D1387757

This copiously illustrated volume is the first systematic general work to do justice to the fruits of recent scholarship in the history of natural history. Public interest in this lively field has been stimulated by environmental concerns and through links with the histories of art, collecting and gardening. The centrality of the development of natural history for other branches of history – medical, colonial, gender, economic, ecological – is increasingly recognized.

Twenty-four specially commissioned essays cover the period from the sixteenth century, when the first institutions of natural history were created, to its late nineteenth-century transformation by practitioners of the new biological sciences. An introduction discusses novel approaches that have made this a major focus for research in cultural history. The essays, which include suggestions for further reading, offer a coherent and accessible overview of a fascinating subject. An epilogue highlights the relevance of this wide-ranging survey for current debates on museum practice, the display of ecological diversity and the environment.

Cultures of natural history

Cultures of natural history

EDITED BY

N. JARDINE, J. A. SECORD and E. C. SPARY

CAMBRIDGE
UNIVERSITY PRESS

Published by the Press Syndicate of the University of Cambridge
The Pitt Building, Trumpington Street, Cambridge CB2 1RP
40 West 20th Street, New York, NY 10011–4211, USA
10 Stamford Road, Oakleigh, Melbourne 3166, Australia

© Cambridge University Press 1996

First published 1996

Printed in Great Britain at the University Press, Cambridge

A catalogue record for this book is available from the British Library

Library of Congress cataloguing in publication data applied for

ISBN 0521 453941 hardback
ISBN 0521 558948 paperback

WV

Contents

Figures

Tables

Notes on contributors

DAVID ALLEN is an Honorary Lecturer in the History of Biology at the Wellcome Institute for the History of Medicine and Co-ordinator of the History of Medicine Programme of the Wellcome Trust. He is the author of three books and numerous journal articles on the social development of natural history, more especially in the nineteenth century. *The Naturalist in Britain: A Social History* (1976) has been a Penguin paperback and translated into Japanese. It has recently been reissued by Princeton University Press.

WILLIAM B. ASHWORTH, JR is Associate Professor of History at the University of Missouri–Kansas City, and Consultant for the History of Science at the Linda Hall Library of Science and Technology, Kansas City. He is interested in interactions between early modern science and the visual arts, and his current book project, 'Emblematic imagery of the scientific revolution', is a thematic study of illustrated title-pages of the seventeenth century.

GILLIAN BEER is the Edward VII Professor of English Literature at the University of Cambridge and President of Clare Hall. Her books and pamphlets include *Darwin's Plots* (1983), *Arguing with the Past* (1989), *Can the Native Return?* (1989) and *Forging the Missing Link: Interdisciplinary Stories* (1992). She is at present writing a study of island narratives.

B. BENSAUDE-VINCENT is Maître de Conférences in History and Philosophy of Science, Paris X University. After recent publications on the history of chemistry (*Histoire de la chimie*, 1993; *Lavoisier*, 1993), she is leading an international research programme – 'Sciences and their publics 1789–1968' – launched by the Centre de Recherches en Histoire des Sciences at the Cité des Sciences et de l'Industrie (Paris). This programme will be developed in a series of comparative historical studies on the popularization of science.

MICHAEL BRAVO is a British Academy Post-Doctoral Research Fellow in the Department of the History and Philosophy of Science at the University of Cambridge, where he lectures and does research on the history of anthropology. Drawing on extensive field-work in the north of Canada, he studies the history and philosophy of ethno-science. His forthcoming book examines early British anthropology seen from the perspectives of both British scientists and the people they encountered.

JANET BROWNE is the author of *The Secular Ark: Studies in the History of Biogeography* (1983) and *Charles Darwin: Voyaging* (1995), and co-editor of *Dictionary of the History of Science* (1981). She recently retired from the Darwin Correspondence project in order to complete her biography of Darwin. She teaches at the Wellcome Institute for the History of Medicine.

HAROLD J. COOK is Professor and Chair of the Department of the History of Medicine at the University of Wisconsin–Madison, and has research interests in early modern European intellectual and social life, with special interest in England and the Netherlands. He has published articles on English and Dutch medicine and natural history, and on the scientific revolution, and two books: *The Decline of the Old Medical Regime in Stuart London* (1986), and *Trials of an Ordinary Doctor: Joannes Groenevelt in 17th Century London* (1994).

ANDREW CUNNINGHAM teaches in the Cambridge Wellcome Unit for the History of Medicine. He is joint editor of a number of volumes, including *Romanticism and the Sciences* (with Nick Jardine, 1990), *The Laboratory Revolution in Medicine* (with Perry Williams, 1992) and *Medicine and the Reformation* (with Ole Grell, 1993). He loves watching gardeners at work in gardens but, to his wife's despair, never shows the faintest inclination to do any gardening.

MICHAEL DETTELBACH is Assistant Professor of History at Smith College in Northampton, Massachusetts, where he teaches modern European history and history of science. At the moment, he is turning his Cambridge dissertation, a biographical treatment of Humboldt and his terrestrial physics, into a book, and is looking into the relationship between German mathematics, the Berlin Academy of Sciences and the Prussian military in the 1820s and 1830s.

JEAN-MARC DROUIN is Lecturer at the Muséum national d'histoire des sciences et des techniques at the Cité des Sciences (La

Villette). He has published *Réinventer la nature: L'écologie et son histoire* (1991). He is currently working on natural history and the public.

PAULA FINDLEN is Associate Professor of History at the University of California–Davis, where she teaches early modern European history and the history of science and medicine. She is the author of *Possessing Nature: Museums, Collecting and Scientific Culture in Early Modern Italy* (1994). Currently she is working on a book entitled *When Science Became Serious*.

MARTIN GUNTAU was, until his retirement in 1992, Professor of History of Science at the University of Rostock in Germany. He has published over a hundred articles, books and edited volumes on historical and philosophical problems in the natural sciences. His works include *Die Genesis der Geologie als Wissenschaft* (1984) and *Abraham Gottlob Werner* (1984).

NICK JARDINE is Professor of History and Philosophy of the Sciences at the University of Cambridge. His books include *The Birth of History and Philosophy of Science: Kepler's 'A Defense of Tycho against Ursus' with Essays on its Provenance and Significance* (1984), *The Fortunes of Inquiry* (1986), *Romanticism and the Sciences*, ed. with Andrew Cunningham (1990), and *The Scenes of Inquiry: On the Reality of Questions in the Sciences* (1991). He is editor of *Studies in History and Philosophy of Science*. His current research projects are on historical consciousness in the sciences and, in collaboration with Alain Segonds, on priority disputes in early modern cosmology.

ADRIAN JOHNS was a Research Fellow at Downing College, Cambridge, between 1991 and 1994, and is currently Lecturer in the History of Science at the University of Kent. He has taught widely in the history of science before the mid-nineteenth century. His own research centres on the social history of early modern natural philosophy, science and medicine, and he is now completing a book on natural philosophy and the history of the book in early modern England.

LISBET KOERNER is Assistant Professor in the Department of the History of Science, Harvard University, specializing in modern German and Scandinavian history of the life sciences. She is currently working on a biography of Linnaeus (forthcoming, Harvard University Press). One of her key interests is the history of how science has been understood to relate to politics and economics.

ANNE LARSEN received her Ph.D. from Princeton in 1993; her dissertation is entitled 'Not Since Noah: The English Scientific Zoologists and the Craft of Collecting, 1800–1840'. Presently she is an independent scholar associated with the Smithsonian Institution, where she continues to investigate the origins and construction of zoological specimens.

LYNN K. NYHART is Assistant Professor in the Department of the History of Science at the University of Wisconsin–Madison, where she teaches courses on the history of modern science, the history of biology and biology and gender. Her book, *The Shaping of Biology: Animal Morphology and the German Universities, 1800–1900*, will be published by the University of Chicago Press in 1995. She is beginning a new project on nineteenth-century zoological research outside Germany's universities and looks forward to visiting numerous German zoos, natural history museums and fisheries research stations in the next few summers.

DORINDA OUTRAM lectures in history at University College, Cork, Republic of Ireland. She was Visiting Scholar at Griffith University, Australia in 1989–90, and Landon Clay Visiting Associate Professor at Harvard in 1992–3. She is the author of *Georges Cuvier* (1984), *The Body and the French Revolution* (1989) and *The Enlightenment* (1995).

DANIEL ROCHE is Professeur d'Histoire Moderne at the Sorbonne, Directeur d'Études at the École des Hautes Études en Sciences Sociales, and Directeur of the Institut d'Histoire Moderne et Contemporaine in Paris. He teaches on the intellectual, cultural and social history of modern France, and his current interests include the cultural anthropology of mobility and the history of equestrian culture in Europe. Amongst his publications are *Le Siècle des lumières en province* (Paris, 1978, new edn 1989); *The People of Paris* (orig. 1981; Leamington Spa, 1987); *Les Républicains des lettres* (Paris, 1988) and *The Culture of Clothing* (orig. 1989; Cambridge, 1994).

MARTIN RUDWICK has taught the history of science at Cambridge, Amsterdam, Princeton and, since 1988, in the Science Studies programme at the University of California–San Diego. He is the author of *The Meaning of Fossils* (1972), *The Great Devonian Controversy* (1985), *Scenes from Deep Time* (1992) and many articles on the history of geology and palaeontology in the nineteenth century.

LONDA SCHIEBINGER is Professor of History and Women's Studies at Pennsylvania State University, and Founding Director of the Institute for Women in the Sciences and Engineering. She is author of *The Mind Has No Sex? Women in the Origins of Modern Science* (1989) and *Nature's Body: Gender in the Making of Modern Science* (1993), along with numerous articles and book chapters. She is currently working on a book entitled 'Women and science: the clash of two cultures'.

ANNE SECORD worked as Assistant Editor of *The Correspondence of Charles Darwin* for the first seven volumes. She is currently a Resident Associate of the Cambridge Wellcome Unit and an external Ph.D. student at London University. Her research focuses on popular, particularly working-class, natural history in nineteenth-century Britain.

JIM SECORD lectures in the Department of History and Philosophy of Science at the University of Cambridge. His interests include the history of the life and earth sciences since 1750, and the relations between science and its audiences. His publications include *Controversy in Victorian Geology: The Cambrian–Silurian Dispute* (1986). He has completed a book on early Victorian controversies about creation and natural law, provisionally entitled *Evolution for the People*.

EMMA SPARY is a Research Fellow at the Department of History, University of Warwick. Between 1992 and 1995 she was a Research Fellow of Girton College, Cambridge, and taught at the Department of History and Philosophy of Science, Cambridge, on natural history in the eighteenth and nineteenth centuries. Amongst her research interests are notions of the natural, the uses of the body, agriculture and food, and forms of association in eighteenth and early nineteenth-century France. Her publications include *Making the Natural Order*, forthcoming.

KATIE WHITAKER is a Ph.D. student in the Department of History and Philosophy of Science at the University of Cambridge. Her dissertation title is 'Curiosi and Virtuosi: Gentlemanly Culture, Experimental Philosophy and Politics in England, 1580–1720'.

PAUL WOOD is an Associate Professor in the Department of History, University of Victoria, Canada, where he teaches courses on the history of science and the intellectual history of early modern Europe. His research focuses on the Scottish Enlightenment, and he has published *The Aberdeen Enlightenment: The Arts Curriculum in the Eighteenth Century* (1993) as well as articles on Scottish science and philosophy.

Acknowledgements

First and foremost, we thank our contributors for the promptness with which they delivered their fine essays and responded to editorial suggestions.

Since the autumn of 1988, the Department of History and Philosophy of Science and the Wellcome Unit for the History of Medicine at Cambridge have sponsored the Cabinet of Natural History (the Cambridge Group for the History of Natural History and the Environmental Sciences). Our plans for the present volume were inspired by the lively sessions of the Cabinet, and we thank all its members, especially its founder members and secretaries: David Allen, Ken Arnold, Henry Atmore, Michael Bravo, Helen Brock, Janet Browne, Andrew Cunningham, Michael Dettelbach, Shelley Innes, Myles Jackson, Adrian Johns, Simon Schaffer, Anne Secord, Ruth Stungo, Jennifer Tucker, Sonia Uyterhoeven, Max Walters, Katie Whitaker and Frances Willmoth.

Sven Widmalm advised on Swedish, and Emma Spary took main responsibility for translating the essays of Martin Guntau and Daniel Roche.

We thank Marina Frasca-Spada, Anne Secord and Paul White for constructive suggestions at various stages in the editing and for putting up with us.

Introduction

I The natures of cultural history

In 1690, the Perpetual Secretary of the Académie Française, Antoine Furetière, included natural objects in his definition of the symbol in the Académie's *Dictionnaire universel*: 'The lion is the symbol of valour . . . the pelican that of paternal love'.[1] At the end of the seventeenth century it was still commonplace for natural objects to be used as symbols for particular human qualities. A century later, however, such readings had little place in learned writing. As the Revolutionary legislator, Constantin-François Chassebeuf de Volney declared: 'Man is ruled by *natural laws . . .* and these laws, *the common source of good and evil*, are not written in the stars, nor hidden in mysterious codes'. But in rejecting symbolic readings of natural objects, Volney by no means denied that human lessons are to be learnt from natural history. Rather, knowledge of 'the nature of the beings which surround him and [of] his own nature' would teach man the 'motors of his destiny' – in this specific context, the true course of Revolutionary political reform.[2]

The period covered by this volume shows many such drastic shifts in the meanings and human significances of natural objects. But there is one constancy, namely, the importance of the roles assigned to natural history in the commonwealth of learning: as a universal discipline, prior to political, social and moral order; as the partner with civil and sacred history in the revelation of the workings of divine providence; as the universal and stable foundation for the transitory and speculative systems of natural philosophy; as the basis for the agricultural, commercial and colonial improvement of the human estate. Today, natural history seems marginal to our concerns, appearing primarily as an amateur, popular, local study. It is the experimental and mathematical sciences to which debates about the 'true' principles of social and mental order appeal, and which serve as a model of expertise and professionalism.

Nature too has been marginalized and devalued. In his *Die Lehrlinge zu Sais* ('The Apprentices at Sais'), Friedrich von Hardenberg, a Saxon mining official, poet and naturalist better known as Novalis, declared that 'The ways of contemplating nature are

innumerable.'[3] The apprentices live in a museum, working under
a master modelled on Abraham Gottlob Werner, director of the
Freiberg mining academy. The work opens with their celebration
of the multiformity of nature: as inexhaustible treasury of phenomena, to be gathered and ordered by the rational philosopher; as
the invincible oppressor of mankind, the nemesis of reason; as a
vulnerable enemy to be enslaved by turning her forces against her;
as prolific mistress of the poet, whose playful creativity he emu-
lates; as ailing victim of naturalists who plunder her; as longed-for
homeland; as wilderness to be tamed and cultivated; as desolation,
prison and slaughterhouse; as hieroglyphic, divine language,
mirror of the soul.[4]

If a single vision predominates in modern Western society, it is
that of a passive and disempowered nature, slave and victim of
human agency. Admittedly, notions of an active nature are not
altogether extinct: there are, for example, many who adhere to
some version of the 'Gaia hypothesis', according to which the Earth
is an organism whose harmonious balance, temporarily disrupted
by humanity, will eventually be restored.[5] But Novalis' multifaceted creative nature, variously well- or ill-disposed towards mankind, is irretrievably lost. The present volume is concerned with
the practices of naturalists from the sixteenth-century revival of
natural history to its late nineteenth-century transformation at the
hands of practitioners of the new biological sciences. Through
these studies, we hope to convey something of the past complexity
and diversity of attitudes to nature.

Histories of natural history

In his *Introductory Discourse on the Rise and Progress of Natural
History* of 1788, James Edward Smith, purchaser of the Linnaean
collections and first president of the Linnean Society of London,
sketched the history of natural history from its conjectural origins
when man lived in 'a state of nature'. Claiming to offer an impartial
account of the merits of past naturalists, Smith tells a story of liber-
ation of natural history from 'superstitious theory', from vulgar
insistence on medical utility and from national prejudice (especially
French). Linnaeus, Smith claims, was the one who first 'supervised
and methodized' the whole of natural history, and subsequent progress has come about through the spread of Linnaean doctrines
and methods.

Certain aspects of Smith's account – its placing of natural history
in the context of a conjectural universal history of mankind, and
its linking of the advancement of learning with the conquest of
superstition – are typical of the Enlightened historiography of his
period. But there are other features that his account shares with

the vast majority of subsequent histories of natural history: the narration of a progress culminating in the present state of the subject;
the emphasis on the discovery by heroic geniuses of doctrines
which anticipate current views; and the explanation of the growth
of knowledge by the appeal to the use of sound methods. Further,
for all their frequent parades of objectivity, they serve to legitimate
particular theoretical concerns and disciplinary factions or to locate
the historian as practitioner within a particular disciplinary tradition. Such epic histories so re-present and re-work past cultures
of natural history as to make them appear directed towards the
writer's own time and culture.

It is now fashionable to dismiss such histories *en bloc* – but this
is surely a mistake. Victor Carus' *Geschichte der Zoologie* (1872),
Julius von Sachs' *Geschichte der Botanik* (1875) and K. A. von Zit-
tel's *Geschichte der Geologie und Paläontologie* (1899), are magisterial histories of zoology, botany and geology which celebrated the
consolidation of these disciplines as triumphant natural sciences in
the new German *Kaiserreich*.[6] But they are also works whose massive and meticulous scholarship laid a foundation for subsequent
interpretations.

The link between upheaval in the life and earth sciences and
concern with their histories was evident also in the 1950s, 1960s
and 1970s, a period of dramatic developments in taxonomy, genetics and molecular biology, many of which challenged the neo-
Darwinian evolutionary synthesis. At the same time the earth sciences were revolutionized by the new plate tectonics. The
appearance of a number of major new historical journals, such as
the *Journal of the History of Biology*, founded in 1967, and *Histoire
et nature*, first published in 1971 under the title *Histoire et biologie*,
marked the upsurge in interest. Works such as Bentley Glass et
al. (eds.), *Forerunners of Darwin* (1959) and Philip Ritterbush,
Overtures to Biology (1964) traced the remote ancestry of modern
disciplines. At a more local level there were many attempts to find
distant anticipations of the latest theoretical innovations. For
example, much discussion of the works of the eighteenth-century
botanist Michel Adanson was generated by claims that he had
invented methods of classification according to the proportion of
matches in characters, all characters being equally weighted –
methods anticipating those of the new computer-based taxonomy
of the 1970s.[7]

It would be a mistake to portray all historiography of natural
history of the first half of the twentieth century as built around
scientific progress and the anticipation of current doctrines. There
are notable exceptions. One is *Form and Function: A Contribution to
the History of Animal Morphology* (1916), by Edward Stuart Russell,
proponent of a holistic, vitalistic and teleological biology. It is

remarkable for the respect it shows for the categories and philosophical assumptions of past naturalists, and for the sympathy with which it expounds past approaches far removed from the main lines of progress as generally understood at the time. Equally striking are the works of Henri Daudin, pupil of the anthropologist Lucien Lévy-Bruhl, and friend of the cultural historian Lucien Febvre. His *De Linné à Lamarck. Méthodes de la classification et idée de série en botanique et en zoologie (1740–1790)*, and *Cuvier et Lamarck. Les Classes zoologiques et l'idée de série animale (1790–1830) (1926–7)*, are notable for the care with which they relate past systems of classification to the specific collecting, horticultural and curatorial activities of the naturalists.

Concern with the history of natural history of another very different sort is a by-product of the rules of botanical and zoological nomenclature established at the International Congresses of Botany, Zoology and Geology in the closing decades of the nineteenth century.[8] Ascertaining the correct names of living beings in accordance with the internationally agreed rules often requires extensive study of the publications and collections of past naturalists. As well as a substantial body of valuable scholarly literature, this professional interest of systematists and taxonomists in the history of their own discipline has produced a series of indispensable reference works – notably, for Britain, Ray Desmond and Christine Ellwood's *Dictionary of British and Irish Botanists and Horticulturalists* (1994), and Frans A. Stafleu et al. (eds.), *Taxonomical Literature* (1976–1988), as well as the *Journal of the Society for the Bibliography of Natural History* (later *Archives of Natural History*), founded in 1936.[9]

Despite these exceptions, the historiography of natural history continued to be dominated by tales of anticipations and progress until quite recently. Since the mid-1960s, however, the emphasis has gradually shifted to approaches that, in Thomas Kuhn's words, 'rather than seeking the permanent contributions of an older science to our present vantage, attempt to display the historical integrity of that science in its own time'.[10] Thus, in the field of history of natural history, a number of works sought to reconstruct the meanings of past theories and systems within the framework of the presuppositions and conceptual categories of their period; notable examples are Elizabeth Gasking, *Investigations into Generation, 1651–1828* (1967) and Mary P. Winsor, *Starfish, Jellyfish and the Order of Life* (1976). Some emphasized the social contexts and uses of the sciences in place of the older 'internalist' chronicles of sequences of discoveries and theories; typical of this reorientation are David Allen, *The Naturalist in Britain. A Social History* (1976), Martin Rudwick, *The Meaning of Fossils* (1972) and Paul Farber, *The Emergence of Ornithology as a Scientific Discipline* (1982).

Others protested at the 'presentism' of histories of scientific progress, with their imposition of current categories, interests and values on past agents; instead, they have sought to understand past agents in their own terms, to reconstruct their mentalities and conceptual frameworks as local and historically contingent creations. In this they have drawn upon the work of anthropologists concerned to do justice to cultural difference, notably Clifford Geertz, *The Interpretation of Cultures* (1973) and, in a quite different vein, Dan Sperber, *Rethinking Symbolism* (1975).

Typical of these new 'anthropological' perspectives in the history of natural history are Wolf Lepenies, *Das Ende der Naturgeschichte* (1976), Scott Atran, *Fondements de l'histoire naturelle. Pour une anthropologie de la science* (1986, of which *Cognitive Foundations of Natural History* (1990) is an amended translation) and Krzysztof Pomian, *Collectionneurs, amateurs et curieux* (1987). In rejecting the traditional narrations of continuity in the growth of disciplines, some have followed Kuhn in claiming priority for local and tacit practices in the formation of the successive and incommensurable 'paradigms' of the sciences. Others have emulated Michel Foucault in focusing on radical discontinuities at the level of discursive practices and institutional regimes. In particular, there has been much debate about Foucault's account, in *Les Mots et les choses* (1969), of an abrupt shift around 1800 of natural historical discourse from static tabulation of the external similarities and differences of plants and animals to dynamic narration of the inner developmental and historical processes of living beings.[11]

Where Foucault emphasized the temporal discontinuities of disciplines, others have attended rather to their spatial and social discontinuities, arguing for the importance of national styles, and of divergences between the metropolis and the provinces, between elites and artisans, between authors and their publics, between men and women. And, as Michel Serres has pointed out, the need for a local historiography is particularly pressing in the case of natural history, given the locality of occurrence of natural objects and the fact that naturalists in different places perceive different natural worlds.[12] Examples of such 'decentred' history of natural history include Bernard Smith's pioneering *European Vision in the South Pacific* (2nd edn, 1985), James A. Secord, *Controversy in Victorian Geology: The Cambrian–Silurian Dispute* (1986) and Mary Louise Pratt, *Imperial Eyes: Travel Writing and Transculturation* (1993).[13]

Cultural history

As a recent reviewer has observed, few historians nowadays 'feel entirely comfortable saying that they *don't* do cultural history'.[14]

The identity of cultural history is hotly contested, and the remarks that follow are unashamedly prescriptive.[15]

Two commitments seem to us to be constitutive of cultural history. First, there is the concern with culture in the sense of that which gives meaning to people's lives, a concern that covers both Matthew Arnold's elitist 'best that has been known and thought in the world' and William Morris's culture 'of the people, by the people, and for the people'.[16] Secondly, cultural history has to deal with culture in the sense of social habits, the totality of the skills, practices, strategies and conventions by which people constitute and maintain their social existences. In both these senses natural history is a cultural phenomenon. For the values which attach to people's lives are informed by natural historical reflections on the place of humans in the natural world; and natural history may provide models for the moral and political order of human society. Further, in so far as natural history is a discipline, it lends itself to treatment in terms of the conventions, skills and strategies – let us call them, collectively, practices – through which knowledge claims have been promoted, secured and defended.

It is, we believe, of the utmost importance for all disciplinary history to do justice to the full range of such practices. Our concern in this book is to illustrate the range and diversity of cultures of natural history over several centuries. Rather than presenting natural historical knowledge as generated by isolated individuals working wholly within the domain of the mind, we wish to portray natural history as the product of conglomerates of people, natural objects, institutions, collections, finances, all linked by a range of practices of different kinds. As a rough and ready guide we may distinguish the following types of practices.[17] *Material* practices are ways of making, handling and transforming things; in the case of natural history they include the gathering, transport and preparation of specimens, the making and distribution of books and illustrations, the performance of experiments. *Social* practices cover the whole range of associations, recruitments, delegations and negotiations; in particular, in natural history, they include the skills of inspiring trust in other natural historians and assessing their trustworthiness, the conventions and strategies relating sponsors and patrons to naturalists and naturalists to informants and assistants, the regulations and routines of behaviour in the institutions of natural history – courts, academies, universities, gardens, museums, laboratories.[18] *Literary* practices are conventions of genre, representation and persuasion; in natural history and other disciplines these include, along with rational argumentation, the gamut of rhetorical and aesthetic forms of persuasion – appeal to historical precedent, to the interest, self-esteem and taste of the reader, for example. *Bodily* practices are forms of bodily, sartorial or gestural self-presentation and normative accounts of physical and emotional experience in response to particular situations; natural historical examples include the legitimation of natural historical enquiry by appeal to the emotional experiences it engenders, or even accounts of the attire most appropriate for the naturalist. Last, but by no means least, there are *reproductive* practices, that is, the means by which skills and knowledge are handed on from generation to generation; with natural history, as with other disciplines, these include not merely procedures of formal instruction, but also the informal ways in which the various practical and social skills of the naturalist are imparted.

Exaggeration of the powers of one of these categories of practices at the cost of others is responsible for certain distortions and excesses. Social practices are, indeed, of primary importance for all cultural historians, for it is they that constitute and maintain society, with its institutions and forms of association; it is they, indeed, that provide the framework for all human activities. Alas, some have gone further, attempting to reduce all practices to social practices, and venturing extreme metaphysical claims, for example, that truth and rationality in the arts and sciences is but a mask and emblem of power, or that knowledge in the arts and sciences is but a projection of social interests. Literary practices are likewise of paramount importance for cultural historians who deal with learned disciplines. Unfortunately, some have reduced all past social and natural worlds to the surfaces of documents, insisting that both the objects and the authors of natural knowledge are mere projections from the flat plane of the textual universe. One route to this fantasy starts by ignoring the material traces of past disciplines – instruments, buildings, specimens – and goes on to claim that since our only access to the past is through texts, texts are the only genuine subjects of historical enquiry. Another, and more insidious, inference moves from the observation that all human practices embody symbolic and conventional elements to the conclusion that all human practices are at bottom linguistic practices. Such 'pantextualism', much in evidence in the 'cultural studies' sections of bookshops, makes the historian both blind and unjust: blind to the social and natural materials represented in texts and involved in their production, and unjust to those silent majorities who never made it into the world of documents.[19]

In contrast with such belated idealisms, a cultural history that attends to the full range of disciplinary practices can, we believe, do justice to the natural, social and textual worlds. By tracking the local and day-to-day routines of past inquirers, such a history can convey aspects of their lived experience. By studying the means by which they sought to resolve questions, it can reconstruct the ranges of questions real for them, their 'scenes' of inquiry.[20]

By reconstructing changes in investigative practices and the ways they were brought about, it can explain the formation and dissolution of disciplines. And by charting the production, distribution and reception of knowledge claims, it can reveal the ways in which the social and natural worlds give rise to their representations and are transformed by those representations.

Scope of the book

The essays in the volume cover the contents and context of natural history from the sixteenth century to the present. The work falls into three sections, starting at the time when the first botanic gardens were being founded across Europe, in Italy, Holland, France and England, as discussed by Cunningham. Whilst natural history flourished in this medical context, as Cook's essay reveals, Findlen and Whitaker show that it also found powerful support in the courtly and gentlemanly cultures of Renaissance and early modern Europe. Ashworth's essay explores the significances attached to natural historical emblems in these different contexts. Natural history was readily assimilated into the gift–exchange society of the wealthy nobles, which encompassed also the objects of civil history, from paintings to coins. Natural historical objects acquired a concomitant value, since they exemplified the rare, representing the new colonial wealth that could be obtained from the exploitation of the exotic. It was the appearance of natural history as a set of practices favoured by court culture which principally ensured the development of natural history as independent from medical and agricultural concerns. As Johns reveals, the development of print culture in the early modern period offered writers on natural history new opportunities, but also new problems, as they struggled to fix a meaning for natural knowledge.

Our second section covers the period from the end of the seventeenth century to that of the eighteenth century, during which time natural history and its practitioners began to acquire autonomy from courtly culture. The establishment of societies and academies, described by Roche, served to provide naturalists, who were often not of noble birth, with a legitimacy independent of their individual position in the early modern patronage society. As collecting became a pastime of the 'enlightened' classes, increasing the rate at which new specimens flooded into European collections, naturalists were increasingly successful in soliciting State funding for their activities. Naturalists linked their activities to fashionable concerns with natural and experimental philosophy, and with the wider Enlightenment movement, so that, as Koerner shows, the question of the practical and/or natural criteria for ordering natural objects became increasingly fraught for naturalists such as Linnaeus. The nature of history, too, was open to debate as naturalists distanced their enquiry from that of civil history, whilst many attacked older sacred histories of nature and the earth's past, as Guntau's essay reveals. The use of the microscope and the development of comparative anatomy also opened new realms of enquiry for natural history, as phenomena such as generation or the operations of the mind became legitimate subjects for natural historical enquiry, as Wood and Spary suggest. But, as Schiebinger's contribution shows, the growing power of naturalists to represent the natural allowed them to make normative claims about social relations. By the end of the eighteenth century, radically new forms and agendas of natural historical inquiry were emerging, particularly in the German lands, as Jardine argues.

In our third section, Outram's essay demonstrates the increasing institutionalization of natural history from the beginning of the nineteenth century, showing how the construction of the naturalist and natural historical practice were shaped by place and resources; and both she and Dettelbach indicate how apparently 'new' kinds of enquiry appearing at this time owe much to the Romantic notion of experimentation and the invention of the 'discovering hero'. The work of Beer and Browne also illustrates the relations between place, language and natural historical knowledge; for them, as for Dettelbach, the problem of natural historical knowledge at a distance is what is at issue. Both Beer and Bravo address the problem of ethnocentrism in early nineteenth-century travel accounts, although in rather different ways, and both examine how far natural historical travellers could mediate between distant cultures and their own. Rudwick's essay, too, considers the problems of representing the distant, whether in space or time, which naturalists confronted in their endeavours to present themselves as experts in a new science of the earth. Such problems were manifest not just in the field, but also in the proximate site for the representation of natural historical knowledge, the collection. Where eighteenth-century collections had been aimed at a single public, differentiated only by degree of knowledge, collections came to be differentially designed during the course of the nineteenth century, with one face for amateurs and another for naturalists. Thus they partook in the shaping of 'the public' and its exclusion from natural historical expertise. The essays of Larsen, Allen, A. Secord and Bensaude-Vincent and Drouin address this shift from different perspectives: those of the elite reading public, of artisans, of women, of private collectors. Gradually, natural historical displays ceased to be the site of active research by naturalists: in many institutions, public and research collections were separated; in others, as Nyhart's chapter reveals, the site of natural historical research shifted from the collection to the laboratory.

The contributions to this volume are not intended to be specific research articles. Our intention is that this should serve as a work of first resort, demanding no previous acquaintance with the literature of the history of natural history. Because the chapters are intended for a general readership, each provides an account of the wider framework of the subject area. However, all the contributions embody at least some of our historiographical prescriptions. All avoid anachronism, respecting the categories of the naturalists themselves. Almost all are focused on the practices of natural history; a number deal with the social, political and moral uses of natural history; and many discuss the settings and contexts of natural history. The need for generality and continuity has inevitably produced a Eurocentric bias and limited attention to local and popular issues. However, decentring is evident in the treatment of developments and receptions of natural history outside the spheres of elites and savants, and in the discussion of the impacts of natural historical exploration on indigenous cultures. Taken as a whole, these essays do, we believe, convey the richness and variety of the past cultures of natural history.

History or hyper-reality?

If a single conclusion is to be drawn from this volume, it is that there is no 'natural' conception of nature, no stable inventory of the products of nature, and no universal register of questions timelessly posed by nature. Rather, the contributions reveal how various are the frameworks that have structured and informed natural historians' dealings with nature, how the boundaries between the natural and the conventional, artificial and social have been continually contested and relocated.

No matter how determinedly we seek out the wild, we cannot hope to escape from our time- and culture-bound ways of seeing and interpreting, to encounter nature prior to all perceptual ordering and judgement. Even so-called 'nature reserves' contain not untrammelled nature, but a managed, culled, restricted nature, where access is controlled and where the observer is constantly guided, so that the supposedly natural spaces are rendered just as much 'hybrids' between the social and natural as those areas that ecologists deplore for the human destruction wrought in them.[21] Even when ecologists and naturalists venture into 'virgin' territory, the object of their observations is not raw nature, but nature measured and graded, classified and tagged, registered and simplified. When we consider our present-day forms of representation – photographic, cinematographic, holographic – it is hard to avoid the feeling that they have an unprecedented tendency to cut us off from the natural world. Such simulacra are unambiguously representational: despite their intensity and reality they bear few, if any, traces of the material objects and the human labour that went into them. Rather, their impact depends largely on their success in conveying the scientific sophistication of their making. The more engaging and convincing the images, the more the viewer is led to take on trust the reliability and authority of scientists and their technologies. In representing nature past and present, such 'hyper-real' images displace and supplant it.[22]

Through historical studies we can hope to regain the natural and human worlds that we are in danger of losing, as we uncover the ways in which humans have worked on natural objects to produce knowledge of nature. Thus we may gain a more critical understanding of our own concerns and dealings with nature. None shall lift the veil of Isis, Goddess of Nature, declared the ancient oracle. But Novalis tells us that one – a historian, no doubt – did succeed, only to see 'wonder of wonders, himself'.[23]

I Curiosity, erudition and utility

2 Emblematic natural history of the Renaissance

Natural history in the Renaissance was an area of study that bore little resemblance to our modern notions of the discipline. Renaissance natural historians had reasons for studying nature and ways of writing about nature that contrast strongly with our own. The Renaissance approach is well worth understanding, since it sheds a great deal of light on Renaissance culture as a whole, but to appreciate it properly we must put aside all preconceptions of what natural history should be and allow ourselves to encounter Renaissance natural history on its own terms. We need to forget everything we know about zoology and comparative anatomy and taxonomy and be willing to entertain approaches that seem to venture far beyond the pale of what we consider science. If we can manage this, however, we will be richly rewarded. The Renaissance view of the natural world was more densely layered and more intricately interwoven than ours, and it can be a great pleasure to reconstruct that view and perhaps dwell within its sight for a brief while.[1]

The best way to begin this reconstruction is to open a volume written in the sixteenth century, by someone who claims to be a natural historian, and that purports to be about the natural world, and read what it has to say. There are a number of such works to choose from, although you could store them all in a modest-sized bookcase. Most of the volumes are large folios, forbidding in their bulk, but that is not surprising since they often survey all of nature between their covers. However, we need not read an entire work, at least not initially. One chapter ought to be enough, provided that we read it all (not just the parts that look familiar) and that we resist the temptation to judge until we are through.

Gesner and the fox

So that we get the most representative picture possible, let us choose a selection from the most widely read of all Renaissance natural histories, the *Historia animalium* of Conrad Gesner (1516–65). It was published in four volumes in Zurich between 1551 and 1558, with the first volume discussing live-bearing quadrupeds, and the succeeding volumes treating egg-laying quadrupeds, birds,

Figure 2.1 Fox. Woodcut from Conrad Gesner, *Historia Animalium Lib. I* (Zurich, 1551), p. 1081.

and aquatic animals.[2] Let us look at the chapter on *vulpis*, the fox, which we encounter near the end of volume one.[3] It is a sizeable essay of sixteen folio pages, and if it were reset to fit into this volume, it would occupy at least sixty pages. One notices, immediately, that it begins with a woodcut illustration, and a most handsome one at that (Figure 2.1). In fact, most of the articles in Gesner's *History* contain illustrations, and their naturalism and attractiveness have been often remarked upon. We will return to say more about these woodcuts, but first it is the text that will occupy our attention.[4]

The chapter is divided up into eight sections, lettered A through H (nearly every chapter, it will turn out, is sectioned in an identical manner). In section A, a brief paragraph, we learn about names: that *vulpis* in French is 'regnard', in English, 'fox', in Dutch, 'vos'; further equivalents are given for a large number of ancient and modern languages. In section B we learn about regional differences in foxes: for example, that foxes in Russia tend to be black rather than red, and those in Spain are often white. Foxes everywhere, however, share one attribute in common: their large bushy tails.

In succeeding sections we learn about the fox's daily habits and movements, its different cries and calls, its relationships with other animals, its diet, its suitability as a source of food or medicine, and more. And as we slowly move through these long paragraphs we gradually become aware of Gesner's basic method for learning about nature: quite simply, he read *books*, untold numbers of books, and he gathered from these texts and assembled on his pages the facts that he thought appropriate. It is not difficult to discover this, because Gesner pairs virtually every fox fact with a name, his source. So we read: 'foxes are very ravenous in Sardinia, *Aelian*', or 'foxes die if they eat bitter almonds, *Dioscorides*'. There are over eighty different authorities cited in the fox article alone. Most of them are ancient authors; some, such as Aristotle and Albertus Magnus, are familiar names, but many more are minor literary figures that would be unknown to anyone except a dedicated specialist. It would seem, then, that Gesner was a competent classical scholar, that he read an astonishing number of obscure books, and that he preferred ancient authority over modern. It is also apparent that, for Gesner, natural history was a discipline forged in the library with the bibliographic tools of the scholar, rather than an observational science built up by a direct personal encounter with nature. To use a term that is commonly applied to those in the Renaissance who believed that the best answers were usually found in the writings of antiquity, Gesner was a *humanist* and, in his eyes at least, natural history was, first and foremost, a *humanist* pursuit.[5]

Fables and folklore

As we continue our reading, another impression becomes inescapable. There are some very odd facts here, and many of them sit very uneasily in the organizational cubbyholes that Gesner has provided for them. For example, section C is purportedly on the habits and voice of the fox. Gesner informs us that the fox's yelp is so distinctive that the Romans invented a special word for it – 'gannire' – and Gesner goes so far as to provide examples from Terence and Plautus as to how this verb was applied metaphorically to men who tend to yelp at others. The same section informs us that the fox does not make its own den but rather relies on one built by the badger, whom the fox drives away by depositing excrement at the entrance. Still in the same paragraph we are told that in northern countries the fox crosses frozen rivers very deliberately, listening carefully for the sounds of cracking ice, and it is so good at finding a passage that other inhabitants of these regions will only cross where the fox does. And it goes on: the fox is often bothered by flies and gnats, but it relieves its torment by taking a mouthful of hay and immersing itself slowly in water until all the insects have taken refuge in the hay, which the fox then quickly releases. The fox will often cover itself in red clay and lie on its back as if dead; crows and ravens, delighting in the death of their enemy, will land on the carcass in triumph, only to be devoured. Sometimes the fox will bury its nose in the mud and raise its tail into the shape of a bird's neck, creating a strange avian form that other birds are compelled to investigate, and they too become fodder for the fox. And still it goes on.

What are we to make of this barage of folktales and myths? Why are such stories here, in a work of natural history? One might choose to believe, as many commentators have, that Gesner was simply a lousy natural historian; that for all his humanistic fervor he patently lacked the common sense to discriminate between fact and fiction. But such a conclusion makes a dangerous presupposition about natural history; it assumes that good natural history consists only of true facts, and that a natural history containing mythical or apochryphal information is somehow inferior. But perhaps Gesner did not feel that way. Perhaps he thought that a proper essay on the fox would include not only information on the fox's name, size and appearance, but also every fox folktale, every vulpine myth, every reynardian legend that has come down to us. Perhaps Gesner believed that such tales reveal to us a great deal about the place of animals in human culture, and that one of the goals of natural history, perhaps the supreme goal, is to understand the intricate web of relationships that interconnect humans and animals. I would venture that this latter intrepretation makes much more sense of Gesner's natural history. Gesner used every available thread because he was trying to weave the richest tapestry possible. And if we can just accept that as a viable approach to natural history, and step back for a moment and behold his creation, we realize that, by those terms, Gesner was eminently successful. The fabric he wove is indeed quite wondrous to behold.[6]

Adages and emblems

And if we can hold on to our new tolerance for a moment longer and return to Gesner's fox chapter, we can now attempt to come to grips with a section that we have avoided until now, the mysterious section H that concludes each and every animal chapter. If earlier sections have perplexed some because they seem to be compartments over-filled with facts of doubtful value, this final section can be downright disturbing, because many will not even recognize the compartment. Gesner does not label his sections except for the letters, but if we had to think of a title for section H, the best might be 'Associations' – all those ways in which animals and their attributes have intruded themselves into our language, literature, and art. For most animals in Gesner's work, H is the largest section of all, and so it is for the fox. We find, for example, a complete listing of fox epithets: crafty, sly, cunning, deceitful, each one supported by a plethora of classical quotations. We find a lexicon of all the meanings that 'foxy' can have as an adjective. Gesner provides us with instances of the fox as metaphor: Christ, for example, called Herod a fox because of the guile he displayed. We learn about the fox as omen: that it is bad luck to encounter a newborn fox on the path. Gesner lists all the appearances of the fox in Scripture, such as Matthew 8: foxes have holes, and birds have nests, but Christ has no place to lay his head. If the fox had appeared in pagan mythology, or in Egyptian hieroglyphics, that would be recounted here (the fox, it so happens, is absent from both, but many other animals had some status as hieroglyphs or mythological symbols and this is duly discussed in the appropriate section H). We are treated to several *pages* of fox proverbs: 'a fox takes no bribes'; 'he is yoking foxes'; and many more.

Finally, after six full pages of fox associations, we encounter, without a word of explanation, the following four-line paragraph. I will translate it in full: 'Mind is worth more than beauty. A fox, entering the workshop of a stage-manager, came upon a smoothly polished mask of a human head, so elegantly fashioned that although it lacked breath, it appeared to be alive in other respects. When the fox took the mask in its paws, it said: What a fine head this is, but it has no brain. Alciati, in his Emblemata.' This would seem a fitting finale, in its strangeness, to what seems to be, initially, a very odd set of passages about the fox and its symbolic meanings.

A modern reader's initial encounter with a Gesnerian section H can be a most unsettling experience. The earlier sections, with their wealth of animal fables and folklore, might be made tolerable even to a critic, for whether you view it as good or bad natural history, it is still natural history. But section H seems to have forsaken natural history altogether. It resembles a vast exercise in philology, linguistics, literary criticism, and biblical exegesis that has little to do with the study of nature. At least in the other sections we can recognize the authorities and admit their relevance: Aristotle, Pliny, Aelian, Dioscorides, and Albert the Great all earned laurels as students of nature. But in section H we encounter Suidas and Planudes, Horapollo and the Greek Anthology, Erasmus and Alciati, most of whom you will not recognize, and none of whom you would expect to encounter in any history of natural history. The easiest way to resolve the paradox of section H is simply to ignore it as inconsequential or idiosyncratic, and that is precisely what most commentators have done. But should we do so? If Gesner saw fit to devote six of his sixteen fox pages to epithets, icons, proverbs, and emblems, should we not allow for the possibility that, in the mid-sixteenth century, knowledge of animal symbolism was considered an essential aspect of natural history?

I would like to suggest that such was the case – that Gesner lived in an age that delighted in the allegorical and the adagial and that regarded symbolic meanings as anything but inconsequential. Since we now live in an era that is profoundly uninterested in and uninformed about proverbs and emblems, entertaining such a possibility takes a considerable leap of faith. But we really ought to make the effort, if we truly wish to understand Renaissance natural history. And it would help if we were better aware of two literary giants of the Renaissance, Erasmus and Alciati, who respectively made proverbs and emblems an integral part of Renaissance culture.

Erasmus and Alciati

Most people are familiar with Desiderius Erasmus as a prominent figure of the Reformation, but it can be a surprise to learn that in his own day he was more esteemed as the compiler of a vast annotated encyclopaedia of proverbs, the *Adages*. First issued in 1500 and greatly enlarged over the next thirty years, the *Adages* in its final form contained nearly four thousand proverbs, each with a running commentary by Erasmus, who was easily in Gesner's league as an indefatigable classical scholar.[7] Not surprisingly, many of the Erasmian proverbs feature animals: 'owls to Athens', 'the tired ox treads more firmly', 'let sleeping dogs lie'. If we browse through Erasmus with an eye focused on foxes, we discover many delightful adages: 'the fox is given away by his Brush [tail]' (a variant on the more familiar proverb, 'the lion is known by his claw'); 'an old vixen is never caught in a trap'; 'the fox knows many ways, the hedgehog one really good one'; as well as 'a fox takes no bribes' and the others that we already encountered in Gesner. And we must admit that the Erasmian proverbs collectively and effectively sum up what the fox has always symbolized in human culture: cleverness, craftiness, and the ability to learn from experience. In Gesner's opinion, and in the view of most of his contemporaries, such proverbs tell us as much, if not more, about the fox as details about the shape of its ears or the size of its litter.

Andrea Alciati is less familiar to most of us than Erasmus, although we have already encountered him in passing, for Gesner ended his fox chapter with a quotation from Alciati's *Emblemata*. The *Emblemata* was a book of emblems, the *first* book of emblems, since Alciati himself invented the genre. As Alciati envisioned it, an emblem is the combination of a short (and preferably obscure) motto, an image (also preferably obscure), and an explanatory epigrammatic poem. The emblem was intended to first mystify, and then delight, as the separate elements of word and icon, assisted by the explanatory poem, combined with illuminating clarity to convey a clever truth. The *Emblemata* was first published in 1531, and it proved to be immensely popular, unbelievably popular, going through many editions and enlargements, and then spawning in turn a host of imitators. By the end of the century there were literally hundreds of different emblem books, containing tens of thousands of emblems.[8] The idea of the emblem fitted perfectly with a Renaissance spirit that treasured symbolic meanings and hidden truths, and it is no exaggeration to call the last half of the sixteenth century the Age of the Emblem, since emblems infiltrated virtually every aspect of Renaissance culture. Natural history was no exception. Many of Alciati's emblems utilized animals. For example, one emblem shows a lynx standing over a newly killed deer while gazing at a flock of sheep in the background. The motto is 'Forgetfulness is the parent of poverty', and the epigram tells us to be careful not to neglect the present while looking for greater things in the future. Similarly, we can find in Alciati an image of hares cavorting about a dead lion, symbolizing courage in the absence of danger, and an emblem of a chameleon, representing flattery, because it can change its colours to fit the situation. Not surprisingly, we can find a fox in Alciati, sitting up and gazing at a human mask that it holds in its hands (Figure 2.2). The motto tells us that the mind is worth more than beauty, and the epigram, which we have already quoted in full as the concluding paragraph of Gesner's fox chapter, might be boiled down to the query: what good is an empty head?[9]

Depending on your point of view, it is either fortunate or unfortunate that Gesner worked at the very beginning of the Age of Emblems, for his only emblematic source was Alciati; had he published his *History of Animals* fifty years later, section H could have been much more swollen with emblematic material. For it is clear that the idea of the emblem captured the very essence of Gesner's view of nature: that the natural world is a complex matrix of seemingly obscure symbols and hidden meanings, which can suddenly become clear in a burst of illumination, if only you view it from enough different angles. Gesner's text was so thorough and so all-embracing because it was necessary to cover all these angles, if one is to ensure complete understanding.

The role of the illustration

I hope that our encounter with one tiny portion of Gesner's text has elicited the outlines of a view of nature that, while different from our own, is nevertheless fully self-consistent, and compatible with the other cultural traditions of the Renaissance that we have discussed. But our picture of Gesner's natural history is still

52   ANDREAE ALCIATI

Mentem non formam plus pollere.

Ingreſſa uulpes in Choragi pergulam,
  Fabrè expolitum inuenit humanum caput,
  Sic eleganter fabricatum,ut ſpiritus
  Solùm deeſſet,cæteris uiuiſccret:
  Id illa cùm ſumpſiſſet in manus,ait,
  Hoc quale caput eſt,ſed cerebrum non habet.

Figure 2.2 Fox emblem. Woodcut from Andrea Alciati, *Emblematum libellus* (Paris, 1534), p. 52.

incomplete, for we have yet to consider some of the truly novel features that Gesner, uniquely, brought to the writing of natural history, and that, ultimately, would have as great an impact as his emblematic view of nature.

The first novelty is the incorporation of a visual image to supplement the text. We tend to take it for granted that books about animals should contain pictures of those animals, but there was very little precedent for this in classical natural history – Aristotle, Pliny, and Aelian were purely textual sources, as was Albertus, and whether in manuscript or printed form their texts were hardly ever accompanied by images of the animals under discussion. Gesner, however, illustrated practically every animal he discussed, using for the most part very fine woodcuts. The printed text had never seen the like of his impish fox or his droll hedgehog (Figure 2.3).

The reasons for Gesner's decision to include not only pictures, but naturalistic pictures, are not immediately obvious. There was

Figure 2.3 Hedgehog. Woodcut from Conrad Gesner, *Historia Animalium Lib. I* (Zurich, 1551), p. 400.

one genre of animal books that had commonly been illustrated, namely the medieval bestiary, and since the bestiary was primarily a book of animal symbolism, and since Gesner was so interested in animal symbols himself, one might wonder whether he found inspiration in one of those beautiful illuminated bestiary manuscripts that were so widely scattered around the libraries of Europe.[10] But the possibility, initially tantalizing, is really untenable. Bestiary illustrations, for all their charm and beauty, were highly stylized, and often unrecognizable as to species; and they usually depicted the animal engaged in the act that gave it importance as a symbol, so that the bear is depicted licking its unformed cubs into shape, and the pelican is shown nourishing its young from its bloody breast. Had Gesner used the bestiary as his model, then surely his fox would have been lying on its back, covered with mud, and enticing the birds from the trees, as it customarily did in the bestiary. Moreover, the bestiary tradition is no help in explaining why Gesner wanted illustrations that were naturalistic.

Another possibility is that Gesner was inspired by the animal images in an early printed book that is better known to botanists, namely the *Gart der Gesundheit* ('Garden of Health') (Mainz, 1485). The *Gart*, as it is usually called, was the first printed book to contain illustrations drawn from life by an artist, and these woodcuts are a notable improvement over the copybook specimens that one usually finds in herbals.[11] The *Gart* is primarily a book of medicinal remedies, and consequently most of the new illustrations are of plants, but there are a few animals of medicinal value that managed to creep in, and these woodcuts, while still relatively crude, belong to a completely different artistic world from their predecessors. The fox of the *Gart* is, I would suggest, a worthy ancestor to Gesner's smirking vixen (Figure 2.4).

Figure 2.4 Fox. Woodcut from *Gart der Gesundheit* (Mainz, 1485), ch. 426.

Vulpis   eyn fuſch   Cap·cccc̃xxvj·
Vlpis latine· Die meiſter ſprechen das diß gar ein bedruͤ
glich dier ſy·wan die hunde es iagen ſo nympt es den ſwantz

A third possibility is that Gesner was influenced by some of the late classical manuscripts that he consulted while compiling his natural history. Late classical manuscripts are quite different in appearance from medieval manuscripts or early printed texts, because the illustrations, when they are present, are quite naturalistic. As one might imagine, such manuscripts are very scarce, but we know that Gesner studied at least one, a Greek manuscript of Oppian's *Cynegetica* (*c.* 217 AD) that is now in Venice. This manuscript is beautifully illustrated, and the representations of animals are quite wonderfully rendered. We know Gesner was taken by the images, because he copied two of them for inclusion in his *Historia animalium*: the hyena and the ichneumon (Figure 2.5). One might well imagine that he found the Oppian manuscript so attractive, with its regular use of lifelike illustrations, that he used it as model for his own work.[12]

However, it is probable that the major inspiration for Gesner's decision to give zoology a visual component came from the botanical revolution of the 1530s. The appearance of the *Herbarum vivae eicones* ('Living Images of Plants') (Strasbourg, 1530) by Otto Brunfels (*c.* 1489–1534) marked a watershed in natural history illustration, since all of the large woodcuts were based on watercolours drawn from life by an artist, Hans Weiditz (before 1500–*c.* 1536), who came out of the studio of the noted naturalistic painter, Albrecht Dürer (1471–1528), of Nuremberg. The superiority of these images, and the value of their accuracy for the physician

DE ICHNEVMONE.

Ichneumonis hanc imaginem, cui parum tribui, ex uetuſto manuſcripto codice Oppiani Venetijs naĉtus ſum.

Figure 2.5 Ichneumon. Woodcut from Conrad Gesner, *Historia Animalium Lib. I* (Zurich, 1551), p. 635. Gesner's caption explains he found this image in a Greek manuscript of Oppian.

trying to identify medicinal plants, was immediately obvious, and before long every printed herbal not only contained pictures drawn from life, but trumpeted this fact as an advertisement from the very title-page.[13] Gesner was a skilled botanist himself, and it may well be that having observed the success of naturalistic images in botany, he simply decided to do the same for books on animals. It is not quite so easy to assemble a portfolio of drawn-from-life images of animals, since many are exotic, and even the local ones cannot easily be persuaded to pose like plants. But Gesner, with the help of friends and correspondents, did manage to collect together the best available images of a large number of animals, and the face of natural history was changed forever as a result.

The problem of newly discovered animals

The second innovative feature of Gesner's approach to natural history, which in some ways is even more surprising than his insistence on accurate pictures, is that he included, and even sought out, information on new animals that were unknown to the great classical authorities. In an adventurer, novelty is often welcome, but in a classical scholar, especially a humanist scholar, it is not common to see a willingness to embrace evidence that undermines the authority of ancient heroes. It is to Gesner's great credit that he made room in the classical menagerie for the exotic and classically unprecedented animals that were beginning to trickle in from the far north, the New World and the East Indies. Granted, there were not many of these in 1551, but Europe had discovered the opossum, the guinea pig, and the bird of paradise, and Gesner included all of these, with illustrations, in his *Historia animalium*, even though he has to break with his carefully worked out format to do so. The opossum, after all, has no classical names, no references in Aristotle or Pliny, no network of fables built around it, no emblematic baggage. But Gesner welcomed the strange beast into his stable anyway.

Contemporary observers

Gesner's welcome of the exotic is the result of a third innovation that is easily lost to view in the glare of his elegant woodcuts and clever proverbs. Gesner was willing to supplement his classical scholarship with observations and stories from contemporary observers. Olaus Magnus was the author of a book on the natural history of Scandinavia; Gesner mined this for the fact that some northern foxes have a black cross-shaped marking on their backs. Sebastian Münster was one of the great geographers of sixteenth-century Germany; Gesner learned from him what the foxes of Germany and Russia look like. Pierre Gilles (1490–1555) was a fellow humanist, who in 1532 published an edition of Aelian's *On Animals*, and who also wrote a book on the names of fish found in the Mediterranean. He was also a traveller who visited Constantinople and wrote a popular account of his visit. Gesner got his tidbit about the fox imitating a bird from Gilles's book on fish (where he had inserted the story as a parenthetical parallel to the fishing frog, who attracts fish with a fleshy worm). The list could go on extensively. This abundance of new information from modern observers is one of the features that distinguishes Gesner's natural history from all of its predecessors.[14]

Now we do have a relatively complete picture of Gesner's natural history. And if we try, perhaps foolishly, to epitomize his achievement, we might say that he established natural history as a humanist discipline, firmly grounded in the writings of antiquity, but that he enriched it with the addition of a whole new world of emblematic and proverbial associations, a new concern for accurate pictures, and the incorporation of contemporary observations and new discoveries. The question we must now address is: what happened to natural history as a result of Gesner's reformulation of the discipline?

The answer, as one might imagine, is complex, primarily because the innovations – emblems and proverbs, naturalistic images, and contemporary observations – lie rather uneasily alongside one another, and each would have the potential to pull subsequent natural history in different directions. Indeed, one can see a conflict arise in the subsequent work of Gesner himself. Gesner continued to gather information on quadrupeds, even after volume one of the *Historia animalium* was published, and this new information appeared as supplements to the later volumes, in separate epitomes that contained just the woodcuts, and finally in a German

Von der Indianischen Mauß.

Ichneumon.

Wo diß thier zů finden.

Figure 2.6 Ichneumons. Woodcut from Conrad Gesner, *Thierbuch* (Zurich, 1551), fo. 115r.

abridgement of all four volumes that was published in 1563. If we follow an animal through these twelve years, we discover that most of Gesner's supplements consist of new or better pictures of the animals. His fox woodcut could not really be improved upon, although he does acquire a picture of the northern cross-bearing fox and includes this in the appendices. His 1551 image of the ichneumon, we recall, was taken from the Oppian manuscript; in 1553 a picture from life was published, and Gesner added this to all later editions (Figure 2.6). The civet had been pretty well represented in 1551, but Gesner was sent two more pictures subsequently, and he prints the better of the two in the 1563 German abridgement. Similarly, Gesner was continually on the lookout for *new* animals. When André Thevet (1504–92) published his *Les Singularitez de la France Antarctique* in 1557, Gesner immediately plundered it for images of the sloth and the toucan, and by 1563 he had added a large contingent of new and exotic animals to his original few. There were many new emblem books published in the late 1550s, but Gesner does *not* seem to have mined them to enrich the emblematic associations of his animals. Apparently the desire to have accurate images, and images of new animals, came to take precedence over the desire to place an animal in a more complex emblematic environment.

Belon and Rondelet

One sees a different trend in the work of two contemporaries, Pierre Belon (1517–64) and Guillaume Rondelet (1507–66). Neither scholar aspired to write a comprehensive encyclopaedia of the natural world, but in the more limited spheres in which they worked – Rondelet on fish, Belon on birds, fish, and exotic animals – each placed great stress on personal observation. Both were trained in humanist natural history, but neither paid much attention in their published works to classical authority, except for Aristotle, and neither had the slightest interest in the emblematic view of nature. But like Gesner, Belon and Rondelet were insistent that narrative be accompanied by pictures 'drawn from life'. In fact, the importance of lifelike image is now so paramount that it is explicitly stated on the title-pages of all their works; Belon's 1553 book on fish carries the sub-title: 'with pictures that are likenesses from life', while Rondelet's 1554 study of marine life carries a similar continuation: 'in which true likeness of fish are given'. One can see why Belon and Rondelet have emerged as heroes to modern zoologists, for their narrow approach to natural history is much more akin to ours than the grand vision of Gesner.[15]

The flowering of emblematic natural history

It would be a great mistake, however, to think that some kind of transformation occurred here – that Gesner's grand emblematic natural history somehow withered in the face of a more observational approach, thanks to Rondelet and Belon, and perhaps even the late Gesner. In truth, it was the Belon/Rondelet vision that faded, for almost a full century, and the emblematic view that flourished. We can see this in the simple fact that it was Gesner's encyclopaedia that was reprinted in 1604 and again in 1617–20 and once again as late as 1669. None of the zoological works of Rondelet or Belon went through subsequent printings. We can see it also in the growing popularity of animal emblems in the last half of the sixteenth century. More and more animals were woven into the emblematic fabric, even some of the newly discovered ones, and the trend was so powerful that by 1595 it was possible to publish a work that would have gladdened Gesner's heart: *Symbolorum & emblematum ex animalibus quadrupedibus desumtorum centuria altera . . .* ('Another Century of Symbols and Emblems Derived from Quadruped Animals') by Joachim Camerarius (1534–98).

Camerarius's book contains one hundred emblems of animals; it was the second of four volumes which altogether contained four

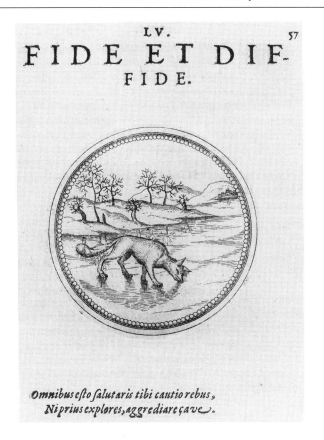

LV.   57

FIDE ET DIF-
FIDE.

Omnibus esto salutaris tibi cautio rebus,
Ni prius explores, aggrediare cave.

Figure 2.7 Emblem of the fox on ice. Engraving from Joachim Camerarius, *Symbolarum et emblematum . . . centuria altera* (Nuremberg, 1595), no. 55, fo. 57r.

hundred emblems of plants, insects, and birds, as well as animals.[16] Each emblem is illustrated with an exquisite engraving, but in addition to the requisite motto and epigram we have a full page of scholarly commentary provided for each emblem. The commentary shows us that Camerarius was as well versed in Aristotle and Gesner as he was in Alciati and Erasmus. Camerarius offered emblems for familiar animals, such as hedgehogs and bears, and for less familiar ones, such as ichneumons, chameleons, and even opossums. The fox has no less than three emblems, and it is instructive that all three are brand new, invented by Camerarius, which certainly shows us the vitality of emblematic natural history in the 1590s. The first (emblem 55) shows a fox crossing a frozen river while listening to the ice, with the motto: 'Trust and mistrust' (Figure 2.7). The story of the vigilant fox, we might recall, is told by Gesner, who in turn found it in Pliny. The second (56) shows a fox sneering at the attempts of two hounds to catch it, and the motto is: 'Security without fear'. The text relates that a pregnant vixen is so clever that it is rarely caught; the story originated in Aristotle, became a proverb in Erasmus, and was of course duly quoted by Gesner. The third emblem (57) shows a badger trying

Figure 2.8 Emblem of the fox and badger. Engraving from Joachim Camerarius, *Symbolarum et emblematum . . . centuria altera* (Nuremberg, 1595), no. 57, fo. 59r.

to get into its den, which is occupied by a fox (Figure 2.8). The motto is: 'What you want, another has'. This emblem, too, was fashioned from a tale, found in Gesner, that goes all the way back to antiquity.

Camerarius's emblem book has been virtually ignored by historians of natural history, but in fact it is very much a work of natural history in the Gesnerian tradition.[17] It is not irrelevant that Camerarius was an extremely competent botanist and published several botanical works that are highly regarded. Yet he clearly felt that the purpose of studying nature was not just to describe and illustrate, but also to create and illuminate and uncover new meanings in the natural world. If Gesner's natural history planted the seeds of the emblematic approach to nature, then the emblem books of Camerarius mark the true flowering of emblematic natural history.

Aldrovandi and the fruition of emblematic natural history

As a final demonstration of the power of emblematic history in the late Renaissance, let us turn to the work of Ulisse Aldrovandi (1522–1605).[18] Aldrovandi taught at Bologna and published the first of what would ultimately be a thirteen-volume encyclopaedia of natural history in 1599; and from the date it is easy to imagine Aldrovandi as several generations removed from Gesner and the other zoologists who published in the mid-sixteenth century. In fact, Aldrovandi was only five years younger than Gesner and Belon, and he came out of precisely the same cultural milieu. He just had the good fortune to live forty years longer than his contemporaries. He was seventy-seven years old when his first volume, on birds, was published, and he lived to see three more volumes printed before his death. Fortunately, his literary executors ensured that his remaining manuscripts were published more or less as Aldrovandi envisioned them. The first quadruped volume appeared in 1616, with another in 1621 and the last in 1637. It is in this later volume, the *De quadrupedibus digitatis viviparis* ('History of Live-bearing Quadrupeds with Claws'), that we encounter, once again, our friend, the fox.[19]

Aldrovandi's article is twenty-eight folio pages long, which makes for an impressive looking package. What is initially most impressive, however, is the illustration: it shows, not the external appearance of the fox, but its skeleton (Figure 2.9). And indeed, the section in which it is embedded is titled 'Anatomica', a term we hardly ever encounter in Gesner. Here at last is a naturalist who is looking beyond surface features for more significant information, such as the details of anatomical structure.

Or so it might seem. Aldrovandi did indeed make many anatomical studies of animals, and he was one of the very first to do so, and woodcuts of skeletons of various specimens are scattered throughout his encyclopaedia. It is quite proper that he be given credit for this. Aldrovandi also introduced a number of new illustrations into his volumes, such as an improved ichneumon (Figure 2.10), and he kept a stable of artists busy furnishing pictures of animals drawn from life. However, it would be a real mistake to conclude that by taking an interest in comparative anatomy, or in improving illustrations, Aldrovandi was somehow turning his back on the emblematic approach as developed by Gesner. One has only to read some of the other section headings of Aldrovandi's fox article to be rapidly disabused of this notion. After shorter sections on Names, Habits, Voice, Food (and Anatomy), we find sections

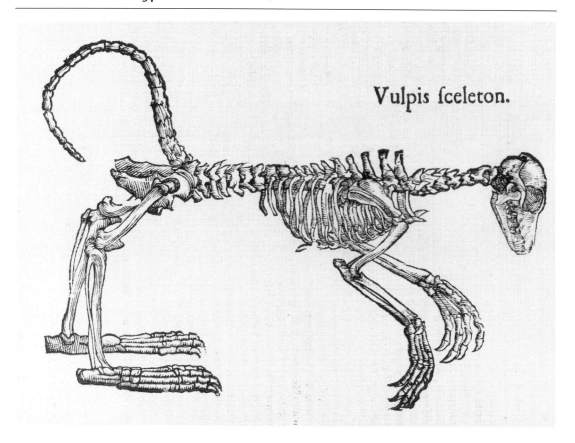

Vulpis fceleton.

Figure 2.9 Fox skeleton. Woodcut from Ulisse Aldrovandi, *De quadrupedibus digitatis viviparis* (Bologna, 1637), p. 198.

on Antipathies and Sympathies, Physiognomy, Epithets, Emblems and Symbols, Fables, Hieroglyphics, Proverbs, Allegories, Morals, Omens, and Symbolic Images, to name just some of the headings.

What Aldrovandi has done is not to break away from emblematic natural history, but rather to expand the emblematic world of animals by adding in all the new emblems, adages, and images that appeared in the half-century after Gesner. In the paragraph on Emblems, for example, we find not only the one Alciati emblem that Gesner had already extracted, but two of the three from Camerarius, as well as two further examples from other emblem books that we have not discussed. Gesner had limited knowledge of Egyptian hieroglyphics, since only the short text of Horapollo was available in 1551, but Aldrovandi had access to Piero Valeriano's prodigious *Hieroglyphica*, an exhaustive commentary on Egyptian symbolism that was first published in 1556, and he drew on it frequently. Aldrovandi also had the advantage of being able to consult Cesare Ripa's *Iconologia*, a fascinating work first published in 1593 that described accepted ways of personifying such abstract concepts as Truth or Nature. Many personifications involved animals, so that Arrogance, according to Ripa, was represented by a woman

Figure 2.10 Ichneumon. Woodcut from Ulisse Aldrovandi, *De quadrupedibus digitatis viviparis* (Bologna, 1637), p. 301.

holding a peacock, while Natural Instinct was to be personified by a youth trying to prevent a weasel from entering the mouth of a toad. Aldrovani was very fond of Ripa's constructions, and in the Personification section of the fox article we learn that a pair of foxes is one of the attributes of Christian faith, where they represent the heretics that must be converted. Aldrovandi could also utilize Giovanni Battista della Porta's *De humana physiognomonia*, published in 1586, which compared animal and human faces and argued that similarities in appearance reveal similarities in character. Porta has a great deal to say about the vulpine personalities of people with long noses and close-set eyes, and if you do not wish to consult Porta, you can find it all neatly digested in Aldrovandi.

Aldrovandi's real achievement, then, was in bringing the emblematic view of nature to fruition. In his many volumes he was able to weave a richer fabric than Gesner ever envisioned, because he had more threads and colours to work with, and if you believe that the goal of natural history is to capture the entire web of associations that inextricably links human culture and the animal world, then Aldrovandi's natural history was remarkably successful.

The collapse of emblematic natural history

Fifty years after Aldrovandi's death, the emblematic view of nature would collapse. Natural history would suddenly take on a more familiar form, as naturalists abandoned the entire associative framework and began to focus on description and anatomical investigation, with the ultimate goal of a natural system of classification.

Conjectures as to the cause of this demise have been offered, but we will not take up the problem here. Perhaps for this occasion we should adopt the position of Michel Foucault, who said that it simply happened; after 1650, people abruptly ceased to think in terms of associations and similitudes as ordering principles of nature and began to look at the world in other ways.[20]

For the student of the Renaissance, the important point is not that emblematic history ultimately disappeared, but rather that for a full hundred years it flourished, and indeed dominated attitudes toward nature. If we wish to study Gesner, or Aldrovandi, or Camerarius, and understand their works in the spirit in which they were written, then we must be prepared to think adagially, allegorically, and analogically. We must see an animal as a symbol, a character, in some greater language of nature. To paraphrase Galileo, if we wish to read the book of nature, we must first comprehend the language and read the letters in which it is composed. In the late Renaissance, that book was written in the language of emblems, and if you do not read that language there is little that will make any sense.

Further reading

Ackerman, James S., 'Early Renaissance "naturalism" and scientific illustration', in Allan Ellenius (ed.), *The Natural Sciences and the Arts* (Uppsala, 1985), pp. 1–17.

Ashworth, William B., Jr., 'Natural history and the emblematic world view', in David C. Lindberg and Robert S. Westman (eds.), *Reappraisals of the Scientific Revolution* (Cambridge, 1990), pp. 303–32.

Copenhaver, Brian P., 'A tale of two fishes: magical objects in natural history from antiquity to the scientific revolution', *Journal of the History of Ideas*, 52 (1991), pp. 373–98.

Delaunay, Paul, *La Zoologie au seizième siècle*, Histoire de la pensée, 7 (Paris, 1962).

Doggett, Rachel, Dunnington, Jean and Miller, Jean, *Fabulous Beasts: Renaissance Animal Lore* (exhibition catalogue) (Washington, DC, 1993).

Findlen, Paula, *Possessing Nature: Museums, Collecting, and Scientific Culture in Early Modern Italy* (Berkeley, 1994).

Foucault, Michel, *The Order of Things: An Archaeology of the Human Sciences* (New York, 1970).

Friedmann, Herbert, *A Bestiary for Saint Jerome: Animal Symbolism in European Religious Art* (Washington, DC, 1980).

George, Wilma, *Animals and Maps* (Berkeley, 1969).

Gmelig-Nijboer, Caroline Aleid, *Conrad Gesner's Historia Animalium: An Inventory of Renaissance Zoology* (Meppel, 1977).

Harms, Wolfgang, 'On natural history and emblematics in the sixteenth century', in Allan Ellenius (ed.), *The Natural Sciences and the Arts* (Uppsala, 1985), pp. 67–83.

Koreny, Fritz, *Albrecht Dürer and the Animal and Plant Studies of the Renaissance*, trans. Pamela Marwood and Yehuda Shapiro (Boston, 1988).

Praz, Mario, *Studies in Seventeenth-Century Imagery*, 2 vols., Studies of the Warburg Institute (London, 1939–47).

Raven, Charles E., *English Naturalists from Neckham to Ray: A Study of the Making of the Modern World* (Cambridge, 1947).

Thomas, Keith, *Man and the Natural World: Changing Attitudes in England 1500–1800* (London, 1983).

3 The culture of gardens

The garden

God once planted a garden eastward in Eden: and there he put the man whom he had formed, to dress it and to keep it. God walked in this garden in the cool of the day. But man ate the forbidden fruit of the tree of knowledge and God expelled him from this garden. Cursed is the ground for thy sake, God said to Adam, in sorrow shalt thou eat of it all the days of thy life; and God sent Adam forth from the garden of Eden, to till the ground from which God had made him.

God never made another garden. In the Christian tradition all other gardens have been made by man seeking, often, to recreate that half-remembered half-imagined Paradise, and to remove God's curse from the ground.

Thus all gardens except that first Eden are the result of human art and skill: they are Nature re-arranged by artifice. Within the garden an aesthetic prevails: beauty is constructed from rearranging Nature. Such rearrangement needs unremitting work; all gardens need the constant nurturing hand of the gardener. The gardener's work preserves their identity as gardens, and keeps them from slipping back into non-gardens. If the garden falls into neglect all the vistas become overgrown, the exotic introductions fail, and the land is 'recolonized' by native plants.

Dirt, says Mary Douglas, is matter in the wrong place. Gardens, one might say, are Nature put into the right place. And that place is an enclosed place. For all gardens are spaces delimited, areas which man has artificially enclosed, with either a physical or visual boundary. The boundary is essential. Even when the garden is intended to portray an unbounded landscape, there are boundaries, visual and physical, between garden landscape and landscape landscape. Within the wall there is order and cultivation, outside there is wildness. Inside there is rationality, care and nurturing, outside there is Nature raw and savage. Inside there is art, outside there is Nature. Brought within the wall, the weak can be made strong by preferential growing conditions; but if a plant should 'escape' over the wall (as gardeners say) and into the world, then what inside the garden had been treasured and pampered, outside its walls becomes a nuisance and a weed.

God and the garden

Gardens are spiritual places, and always have been, in both the Western and Eastern garden traditions. *Paradise* is a word adopted into Greek from old Persian, where it meant 'an enclosed park or garden'. This equation is very important. Paradise is a garden: a garden is a paradise. We can make a paradise on earth by making a garden. And the central function of this paradise is spiritual, primarily that of contemplation and spiritual renewal; and it is also a place for mystical thinking. A garden is a place of peace and harmony, a place to promote and reflect spiritual well-being, a place where there is refreshment for the soul.

What is most interesting about this from a modern perspective, and particularly from the perspective of the naturalist, is that gardens and gardeners were not particularly concerned with plants, which is probably the first thing that we think of today with respect to gardens, with our passion for planting in our own little paradises.

The first garden in the Judaeo-Christian tradition was a Paradise, and an enclosed one: the garden or paradise of Eden. Adam and Eve were thrown out of this garden, and the attempt to get back to it can, perhaps, be characterized as the point of Christianity. The only things which the Garden of Eden contained, that are mentioned in Genesis, are a *river* to water it, which divides into four streams as it leaves the garden, and *trees*: 'all trees pleasant to look at and good for food'. No other plants are mentioned.

The essential enclosedness of gardens is particularly important in the Christian tradition with respect to the cult of Mary. For the garden itself, as well as certain flowers (the lily and the rose), came to represent Mary and her virtues. St Bernard of Clairvaux succeeded in showing that the unmistakably sexual garden imagery of the erotic poem 'The Song of Songs' (or Song of Solomon) in the Bible, actually referred to and celebrated asexuality: it was a paean to the chastity and purity of the Virgin Mary. This was the kind of expert exegesis which marked Bernard out clearly as a saint. Indeed, St Bernard managed to show that the Virgin was herself the 'garden close-locked' or *hortus conclusus* referred to in the Song of Songs. (In the Authorized Version it is 'a garden enclosed is my sister, my spouse; a spring shut up, a fountain sealed'.) As a consequence, much of the visual representation of gardens which we have from the Middle Ages is a product of this cult of Mary and her virginity, showing her *as* 'a garden close-locked' *inside* 'a garden close-locked' (Figure 3.1). This was, of course, centuries before Linnaeus joyfully

Figure 3.1 The Virgin Mary, 'a garden close-locked' inside 'a garden close-locked'. From a *Livre des chants royaux* (15th century, Paris), Bibliothèque Nationale, MS. fr. 145.

showed that plants have sex, so the garden could quite reasonably be thought of as a sex-free zone.

In the Islamic tradition the garden as Paradise is, perhaps, even stronger, since the Koran describes heaven as a garden, a state of blessedness. The reward of the faithful will be to dwell in a garden with 'spreading shade . . . fruits and fountains and pomegranates . . . and cool pavilions'. Again, flowing water is presented as an important element of a garden, but as in the Judaeo-Christian Eden, plants (as opposed to trees) are not particularly important.

Islam was spread to the Persians, and reciprocally the Muslims adopted the Persian style of garden: so an Islamic garden is enclosed, it has flowing water, usually in water-courses, it has trees and aromatic plants, and often has buildings within it. The garden is often seen as the 'soul', compared to the 'body' of the building that it accompanies. When the Moguls went conquering to India, they took this model of Paradise garden with them: as a nomadic people they often established their tents within such gardens. Surviving examples of Islamic Paradise gardens include, in India, the Taj Mahal, the Red Fort in Delhi and, in Spain, the gardens of the Alhambra in Granada.

Philosophy and religion have always been close, separating in the West into contrasted spheres only in the last couple of centuries. The philosopher, the seeker of wisdom, has also customarily found his spiritual solace in the garden. In Antiquity the philosopher Democritus used his garden to practise philosophy and go mad in, and Hippocrates supposedly found him there 'sitting upon a stone under a plane tree, without hose or shoes, with a book on his knees, cutting up several beasts, and busy at his study'.[1] Much later, in 1625, another famous philosopher, Lord Chancellor Bacon (Francis Bacon) in England was to write an essay on gardens-fit-for-a-king, and created a more humble one himself. A garden, Bacon wrote, 'is the purest of human pleasures. It is the greatest refreshment of the spirits of man'.

Power and the garden

This spiritual purpose, both religious and philosophical, was a primary purpose in creating gardens, and it was a guide to what a garden should comprise and how it should be constructed and used. This spiritual dimension continued into Renaissance gardens and it was explicit in much discussion and planning of such gardens. Traditionally, historians have paid most attention to the appearance of new, additional motivations, interests and uses for gardens in the Renaissance, but it should be remembered that the spiritual role underlay all these innovations and outlasted some of them, and spiritual refreshment is still a role of gardens today.

Two new kinds of garden were created in Renaissance Europe. Both exhibited a new enthusiasm for gardens as such, and this enthusiasm produced gardens on a new scale, in great numbers, and with new roles. Both new kinds of garden were attempts in some measure to recreate, to give new birth to, ancient practices. The practices being 'reborn' were those of the ancient Greeks and Romans, but as the ancient Greeks seemed to have had little interest in making gardens, attention was much more on the garden world of the ancient Romans. Literary evidence was the main source: garden creators had to read Latin.

The innovations that were made in attempting to copy the Ancients included the laying out of gardens using geometric shapes: perfect circles enclosing squares, enclosing circles, enclosing squares. Geometry and proportion ruled, harmony prevailed. Classical-style buildings were erected in the garden. Classical statuary, both genuine and faked, was introduced into the garden in order to represent the classical gods and myths, as the Greek Ancients themselves had done in their groves. Fountains had adorned the gardens of the ancient Romans, and they had created grottoes and pergolas. The Ancients had had terracing in their gardens and theatres. It was learnt that they had planted trees in rows, had used grafting to create better trees, had employed topiary, and that they loved certain trees – the elm tree, the plane tree for shade, and the bitter-orange tree (*melangolo*) for ornament. All these were adopted into the new Renaissance gardens. And these new gardens were sometimes also constructed on the best understanding of Platonic philosophy, and were intended to reflect cosmological harmony.

The ancient Roman fascination with topiary, for instance, the Renaissance garden-makers learnt from the letters of Pliny the Younger, who had described his two gardens in letters to his friends in about 100 AD. His Tuscan villa garden was full of box clipped into animal shapes:

In front of the colonnade is a terrace laid out with box hedges clipped into different shapes, from which a bank slopes down, also with figures of animals cut out of box facing each other on either side. On the level below there is a bed of acanthus so soft one could say it looks like water. All round is a path hedged by bushes trained and cut into different shapes, and then a drive, oval like a racecourse, inside which are various box figures and clipped dwarf shrubs. The whole garden is enclosed by a dry-stone wall that is hidden from sight by a box hedge planted in tiers.[2]

This Tuscan garden of Pliny also had plane trees and gentle fountains.

The first of the new kinds of garden in the Renaissance was the princely garden for the attention and the admiration of other people, rather than kept closed (*conclusus*) for the personal spiritual refreshment of its princely owner and his friends. Like the palaces that they were concurrently constructing for themselves, these gardens were intended to say something about the power and importance of their princely creators.

It was a novel thing in early modern Western Europe to use a garden as a weapon in the game of social and political power. In accordance with the general Romantic characterization of the Renaissance, the princely gardens have usually been seen by historians as 'gardens as works of art', with the implicit assumption that a 'work of art' speaks for itself about the social claims of its owner. But it should be rather surprising that a garden, *as a garden*, could have been made into an effective way of making a statement about social standing and political ambition, especially given its history as a spiritual retreat. The first requirement in order for the garden to fill such a role was, naturally, that it had to be taken out of the closet: the enclosed garden had to be opened up to a wider audience. But in itself this was not enough; presentational skills were needed, and this is where the architect was brought in.

These power gardens were begun in the mid- to late-fifteenth century in Italy, by the rulers of the richer and more powerful city-states. Rome and Florence were the most important early centres. I shall concentrate here on one particular garden as exemplary of certain features of Renaissance princely gardens, one which has been described as one of the three most influential gardens in European history.[3] It is also a particularly good one to choose as a source of the view of the garden as power dressing, since it concerns the political ambitions of one of the greatest of earthly princes of the Renaissance, the Pope. Moreover, it was two gardens in one.

These gardens were created by Pope Julius II (pope from 1503 to 1513), who was taking up in its entirety the scheme of an earlier Pope, Nicholas V (1447–55), to create an *instauratio Romae*: 'The whole Vatican area was to form the imperial palace of the new Julius Caesar, *imperator* and *pontifex maximus*, an explicit celebration of the pope-emperor'.[4] The important garden innovations were part of this imperial scheme of Julius II and, naturally enough, as they were intended to achieve imperial goals, so they were constructed on an imperial scale. The Renaissance garden, in its origin, was an instrument to make claims about power, as well as classical learning.

The two gardens that bore this weight of imperial ambition were part of the Belvedere Court, a vast, multilevel construction connecting the papal palace to the Villa Belvedere up a little hill.[5] It

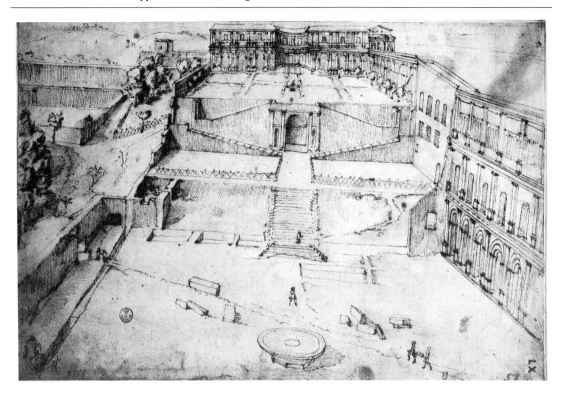

Figure 3.2 Bramante, Belvedere Court in the Vatican Palace – original drawing, viewed from the primary viewpoint in the papal apartments. The important role of staircases, and the minimal role of plants, is clear from this sketch.

was designed by the papal architect, Donato Bramante (1444–1514), first architect of the new St Peter's. An enormous space, over 300 yards long and almost 100 yards wide, was designed with a great empty court at the lowest level serving as an open-air theatre, connected by spectacular stairways to two higher levels of formal gardens with box-outlined parterres and fountains (Figure 3.2). Following Bramante's cue here, grand staircases were to be a leading feature of many later formal gardens. The point of this arrangement was to create a vista: the stairs and terraced gardens created a view which allowed the eye to see the whole garden as one great construction, and allowed the viewer to be impressed by the wealth, taste and learning of the owner of the garden. In the case of the Belvedere, the whole great court with its two levels of garden was designed to be seen at its best from one particular viewpoint in the papal apartments. To achieve this aim of making a visually dominant viewpoint, Bramante employed in his construction that Renaissance innovation in the visual arts – linear perspective – thus making the whole garden look like a Renaissance painting and enable it also to be experienced like one. This is where it is important that this and other gardens of the rich and famous were designed by architects, those practitioners of a classical art of great importance as an art of persuasion, itself being resuscitated

according to ancient models and standards at this time. For architects design deliberately to achieve particular visual effects, in
order thereby to affect the emotions and opinions of the viewer
and convey messages to him or her, in this case messages about
the Pope's cultivated appreciation of Antiquity, his claims to
imperial power and the proper central role of the Pope in the
modern, sixteenth-century, world. Although such architect-
designed gardens, like the Belvedere Court, often include many of
what are, in essence, *horti conclusi*, these enclosed gardens have
now been brought out into the open, and the intention is always
that the eye of the spectator should see beyond the separate
enclosed-gardens-without-walls, and be led to the grand view or
vista of the garden as a whole. The garden was now part of a bigger
picture.

But there was another garden in the Belvedere Court, of equal
importance and influence. This was a special enclosed garden,
though on a grand scale (Figure 3.3). The distinctive thing about
this garden was that in it ancient (that is, classical) sculpture was
displayed. A Venetian visiting this garden in 1523 wrote:

One enters a very beautiful garden, of which half is filled with growing
grass and laurels, mulberrys, and cypresses, while the other half is paved
in squares of tiles, laid end on end, and in every square a beautiful
orange tree grows out of the pavement, of which there are a great many,
arranged in perfect order. In the centre of the garden there are two
enormous men in marble; one opposite the other, twice life size, who
lie in sleeping position. One is the Tiber, the other the Nile; very ancient
figures, and two fountains issue from them. At the main entrance to
this garden on the left there is a sort of chapel built into the wall where,
on a marble base, stands the Apollo, world famous, a very beautiful and
worthy figure, life size, of the finest marble. Somewhat further on . . .
is the Laocoon, celebrated throughout the world . . . Not far from this
. . . there is a lovely Venus of life size, nude, with a bit of drapery
over her shoulder which partly covers her genitals: as beautiful a piece
as it is possible to imagine.[6]

This garden was making many statements at once. It was displaying outstanding Greek statuary from the Roman imperial past
as trophies of the new imperial claimant, the Pope; it was simultaneously stepping beyond the tradition of the ancient Roman
garden back to the Greeks, by adopting the ancient Greek practice
of placing statuary in the groves where philosophers walked and
talked.

This was the first of all the many statuary gardens in which
antique statuary and inscriptions were displayed among the laurels
and the bitter-orange trees. But it was never to be excelled. For
the statues displayed in this garden were some of the most important and beautiful of all statuary that has been recovered from the

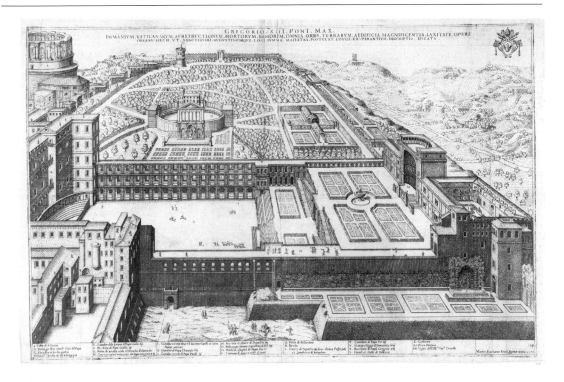

Figure 3.3 The garden and sculpture garden in Bramante's Belvedere Court, in the Vatican Palace. The court is here seen from the side, which is not a viewpoint intended by Bramante, and is potentially confusing. The enclosed sculpture garden is at the very top of the court, and is the walled structure on the right of the picture, letter 'I': the 'Giardino secreto doue e il lacoonte l'apollo et altre statue anticke'. A reclining river god, either the Tiber or the Nile, can just be discerned in it. The main entrance to this sculpture garden was up Bramante's famous circular staircase, which is within the tower at the bottom right of the picture. The papal apartments are on the left of the picture, and the intended vista of the court was from the left of the picture toward the right. The various small papal gardens and wood in the background are not part of the Belvedere court, or of the intended vista. From Antoine Lefrery, *Speculum Romanae Magnificentiae* (Rome, 1574).

ancient world, and were to include what is possibly the greatest of all antique statues, the Belvedere torso – which Michelangelo used as a motif in his paintings for the Sistine Chapel, and of which he supposedly said, 'Truly this was created by a man who was wiser than Nature! Pity that it is a torso.' These statues had been discovered in Rome in the course of the developments that the papacy was making in order to rebuild the city as a renewed imperial centre under the rule of the Pope.

The Belvedere Court was never fully finished according to Bramante's designs, and a later Pope, Pius V (1566–72), removed the statues from the statue garden, while at the end of the sixteenth century another Pope was to appropriate the higher-level garden of the court to build the Vatican library. But the influence of these two Belvedere gardens was nevertheless immense, with respect to the purpose, scale and content of princely gardens.

The great princely gardens of Italy, France and England of the sixteenth and seventeenth centuries all build on these features, adapted to the local terrain and circumstances. Such grand perspectival vistas came to be typical of many subsequent gardens of Rome, especially those built by the numerous resident cardinals, and of other princely gardens of Italy. The French grand garden, by contrast, was generally on flatter ground, so it usually had lakes where the Italian gardens had fountains, and the garden was designed to be seen from the grand house or a raised walk or terrace, with beds one looked down on. Versailles, with its great vistas and avenues, is such a garden on the very grandest scale. The patterned floral parterre, or bed, became the fundamental item of a French garden, and as its intricate pattern could be made equally well and more conveniently with coloured sands and tiles, the parterres in the grand gardens came at one time to have no flowers in them at all.[7]

Plants and the garden

The classical gods had now been introduced to the Paradise of the Christian tradition. The first great fashion of garden ornamentation of the sixteenth century had been for classical sculpture; the second was for increasingly extravagant and impressive fountains, and here the Tivoli gardens constructed by the Cardinal of Ferrara, Ippolito II d'Este, take the laurel and spectacularly outdo the ancient Romans at their own art of garden fountains. But still, even toward the end of the sixteenth century plants and flowers – their exotic allure, their variety of colour and shape, their beauty – were not particularly important. It was not until the seventeenth century that 'the gardener came into his own as master of the garden, diminishing the role of the fontaniere'.[8] Thus, perhaps surprisingly, the introduction of new plants into Europe through the voyages to the New World, had not in itself automatically led either to the creation of these princely gardens, or to their transformation into plant gardens.

The primary way in which gardens became full of plants was through the establishment of 'physic' or 'botanic' gardens. These gardens were first established in medical faculties of universities, and their initial aim was to improve the education of future doctors in the knowledge of the plants which made up the basis of most drugs. The first such gardens were created in the 1540s in Padua and Pisa, the universities of Venice and Florence respectively, and sometimes with a lectureship in the 'simples'. Shortly after,

Figure 3.4 The Padua physic/botanic garden today: the garden is still on the original site and the beds are laid out in geometric patterns. The circular shape was thought most perfect for a garden, but in practice it gave no room for expansion, and Padua was one of the very few places where it was adopted.

gardens were established in Florence, Pavia, Ferrara, Bologna, Rome, Messina and other places in Italy.[9] Elsewhere gardens were established more slowly: at Montpellier in 1598, in Paris (the Jardin Royal des Plantes Médicinales) in 1640. In Britain the Oxford Physic Garden was founded by a private benefaction from Sir John Danvers in 1621; the Chelsea Physic garden in London in 1673 to teach apothecary apprentices; the Edinburgh garden (now covered by Waverley Station) in the 1670s by two concerned doctors; the Cambridge garden not until 1763, though there had been earlier attempts.

The physic gardens were Renaissance gardens in two ways. First their arrangement was on the same principles as other gardens, and geometrical in layout. Padua, for instance, is a circle with squares inscribed, and circles within the squares; Oxford is square, subdivided into squares. Second, their creation was linked to the attempt to recover something ancient: the medical plant lore described in the works of the ancient Greeks who wrote on plants and medicines: Dioscorides, Theophrastus and Galen.

Medicine ('physic') was the occasion for the making of these gardens, but their flourishing stemmed from the fact that they came to be run by people passionate about plants: 'botanists' in the root meaning of the Greek term, plant enthusiasts. A 'botanic' garden is a plant garden. In most physic gardens we find that the leading spirit was someone with primarily botanical interests, such as Luca Ghini in Pisa or Ulisse Aldrovandi at Bologna. Hence these gardens were used as a sort of depot for the collection,

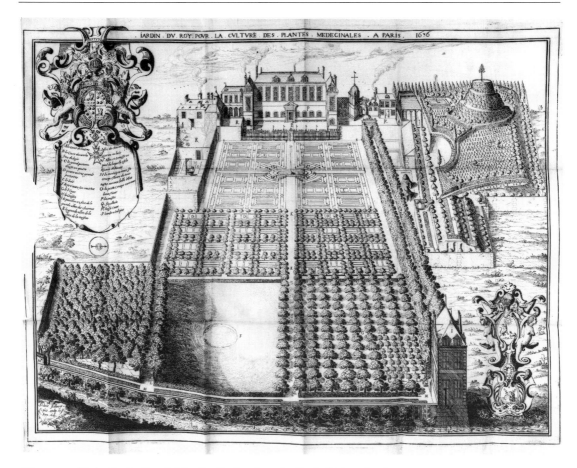

Figure 3.5 Paris: Jardin Royal. The Jardin Royal, or Jardin du Roi, or Jardin des Plantes, was founded in Paris in 1640 with royal support, as a public pleasure garden as well as a physic garden. The moving spirit, Guy de la Brosse, was a physician in ordinary to Louis XIII. Lecturers were endowed in the garden on royal funds, and it was possible to obtain a virtually complete medical education there free. In the course of the French Revolution it became the Museum of Natural History (Muséum d'Histoire Naturelle), and in that guise still exists on its original site. From Guy de la Brosse, *Description du jardin royal des plantes médicinales* (Paris, 1636).

storage and distribution of new plants, and they were centres of correspondence and exchange networks. The aims of the keepers of these gardens (in contrast with those of the medical faculties who paid for the upkeep) came to be to use these gardens as places in which to assemble and culture all the plants in the world; the number of plants being, of course, constantly increased by introductions from the New World. In this sense the gardens become living catalogues of plants, living catalogues of Creation. And in these Gardens of Eden, these plant paradises, the botanists tried

Figure 3.6 The growing and preparation of medicinal herbs, from the physic garden to the patient; the distilling process, to extract the medically active essences of plants, is also illustrated. Woodcut from the title-page of Eucharius Rösslin, *Kreuterbuch* (Frankfurt, 1536).

to find the 'natural order' existing among plants: the natural relationships, the families in which God had created them.

This activity in growing new plants in the garden was reflected in print, and a number of large illustrated books on plants were published during the sixteenth century and found a ready market: such as Otto Brunfels' *Herbarum vivae icones* ('Living Pictures of Plants') (1530); Leonhard Fuchs' *De historia stirpium* ('On the History of Plants') (1542); Pietro Andrea Mattioli's *Commentarii* on Dioscorides (1554); and Rembert Dodoens' *Cruydeboek* ('Herbal') of 1563, which was pillaged by John Gerard for his *Herball, or Generall Historie of Plantes* (1597) (Figure 3.7). The book acted as a garden: around 1649 a Cambridge medical student was advised to read John Parkinson on plants (either his *Paradisi in sole paradisus terrestris. Or, a Garden of All Sorts of Pleasant Flowers which our English Ayre will permitt to be noursed up* (1629), or his *Theatrum Botanicum: The Theater of Plantes* (1640)), for Parkinson

does not only outvie Gerard in the number, but also in the method of his plan, sorting and bedding them so neatly together in divers knots and tribes, that you enter into him, as in to a fair well ordered garden, where you may readily pick out which herb you will.[10]

Figure 3.7 Part of the title-page of Gerard's *Herball*.

The book is a garden: the garden is equally a book in which God's planning of the natural world can be uncovered. Indeed, as the botanists came up with ever new versions of the 'natural order', so the gardeners would have to dig the plants up and rearrange them in the beds. Once again the gardeners turned to Latin, not this time to read the Ancients on gardens, but to name coherently and systematically, in a common and neutral language, the residents of the garden, the plants; thereby they created botanical Latin, a language the ancient Romans had never known.

The range and variety of plants increased dramatically through the use of the physic gardens, and catalogues were sometimes published which reveal the extent of the living collections at certain dates. From the catalogue of the Padua Garden, we know that in 1644 it had

11 varieties of anemone, 10 of geranium, 13 different irises, no less than 15 varieties of malva (mallow), three of the 'marvel of Peru' . . . 13 different narcissi, nine orchids, three water lilies, five peonies, eight poppies, perhaps as many as 20 varieties of the ranunculus or buttercup family, and 11 roses.[11]

Some people came to see the beauty of the garden as depending largely on the variety and colour of flowering plants,[12] especially those introduced from abroad – 'outlandish' plants, as John Parkinson called them. Flowers were all the fashion in many private gardens of the late sixteenth and seventeenth centuries all over Europe, as is evident from Sir Thomas Hanmer, writing around 1659 in England:

As soon as peace hath introduced plenty and wealth into a country men quickly apply themselves to pleasures, and by degrees endeavour to improve them to the height. Amongst the innocent ones, persons of quality and ingenuity have in all ages delighted themselves with beautiful gardens, whose chiefest ornaments are choice flowers, trees and plants; after which rarities the French and Dutch have for some years been most diligent enquirers and collectors, as the Italians and Germans before them, and since our late war many of this nation. The rich among us now are not satisfied with good houses and parks, or handsome avenues and issues to and from their dwellings, their ambition and curiosity extends also to very costly embellishments of their gardens, orchards and walks, and some spare no charge, amongst other things, in procuring the rarest flowers and plants to set them forth withal.[13]

The flower which attracted most passionate attention was the tulip, introduced from Turkey in the mid-sixteenth century; there was an extraordinary episode of 'tulipomania' in Holland from 1643 to 1647 when, after a frenzy of betting on the colours and patterns of flowers that would grow from newly bred bulbs, there was then a 'bubble' of financial speculation in bulbs, and when this burst many people were left heavily in debt.[14]

The plantfulness of gardens was not to last: both the French formal garden and, later, the 'landscape garden', made flowers and plant variety less important than the grand vista. The passion for planting flowers in gardens does not seem to have returned until the nineteenth century.

Cultivating one's garden

Finally we turn to two other senses of 'the culture of gardens': garden culture, and garden cultivation. That is to say, on the one hand the role of gardens as places to practise social cultivation, and on the other the practical cultivation or horticulture needed to keep the vistas visible and to make the plants and trees grow where art required them.

We often know, or can discover, what garden owners and planners were trying to achieve with their gardens, in terms of effect. But of their actual use we know almost nothing systematically. Yet it is clear that the garden was a most important social location at certain periods for both private and public purposes: for meetings, plays and picnics, for discussing philosophy, war, peace and love, for trysts and quarrels, and, best of all, for firework displays. The gardens of sixteenth-century Rome, for instance, were used for many kinds of entertaining: jousting, banquets, pageants, masques, philosophical discourses, poetry recitals, plays and concerts.[15] The gardens of England in the late sixteenth century were where lovers met secretly, to the disapproval of more puritan observers:

In the fields and suburbs of the cities they have gardens, either paled or walled round about very high, with their arbours and bowers fit for the purpose . . . And for that their gardens are locked, some of them have three or four keys a-piece, whereof one they keep for themselves, the other their paramours have to go in before them, lest haply they should be perceived, for then were all their sport dashed . . . These gardens are excellent places, and for the purpose; for if they can speak with their darlings nowhere else, yet there they may be sure to meet them and to receive the guerdon [reward] of their pains: they know best what I mean.[16]

When social events happen in a garden, the garden gives them added meaning, or different meaning, and makes possible types of social negotiation and relation not possible within the different, and more stationary, geography of a house or hall. A walk in a garden can lead to contrived encounters. One can hide in the garden or meet secretly in the dark between the hedges; one can whisper without being overheard. And gardens are, after all, deliberately constructed to create mood: this mood colours and shapes what does and what can go on between people in a garden. A social history of garden use, if we had one, would be very revealing of attitudes and behaviour at different times.

And last of all there is the history of the culture or practical cultivation of gardens, the history of horti-culture, of the 'low culture' of how gardens were actually cultivated, and by whom, the

Martius, Aprilis, Maius, sunt tempora ueris ·   **VER** *Pueritię compar*   *Vere Venus-gaudet florentibus aurea sertis ·*

Figure 3.8 'Spring': a Dutch garden scene of 1570; engraving of a picture by Pieter Bruegel the Elder. The spring garden tasks are being carried out to the instructions of the mistress of the house by garden labourers of both sexes, who are mostly portrayed as faceless.

history of husbandmen or practical gardeners and their knowledge and skills. Here the historians have not been so busy as they have been in investigating the world of garden owners. We lack a history of garden*ing*: a history of skills in the garden, of tacit knowledge, a history of how it was made possible to grow all the wrong things in all the wrong places all of the time, a history of all the practical knowledge without which none of the plans for gardens could have been carried into effect. Although some cardinals and princes believed that it was good spiritual exercise to put their own hand to the planting,[17] and there were passionate gentlemen gardeners at times, as for instance in mid-seventeenth-century England and the Netherlands, yet the back-breaking, routine work of gardening in all weathers fell not to the owners of grand gardens but to their servants.[18] The people with these skills were, for the most part, people who did not write their knowledge down, or did not know how to write at all. Even a garden technician who rose to the highest level of theory in garden planning such as André Le Nôtre (1613–1700), 'Gardener of Kings, and King of Gardeners', who designed and supervised the construction of the Versailles gardens for Louis XIV of France, and who is universally recognized as a genius at garden design, did not write anything down about his craft. However, Le Nôtre was very aware of the importance of his (practical) gardening skills to his achievement in the high theory of making gardens. In a famous moment, when the king ennobled him, Le Nôtre chose for his coat of arms three snails and a cabbage, and when Louis XIV expressed astonishment, Le Nôtre said, 'Sire, would you have me forget my spade? Think what I owe it!'

It is certainly the case that some history of practical gardening could be written, paralleling the history of theoretical gardening that we already have. A few practitioners did write down something of the skill and craft of gardening, and those writing in English included Thomas Hill, with his gardening books of the 1560s and 1570s, such as *The Gardener's Labyrinth*, and Phillip Miller, 'Gardener to the Botanic garden at Chelsea', with his very popular *Gardener's Dictionary*, published from the 1720s. From such works we can get a glimpse of the endless clipping, manuring, planting, propagating, replanting and weeding that a formal garden required, of the advance planning to make the garden immaculate and perfect for whenever its master should visit it, and of the trades which grew up to supply the demand for seeds and plants.[19] Practical gardeners of the past are indeed visible if not yet audible, for they can be seen getting on with the job in countless drawings and paintings of royal and noble gardens – though, as they were 'only' servants engaged in a mechanic art, they are usually portrayed as faceless men and women: anonymously. But there could be joy and spirituality in such work for the labourers, as well as sweat. Thomas Hill, a lover not just of gardens but of plants and practical gardening, wrote in 1577:

The life of man in this world is but thraldom, when the Sences are not pleased and what rarer object can there be on earth . . . than a beautifull and Odoriferous Garden plot Artificially composed, where he may read and contemplate on the wonderfull works of the great Creator, in Plants and Flowers; for if he observeth with a judicial eye, and serious judgement their variety of Colours, Sents, Beauty, Shapes, Interlacing, Enamilling, Mixture, Turnings, Windings, Embossments, Operations and Vertues, it is most admirable to behold, and meditate upon the same.[20]

Further reading

Clifford, Derek, *A History of Garden Design* (London, 1962).

Comito, Terry, *The Idea of the Garden in the Renaissance* (New Brunswick, NJ, 1978).

Goody, Jack, *The Culture of Flowers* (Cambridge, 1993).

Harvey, John, *Medieval Gardens* (London, 1981).

Hill, Thomas, *The Gardener's Labyrinth* (orig. 1577), ed. with an introduction by Richard Mabey (Oxford, 1987).

Hobhouse, Penelope, *Plants in Garden History* (London, 1992).

Lazzaro, Claudia, *The Italian Renaissance Garden* (Yale, 1990).

Moynihan, Elizabeth B., *Paradise as a Garden in Persia and Mughal India* (London, 1980).

Prest, John, *The Garden of Eden: The Botanic Garden and the Re-Creation of Paradise* (Yale, 1981).

Reeds, Karen, *Botany in Medieval and Renaissance Universities* (New York, 1991).

Woodbridge, Kenneth, *Princely Gardens: The Origin and Development of the French Formal Style* (New York, 1986).

4 Courting nature

Between the mid-fifteenth and the mid-seventeenth centuries the study of nature enjoyed a remarkable resurgence within Western Europe. Natural history as a form of writing had been codified by the ancients in the Greek philosopher Aristotle's (384–322 BC) works on animals, his disciple Theophrastus' (*c.* 371–286 BC) works on plants and stones, the physician Dioscorides's (*c.* 40–80 AD) *Materia medica*, and the Roman encyclopaedist Pliny's (22/23–78 AD) *Historia naturalis*; but natural history as a discipline was an early modern invention.[1] Before the sixteenth century no university in Europe had hired professors to teach natural history or funded botanical gardens. While late medieval princes, such as the Emperor Frederick II in mid-thirteenth-century Sicily and the Duke of Burgundy, Jean de Berry (1340–1416), included natural objects among their coveted possessions, none of them considered natural history an important enough subject to devote vast quantities of resources to it. Few patricians boasted museums of natural history in their homes or private botanical gardens on their estates, as they were to do with increasing regularity in the sixteenth and seventeenth centuries. By the 1470s the first printed Latin translations and re-editions of the works of Aristotle, Theophrastus, and Pliny had appeared but only a handful of scholars engaged directly with the content, let alone attempted to write their own natural histories. With the exception of Frederick II, few princes made scientific activities a courtly pursuit; with the exception of Albertus Magnus (*c.* 1200–80), almost no natural philosophers considered natural history a subject worthy of their attention. Natural history languished just as Pliny had left it: unwieldy, encyclopaedic, prone to anecdote, and highly unfashionable. As the Bolognese naturalist Ulisse Aldrovandi (1522–1605) remarked in 1585, it was a 'faculty buried for so many hundreds of years in the gloom of ignorance and silence'.[2]

Despite the great vogue for bestiaries, herbals, and lapidaries, the naturalistic poetry of the twelfth-century writer Alain of Lille, and the detailed commentaries of Albertus Magnus on Aristotle's scientific writings, the scholastic and courtly worlds of the later Middle Ages exhibited very little interest in the comprehensive

description of nature that we most commonly associate with the
discipline of natural history.[3] Instead, nature was primarily a subject for allegory and commentary or for medicine. Equally important, there was a vast difference in the way in which inhabitants
of these diverse environments – the cathedral schools, the universities, and the courts – perceived nature. Each developed a particular form of discourse suited to its immediate context. In contrast,
the study of nature during the early modern period linked the
learning of the schools to the culture of the humanist academies
and the display of the courts. Central to the erosion of these barriers was the new image of the naturalist, who was increasingly
perceived as someone who could move readily between these different worlds.

The revival of natural history was due as much to the new social
circumstances of the naturalist as to the reverence for ancient learning and the voyages of exploration and discovery. By the mid-
fifteenth century humanists with an interest in natural history
could be found at leading Renaissance courts. Nicolò Leoniceno
(1428–1524), the physician who initiated the first full-scale critique
of Pliny with his *De Plinii . . . erroribus* ('On the Errors of Pliny')
(1492), served Borso d'Este in Ferrara and was a guest of Lorenzo
de' Medici when he travelled to Florence in 1490. Respondents
to his attack on Pliny, such as the lawyer and Gonzaga courtier
Pandolfo Collenuccio and the Venetian patrician, Ermolao Barbaro
(1444–93), were also members of the humanist elite who valued
the learning of the Greco-Roman world as worthy of recuperation
and imitation.[4] Thus it was in the society of educated professionals,
patricians who viewed learning as an appropriate leisure activity,
and rulers who financed the activities of scholars that an intense
interest in the accurate depiction and description of the natural
world first arose.

The naturalist and the prince

The academic community that revived natural history self-
consciously presented it as a form of enquiry designed to appeal
broadly to different sectors of society, most particularly the urban
patriciate and nobility. In his 1543 lecture inaugurating his position
as the first professor of natural history at the University of Ferrara,
Giuseppe Gabrieli (1494–1553) aptly summarized the new social
parameters of his discipline. Natural history, he proclaimed, was
appropriate 'not only for humble and lowly men, but people
coming from every social class conspicuous for political power,
wealth, nobility and knowledge such as kings, emperors, princes,
heroes, poets, philosophers, and similar men'. Reflecting on the
uneasy status of his chosen field of study, Gabrieli lamented the

'unhappy decadence of our times' that led princes to neglect to support naturalists and led physicians to shun the study of nature as an ignoble art. He responded to this problem by ennobling natural history, thereby making it worthy of his audience. Contrasting the study of nature with other sciences, he exclaimed, 'This is the only science of divine origins, the only one given to men by the Gods'. Exhorting his fellow Ferraresi to emulate the munificence of their rulers, Gabrieli lavished praise upon the d'Este princes whose patronage of men such as Leoniceno had allowed natural history to 'raise its head from the most profound darkness'. Such praise was pointedly directed at Ercole II who was undoubtedly in the audience, along with most of the courtiers in Ferrara, to hear the city's first official lesson on nature.

The rulers of Ferrara were among the first princes who brought 'men of great and diligent erudition [to their court] and supported them at great expense . . . in order to revive this knowledge of plants'.[5] Like many early modern rulers, they increasingly came to see natural history as a noble, pleasurable, and useful subject, worthy in every respect of their patronage. Similarly, the physicians who so disparaged the empirical study of nature were soon persuaded that natural history had a legitimate place in the medical curriculum, in part because their employers seemed so willing to finance professorships and research facilities (museums, anatomy theatres, and botanical gardens) in this field and eagerly hired physicians with an expertise in natural history to fill the most important positions in the cities and at the courts. In the half-century preceding Gabrieli's inaugural lecture, humanist princes and their scholars throughout Europe gradually integrated natural history into court culture. During the middle decades of the fifteenth century, such popes as Nicholas V and Sixtus IV financed lavish editions of Aristotle and Theophrastus, meticulously prepared by Theodore Gaza (c. 1400–75). The Medici Pope Leo X further placed the papal imprimatur upon natural history when he appointed Giuliano da Foligno to the first professorship in natural history in 1513 and, shortly afterward, displayed a stuffed rhinoceros in the papal palace.

By the time Gabrieli composed his oration, princes competed openly among each other to establish the most extensive research programme in this area, particularly in Italy where courts abounded and where natural history first experienced its intellectual revival. The very year that Ercole II d'Este appointed Gabrieli to the chair in 'medicinal simples', as natural history was often called, Cosimo I de' Medici, the first Grand Duke of Tuscany, wooed the eminent naturalist Luca Ghini (c. 1490–1556) away from Bologna to accept the first chair at the University of Pisa and manage its new botanical garden. In a relatively short period of time, humanists who initially considered an interest in nature to be marginal to their scholarly pursuits found themselves studying nature full time, largely because that study was supported by patronage that began in the courts, and moved outward to include the universities and the cities.

As a scientific discipline, natural history had certain distinct advantages that facilitated its introduction to the Renaissance courts. The popularity of bestiaries and lapidaries among the late medieval nobility provided an important precedent for the intro- duction of natural histories, particularly ones filled with elaborate allegorical imagery like the work of Aldrovandi and Joachim Cam- erarius the Younger (1534–98) that defined the 'emblematic world view'.[6] Furthermore, Aristotle's success in gaining the patronage of Alexander the Great and Pliny's attentiveness to the tastes of the Roman patriciate made them reasonable prototypes for the naturalist as courtier. Aldrovandi alluded to this explicitly when he chose to portray himself giving his *Ornithologia* (1599) to Pope Clement VIII, surrounded by images of Aristotle and Pliny offer- ing up their works to Alexander the Great and the Emperor Vespasian (Figure 4.1). Natural objects as well as books about nature became popular at court. Menageries, gardens, and museums all appealed to an audience that cultivated the exotic and the beautiful and thrived upon display. Rulers from the Medici to Louis XIV, regardless of their interest in natural history as a disci- pline, loved the aesthetics of nature. The French surgeon Ambroise Paré's (1510–90) characterization of Charles IX as some- one who delighted in 'large, weighty, and monstrous things' applied to most early modern monarchs. We have only to recall the rhinoceros that Leo X brought to Rome or the elephant that Louis XIV commanded his academicians to dissect in 1683 to real- ize how apt this description is.[7]

Unlike the traditional components of natural philosophy, natural history was tactile and visual, and required no specialized knowl- edge in order to participate. The professional naturalists who regu- larly botanized throughout the world were accompanied not only by assistants who collected and recorded what they observed but also by wealthy patricians who could imagine no greater pleasure than an excursion into the countryside.[8] The ability of naturalists to arrange objects in pleasing ways in their museums and the skill with which artists captured nature's contours on canvas served to heighten the aesthetic pleasure of viewing nature. Court artists such as Jacopo Ligozzi in Florence, Teodoro Ghisi in Mantua, and Giuseppe Arcimboldo in Prague made naturalistic illustrations their speciality. The Medici grand dukes were so enamoured of Ligozzi's skill in this area that they willingly lent him out to

Given the incorporation of artistic and literary skills into natural history, it is not surprising that the courts most known for their cultivation of learning and art provided crucial support for the revitalization of natural history. In Florence, the Grand Dukes Francesco I and Ferdinando I, sons of the Cosimo I who had established natural history in Pisa, expanded Medici support. Francesco I invited Aldrovandi to view his private collections and encouraged the *Natural History* that Aldrovandi claimed would rival Aristotle's and Pliny's; Ferdinando I established a natural history museum in the Pisan botanical garden in the 1590s and sent his court botanist to Crete to explore the flora and fauna there. The Medici grand dukes continued to be so diligent in sponsoring natural history that in 1697 the court botanist Paolo Boccone (1633–1704) remarked that all the courtiers in Florence studied nature 'in imitation of their ruler'.[9]

Besides Florence, Rome, Prague, and Vienna were the principal courts in which natural history flourished. Within the papal state, the Sack of Rome in 1527 precipitated a temporary demise in such

Figure 4.1 Ulisse Aldrovandi giving his *Ornithologia* (1599) to Clement VIII. From Ulisse Aldrovandi, *Ornithologia* (Bologna, 1599).

Aldrovandi to expand his choice of subjects through access to the Bolognese professor's museum.

Figure 4.2 A botanical expedition in the sixteenth century as depicted by Gherardo Cibo (1512–1600).

activities, though the annexation of Bologna allowed the popes indirectly to take credit for the thriving community of naturalists clustered around that city's university. Encouraging the support of natural history at the University of Bologna to rival the activities in Pisa and Padua, the early modern popes reminded the upstart Medici and the 'godless' Venetians that they were the supreme patrons of learning in Europe. By the mid-sixteenth century calmer political waters afforded the papacy the opportunity to make Rome a centre for natural history again. Within a few decades, some of

Figure 4.3 Michele Mercati's mineralogical museum in the Vatican, off the Belvedere courtyard. From Michele Mercati, *Metallotheca* (Rome, 1717).

the best naturalists in Italy came to Rome as papal physicians. During his tenure as papal physician in the 1550s, Ippolito Salviani (1514–72) published his remarkable *Aquatilium animalium historia* ('History of Aquatic Animals') (1554–8), dedicated to Pope Paul IV and with introductions by Pope Julius III, the Holy Roman Emperor Charles V, the King of France Henri II, and the Grand Duke of Tuscany Cosimo I. In the 1560s Michele Mercati (1541–93) was invited to supervise the botanical garden and create a mineralogical museum, known as the *Metallotheca*, that opened up into the famous Belvedere courtyard of the Vatican (Figure 4.3). Naturalists such as Aldrovandi visited this museum and no less a figure than Mercati's mentor Andrea Cesalpino (1519–1603), known for his precise classification of plants and a description of the circulation of the blood that prefigured William Harvey's, was wooed away from the Medici stronghold in Pisa to join the community in Rome.

In Prague and Vienna a succession of Habsburg emperors searched the length and breadth of the Holy Roman Empire to bring Europe's best naturalists to the court. In quick succession,

Pietro Andrea Mattioli (1500–77), author of a famous translation of and commentary on Dioscorides' *Materia medica*, and the Flemish naturalists Rembert Dodoens (1517–85) and Carolus Clusius (1526–1609), both known for their accurate work in plant classification, arrived. Mattioli's rise to fame illustrates well how the munificence of princes facilitated the study of nature. Shortly after the publication of his Italian translation of Dioscorides in 1544, which he dedicated to the Archduke Ferdinand, Mattioli confessed to Aldrovandi, 'I am hoping for some notable reward'. Within a year, he was appointed imperial physician, a position he enjoyed under Ferdinand and his successor, Maximilian II. In Italy Mattioli had been one of many town physicians; in Prague and later Innsbruck he had artists, engravers, printers, and translators at his disposal. During the 1560s his 'Dioscorides', as contemporaries called it, grew from a local project into a work of international renown. Under imperial patronage, Latin, German, and Czech translations appeared, and illustrations were made to accompany the text. By 1565 it had become a lavish work of some 1,500 pages.[10] Rarely had a career depended so much on one book. Mattioli became the envy of other naturalists such as Aldrovandi and Giovan Battista della Porta (1535–1615), both of whom tried unsuccessfully to interest the Habsburgs in supporting their natural histories.

As the above examples suggest, naturalists had much to gain by arriving at court, though certainly this was not the only environment in which to study nature. An appointment as court physician, imperial 'herbalist and plant collector' in the case of Clusius, royal surgeon in the case of Paré, or royal cosmographer and 'overseer of the Royal Collection of Curiosities at Fontainebleau' in the case of André Thevet (1504–92), allowed a naturalist unique access to resources vital to the global rewriting of natural history then underway.[11] Enhanced status and greater visibility made other naturalists more likely to share information and specimens with someone under princely sponsorship, thereby increasing a naturalist's authority as a commentator on nature. When Mattioli moved to the imperial court, he expanded his knowledge of nature from north-central Italy, the region where he had lived and worked, to the entire breadth of the Holy Roman Empire. Similarly, the Spanish physician Francisco Hernández (1517–87) extended his reach from Spain to the farthest reaches of the Spanish empire – the Indies – in his capacity as court physician, while naturalists under papal sponsorship commanded the resources of the Catholic missionaries scattered throughout the world. By making natural history a 'princely thing', a phrase Ferdinando I de' Medici used to persuade his brother Francesco I to fund Aldrovandi's *Ornithologia*, the first generation of naturalists ensured themselves some measure of prestige and financial support, and used their access to the courts to collect the materials necessary to examine nature.[12]

The naturalist and the city

While a select group of naturalists made their careers at court, the majority of their colleagues worked within a predominantly urban setting. Despite the financial support Aldrovandi received from many Italian princes, he remained in Bologna, employed by the city as a university professor, curator of the botanical gardens (f.1568) and *protomedico* (a post from which he supervised the natural ingredients placed in medicines). Many of his contemporaries – most famously Conrad Gesner (1516–65) in Zurich and Leonhart Rauwolf (1535–96) in Augsburg – were employed as town physicians; other, like Francesco Calzolari (1521–1600) in Verona, famous for his botanical excursions to Monte Baldo, and Ferrante Imperato (1535–1601) in Naples, well known for his careful work on fossils in his *Dell'historia naturale* (1599), made a living as apothecaries. Even naturalists who arrived at court did not always stay. Both Dodoens and Clusius, luminaries of the Habsburg court in the 1570s and 1580s, chose to return to the Netherlands when the University of Leiden offered them positions.[13]

Natural history quickly became popular with the urban patriciate in the sixteenth century. It fitted well within the marketplace culture of the city, where goods were bartered and sold. In numerous cities, clerics, lawyers, physicians, and other notables cultivated gardens and museums and participated in the endless cycle of visits, exchanges, and demonstrations that defined the world of collecting. By the time the Swiss physician Thomas Platter visited the home of Sir Walter Cope in London in 1599 to see the rooms 'stuffed with queer foreign objects', museums of natural history abounded on the continent, among them his brother Felix's collection in Basle.[14] The most famous Renaissance 'theatres of nature' belonged to such professional naturalists as Calzolari, Imperato, Mercati, and Aldrovandi (who with evident pride described his museum in Bologna as the eighth wonder of the world). By the seventeenth century the ownership of such collections had become more diffuse, as natural history increasingly became integrated into the leisure activities of the urban elite. Significantly, only four of the twenty-six letters comprising Paolo Boccone's *Osservazioni naturali* (1684) were addressed to men with a professional interest in nature; the remainder were directed to leading citizens of various cities. The great collections of this epoch were maintained by clerics such as the Jesuit Athanasius Kircher (1602–80), founder of the Roman College museum, and Manfredo Settala (1600–80) in Milan; by lawyers such as Nicolas-Claude

Fabri de Peiresc (1580–1637) in Aix; and by the members of fledgling scientific societies like the Academy of the Lynx-Eyed in Rome, the Royal Society in London, and the Paris Academy of Sciences.

Unlike John Tradescant the Elder (1570/75–1638), who collected nature as part of his professional duties as gardener first to Robert Cecil, Earl of Salisbury, and then to George Villiers, Duke of Buckingham, the gentlemen naturalists of this later era defined themselves as *virtuosi*, men of leisure whose curiosity knew no bounds. These were the men who rushed to buy tulips, puzzled over fossils and luminescent stones, and exclaimed over the Danish anatomist Nicolaus Steno's (1638–1686) skilful dissections in Leiden, Paris, and Florence. 'This Monsieur Steno is the rage here', wrote one provincial physician during a trip to Paris in 1665. 'Neither a butterfly nor a fly escapes his skill. He would count the bones in a flea – if fleas have bones.' Only a few months later, the same person attempted to replicate 'the style of Monsieur Steno' in the French town of Caen for the local noblemen and noblewomen, eager to see nature's latest curiosity.[15] So quickly did the urban elites in various cities inform each other of Steno's skill, that long before he arrived in Florence in 1666 the Medici court naturalist Francesco Redi (1626–97) was already anticipating the pleasures of dissecting animals from the grand-ducal menagerie with him.

Studying nature appealed to urban elites as well as courtiers primarily because it made nature a part of a nascent consumer culture. For patricians who collected art, furniture, and books, the possession of natural objects was yet another means of displaying one's accumulated wealth. When Elizabeth I's surgeon, George Baker, visited John Gerard's (1545–1612) garden in Holborn he wondered 'how one of his degree . . . could ever accomplish the same.'[16] The more exotic the objects in a museum, the more the owner could boast about the dangerous voyages undertaken, the contacts in every corner of the world, and the care employed to achieve the desired result. No expense was spared in the re-creation of nature in microcosm.

Patricians particularly delighted in novelties. New objects and new experiences, all of which collecting offered in plentiful supply, featured prominently in their leisure activities. While professional naturalists benefited from expanded lay interest in nature, they occasionally lamented the faddishness of it all. In 1594, for example, Aldrovandi recalled with nostalgia the excitement that his planting of a 'Peruvian chrysanthemum', the first sunflower seen in Italy, had occasioned in Bologna (Figure 4.4). 'It was visited by all the gentlemen and gentlewomen for the size and beauty of its flower. But now it is so vulgar that no one cares about it.' In

Figure 4.4 Ulisse Aldrovandi's 'Peruvian chrysanthemum' (sunflower). From Biblioteca Universitaria, Bologna, *Aldrovandi. Tavole di piante*, vol. I, p. 75.

similar fashion, Gerard noted in his famous *Herball, or Generall Historie of Plantes* (1597) that goldenrod no longer elicited sighs of appreciation from the English nobility since it had become common.[17] Professional naturalists continued to find even the most ordinary aspects of nature important to their work, but they catered to an audience who subsumed nature within the category of wonder.

The activities surrounding the study of nature further integrated it into the social world of the urban elites. Since individual naturalists rarely had the resources to create more than modest museums, the most successful collectors depended upon gifts to make their encyclopaedias of nature complete. Echoing the list of potential connoisseurs of natural history that Gabrieli articulated in his 1543 oration, Aldrovandi thanked the 'famous Cardinals, Bishops,

Archbishops, Dukes, Barons, Counts, Doctors, and Scholars' who had contributed the objects which he classified in his unpublished *Methodus fossilium* ('Method of Fossils'). Similarly, the apothecary Imperato felt that he could not have completed his natural history without the help of 'most of the virtuosi in Europe.'[18] While popes and princes offered up lavish gifts worthy of their status as patrons – in 1635 Tradescant acquired mementoes of Henry VII and Henry VIII from the reigning Charles I – naturalists, travellers, and the simply curious provided the ordinary stuff that formed the core of such collections.

As patricians travelled with greater frequency, both within Europe and beyond, museums became a privileged site of sociability. In these settings urban elites met and conversed upon a wide variety of topics, exchanging news as they would later do in Theophraste Renaudot's weekly meetings at the Bureau d'Adresse in early seventeenth-century Paris or in the eighteenth-century coffee-houses. Subscribing to Pliny's encyclopaedic definition of natural history as everything in the world worthy of memory, collectors made the museum an ideal setting in which to converse.[19] The infinite number of curious objects displayed guaranteed an endless array of topics. Aldrovandi's and Gesner's lengthy Latin tomes memorialized the culture of erudition, but more popular volumes such as Francesco Imperato's *Discorsi intorno a diverse cose naturali* ('Discourses on Diverse Natural Things') (1628) or Redi's *Esperienze intorno alla generazione degl'insetti* ('Experiments on the Generation of Insects') (1668) captured the essence of these dialogues about nature. While the virtuosi of Naples enthused about pygmies, crocodiles, mandrake roots, tarantula, fossils, and other natural marvels (Figure 4.5), their counterparts in Florence enjoyed Redi's nauseating experiments with vipers, scorpions, and the rotting carcasses which proved that spontaneous generation existed only in the mind of Aristotle and not in nature. The point of it all was not solely (or even primarily) to determine any truth about nature, but to fuel the conversations of the virtuosi who continued to exclaim over these things.

By the late sixteenth century few members of the urban elite had neglected to enter a museum. Naturalists such as Gesner and Aldrovandi, ever mindful of the legacy they would leave to posterity, kept records of their visitors. Gesner's *Liber amicorum* ('Book of Friends') recorded the names of 227 scholars who had visited him between 1555 and 1565. Aldrovandi collected the names of all the visitors whom he felt shed honour upon his enterprise in two books: the *Liber in quo viri nobilitate honore et virtute insignes, viso musaeo . . . propria nomina ad perpetuam rei memoriam scribunt* ('Book in Which Men of Extraordinary Nobility, Honour and Virtue Who have Seen the Museum . . . Write Their

Own Name') and *Catalogus virorum, qui visitarunt Musaeum nostrum* ('Catalogue of Men Who Have Visited Our Museum'). While the former contained only the names of the most illustrious visitors, including the 1635 entry of one Cardinal Ubaldo, the papal legate who described the museum as the work of 'another Noah' (Figure 4.6), the latter arranged entries for 1,579 people 'according to the order of dignity, study and profession' and by city or region. Many of Aldrovandi's visitors came from the courts – for example, the Duke of Mantua, Rudolf II's councillor and the 'Carver for the King of Poland'. Others were fellow naturalists and collectors such as Camerarius and the young Caspar Bauhin (1560–1624), whose *Pinax theatri botanici* ('Index to the Theatre of Botany') (1623) was one of the most important contributions to taxonomy in the seventeenth century. However, the vast majority of visitors were patricians who viewed nature as a continuous spectacle placed in front of them by the humanists who revived the empirical study of nature.[20]

Naturalists collected nature to increase their store of observations

Figure 4.5 The museum of Ferrante Imperato in Naples, 1599. From Ferrante Imperato, *Dell'historia naturale* (Venice, 1672 edn.).

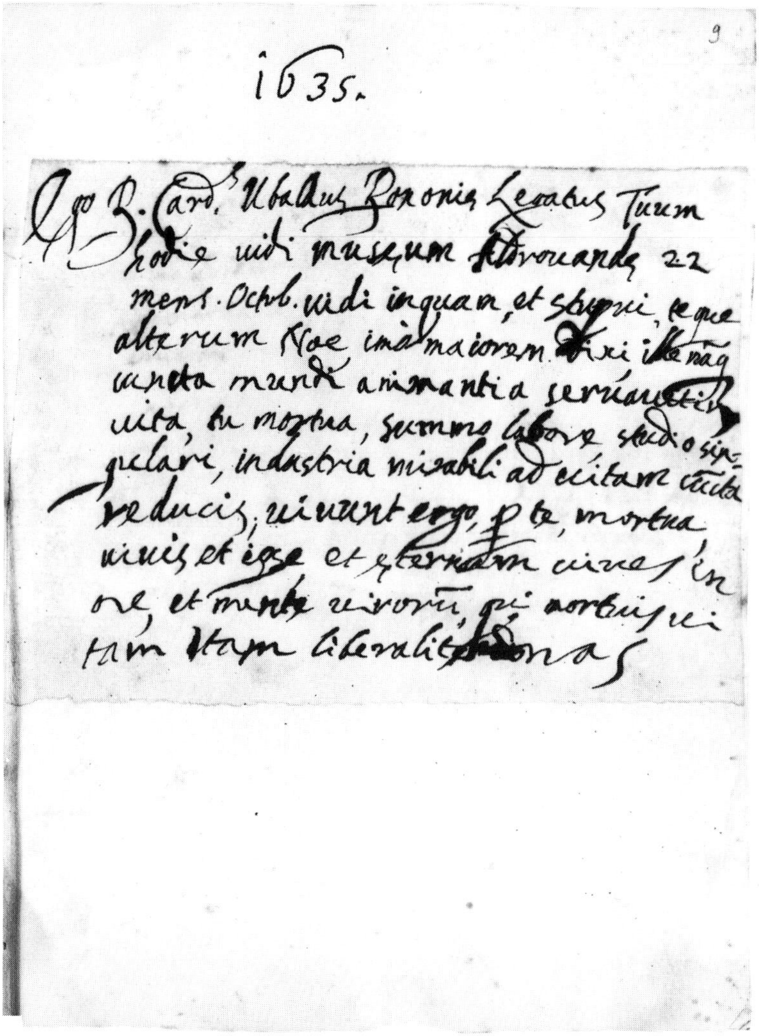

Figure 4.6 Page from Ulisse Aldrovandi's 'Book in Which Men of Extraordinary Nobility, Honor and Virtue Who Have Seen the Museum . . . Write Their Own Name'.

and refine their statements about the essential structures that composed the natural world, but they also derived great prestige from these activities.[21] The increased mobility of the upper classes privileged travel and collecting as activities that linked the European cities. By the middle of the seventeenth century Kircher rightfully claimed that no trip to Rome was complete without a tour of the Roman College museum. Few people went to London without seeing Tradescant's 'ark' in Lambeth or the Royal Society's 'repository'. Collecting allowed naturalists to expand their range of acquaintances, coming into contact with virtually every sector of society, from the lowly fishermen who sold them oddities in the market-place to the princes who paid handsomely to see such things. The naturalists who had begun their careers with the scholarly task of correcting the errors of Pliny or completing the work of Aristotle, soon found themselves in possession of the world.

The naturalist and the world

Natural history intrigued princes and merchants because it was a tangible sign of their ability to roam the globe. Whether a naturalist travelled like Rauwolf, who spent 1573–6 in the Far East, or stayed at home like Aldrovandi, his success depended upon his ability to incorporate the novelties of the New World, Africa, and Asia into his findings. The new interest in material culture engendered by the voyages of exploration contributed greatly to the development of natural history. In cities such as Seville, London, and Amsterdam naturalists enjoyed the support of patrons who saw a certain profit in collecting data from the Americas. Between 1550 and 1650 entrepreneurs in these different centres launched three of the greatest projects of that period: the natural histories of Mexico, Virginia, and Brazil.

Philip II had already exhibited a personal interest in medicinal simples by the time he commissioned Hernández to spend six years in Mexico gathering its flora and fauna. He had sponsored a botanical garden as part of the royal pharmacy at the Escorial, purchased the collections of a Seville physician and a royal apothecary in Madrid, appointed Honorato Pomar (professor of medicine at the University of Valencia) as court naturalist, and sponsored a 1555 Spanish translation of Dioscorides. In 1570 he appointed Hernández, already a court physician, '*protomédico general* of our Indies, isles and *tierra firme*', requesting that he depart for New Spain immediately. Between 1571 and 1577 Hernández and the artists and assistants who accompanied him collected thousands of plants, animals, and minerals, and interviewed numerous natives about their medicinal uses. By the time Hernández returned to Spain, he had sixteen precious volumes of notes, specimens, and illustrations.[22] Despite his efforts the majority of his work never appeared in print. Philip II decided to publish a more modest natural history of Mexico, leaving the majority of Hernández's work in storage in the Escorial where it was lost during a fire in 1671. After Hernández's death in 1587, the king entrusted a Neapolitan physician with the project. It languished in Naples for several decades until members of the Accademia dei Lincei ('Academy of the Lynxes') brought it to Rome. In 1648 the definitive edition of the *Rerum medicarum Novae Hispaniae Thesaurus* ('Treasury of Medicinal Things from New Spain') appeared, dedicated to Philip IV. But by then the Spanish were no longer the powerful rulers they had once been and no one in the Escorial seemed to really care.

In 1585 Sir Walter Raleigh launched a second expedition to Virginia, with the blessing of Elizabeth I. On this trip he took three scientifically skilled men with him: the astronomer Thomas Harriot, the metallurgist Joachim Ganz, and the artist John White (*c.* 1540/50–1606). One of Harriot's and White's principal tasks was the collection and illustration of North American specimens to publicize the English colony at home. When White returned to England in 1587 he discussed his botanical findings with John Gerard, who published some of them in his *Herball*, and shared a few insect drawings with Thomas Penny (d. 1589) who included them in his *Theater of Insects*, published by Thomas Moffet in 1634. Other illustrations found their way into Walter Bigges's *Expedition of Francis Drake* (1588), Harriot's *A Brief and True Report of the New Found Land of Virginia* (1590), and Edward Topsell's *Historie of Four-Footed Beastes* (1607).[23] The promised natural history of Virginia, however, never appeared: White was increasingly busy in his role as governor of Raleigh, Virginia; Harriot had returned to England; Raleigh's financial situation was always shaky; and the Queen continued to be ambivalent about her colony. The enthusiasm with which patrons initiated these collecting projects did not always last. By the seventeenth century only the objects displayed in Tradescant's museum gave testimony to the Virginia project. The British empire had been established but its natural history had yet to be written.

The same year in which Hernández's natural history of Mexico appeared, Georg Markgraf's (1610–43) *Historia naturalis Brasiliae* ('Natural History of Brazil') (1648) appeared. Markgraf had been employed by Prince Johan Maurits of Nassau to gather information on this region during Maurits's tenure as Governor-General of Brazil for the Dutch West India Company. In Brazil Johan Maurits had established a zoo, botanical garden, and museum, and employed several artists to produce over one thousand illustrations for the natural history. In many respects, the *Natural History of Brazil* was a more successful project than those of Mexico and Virginia, simply because the success of its patron depended upon its completion. Unlike Philip II and Elizabeth I, Johan Maurits was a minor prince with little land and no power to his name. Seven years in Brazil provided him with a unique opportunity to collect and create artefacts to amuse the princes of Europe. When he returned to the Netherlands he spent the next few decades distributing lavish gifts – 'documents, paintings and parrots', as one contemporary reported – to the kings of France and Denmark, the anatomical theatre in Leiden, and the Elector of Brandenburg who awarded him with the title of imperial prince in return. Similarly, the *Natural History of Brazil* became a gift that he could distribute to gain further favours.[24]

Natural history fitted well within the images of empire promoted by early modern rulers and the majority of naturalists benefited from these ambitions, directly or indirectly. As the above examples suggest, however, it was much easier to accumulate objects for the new history of nature than to see their description into print. When Gesner died only his histories of animals had been printed; Aldrovandi published four works in his long life, leaving hundreds of manuscripts to gather dust in his museum. The expense of Joseph Pitton de Tournefort's path-breaking *Elémens de botanique* (1694) effectively curtailed the publication of the Paris Academy of Sciences' history of plants. The problem was not simply the cost of financing these books but also the attitude of patrons towards the printed word. Museums and botanical gardens fitted readily into court culture and urban life, making natural history a fashionable pursuit among the upper classes. In contrast, the lengthy humanist encyclopaedias of nature served the community of professional naturalists more than the interests of their patrons.

Both the culture of display and the culture of erudition shaped natural history as a discipline. With the decline of the courts in the eighteenth century natural history moved into the salons, coffee-houses, and provincial academies. Enthusiasts flocked to the public lectures in the royal gardens in Paris and eagerly consumed the more accessible natural histories, written in vernacular languages, that introduced them to new taxonomic systems. While the social world in which natural history first emerged did not last, certain practices associated with it continued. Naturalists such as Buffon were concerned with expanding their audience, just as Renaissance scholars such as Gabrieli had worried about finding an audience for their work. When the Swedish naturalist Carolus Linnaeus reformulated the study of plants in the early eighteenth century by renaming them, he did not neglect to celebrate the names of worthy patrons who advanced the study of natural history. They, like the 'Fathers of Botany' – naturalists whose contributions Linnaeus deemed significant – were to be immortalized in binomial nomenclature. 'Those Emperors and Kings shall enjoy everlasting remembrance who have founded Public Gardens, endowed professional chairs of Botany, sent Botanists in the interests of science into foreign lands', he wrote in his *Critica botanica* (1737).[25] Patrons were just as important as naturalists; without them, the study of nature could not progress. Situated between the Renaissance court and the Enlightenment salon, between the universities and the academies, the study of nature developed into the discipline of natural history.

Further reading

Eamon, William, 'Science and popular culture in sixteenth century Italy: the professors of secrets and their books', *Sixteenth-Century Journal*, 16 (1985), pp. 471–85.

Findlen, Paula, 'Controlling the experiment: rhetoric, court patronage and the experimental method of Francesco Redi', *History of Science*, 31 (1993); pp. 35–64.

 Possessing Nature: Museums, Collecting, and Scientific Culture in Early Modern Italy (Berkeley, CA, 1994).

Houghton, Walter E., Jr., 'The English virtuoso in the seventeenth century', *Journal of the History of Ideas*, 3 (1942), pp. 51–73, 190–219.

Impey, Oliver, and MacGregor, Arthur, (eds.), *The Origins of Museums: The Cabinet of Curiosities in Sixteenth- and Seventeenth-Century Europe* (Oxford, 1985).

Kenseth, Joy (ed.), *The Age of the Marvelous* (Hanover, NH, 1991).

Lux, David, *Patronage and Royal Science in Seventeenth-Century France: The Académie de Physique in Caen* (Ithaca, NY, 1989).

MacGregor, Arthur (ed.), *Tradescant's Rarities: Essays on the Foundation of the Ashmolean Museum 1683 with a Catalogue of the Surviving Early Collections* (Oxford, 1983).

Pomian, Krzysztof, *Collectors and Curiosities: Paris and Venice, 1500–1800*, trans. Elizabeth Wiles-Portier (London, 1990).

Sarton, George, 'Natural history', in his *Appreciation of Ancient and Medieval Science During the Renaissance* (Philadelphia, 1955), pp. 52–132.

Siraisi, Nancy G., 'Life sciences and medicine in the Renaissance world', in Anthony Grafton (ed.), *Rome Reborn: The Vatican Library and Renaissance Culture* (Washington, DC, 1993), pp. 169–97.

Stroup, Alice, *A Company of Scientists: Botany, Patronage, and Community at the Seventeenth-Century Parisian Royal Academy of Sciences* (Berkeley, 1990).

5 The culture of curiosity

The study and collection of natural objects in the seventeenth century was undertaken as part of a broad interest in rarities and wonders of all sorts, natural and artificial. Natural histories such as Robert Plot's *Natural History of Oxfordshire* (London, 1677) included treatments of many not strictly natural phenomena such as the houses of the local gentry and the antiquarian remains of the area. Similarly, antiquarian histories such as Thomas Fuller's *History of the Worthies of England* (London, 1662) included such natural subjects as natural wonders and medicinal herbs. Both natural and artificial rarities, or 'curiosities' as they were called, filled the collections that were eagerly formed by gentlemen and scholars who described themselves as 'curiosi' or 'virtuosi'. My discussion of these curiosi will concentrate on England where the culture of curiosity was particularly pronounced.

Curiosi were aristocrats, gentlemen and aspiring gentlemen, dispersed through the counties of England in their homes in the summer, but converging on London in the winter where they attended meetings of the Royal Society. Predominantly landowners, they also included clergymen, lawyers, university men, physicians, wealthy merchants, and apothecaries. Curiosity was considered an important attribute for an accomplished gentleman to possess. It was an attitude of mind involving a fascination and admiration for the rare, novel, surprising, and outstanding in all spheres of life. Young gentlemen were trained in curiosity by making the grand tour on the Continent. Often accompanied by a tutor, they followed standard routes to view the many curiosities described in their guidebooks. They returned to England as fully-fledged curiosi, bringing back rarities of nature and art which formed the basis of their collections of curiosities. They settled into the life-style characteristic of curiosi, travelling to seek out and view rarities, displaying rarities in their houses, gardens, and estates, and visiting each other to view and discuss these rarities.[1] These curiosi formed an educated culture whose outlook on the world must be understood if we are to appreciate how natural objects were studied in the period.

Curiosity and wonder

Curiosity in the seventeenth century played the same role as would the sublime in the eighteenth century: it was the standard of appreciation of nature and art.[2] Appreciation of the sublime was to involve astonishment, awe, and terror, felt in the face of vastness, power, and grandeur. Curiosity, on the other hand, involved wonder and admiration at whatever was rare or outstanding, whether in size, shape, skill of workmanship, or in any other respect. Such rarities formed the curiosities whose unusual and outstanding qualities curiosi admired and wondered at.

Thus rare exotic animals and birds were curiosities which produced wonder: the first rhinoceros to be brought to England was regarded as 'a very wonderfull creature' for its unusual appearance, size, and strength. A curious waterfowl produced wonder by its remarkable eating habits: 'It would eat as much fish as its whole body weighed, I never saw so unsatiable a devourer, I admir'd how it could swallo[w] so much & swell no bigger: I believe it to be the most voracious creature in nature, it was not biger [sic] than a More hen'.[3]

Natural curiosities also produced wonder by their beauty and the apparent art of their contrivance. Butterflies in particular were wonderful for their 'curiously variegated Wings, admirably beautiful for their Colours or Texture' and shells were admired as 'engraved and painted with various Colours and Figures', 'curiously striated, with transverse Lines', or 'very curiously variegated, with triangular Figures white upon black'. The similarity of nature to art could become so wonderful as to produce incredulity, as happened in the case of the Chinese chair, one of the greatest curiosities of the Royal Society's museum. Supposedly made from the natural growth of a root, it led one viewer to conclude that 'I cannot possibly believe that art did not come to assist, so elegantly is it carved'.[4]

Natural features of the countryside were also curiosities to be appreciated with wonder. The hills of Derbyshire were appreciated as the 'most prodigious high mountains as ever my eye beheld', and one of the caves there was 'esteemed one of the wonders of England' because of 'the vast largeness of it' which made it a true rarity. At Knaresborough in Yorkshire, a petrifying spring, another rare natural feature, was a curiosity considered 'admirable'.[5]

Experiments and mechanical inventions were also seen as wonderful curiosities. Thus the virtuoso Robert Boyle (1627–91) reported experiments which inspired wonder in his 'Tracts of A Discovery of the Admirable Rarefaction of the Air' and 'The

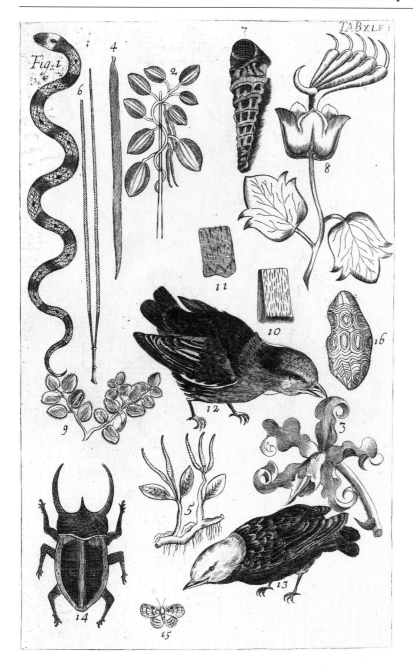

Figure 5.1 A selection of natural curiosities. From James Petiver, *Gazophylacium naturae et artis* (London, 1702–6).

admirably Differing Extension of the same Quantity of Air rarified and compressed'. Visitors to the Royal Society wondered at the experiments which they saw there: the Duchess of Newcastle (1624?–74), writer of poems, plays, and works on natural philosophy, was 'full of admiration' when she visited the Society and was shown some 'fine experiments . . . of Colours, Loadstones,

Microscope, and of liquors: among others, of one that did while she was there turn a piece of roast mutton into pure blood – which was very rare'.[6]

Unusual human beings were also appreciated as wonderful curiosities. A 'native Irishman, Edmund Mallory . . . two yards and a half tall' was considered 'a wonderful sight', and the Cumbrian clergyman, botanist, and antiquarian William Nicolson (1655–1727) admired the 'wondrous feats performed by one John Valerius, a German, born without Arms' whose curious drawings were displayed at the Blue Boar in London.[7]

Houses and their gardens could be rare curiosities and so inspire wonder. The virtuoso John Evelyn (1620–1706), visiting the house of Lord Sunderland, found that 'above all are admirable & magnificent the severall ample Gardens furnish'd with the Choicest fruite in England, & exquisitely kept: Great plenty of Oranges, and other Curiosities: The Parke full of Fowle & especialy Hernes [herons]'. Similarly, the Yorkshire antiquarian and collector Ralph Thoresby (1658–1725) admired 'the greatest house in England, viz. Audley-end, a vast building, or rather town walled in; it is adorned with so many cupolas and turrets above, walks and trees below, as render it a most admirably pleasant seat'.[8]

Even ordinary objects could become noteworthy curiosities if they were associated with strange and wonderful stories. An apparently unremarkable fly in Ralph Thoresby's collection was in fact a rare curiosity because of its wonderful history: it was 'sent me by the Reverend Mr. Hall of Fishlake, An. 1699 with this remarkable Account, That in May the same Year, at Kerton in Lincolnshire, the Sky seem'd to darken North-Westward . . . as though it had been with a Shower of Hailstones or Snow; but when it came near the Town it appeared to be a prodigious Swarm of these Flies, which went with such a Force towards the South-East, that Persons were forced to turn their Backs of them, to the Wonder of those that were abroad and saw them'.[9] The fly was a curiosity which would inspire wonder in visitors to Thoresby's collection when they were told its strange history.

Curiosities could also inspire wonder by their richness and cost. In the Jewel House of the Tower of London, William Nicolson

---

Figure 5.2 An admirable artificial curiosity: Enston Waterworks formed by Thomas Bushell Esq. from 'a Rock so wonderfully contrived by *Nature* her self, that he thought it worthy of all imaginable advancement by *Art*'. A house with a banqueting room looks out over the 'Ingenious Contrivances' of the waterworks, including a vertical 'Column of water rising about 14 foot, designed to toss a Ball', streams of water which 'sportively wet' any person on the island, and two spouts designed to wet the back and legs of the curioso as he retreated over the bridge. From Robert Plot, *Natural History of Oxfordshire* (Oxford, 1677).

admired 'The Rich Crown of State . . . in which there's a large Emerald (green) of 7 Inches round; the finest Pearl in the world, pawned by King Charles II to the Dutch for 40000£ and a Rubie (given by the Jewes of London to the late King James, when he was Duke of York) of an inestimable value'.[10]

The wonder felt by curiosi was quite unlike the awe and amazement with which the sublime was later to be appreciated. The sublime inspired the strongest emotions which the mind was capable of feeling. In this delightful state of astonishment, as Edmund Burke described in his *Philosophical Enquiry into the Origin of our Ideas of the Sublime and Beautiful* (London, 1757), the mind is overcome and made incapable of reasoning.[11] The state of wonder caused by curiosities in the seventeenth century was completely different from this: curious wonder was reasoned and articulate. Curiosi understood what they found wonderful and expressed this in their writing and conversation. Exact descriptions of curiosities were recorded which stated precisely what about them was considered wonderful. Thus when John Evelyn viewed a rhinoceros he noted in detail all its outstanding and wonderful features, forming a reasoned and articulate expression of wonder:

It more ressembled [*sic*] a huge enormous Swine, than any other Beast amongst us; That which was most particular and extraordinary, was the placing of her small Eyes in the very center of her cheekes and head, her Eares in her neck . . . her Leggs neere as big about as an ordinarie mans wa[i]st . . . but what was most wonderfull, was the extraordinary bulke and Circumference of her body, which though very Young . . . could not be lesse than 20 foote in compasse: she had a set of most dreadfull teath, which were extraordinarily broad, and deepe in her Throate . . . but in my opinion nothing was so extravagant as the Skin of the beast, which hung downe on her ha[u]nches . . . loose like so much Coach leather . . . and these lappets of stiff skin, began to be studdied with impenetrable Scales, like a Target of coate of maile, loricated like Armor . . . T'was certainly a very wonderfull creature, of immense strength in the neck, and nose especially.[12]

As part of this articulate expression of wonder, curiosities were compared so as to reveal what was really outstanding and rare. The German traveller Zacharias Conrad von Uffenbach, who visited England in 1710, described Hans Sloane's collection of corals as 'especially charming' since the corals 'were not only of unusual size but also quality', but he criticized the 'great cornua Hammonis' in John Woodward's collection since 'their size did not equal those we saw in Limburg at Herr Reimer's'. The clergyman John Covel (1638–1722), in his northern travels, was 'particularly pleased with Gingling-Cove and Reeking-Cove, near Ingleton, which (he saies) outdoe Oakey-Hole in Somersetshire and all the wonders of the Peak'. At Kensington Palace William Nicolson 'spent two

Hours in walking about the fine Garden, Wilderness and Green-House' and viewed the 'Queen's Dressing-Room hung with Neddle-work, in Satin . . . And the great Gallery stored with excellent Pictures' concluding with a comparison, that the 'whole [was] much Superiour to the Palace at St. James's'.[13]

In addition to detailed description and comparison, curious wonder also essentially involved the attempt to understand its object. Wonder occurred in the face of a phenomenon which was not understood and led the curioso to speculate and philosophize on the causes of curiosities. Thus when Ralph Thoresby went to see some wonderful corn which had been 'rained down the chimney upon the Lord's-day seven-night', he was led to consider 'What it may signify, and whether it doth proceed from natural causes . . . or preternatural'. Similarly, when the antiquary and collector Lord Coleraine (1636–1708) described the remarkable burning of a haystack at Bath to the curious collector William Courten (1642–1702), his admiration at the monstrous occurrence led him to philosophize on its cause:

a Great Hay stack att the Bath . . . takeing fire of ittself about August last indangerd the whole Towne, allarmd the inhabitants, occasiond much discourse & admiration att the flying out of itt in great peices [*sic*] of this (then inflamed matter) with a Crackling Noyse, & desperat[e]ly scalding while itt burnt. Noe doubt butt ye Tendrest herbs & ye dews wch fall upon y^m too are much impregnated with the sulphureous Attomes ariseing from ye Springs Thereabouts, so y^t 'tis no wonder y^t ys Grass at ye Bath being layd up not th[o]roughly dry should from the abundance of those minerall exhalations (both in & about itt) take a more y^n ordinary heate wch did att once bake itt hard & light itt, as a Cake of Seacole is accended & concocted.[14]

This reasoned articulate wonder which sought understanding was contrasted by curiosi with speechless astonishment in which the mind was baffled and brought to a stand, unable to exercise its judgement. True wonder was felt by the Christian curioso who observed 'the true Works and wonderful Contrivances of the Supreme Author'. Unreasoning astonishment in which causes were not understood was seen as the fate of the non-Christian whose superstition led him 'to think all strange things Supernaturall' and not to distinguish the wonders of nature from genuine miracles: 'To behold a Rainbow at night' was a curiosity to be wondered at, but was 'no prodigie unto a Philosopher'. Wonder was a religious activity: whereas vain amazement led to superstition, just admiration of God's works was a religious duty, and the viewing of natural curiosities led to admiration of the wisdom and power of God. Thus Ralph Thoresby, wondering at the sights which he saw in the Peak District, was led to admire God's power: 'God, who is truly Θαυματουργος, the only worker of wonders, has more manifested his might in this than in any other county in England, such the heaps of wonders therein'. Curious natural philosophy dealt with the true wonders of God's works and was 'Next to Gods Word . . . the most Soveraign Antidote to expell the poison of Superstition; and . . . the most approved food to nourish faith'.[14]

The need to experience Christian wonder and avoid vain superstition led curiosi to have a serious concern with the accuracy of reports of curiosities: only reports which were ascertained to be true could produce Christian wonder at the real works of the Creator. Accounts of curiosities thus had to conform to stringent standards of accuracy. They had to be 'particular', describing all the details of the phenomenon minutely and circumstantially. They had to show care and diligence which ensured their reliability.[16] This accurate reporting of wonders formed a new style of natural history, characteristic of the period.

Curious natural histories

Natural histories were written within this culture of wondering curiosity. They described curiosities, treating novelties, rarities, and wonders, and ignoring the common and ordinary. This bias can be clearly seen in the *Natural History of Oxfordshire* written by Robert Plot (1640–96), a commoner of University College, Oxford, who was to become the first custodian of the Ashmolean Museum and professor of chemistry at Oxford in 1683. Plot's treatment of plants and animals was not a complete catalogue including the common, but described only 'such, as either have not been noted before, are very unusual, or have somthing [*sic*] extraordinary attending them' (p. 175). The other chapters of his natural history similarly concentrated on rare and wonderful phenomena.

Natural histories of a particular geographical region were a popular genre in the period, and many curiosi wrote local natural histories modelled on Robert Plot's *Natural History of Oxfordshire*. Plot himself wrote *The Natural History of Stafford-shire* (Oxford, 1686). The Wiltshire gentleman and antiquarian John Aubrey (1626–97) wrote *The Natural History of Wiltshire* (written between 1656 and 1691; published London, 1847, ed. J. Britton) and, having been empowered to survey Surrey by a licence from the royal cosmographer John Ogilby, performed a perambulation of that county in 1673 which resulted in *The Natural History and Antiquities of the County of Surrey* (published posthumously in London, 1718–19). The Lancastrian physician Charles Leigh (1662–1701?) wrote *The Natural History of Lancashire, Cheshire, and the Peak, in Derbyshire* (Oxford, 1700), and *The Natural*

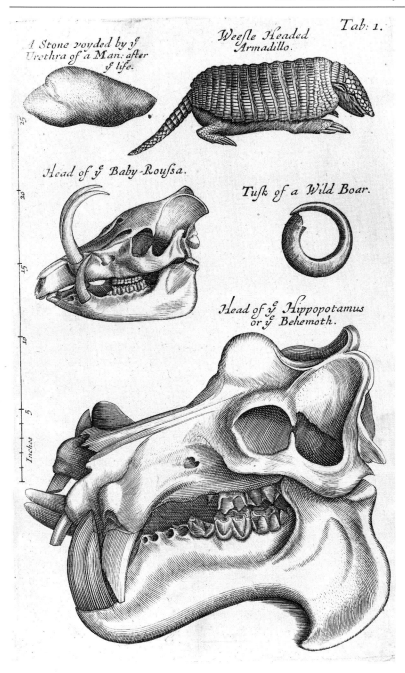

Tab: 1.

A Stone voyded by y^e Urethra of a Man: after y^e life.

Weesle Headed Armadillo.

Head of y^e Baby-Roussa.

Tusk of a Wild Boar.

Head of y^e Hippopotamus or y^e Behemoth.

Inches

Figure 5.3 An illustration in the curious accurate style showing specimens in great detail. The provision of a scale increases the impression of accuracy and also enables the viewer to wonder at the contrast in size between the hippopotamus and the armadillo. From Nehemiah Grew, *Musæum Regalis Societatis* (London, 1681).

History of Northamptonshire (London, 1712) was provided by John Morton (1671?–1726), a Northamptonshire clergyman, naturalist, and antiquarian.

All of these local natural histories concentrated, in true curious style, on rarities and wonders, taking curiosities of all sorts as their subject matter without strict specialization in natural phenomena. They treated all the topics of interest to curiosi: mineral waters; formed stones; plants; animals; notable experiments; mechanical inventions; unusual methods used in agriculture, mining, and manufacture; the houses, gardens, and estates of the gentry and nobility; and antiquities. The rare and fine of all of these subjects formed the interest of curiosi and the subject matter of natural histories.

In describing wonderful curiosities, natural histories adopted the new style of accurate reporting which demonstrated the truth of the wonders which they described and thus fitted them to be objects of the true Christian wonder of the curioso. The natural histories of Plot and others were full of matter-of-fact reports of detailed observations which involved great circumstantial detail demonstrating their accuracy. Measurements were frequently reported, providing further proof of accurate reporting, and illustrations were made as realistic as possible. This concern for detailed accurate description distinguished curious natural historians from their sixteenth-century predecessors whose works had not conformed to these standards of description and who were now criticized as insufficiently careful, or 'curious', in their observation.[17]

Natural histories were written in the 'miscellaneous' or 'essay' style as a further device to cause wonder in the reader. This style avoided pedantry and was seen as particularly suitable for the expression of the wit of a gentleman.[18] It was characterized by variety and contrast: very varied topics were treated without any systematic organization and these different discussions were closely juxtaposed so as to produce striking contrasts. Thus, for instance, Robert Plot in his *Natural History of Oxfordshire* turned from the dramatic death of a scholar of Wadham College struck by lightning to support John Beale's suggestion that the weather be recorded every day so as to learn 'how far the positions of the Planets, or other symptoms or concomitants, are indicative of weathers'. Changing the subject again, he proceeded to describe in detail 'tryals' made on echoes to see whether they differed by night and day, or with a pistol or the human voice. He emphasized the contrast between these different subjects by introducing this last discussion with 'Next the Tragedies . . . it will not be amiss to present the Reader with some of the sports of Nature, and entertain him awhile with the Nymph Echo' (pp. 5–7). All these different discussions were contained in just two pages which served to emphasize the contrast between them.

In adopting the miscellaneous style of writing, involving variety and contrast, writers of natural histories encouraged wonder in their readers. Variety and contrast were the principal techniques used by poets and painters in the period to cause delight and wonder, which were seen as the aim of poetry and art.[19] Thus natural histories were written in a style designed to cause wonder in the reader. They provided rare and surprising material for the wondering speculation and conversation of their curious readers. The accurate style of their descriptions ensured that their reports were true and thus that the wonder was genuine.

Curious collections

Curiosi collected rarities and curiosities avidly, filling their houses and gardens with them. From the hot house in the garden, full of rare exotic plants, to the gallery of paintings in the house, via the curious echo on the stairs, the whole house and garden of a curioso was, like that of the Norwich physician Sir Thomas Browne, 'a Paradise and Cabinet of rarities'.[20]

In addition to the curiosities spread through the house and garden, curiosi generally had a room or sequence of rooms in which their curiosities were concentrated. Here curiosities could be viewed by visitors and form the subject of their curious conversations. These collections or museums had great variety of content: rarities of all sorts, natural and artificial, were included without great specialization. The collection of Ralph Thoresby, described in his *Musæum Thoresbyanum* (London, 1713), is typical in its broad range of curious contents. There were more than two thousand coins and medals, Hebrew, Greek, Roman, British of all ages, and European. The human rarities included a fragment of an Egyptian mummy and the hand and arm of the Marquis of Montrose which 'seems really to have been the very Hand that wrote the famous Epitaph . . . for K. Charles 1st'. The collection contained wonderful monsters such as a 'young Cat (littered at Leedes) with Six Feet and Two Tails having two distinct Bodies from the mid Back'. There were specimens from exotic animals, such as the 'Foot of a great White Bear, eight Inches broad' and the 'Pizle [penis] of a Whale, in Length a Yard and a Quarter', as well as a crocodile almost six feet long. There were shells and butterflies remarkable for their beauty, 'very rare exotic Plants', 'Manna gathered in the Wilderness, where the Children of Israel travelled' and a 'Fragment of the Royal-Oak at Boscobell, where King Charles II was miraculously

Figure 5.4 The collection of Olaus Worm is typical in displaying variety and contrast in the arrangement of the specimens. From the ceiling hang a canoe, a bear, and various fishes and birds. The contents of the shelves on the right include a globe, statues, fish, antlers, birds, corals, stones, fruits, and a zither. From Olaus Worm, *Museum Wormianum seu Historia Rerum Rariorum* (Leyden, 1655).

preserved'. The collection contained fossil shells, beautiful agates, and mathematical instruments which included a telescope, two globes, and an ivory multiplication table. There was an Indian tomahawk and a Turkish scimitar, a 'Tooth-brush from Mecca' and a pair of gloves which had belonged to King James I. There were statues, seals, 'Heathen Deities, Amulets, Charms', and 'Matters relating to the Romish Superstition', including a 'Surprizing Representation of the Trinity'. There were paintings, prints and maps, ancient manuscripts, early printed books and rare editions of the Bible. The collection also contained many Roman antiquities including altars, bracelets, mosaic pavements, urns, lamps, and bricks.

Although catalogues of collections presented their curiosities arranged into kinds, the collection itself was governed by no such systematic classification. Natural and artificial objects were displayed together. Shells, formed stones, medals, and corals might all be contained in the drawers of the same cabinet and animals from all over the world were hung together on the ceiling. The curiosities were crammed into a small space where they filled the walls and ceiling, as well as the drawers and shelves of cabinets.[21]

This close juxtaposition of very different things crammed together in a confined space was the desired effect sought by curious collectors. Widely diverse objects were brought into close proximity in collections so that their variety was emphasized and the contrast between them could be better appreciated. This contrast and variety produced wonder in the viewer. Thus William Nicolson felt wonder at the famous collection of the London physician Hans Sloane (1660–1753) in 'four large Rooms crammed with so much variety': 'The collection indeed, since the Accession of the whole Stores of the late Mr. Charlton's Rarities of all kinds and Mr. Dendridge's Insects, Dr. Plukenet's dryed Plants, &c. is wonderful'.[22]

Collections were designed to provoke wonder in the viewer. They contained the rarities which curiosi found wonderful and were arranged in such a way that the variety and contrast of the rarities were displayed and so inspired wonder. The systematic arrangement of objects into kinds according to some method of classification, which was adopted in eighteenth-century natural history collections, would have been inappropriate in a seventeenth-century curious collection, since an object surrounded by similar but slightly different species would have seemed unsurprising and ordinary. Instead, specimens were arranged to display their variety and contrast, and thus cause wonder. Collections were sites for admiration at the rarities of the world and wondering conversation between curiosi, and their arrangement and contents were designed for this purpose. A curious collection was 'a repository of rare and select objects of natural history and art so curiously and elegantly arranged' that wonder and admiration were provoked.[23]

This artful arrangement of curiosities in the collection or museum made it the best place for viewing curiosities. Here the strangeness of curiosities was displayed by the close juxtaposition of contrasting specimens, whereas in nature no such contrast could be readily made. The rarity and variety of curiosities could thus be better appreciated in a collection than in nature, and curiosi thought that curiosities were improved by being placed in a well arranged collection: Lord Coleraine admired William Courten's ability to 'bring home whatsoever you found good, & make itt better by yo[r] ranging itt so judiciously in your Apartements'.[24]

In a collection objects were brought together from all over the world and displayed in a single place where the whole of Creation could be seen in a single glance. In thus bringing the curiosities from the whole world into one place, the collector recreated Paradise before the animals were dispersed at the Fall, and he and his visitors could view the whole world just as Adam had done after the Creation. This idea of the collection as a recreation of Paradise can be clearly seen in the anonymous 'poem occasion'd by the viewing Dr. Sloans musæum London Dec: 1712'.[25] The poem begins with a description of Adam viewing Creation in Paradise:

> When the fi[r]st man in Paradise had place
> He lookd around and viewd all natures face
> The world had gather'd to its ma[s]ters view
> Its severall kind each wondrous creature knew.
> Here Quadrupeds in all their shap[e]s surround
> Their knowing Lord and Reptiles there abound
> Obsequious fishes press to touch the Shoar
> And all its birds the Airy Region bore
> Its vegetable world the Earth sustain'd
> and various mineralls its womb containd.

The poet recounted how Adam had named the creatures in Paradise and proceeded to compare his own experience in Sloane's collection to Adam's:

> The admiring Lord ye wide Creation knew
> and gave to ev'ry part its name anew.
> Thy crowded world thus do I now survay
> wishing with wondring Eyes for longer day
> Thus while I hear thee name each beauteous part
> Admire the maker's and the owners Art.

The poet viewed the collection, Sloane's world, and wondered at it in the same way as Adam had admired Creation. Sloane himself was likened to Adam naming the creatures.

Sloane's collection was the world collected in one place, hidden indoors unseen by the sun, but seen by the poet with such enjoyment and wonder that he wished that time would stand still. God had spent six days creating the world which was collected here, and the poet proposed that six days could well be enjoyed in viewing it. This wondering viewing of the world was to continue on the seventh day of the week: the first six days were 'enjoy'd' in wondering at God's work, but the Sabbath by contrast was to be 'spent' in this way as a religious duty:

> Phoebus the world to which so fast you fly
> Collected here you pass regardless by
> Time has forgot for scenes of blood to go[26]
> And sure it might for scenes of knowledge too

If six whole days yᵉ new born world employ'd
Six might in viewing thine be well enjoy'd
Spent as seventh too by heavens Command
In wondring at yᵉ great Creators hand
Here all his works in beauty rang'd appear
If theres a paradise on Earth tis here.

Sloane's collection was further likened to Paradise in its preservation of plants by drying. The world after the Fall had seasons in which plants withered, but Sloane's art preserved the colours of plants throughout the year:

In Nature plants decay'd and with[e]red seen
here kinder art makes each an Ever green.

Collections of dried plants provided a 'Hortus Hyemalis' where the plants kept their colour regardless of the season. This meant that they could be viewed in flower even in the winter, just as in the Garden of Eden where it was believed that there had been no seasons, and the plants and trees had borne fruit and flowers continuously.[27]

The collection was presented as a terrestrial paradise where the collector's industry had brought together curiosities from all over the world and his art had arranged them beautifully. All of Creation could be seen in a glance, as Adam had done before the creatures were dispersed at the Fall. The visitor felt wonder at the works of God, just as Adam had done, and admired the collector's art in collecting and preserving the objects, as well as the art of God who had originally made them.

This recreation of Paradise in the collection made hazardous and difficult travel no longer necessary. Products from all over the world were brought together so that the curioso could visit nature by visiting the collection:

No more the Traveller from pole to pole
Shall search the seas or round the globe shall rowll
Safe from the dangers of the deep may be
and visit nature while he visits thee
Here Lappland Bears with Borneos Quantury[28] meet
Those guests of Ice These once dissolv'd in heat
Whales from the north come down to visit day
and flying fishes meet them in their way.

Since collections were the ideal place for viewing curiosities there was no incentive to study objects in their natural environments which only tended to diminish their curiosity. The activities of the curious centred on the collection where the rarities of the world were brought together and could best be viewed and discussed with wonder. Thus when Ralph Thoresby visited William Nicolson in 1694, the two men 'presently retired from the company to his museum, where he showed me his delicate collection of natural curiosities . . . [and] some coins and medals'. When they later went out to visit the local sights, they soon cut short their walk, 'longing to be again in that little paradise, his study'.[29]

The curious way of life centred on the study or library. Here the gentleman curioso collected together the rarities and wonders of the world. Here he and his friends met to admire the works of nature and art and to examine his many books and manuscripts describing natural and artificial curiosities. Here they conversed in curious style, expressing their wonder and admiration, philosophizing on causes and telling the stories which made objects curious.

Further reading

Houghton, Walter E., Jr., 'The English virtuoso in the seventeenth century', *Journal of the History of Ideas*, 3 (1942), pp. 51–73, 190–219.

Hunter, Michael, *John Aubrey and the Realm of Learning* (London, 1975). *Science and Society in Restoration England* (Cambridge, 1981). *Establishing the New Science: The Experience of the Early Royal Society* (Woodbridge, 1989).

Impey, Oliver and MacGregor, Arthur (eds.), *The Origins of Museums: The Cabinet of Curiosities in Sixteenth- and Seventeenth-Century Europe* (Oxford, 1985).

Levine, J., *Dr. Woodward's Shield: History, Science and Satire in Augustan England* (Berkeley, 1977).

MacGregor, Arthur (ed.), *Tradescant's Rarities: Essays on the Foundation of the Ashmolean Museum* (Oxford, 1983).

Pomian, Krzysztof, *Collectors and Curiosities: Paris and Venice, 1500–1800*, trans. Elizabeth Wiles-Portier (Cambridge, 1990).

Prest, John, *The Garden of Eden: The Botanic Garden and the Re-Creation of Paradise* (New Haven, 1981).

Schnapper, Antoine, *Le géant, la licorne, la tulipe: Collections et collectionneurs dans la France du XVIIe siècle* (Paris, 1988).

6 Physicians and natural history

For reasons pertaining to both the 'art' and the 'science' of medicine, large numbers of physicians contributed to natural history in early modern Europe. For their *ars* – the skill or method of treating disease – they needed to know about the uses of plants, animals, and minerals. For their *scientia* – the knowledge of health and disease – physicians shifted the foundations of medical learning from philosophical disputation to investigations of nature. Early modern physicians increasingly came to believe that pursuing 'matters of fact' rather than correct reasoning would provide the most certain basis for both their art and science. Both their professional concerns and intellectual outlook, therefore, caused many physicians to take a deep interest in helping to develop natural history.

While there were many kinds of people who practised medicine – and who did so charitably, part time, or as the mainstay of their living – by and large the physicians were the only ones who had formal degrees from universities. One should allow some exceptions for surgeons, who in some places (notably France and Italy) could enrol in university faculties of medicine. But in general, it was the physicians who were distinguished from all others by the fact that they possessed the medical doctorate, the MD. Moreover, of the four university faculties – arts, law, theology, and medicine – only medicine gave students what we might call a graduate training in natural science (to use somewhat anachronistic terms). Many other people might pick up an excellent knowledge of nature from their studies in the arts faculties, or from reading and discussing matters with private tutors, or even on their own; but physicians who possessed an MD had been formally certified as men (the universities then excluding women) who were well grounded in natural philosophy. The possession of the MD testified as no other certificate could that the physician understood the intellectual foundations of 'physic', a term derived from the Greek word for nature, φύσισ.

Whether with regard to their study of the foundation of health and disease or their cultivation of the art of treating disease, physicians possessed a keen professional interest in all developments related to a knowledge of nature. One late seventeenth-century

English physician, probably Christopher Merrett (1614–95), neatly summed up this concern when he wrote that: 'The word Physician, derived from the Greek φυσικόσ, is plainly and fully rendred by the word *Naturalist*, (that is) one well vers'd in the full extent of Nature, and Natural things; hereunto add the due, and skilful preparation and application of them to Mens Bodies, in order to their Health, and prolongation of Life, and you have a comprehensive Definition of a Physician.' Such a physician-naturalist, Merrett continued, needed to possess a studied and experimental knowledge of many things: vegetables and other simples (such as waters, oils, spirits, and salts), diseases, medicines, manufactures, minerals, animals, earths, stones, etc. Only such a person 'is able to advise fitting Diet, and Remedies, at all times and places'.[1] While few physicians could hope to live up to Merrett's ideals, many found themselves at least dabbling at being naturalists, since both their art and their science demanded it.

The search for medicinals

When it came to the art of treating diseases, the physicians of the sixteenth and seventeenth centuries both developed a more exacting knowledge of ancient medicinal plants, animals, and minerals, and adapted a host of new remedies to their practices. The medieval and early modern European pharmacopoeia borrowed very heavily from ancient sources. The most respectable sources for both simple and compound medicines, such as the so-called *Circa instans* of Platearius (mid-twelfth century) or the *Herbal* of Rufinus (thirteenth century), elaborated and expanded on the work of Pliny the elder, Dioscorides (both of the first century CE), and other ancient authors.[2] These medieval works continued to be studied and used right through the sixteenth century, and sometimes even thereafter. But the fifteenth-century intellectual movement in Western Europe known today as 'humanism' focused increasing attention on the ancient sources. By the later fifteenth century increasing numbers of humanists could handle ancient Greek, and even many with only Latin were involved in an energetic search for ancient manuscripts; moreover, following the appearance of humanist printers, both ancient Greek and Latin authors began to appear in printed editions based on very old manuscripts, making it possible for numerous learned physicians to scrutinize the original sources. This scrutiny of the ancient sources was spurred on by the common supposition that the most ancient knowledge of nature had been the best, it having decayed over the centuries.

Many early modern physicians therefore wished to return to the original sources for their medicines, in the belief that the ancients had possessed excellent remedies, the knowledge of which had been lost or corrupted over the centuries. To obtain the health-giving effects of ancient wisdom, many physicians felt compelled to re-examine classical prescriptions. The philological revolution of the late fifteenth and early sixteenth centuries – which brought into being both a renewed understanding of ancient languages and the development of new palaeographical techniques – made such a reconsideration of the earliest medical texts possible. As well-trained philologists, the medical humanists put their skills to work in arduous attempts to identify properly the classical simples (that is, the individual ingredients used in medicines). But medical humanists quickly found that this work on ancient texts required long labours not only in the library, but in the garden and field as well.

For instance, countless people were involved in the search for the true theriac. The medicinal compound 'theriac' (the word being derived from the Greek word for a wild or venomous animal, and vernacularized in English as 'treacle') was thought to be a wonderfully effective antidote to all poisons and a general preservative against most diseases. A number of ancient authors before Galen (second century CE), and Galen himself, recommended the medicine and listed its ingredients. But theriac had an enormously complex formula; some of the ancient recipes for theriac compounded up to eighty-one simples. By the sixteenth century the annual mixing of theriac in certain cities was done with great ceremonial pomp in the midst of grand festivities, presided over by physicians and magistrates, and finally doled out to local apothecaries and merchants for sale elsewhere. 'Venetian theriac' became particularly well regarded, being distinguished by name in England and throughout Europe. Humanist physicians spent great efforts not only in determining the true directions for compounding theriac, but in attempts to identify accurately the true simples from which the ancients had made it. Some moaned that many of the classical simples were clearly missing in the mixtures sold as theriac in their own day, while others noted that many of the simples used in the modern compound were not definitively the same ones used by the ancients. Physicians therefore struggled to identify correctly the simples mentioned by the ancients by going back to the original sources and attempting to link precisely each Greek term with its modern equivalent. To do so took not only skill with languages and palaeography, but an ability to scour the markets, gardens, and fields for possible ingredients, bringing to bear a knowledge of botany, zoology, and mineralogy. The resulting arguments about the ingedients of theriac might make or break a physician's reputation. The intensive search for the authentic ancient remedies helped to bring about 'a quiet revolution in simples'.[3]

An example of changes in the knowledge of one much sought

after simple described by the ancients – and included in theriac –
is the case of rhubarb. True rhubarb roots worked, it was thought,
by purifying the humours via a gentle purge, having astringent as
well as cathartic effects. But rhubarb's place of origin in the East,
as well as an accurate description of the living plant, were both
uncertain. Humanist physicians did determine that many of the
roots sold as rhubarb in the markets were not the real thing. Many
of these plants, which seemed related to true rhubarb, were native
to the Near East and were consequently cheaper and more
common, but they had quite inferior medicinal effects. The investigations of sixteenth-century herbalists and merchants finally settled
the origin of the true rhubarb as somewhere in China, but it was
not for many long decades to come that an accurate botanical
description of the true rhubarb could be had, much less a secure
supply of the medicine.[4] Further vigorous arguments abounded
about which were the truly good serpent skins, bezoar stones, bits
of mummy, amber, and so forth.

The concern to find authentic ancient medicines also contributed
to the attempts of Europeans to find new trade routes to the East.
Since the most famous classical pharmaceutical authors, such as
Theophrastus and Dioscorides, had lived in the eastern Mediterranean, and had access to simples brought from much further east,
the importation of drugs (often included under the general term
'spices') became a necessity for those who wished to use ancient
remedies. The late medieval pepper and spice trade with the
Levant, especially with Beirut and Alexandria, brought great riches
to the Venetians, Genoese, and Catalans. But the Europeans of the
Atlantic seaboard began their rapacious seaborne enterprises not
only in search of gold and other Christians to help in the struggle
against Islam, but in search of new routes for importing Eastern
spices. First the Portuguese and Spaniards, and then the Dutch
and English, forced their way into the south and east Asian trade
in spices over the course of the sixteenth and seventeenth centuries
by finding and sailing the route around Africa, and so partially
short-circuiting the overland caravan trade to the Mediterranean.
By the early seventeenth century, as Europeans came to control
not only the carrying trade between Asia and Europe but the
regions and people where the spices were produced, the amount
of imported spices grew rapidly, while the prices began to plum-
met.[5] Large quantities of the produce of Asia appeared in European markets, including a tremendous variety of simples that had
been known to the ancients but which had been infrequently available to Western Europeans.

Like the investigations of ancient texts, however, the importation
of ancient remedies brought unintended consequences. For
instance, Columbus' belief that by travelling west he had found the

eastern source of classical simples such as rhubarb and cinnamon eventually proved wrong. But the growing recognition that the western lands were a new world rather than the eastern part of the old did not mean that Europeans gave up the quest to import medicines from there; it only meant that they would have to search out medical uses for plants about which they had previously known nothing. Soon, other botanicals – usually adopted from local custom – began to find their way to Europe in increasing amounts. Most dramatically, tobacco became a successful new import that caused considerable controversy among physicians about whether the supposed medicinal benefits of smoking it outweighed its drying stinks. At the same time, European sugar cultivation – important for medicinal compounds as well as confections – spread with the help of slave labour from the Mediterranean to the Canary Islands and then to the West Indies. And other new addictions came from Asia and north Africa in the seventeenth century, sold first as healthful medicinal drinks: chocolate, tea, and coffee.[6]

Indeed, by the mid-seventeenth century both the English and Dutch East India Companies were making a habit of introducing exotic botanicals to the European market by publicizing their medicinal uses. For instance, a series of short pamphlets running from 1672 to 1695 attributed to one John Peachey (apparently published at the behest of an importer), introduced English readers to a host of new medical simples: Molucco, Virginian, Maldiva and Malabar nuts; Angola, Barbado, Russia and Mexico seeds; roots called nean, ipecacuanha, serapies (or salep), and casmunar (or, in another version, cassummuniar); calumba wood and a wood called lignum nephriticum; and bananas, Bengala bean, Bermuda berries, perigua, cassiny, cylonian plant, blatta Bizantina, and serpent stones.[7] The Dutch physician Cornelis Bontekoe (1640–85) earned the epithet 'the tea doctor' because of his book on the healthful effects of tea, recommending eight to ten cups a day at a minimum, and fifty to two hundred a day as reasonable; rumours abounded (but have so far remained unproven) that he had close connections to the major importer of tea, the Dutch East India Company.[8]

Among the most popular of the new medicinals were guaiacum and cinchona bark. Guaiacum was brought to Europe very soon after Columbus founded Santo Domingo in the mid-1490s, being a remedy used by the indigenous people there to treat syphilis; it was first popularized in Europe by Ulrich von Hutten's *De guiaci medicina et morbo Gallico* ('On the Medicine of Guaiacum and the French Disease') (1519). Also known as 'holy wood', legendary stories developed about how it came to be used by Europeans, associating it with Apollo and a holy celebration on Hispanola in which those infected with syphilis were cured by water sprinkled from the branches of a guaiacum tree. Europeans used fine shavings taken from the wood, which they boiled for a long time in water, then drinking the cooled infusion; at the same time the patient was put on a strict regimen. The great banking house of Fugger, among others, became deeply involved in the trade in guaiacum wood. As people in Europe learned to identify the true guaiacum from the texture of the bark and the colour of the wood, whole logs became the preferred form in which the wood was shipped.[9] The other extraordinarily successful new import, cinchona (also known as 'Jesuit's bark', or simply 'the bark'), was first imported from Peru around the early 1650s. It also quickly took on a legendary history to explain why Europeans adopted it from the local people (being attributed to the healing of the Countess of Chinchon), and was soon being marketed by Jesuit pharmacies and others. The bark, again taken in an infusion, worked to rid people of intermittent fevers quickly, without purging. (From it, many long decades later, quinine was eventually derived.) By the later 1650s, wealthy patients were demanding this very expensive medicine.[10]

Because of the intense pursuit of a healing knowledge of plants, animals, and minerals, by the 1530s, at places like Pisa, Padua, and Montpellier, medical professors were eager to show students examples of the simples used in their medicines, a knowledge that doctors had previously been expected to learn on their own. Padua and Pisa established botanical gardens in the mid-1540s, and the university of Bologna laid down a major botanical garden about 1567. Other medical faculties throughout Europe followed suit. These gardens quickly became places for the cultivation of beautiful and exotic, as well as useful, plants, being true botanical gardens, not just physic gardens. The master gardeners of Europe developed techniques for wintering over non-native and sensitive plants from around the world. By the later sixteenth century, every university that seriously pretended to an excellent reputation for medical teaching had established a botanical garden, while many physicians, apothecaries, and aristocratic scholars had their own private gardens. By the mid-sixteenth century, too, the practice of pressing and drying plants on paper for the purposes of identification (the making of herbaria) – a practice popularized if not invented by Luca Ghini, the first professor of simples at Bologna and first director of the botanical garden at Pisa, in the 1530s and 1540s – made botanical study something that could be done all year long.[11]

At about the same time, some groups of physicians began to compose lists of the simple and compound medicines that they considered useful, and with the help of municipal governments, to limit the dispensing of apothecaries to the ingredients given in these pharmacopoeias. Institutional struggles within the medical community made the physicians keen to retain their superiority over other practitioners. Physicians generally argued that by virtue of their university education they had the ability and right to practise any branch of medicine, and that surgeons and apothecaries ought to be supervised by them. In many printed works, from little pamphlets to large books, physicians stated that the public was being abused by drug sellers, whether these be uneducated empirics or apothecaries, so that the public good would best be served by making all practitioners subject to the oversight of the physicians. But as asserted by one very eminent learned physician, Jean Fernel (*c.* 1497–1558), this imperial view made it necessary for the physician to be better educated in medicinal simples than even the apothecaries: 'The knowledge, collection, choice, culling, preservation, preparation, correction, and task of mixing of simples all pertain to pharmacists; yet it is especially necessary for the physician to be expert and skilled in these things. If, in fact, he wishes to maintain and safeguard his dignity and authority among the servants of the art, he should teach *them* these things.'[12] Increasingly, views like Fernel's had legal power, for many municipalities began to enact laws restricting the local apothecaries to the official pharmacopoeias. The local physicians determined the proper and true simple and compound medicines, limited the apothecaries to dealing in those medicinals alone, and made periodic inspections of the shops to make sure that the ingredients were fresh and proper and that only the medicines listed in the pharmacopoeias were being dispensed. The first official dispensatories were composed in Venice (1496) and Florence (1498), with the first published municipal pharmacopoeia appearing in Nuremberg (1535), followed by ones in Basel (1555), Mantua (1559), Antwerp (1561), Augsburg (1564), Cologne (1565), and Bergamo (1580). London got its first official pharmacopoeia in 1618, and Amsterdam in 1643. While the physicians who wrote such pharmacopoeias staked legal claims of professional authority to their supreme knowledge of medicines, they also reinforced the pressures on other physicians to study simples diligently.

Because of both their classical education and their daily concerns, then, physicians took the lead among those who worked to identify accurately the simples used in medicines, especially the botanicals. Consequently, it is no surprise that the greatest sixteenth-century herbalists were physicians: Otto Brunfels (*c.* 1489–1534), Hieronymus Bock (1498–1554), Leonhard Fuchs (1501–66), Pietro Andrea Mattioli (1501–77), Guillaume Rondelet (1507–66), Francisco Hernández (1514–78), Rembert Dodoens (1516–85), Ulisse Aldrovandi (1522–1605), Andrea Cesalpino (1525–1603), Carolus Clusius (1526–1609), Matthias de l'Obel (1538–1616), and Kaspar Bauhin (1550–1624).

Figure 6.1 Depiction of purple violet, from Leonhard Fuchs, *De historia stirpium* ('On the History of Plants') (1542). The draughtsmen for the illustrations were Heinricus Füllmaurer and Albertus Meyer, and the engraver Vitus Rodolphus Specklin. Fuchs's text drew mainly from ancient authors, although it is not as thorough as others (such as Otto Brunfels's *Herbarum vivae eicones* ('Living Images of Plants') of 1530); but the beautiful woodcuts of 500 plants, each occupying an entire folio page, and drawn in outline with roots, branches, leaves, and often flower and fruit depicted, are remarkable.

Intellectual commitments

But the intellectual world of the early modern physicians saw developments that made them deeply interested in natural history not only because of the importance of simples to their *ars*. Natural history became central to the preoccupations of physicians concerned to establish a firm foundation for their *scientia* as well.

For many reasons the later fifteenth, sixteenth, and seventeenth centuries saw multiple attacks on established university philosophy. These attacks came from both inside and outside the academy. The recovery of a host of new ancient texts made the natural philosophical views of many kinds of non-Aristotelians seem plausible, so that Hermeticism, Neo-Platonism, Epicureanism, Stoicism, and many other philosophical schools developed powerful supporters, while Aristotelians themselves came in an ever greater number of varieties. The bloody religious warfare and search for heresy consequent upon the Reformation and Counter-

Reformation also gave some people pause about how firmly one could or ought to defend a particular view, helping to spread an interest in philosophical scepticism and fideism. And many modern thinkers elaborated their own views of nature and developed adherents: Paracelsus (*c.* 1493–1541), Petrus Ramus (1515–72), Francis Bacon (1561–1626), Galileo Galilei (1564–1642), René Descartes (1596–1650), and many others. Most of the principles by which nature had been thought to operate consequently came under attack by one school or another: matter–form theory, the four elements, the four qualities, the temperaments and humours, even the three principles of the Paracelsians, to mention only a few.

If certainty in knowledge no longer stemmed from the principles of nature, many argued, one could shift ground and begin with a knowledge of the experience of nature, the elements of which constituted 'facts'. The sense of a fact as something that had 'really occurred . . . hence a particular truth known by actual observation or authentic testimony, as opposed to what is merely inferred', was new to the seventeenth century. (The previous meaning of the word, still the most common meaning in the sixteenth and seventeenth centuries, pointed to a deed, since the Latin *factum*, from which the French took *fait* and English also took 'feat' as well as fact, meant a thing done.) To ascertain the real occurrences of nature – the facts – was, however, no easy task, and meant relying on the testimony of the fragile and misleading five senses, making this kind of knowledge of nature more probabilistic than the certain knowledge derived from authentic principles.[13] Nevertheless, Sir Francis Bacon (1561–1626) became famous in later generations for making trials of experience, or 'experiments', into one of the foundation-stones of his natural philosophy. A century later, Dr Hans Sloane (1660–1753), President of the Royal Society and of the Royal College of Physicians, wrote that 'matters of fact' were the very essentials of natural knowledge, and far more certain than other means of knowledge: 'the Knowledge of *Natural-History*, being Observations of Matters of Fact, is more certain than most Others, and in my slender Opinion, less subject to Mistakes than *Reasonings*, *Hypotheses*, and *Deductions* are . . . These are things we are sure of, so far as our Senses are not fallible; and which, in probability, have been ever since the Creation, and will remain to the End of the World, in the same Condition we now find them'.[14] Since the Latin term *historia*, meaning a 'narrative, account, tale, story',[15] derived from the Greek ἱστορία, meaning 'a learning by enquiry', or 'the knowledge or information so acquired' (or more generally 'an account of one's inquiries, a narrative, a history'),[16] natural history came to mean an account of nature based upon information acquired by enquiry, especially through observation (since sight was held to be the least fallible of the senses). By Sloane's time, the units of that enquiry constituted the 'facts'.

Physicians were among those who fought for these changes in epistemology, holding up as an example the ancient physician Hippocrates (*c*.460–370 BCE). Hippocrates, it was thought, had pursued close and detailed investigations into nature without hypothesizing. Upon these foundation stones learned medicine and speculative natural philosophy had been built. If one needed a reform of medical and natural philosophical knowledge, then one ought to start over by following the example of Hippocrates, and observe closely. This view of Hippocrates went back to the early sixteenth century, to Paracelsus and to the revivers of Hippocratism in the medical faculty of Paris and elsewhere. Hippocrates the original, unprejudiced investigator and recorder of nature, rather than Galen the theoretician, gradually became the best ancient physician for students to imitate.[17] Physicians therefore took their place among the earliest university-educated men to develop a deep appreciation for natural history as the rock upon which natural knowledge in general could be built. Consequently, their commitment to natural history went well beyond its practical uses to medical practice. Many developed a deep and abiding commitment to natural history as the one and only true path to a knowledge of nature. For instance, the English physician Martin Lister (1639–1712) had no particular medical ends in his investigations of English spiders (Figure 6.2), while the Dutch Jan Swammerdam, who received a superb medical education in Amsterdam and Leiden, managed to devote himself full time to natural history rather than having to practise, by continuing to live in his father's house.[18] They, and others like them, had imbibed a commitment to natural history as the best way of knowing.

But by the later sixteenth century, when physicians took up the study of natural history, they brought two investigative techniques along with a search for experiences and observations: they made anatomy and chemistry integral parts of their project. As virtually all surveys of the 'scientific revolution' acknowledge, a knowledge of human anatomy moved from a philological concern to know the ancient anatomical texts accurately, to an awareness that Galen and others had sometimes been wrong about the details, to a quick advancement into a very complex view of the structure of the body. Andreas Vesalius (1514–64), a professor of anatomy and surgery at Padua, has become the most famous, but is by no means the only important anatomical investigator of the sixteenth century. By the seventeenth century the notion that a firm grasp of the details of the physical structure of a creature would help to explain it pervaded medical culture. Physicians took to dissecting just about

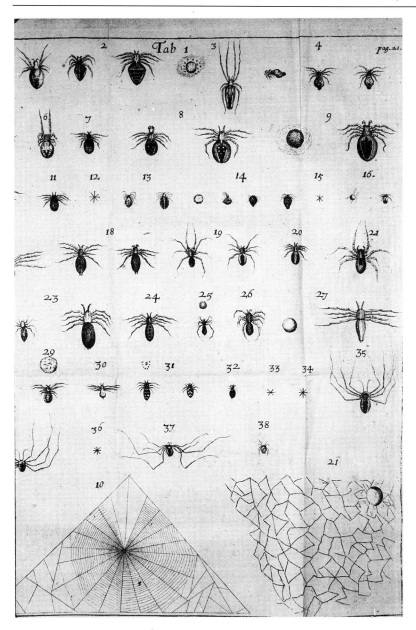

Figure 6.2 English spiders ('De Araneis Octonoculis'), from Martin Lister, *Historiae animalium Angliae* (1678). Note that some of the forms of the webs are depicted, as well as the spiders. The book was composed while Lister was practising in York; after his move to London in 1683 he became an active participant in the Royal Society. He is notable for his work on molluscs as well as 'insects' and medicine. Lister's wife and daughter made many of the drawings he used in his books.

any animal or plant if given the opportunity, in order to analyse them. By the seventeenth century, too, increasing numbers of physicians had taken up chemistry as an analytical tool as well as a means for inventing new remedies. While Paracelsus and others had combined a chemical outlook with other challenges to medical orthodoxy, so that sixteenth-century physicians tended to think of iatrochemists as dangerous quacks, chemistry proved so useful a tool that physicians gradually adopted it – although in doing so they tended to adopt other sorts of explanations for chemical transformations than Paracelsian ones.

By the middle of the seventeenth century, then, physicians had not only become enthusiastic natural historians, they had also combined their investigations into creatures and natural objects with anatomical and chemical analyses. The University of Leiden, for instance, became famous for its medical teaching. There the professors took students into clinics to show them cases of disease; taught them methods of chemical analysis; dissected energetically both privately and publicly; and made sure that the botanical garden, herbaria, cabinets of curiosities, and collections of preserved specimens got heavy use. 'I was much pleasd with a sight of their Anatomy Schole, Theater & Repository', wrote the English visitor to Leiden, John Evelyn (1620–1706). It 'is very well furnish'd with Naturall curiosities; especially with all sorts of Skeletons, from the Whale & Eliphant, to the Fly, and the Spider'. The garden, too, 'was indeed well stor'd with exotic Plants'.[19] Leiden's curriculum had in large part been modelled on Padua's, where anatomy, botany, and the clinic (if not chemistry) had become central to medical teaching in the mid-1560s.[20] The medical faculties of Montpellier and Basel, too, combined chemical, anatomical, and botanical teaching.[21]

The combination of natural history with chemistry and anatomy flourished not only in medical schools, but in the new institutions established in the seventeenth century to investigate nature. The chemical physician Guy de la Brosse (c.1586–1641) became the first instructor at the Jardin du Roi in Paris (later known as the Jardin des Plantes) under the patronage of Cardinal Richelieu: he was both an excellent botanist and excellent (Paracelsian) chemist, teaching both at the Jardin. Richelieu also supported the work of the chemist and physician Théophraste Renaudot and his Bureau d'Adresse, which was in part a forum for public discussions of natural knowledge. Thirty years later, during the 1660s and after, at the Paris Academy of Sciences the botany programme involved not only growing, collecting and illustrating exotic and useful plants, it also involved analysing them anatomically and chemically.[22] In London, too, the Royal Society for the Advancement of Natural Knowledge (as its charter styled it) undertook anatomical and chemical studies as part of its ordinary investigations into nature; the *Philosophical Transactions*, which regularly reported on work of interest to the Society, published many such accounts. Many years ago Raymond Phineas Stearns attempted to break down the papers and experiments done at the Royal Society during its first century of existence; if we take his figures, about 43 per cent of the 'experiments' concerned 'medical sciences', natural

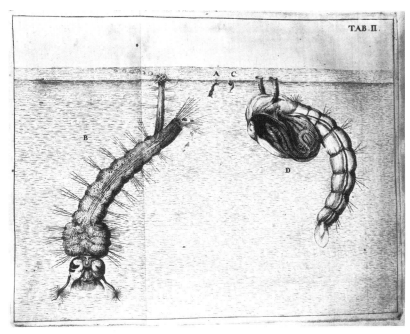

Figure 6.3 Mosquito larvae, from Jan Swammerdam, *Historia insectorum generalis* (1669). Swammerdam's great ingenuity in developing instruments and techniques to study the smallest details of natural things, especially their anatomy, is truly remarkable. He not only worked with powerful single-lens microscopes, but developed tiny tools in order to explore the anatomy of insects, finding (for instance) the wings of a butterfly in the body of a caterpillar. He worked alone and with other Amsterdam physicians.

history, and chemistry, while about 44 per cent of the papers read spoke to natural history or 'medical science'.[23]

Because natural history had become the keystone of both the physicians' *ars* and their *scientia* by the mid-seventeenth century, physicians made up a large proportion of the members of these early modern 'scientific' societies. For instance, Charles Gillispie noted long ago that in the early Royal Society of London, physicians were 'both quantitatively and qualitatively . . . instrumental in implementing the "new philosophy"' (although his ideas about 'science' as mainly physical science caused him to treat their interest as an avocation unrelated to medicine). He noted, too, that in 1664 physicians occupied five of the seven teaching posts at Gresham College (where the Royal Society met), while by his count fourteen of the thirty-seven members of the Royal Society in 1662 (or 38 per cent) were physicians.[24] Marie Boas Hall pointed out that it was 'clever young physicians who, about 1645, began those meetings at Gresham College which were the seed from which the Royal Society was to grow', while A. Rupert Hall noticed that 'at least a fifth' of the members of the Royal Society to 1687 'had some sort of connection with medicine'.[25] More recent analyses of the membership of the Royal Society by Michael Hunter have reiterated the numerical and intellectual importance of physicians to the new philosophy in England. One of the nine classes of members into which Hunter divides the Fellows is that of 'medical practitioners', which includes court and university physicians but excludes

Figure 6.4 Heading of the chapter on the octopus, from Guillaume Rondelet, *Libri de piscibus marinis* (1554–5). Rondelet (1507–66) held a regius professorship in medicine at Montpellier from 1545 until his death. An intensely active anatomist, and an author of several medical works, his fame in subsequent generations has been mainly for his massive work on fish, which remained authoritative for over 100 years. His influence in inspiring his students and other scholars to take up natural history investigations makes him one of the most important naturalists of the sixteenth century.

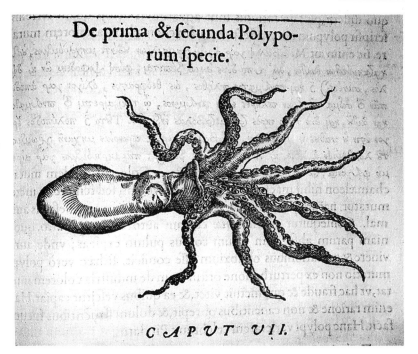

apothecaries and people with MDs who did not practise medicine (like Sir Christopher Wren). Of all the elected fellows from 1660 to 1700, 16 per cent fall into this category (one of the two largest groups); an unusually high 81 per cent of them showed at least some active participation in the affairs of the Society (the greatest activity of any group). They and the scholars and writers together made up between one-third and one-half of the Society's active core.[26]

Physicians composed such a quantitatively and qualitatively significant core of promoters of the 'new philosophy' for two reasons: natural historical investigations had become central to both their *ars* and *scientia*, causing many physicians to be so deeply taken by these studies that they promoted natural history not only vocationally but as an avocation, too; and natural historical investigations had become part of the core studies of nature during the period. It is to do no injustice to the importance of mathematical reasoning, or Cartesianism, Newtonianism, and other natural philosophies, to say that for contemporaries, other kinds of investigations of nature were at least as important as they. Medicine and natural history constituted, in fact, the 'big science' of the early modern period, soaking up enormous sums of money and energy contributed by countless people.[27] The physicians, who were among those leading the pursuit of a knowledge of natural things and events, came to find more intellectual stability in 'matters of fact' than in controverted principles of nature. As a result, physicians were not only crucial to the promotion of natural historical studies as both an activity and an intellectual outlook, they also took an active role in the development of the new philosophy.

Further reading

Arber, Agnes, *Herbals, their Origins and Evolution: A Chapter in the History of Botany, 1470–1670*, 2nd edn., (Cambridge, 1938).

Cook, Harold J., 'Physick and natural history in the seventeenth century', in Roger Ariew and Peter Barker (eds.), *Revolution and Continuity: Essays in the History of Philosophy of Early Modern Science* (Washington, DC, 1991), pp. 63–80.

'The new philosophy in the Low Countries', in Roy Porter and Mikuláš Teich (eds.), *The Scientific Revolution in National Context* (Cambridge, 1992), pp. 115–49.

Foust, Clifford M., *Rhubarb: The Wondrous Drug* (Princeton, 1992).

French, R. K. and Wear, Andrew (eds.), *Medical Revolution in the Seventeenth Century* (Cambridge, 1989), esp. ch. 6, 9, and 10.

Lindberg, David and Westman, Robert (eds.), *Reappraisals of the Scientific Revolution* (Cambridge, 1990), esp. ch. 7, 8, and 10.

Porter, Roy, *Health for Sale: Quackery in England 1650–1850* (Manchester, 1989).

Reeds, Karen Meier, *Botany in Medieval and Renaissance Universities* (New York, 1991).

Schivelbusch, Wolfgang, *Tastes of Paradise: A Social History of Spices, Stimulants, and Intoxicants*, trans. David Jacobson (New York, 1992).

Wear, Andrew, French, R. K., and Lonie, I. M. (eds.), *The Medical Renaissance of the Sixteenth Century* (Cambridge, 1985).

Webster, Charles, *The Great Instauration: Science, Medicine and Reform 1626–1660* (New York, 1975).

Webster, Charles (ed.), *Health, Medicine and Mortality in the Sixteenth Century* (Cambridge, 1979), esp. chs. 9 and 10.

7 Natural history as print culture

With the invention of the printing press in the mid-fifteenth century, the making, distribution and use of printed materials soon became a key part of natural history. As well as transporting and collecting natural objects themselves, practitioners attempted to make and transport printed records of them. Some of the resulting artefacts have remained at the centre of historians' attention ever since. It is hardly an exaggeration to say that the books produced by such figures as Gesner, Aldrovandi, Ray and Buffon have come to constitute the historical identity of natural history.

We must, then, attempt to understand these artefacts. We need to know how and why they were made, how they reached their audiences, and how they were put to use. These are not straightforward questions, and answering them may mean reaching some rather unexpected conclusions. In particular, secure in our own experience of print as providing for reliable communication to large and dispersed audiences, we all too readily consider its history in terms of *fixity* and *quantity*. Printing was unique, we may say, because it caused a vast increase in the numbers of books available, and enabled the stabilization of reliable texts in multiple copies; other consequences then flowed from these. Many historians have indeed predicated their accounts of printing on these elements, and their importance is beyond doubt.[1] Yet a close look at an activity like natural history soon shows that too easy an acceptance of quantity and fixity as foundational can in other respects be rather less satisfying. Above all, it means that those elements themselves remain beyond analysis, since they are regarded as definitive of printing itself. Because of this, such an approach tends to underplay the complexity of the social mechanisms through which early printed books could be not only made, but warranted by their makers as reliable. Nor can it easily comprehend the different meanings accorded them by different audiences. Yet these complications loomed large in the lives of natural historians, and it seems perverse to dismiss them as peripheral. Recently, then, historians have rediscovered these problems, by restoring attention to the work involved in making and reading printed materials. Far from the meaning of a book being fixed by the printing press, they reveal it as something arrived at constructively, by different communities pursuing interpretative conventions specific to their time and place. Their counter-intuitive conclusion is that quantity and fixity are far from self-evident or necessary concomitants of the printing press.

Early printed books existed in, and were constituted by, various complex and dynamic regimes of practice. To understand them – and understanding them is essential to understanding the cultures of natural history – now means comprehending in a single narrative a series of connected worlds: of book traders and readers, of creative collectives and individual authors. How can this be done? One approach has been to look, not at the sheer numbers of copies made and distributed, but at the fate of a single book through its appropriation by different communities. Roger Chartier calls this the 'object study'.[2] Such a strategy allows for a finer appreciation of the subtleties involved at every stage. A related approach will be followed in this chapter. It will construct an object study, of a sort; but it will do so by making use of a mode of communication employed in the early modern era itself.

Seventeenth-century England knew a literary genre, minor at the time and since forgotten altogether, which, for want of a better name, we may call 'sighs literature'. A work conforming to its conventions would typically be biographical, or more specifically autobiographical, in nature. It would be a designedly moral, justificatory, legitimizing autobiography, designed to vindicate its writer from some terrible slur, or to make a moral example of his or her life. To that end, at every point in that life when a major event had occurred, it would break off into an italicized digression called a 'sigh'. These 'sighs' allowed the writer to interrupt the narrative as if by divine *fiat*, in order to articulate its full moral implications. Scripture provided ample resources for such *cris de cœur*. The sighs of troubled cavalier Richard Atkyns (1615–77), to name one such seventeenth-century writer, were steeped in the language of Job and the lamentations of Jeremiah. The result, then, was a bifurcated narrative, marrying the ideal to the real, the general to the particular, and the providential to the contingent. It was a peculiar way of writing, but it had its uses – even though Atkyns himself died in debtors' prison, signally unvindicated.

Without wishing to advocate its revival to any wider extent, this chapter appropriates the most prominent features of 'sighs literature'. It does so in order effectively to discuss the making, distribution, structure and use of printed books in early modern natural history. Sighs literature offers a peculiarly useful structure for such a discussion. For one thing, it jolts us sharply out of our own, twentieth-century, experience of print – the experience which invests fixity and quantity with all the authority of self-evidence.

It does so, moreover, by allowing us to present a credible account of how printed books might act in natural historical communities, directly alongside a critical analysis of that very account. Were the historical impact of print really as simple as we often assume, the following narrative could proceed uninterrupted. The aim in showing that it cannot – that it must be interrupted at every stage by 'sighs' – is to encourage us to question that implicit faith, and to articulate the hidden historical foundations both of print culture and of the enterprise of natural history. In the story which follows, then, it may be best to bear in mind one final counter-intuitive point: for once, the sighs might just be everything.[3]

The Vindication of the Reverend and Learned Doctor *Peter Heylyn* his *Cosmographie*, with certain SIGHS at the end of every *CHAPTER*

The history I have in mind is not that of a person; it is the history of a book. The work I am going to trace is the *Microcosmus*, latterly the *Cosmographie*, of Peter Heylyn (1599–1662). This was one of the most frequently reprinted works of its type in the seventeenth and early eighteenth centuries, and its history is unusually revealing (Table 7.1).

Peter Heylyn was a major religious polemicist, scholar and historian of his time. As his son-in-law affirmed, he was no 'ordinary common Clergy-man', but a favoured agent of the ecclesiastical and political authorities, 'singularly well acquainted . . . with the principal motions and grand Importances in his time both of Church and State'. A staunch supporter of William Laud, the latest Archbishop of Canterbury to be beheaded, Heylyn incurred particular odium among those he called 'the rabble' for his vigorous defences of high-church priestcraft. For all that his contemporary reputation was that of a Laudian militant, his *Cosmographie* left Heylyn's other works far behind when it came to the numbers of editions printed and the longevity of its influence.

The enterprise of cosmography is no longer pursued as avidly as it was in Heylyn's day. Of what, then, did it consist? The short answer is that it consisted of all knowledge about the world. As such, cosmography was essentially universal: it would perhaps be easier to list those branches of knowledge which it did *not* include. Heylyn himself described the field as providing a 'universal comprehension of *Natural* and *Civil* Story', by combining three main approaches. From '*Natural History* or *Geographie*', it treated 'Regions themselves, together with their Sites, and several Commodities'; from civil history, it discussed 'Habitations, Governments, and Manners'; and from 'the *Mathematicks*', it encompassed 'the *Climates* and Configurations of the Heavens'. It was also as

Table 7.1. *Peter Heylyn*, Microcosmus *and* Cosmographie, *1621–1703*; *Edmund Bohun/John Augustine Bernard*, Geographical Dictionary, *1688–1710*

1621	*Microcosmus, or a little description of the Great World*	Oxford: J. Lichfield and J. Short	4°
1625	Μικρόκοσμος ... *Augmented and revised*[a]	Oxford: J. Lichfield and W. Turner; sold by W. Turner and T. Huggins	4° in 8s
1627	... Third edition, revised[b]	Oxford: J. L[ichfield] and W. T[urner], for W. Turner and T. Huggins	4° in 8s
1629	... Fourth edition	Oxford: W. T[urner] for W. Turner and T. Huggins	4° in 8s
1631	... Fifth edition	Oxford: W. Turner	4° in 8s
1631	... Another issue[c]	Oxford: for W. Turner and R. Allott [London]	4° in 8s
1633	... 6th edition	Oxford: [W. Turner] for W. Turner and R. Allott	4° in 8s
1636	... 7th edition	Oxford: W. Turner; sold [by M. Allott, London]	4° in 8s
1639	... 8th edition	Oxford: W. Turner	4° in 8s
1639	... variant[d]	Oxford: W. Turner	4° in 8s
1652	*Cosmographie*	London: for Henry Seile	fol.
1657	... 2nd edition	For Henry Seile	fol.
1665	... 3rd edition	For Philipp Chetwind	fol.
1666	... 'Third' edition	For Anne Seile	fol.
1666	... Anr. edition	For Philip Chetwind	fol.
1669	... Anr. edition	For Anne Seile; sold by George Sawbridge, Thomas Williams, Henry Broom, Thomas Bassett, Richard Chiswell	fol.
1669	... Anr. edition	For Philip Chetwind	fol.
1674	... Anr. edition	For Anne Seile and Philip Chetwind	fol.
1674	... Anr. edition	For Philip Chetwind and Anne Seile	fol.
1677	... Anr. edition	By A.C. for P. Chetwind, and A. Seile; sold by T. Basset, J. Wright, R. Chiswell, T. Sawbridge	fol.
1682	... Anr. edition	For P.C., T. Bassenger, B. Tooke, T. Sawbridge	fol.
1703	*Cosmography* '7th' edition, 'Improv'd ... by Edmund Bohun'[e]	For Edw. Brewster, Ric. Chiswell, Benj. Tooke, Tho. Hodgkin, and Tho. Bennet	fol.
1688	*A Geographical Dictionary*	For Charles Brome	8°
1691	... 2nd edition	For Charles Brome	8°
1693	... 3rd edition	For Charles Brome	fol.
1695	... 4th edition	For Charles Brome	fol.
1710	... '4th' edition[f]	For R. Bonwicke, W. Freeman, T. Goodwin, J. Walthoe, M. Wotton, [and 5 others]	fol.

Notes: [a]Sometimes lacks table of climes (as do next 7 issues). [b]Sheets vary: corrected in press. [c]Cancel title-page. [d]Misprinted title page: '1939'; Sheets sometimes mixed with those of 7th edition. [e]Discontinuous pagination; printed by subscription (*Proposals* circulated separately). [f]Discontinuous pagination.

Sources: A. W. Pollard and G. R. Redgrave (2nd edn., W. A. Jackson, F. S. Ferguson and K. F. Pantzer), *A Short-Title Catalogue of Books Printed in England, Scotland and Ireland, and of English Books Printed Abroad, 1475–1640* (3 vols. London, The Bibliographical Society, 1986–91).
D. Wing, *A Short-Title Catalogue of Books Printed in England, Scotland, Ireland, Wales, and British America and of English Books Printed in Other Countries, 1641–1700* (2nd edn. 3 vols. New York, Modern Language Association of America, 1972).
F. J. G. Robinson, G. Averley, D. R. Esslemont, P. J. Wallis, *Eighteenth-Century British Books: An Author Union Catalogue* (5 vols. Newcastle, University of Newcastle/Folkestone, Dawson, 1981).

omnivorous of sources as it was ambitious of scope. Heylyn thus chose epigraphs for his book to signal that it employed at once the classics, Scripture and natural history: on its frontispiece and title-page were inscribed the testimonials of Virgil, St Paul and Pliny. In fact, Heylyn selected the latter's *Natural History* to exemplify the entire enterprise of cosmography.

What, specifically, was his book *for*? Heylyn himself made it clear that although he might play 'the parts of an *Historian* and *Geographer*' when writing it, he remained primarily a '*Church-man*'. His purpose was to argue for a particular ecclesiastical polity: that of episcopacy. The rule of bishops, he alleged, was the only form of Church governance established by God. In the years immediately preceding the Civil War, a conflict which has been called the last of the Wars of Religion, this was a highly contentious claim. That was why it was both proper and necessary for a churchman like Heylyn to be writing such an exhaustive work. He believed that he could prove his case beyond doubt by grounding it in an authoritative survey of the natural, political and historical knowledge of his time. We can summarize the project, in fact, by reference to the imperatives of Heylyn's most powerful ecclesiastical patron. Laud was championing a notoriously vigorous view of governance which he called 'Thorough', and which he hoped would revitalize Charles I's realm. It would not be inappropriate to see his client's cosmography as 'thorough' natural history.

For the origins of the *Cosmographie*, though, we must go back to Heylyn's early experiences at university. He graduated at Oxford in 1617. It was customary for graduands to read a series of lectures, and he chose geography for his subject. The lectures were a great success. Heylyn followed 'a *new Method* not observed by others', and managed to display himself 'a good Philosopher as well as Geographer'. So impressed was his audience that he was soon offered a Fellowship. He then decided to publish the lectures. Printed and bound, the first copies began to reach their readers on 7 November 1621.[4]

1st Sigh.—— *'Lord, thou art he that took me out of my Mothers Womb!'*

It can be all too easy to assume that the writing and publishing of substantial tomes about the natural world is self-evidently a worthwhile activity. Yet Heylyn went through a remarkably convoluted process in order to establish the legitimacy of his actions. First, the initiative appeared not to have come from Heylyn himself; friends had urged him to publish the lectures. Heylyn then revised them – a process taking two months of hard work – obtained assent from his father for their printing, and had them perused by yet more 'Learned Men'. Only after receiving their approval did he pass the work on to printer John Lichfield. Why such a complex manoeuvre? The answer had to do both with his good name and with the fundamental origins of knowledge. In early modern England, the former was put at risk by too insistent a claim to personal originality, especially when that claim was advanced by the notoriously melancholic and antisocial figure of the university don. A gentleman must appear to shun the braggadocious connotations of authorship. This matters greatly, because Heylyn's credit as a proposer of true knowledge – and thus the reception accorded his book – depended on his reputation as a gentleman.

But in Heylyn's post-lapsarian world – a world irremediably tainted by Adam and Eve's expulsion from Eden – a bigger danger loomed. Perhaps it was religious lèse-majesté, as well as bad manners, to claim personal originality. So Heylyn himself maintained that knowledge did not come from the individual, but from God. Why did he publish? Because, he replied, 'The Lord God brought [the knowledge] to me'.

This, the first complication to any simple story which would leap directly from discovery to published work, is thus a fundamental one. The question of authorship is not trivial. Much was invested in Heylyn's decision whether to publish, and such reasoning has its own complex history.[5]

Who were the readers of this work? Records are scanty. Yet Heylyn's friends were able to claim that copies of his book were 'bought up by Scholars, Gentlemen, and almost every Housholder', and mainly 'for the *pleasantness* of its reading'. Before long, they alleged, 'scarce any Scholars Study' was without its copy. That there is some truth to this may be indicated by the fact that it (or its successor) appeared in the auction catalogues of many deceased scholars' libraries later in the century. The repeated reprinting of the book also testifies to its continued value: it was reissued at least eight times by 1639.

We can, however, identify two readers, and readers who stand out in importance: King James I and the then Prince of Wales, Charles. *Microcosmus* bore a dedication to Charles, and Heylyn presented the first copy to him in person. But unfortunately for Heylyn, the Prince's evident approval was countered by an explosive reaction from his father. Presented with the volume by John Young, Dean of Winchester, the King was seen to 'peruse' it curiously. This, the second printing of *Microcosmus*, included for the first time a detailed 'Table of the Principall things herein contained', and James turned straight to this table. There he read the first entry, which happened to concern the vital courtly subject of heraldry. '*ARMES*', it said: 'why in the same Eschocheon those of *England* give place to *France*, 490'. Turning to page 490, the King read that the reason for this subordination of honour was simply 'that *France* is the larger & more famous kingdome'. He flew into a rage. In the full heat of his 'Anger and Passion', James called for this impudent book to be searched out and suppressed. Heylyn seemed about to fall victim to the court's power over the world of printing – to what, in our own day, we might call censorship.[6]

2nd Sigh.——   *'But Lord, what dangers and mischiefs is Man subject to! There is but one way to Live, and hundreds of ways to be depriv'd of Life'.*

Despite his precautions, Heylyn's good name was now in peril. He had to tread carefully. Fortunately, the content of his book had been constructed to be helpful here. Mario Biagioli's recent studies of Galileo suggest that at court a figure like Heylyn worked to 'efface' himself as an author, so as to represent the prince as the source of all knowledge, and himself as a conduit channelling celebrations of that fact. Microcosmus *could thus scarcely offer to add anything to Charles's knowledge of, or power over, the world. Both, axiomatically, were already perfect. Heylyn was a skilful enough participant in the affairs of the court to appreciate that much. It was thus to Charles as 'the greatest and most accomplished traveller' of all that Heylyn presented his own 'little World'.* Microcosmus *could not augment his knowledge; it could only act as an 'abreviarie' of knowledge he already possessed. But Heylyn's book was now an element in court culture: that was what it was* for, *and that culture, with its intricate protocols of presentation, perusal and patronage, would now decide what it* meant.

James's response reveals something of the attendant perils, and is highly suggestive of the politics of royal reading. For contemporaries attributed his immediate discerning of the offensive passage, not to adroit use of the index, but to 'the Kings peircing Judgment'. The royal nous *had immediately 'spyed out a fault, which was taken no notice of by others'. His perspicacity thus constituted yet more support for the notion that 'God always endows Kings his Vice-gerents with that extraordinary gift, (the Spirit of discerning) above other mortals'. The King possessed all the knowledge, all the skill; his was the discerning, discriminating power. This was the inseparable counterpart of Heylyn's representation of the origins of knowledge as divine. If knowledge came from God, then God's representative on Earth was its proprietor.[7]*

Told of this calamity, Heylyn flew into a panic. He risked losing the King's favour. What to do? Young advised him to go straight to court himself, and appeal to Charles's patronage; but, unsure whether Charles, too, might not have been offended, Heylyn dithered. Lord Danby eventually supplied the answer. Heylyn was to write a letter of 'Apology and Explanation', and transmit it to the King. This letter would explain what Heylyn had really meant to say in the offending passage, and reattribute the undeniably offensive wording to another agent. To restore his good name, then, Heylyn must establish an alternative reading from the literal one. 'Thus Mr. *Heylyn* was the interpreter of his own words'.[8]

'Lord, what obdurate heart is it that will not be concern'd in putting   *3rd Sigh.——*
Soul and Body into the Hands of Strangers, and contrary minded Men!'

Heylyn had to seize back the 'soul and body' of his words. His tactic of providing an 'interpretation' was a well-recognized one. Moreover, in Charles's court Heylyn himself was to become a peculiarly valued 'interpreter' of books. So in 1633–4, it was he who was entrusted to 'collect' scandalous passages from William Prynne's puritan attack on stagecraft, Histriomastix, *'reduce them into method', and provide Laud with evidence against Prynne for his fateful trial at Star Chamber – a defining moment before the Civil War.*

The point is one of the classification *and* use *of books, and thereby impinges on notions of their* influence. *Opponents of such 'interpreting' procedures objected that with them in play one could no longer even 'tell what a libell is'. Laud's lackeys, they claimed, could 'transubstantiate' any books at all into libels. Mercator's* Atlas *itself was banned at this time, on account of its 'sundry scandalous and offensive passages'. We thus need to appreciate the active processes of adjudication by which natural history books attained their meanings in sites of reception such as the royal court. Courtiers often read a text piecemeal, for isolated phrases, rather than pursuing an extended argument from beginning to end.* Microcosmus *was constructed out of, and for, this sort of reading.*[9]

Heylyn's letter reached James, and succeeded in mitigating his wrath. The 'interpretation' it offered therefore bears repetition. 'The burden under which he suffered', Heylyn maintained, 'was rather a mistake than a crime, and that mistake not his own, but the Printers.' They had printed *is* instead of *was* at the crucial point. This was a convincing tactic, since, as Heylyn was able to remind the King, it was 'most ordinary in them to mistake one word for another'. The printing-house, then, had been the place at which those seditious words had been created, not the scholar's study.[10]

'Lord, how are they increased that trouble me! Yet let them not Tri-   *4th Sigh.——*
umph over me, nor be believed when they speak Scandalously against me; but let the mischief of their own Lips fall upon them'.

Heylyn's strategy is revealing of the practical work which had to be done if a book – especially the sort of elaborate volume characteristic of natural history – were to be successful. To recover that work, we need to look to the printing house [Figure 7.1]. Its culture can be reconstructed by attending to the testimony of printers themselves – from 'their own Lips', as it were.

A printer did not simply reproduce a manuscript slavishly. A substantial degree of interpretative autonomy was expected of him. A good compositor must 'read his Copy *with consideration', manuals advised, 'so he may get himself into the meaning of the* Author'. *He must then use typography to render that meaning clearly for 'the capacity of the* Reader'. *That is, he must not only interpret the author's meaning but also anticipate readership. But printers had to work quickly. Rather as a typist does today, they seem to have worked by the syllable, word, or phrase, rather than by mentally registering individual letters. As a result, their errors were often multi-character ones, which could 'corrupt & pervert the sence'. The most common of all were monosyllabic; the worst appeared in the early 1630s, when the King's Printers themselves omitted the word 'not' from the Seventh Commandment and exhorted Charles's subjects to commit adultery. Heylyn's attribution of a monosyllabic error to his printers was, then, believable. But he himself admitted that 'if it had been of a higher Crime than of a Monosyllable, it had not been pardonable'.*

Many houses employed a corrector. But good paper was too expensive to throw away casually, so books would be made up of sheets in different states of correction. In consequence, no two copies of a given 'edition' need be the same: there are no identical copies of the first Shakespeare folio, for example. Rooted in the practice of printing, this conclusion is problematic for our assumption that printing 'fixes' texts. Heylyn's book(s) provide a good example: there is no definitive text of Microcosmus.[11]

Heylyn had satisfied the King. Knowing that he could not afford to make a similar mistake again, though, he resolved to remove the passage altogether from future editions. Meanwhile, his reputation at court restored, Heylyn went to France for five weeks. While there, he penned an account of his travels. The resulting text could probably have laid James's fears to rest by itself, such was the virulence of its Francophobia. In the event, it merely caused Heylyn more trouble. Lent to various 'friends', more and more copies were written out by hand, until eventually one of the recipients passed his or her manuscript on to a printer. There soon appeared a published version, 'printed', Heylyn noted with dismay, 'by a false *Copy*, full of gross Errors and insufferable mistakes'. If he had managed to evade responsibility for one result of the printer's craft, he could surely not repeat the achievement for another. Heylyn was forced to print his own text as quickly as he

Figure 7.1 The printing house. From S. Ampsing, *Beschryvinge ende lof der Stad Haerlem in Holland* (Haarlem, 1628), facing p. 392. Bodleian Library, classmark Douce A.219.

could, in order to counter this unauthorized version. So it was that his French travel narrative came to be printed with his *Cosmographie*.[12]

5th Sigh.——   *'And now, good God! what a golden opportunity have I lost!'*

Hoping to re-establish his credentials, Heylyn had instead fallen foul of yet another aspect of printing and bookselling which throws common-sense notions into doubt. Unauthorized printing was by all accounts common. Only in the late eighteenth and nineteenth centuries, with the advent of comprehensive mechanisms for the location and safeguarding of literary property and originality, and with the arrival of the steam press, could a modern realm of authorship develop. Meanwhile, an unproblematic publication process was the exception, not the rule. To produce a book was an achievement. *And (sigh!) it was an immensely hard and complex one. Fixity had to be fought for. Figures like John Ray (1627–1705) and Robert Plot (1640–96) devoted themselves to such labour night and day – it was essential to the development of natural history. Yet it remains invisible to conventional historical treatments, which begin by classifying it as incidental. The problem which Heylyn faced here – of 'piracy', as contemporary writers called it – contravenes all the most basic values of our own print culture. It thereby throws their self-evidence into doubt, and serves to focus our attention on their historical development.*

Thanks to such efforts, Heylyn certainly became a recognized authority. The court thought him qualified to pronounce on matters of quite recondite mathematical and natural knowledge. When, for example, one Captain Nelson appeared, claiming to have solved the longitude – the key problem of the mathematical sciences – Heylyn was chosen to adjudicate. On his verdict, he was told, rested 'the credibility of the *phaenomenon*'.[13] But this reputation took a fall with the descent into civil war. In January 1640 Parliament summoned him to appear in answer to charges levelled by his old victim, William Prynne, now newly released from prison and the darling of the London crowds. Although exonerated by the hearing, Heylyn found that 'such as took up matters upon trust and hear-say' now 'looked on me as a person forfeited, and marked out for ruin'. This was epitomized for him by an enigmatic experience which occurred as he left Westminster after the hearing. As he walked towards White-hall, a mysterious gentleman stepped into his path and forced him into the road. Laughing scornfully at Heylyn, he hissed, '*Geographie is better than Divinity*', and disappeared into the throng. Heylyn never found out his identity. But the moment haunted him for the next decade. What had this man meant?[14]

'Lord, thou hast made mine Enemies rejoyce over me, and laugh me   ——*6th Sigh.*
to scorn!'

What was this mysterious Londoner's meaning? Heylyn himself was unsure. None the less, his intervention may stand for the fact that printed books do not come with certificates of veracity or instructions for use; they are subject to practices which are necessarily beyond their own stipulation. Heylyn could not control what readers in this newly disordered realm made of his Microcosmus. *All the more so since the book ran through eight editions, six of them unauthorized. That was probably why the passage which had so offended King James was never actually obliterated, despite Heylyn's wish. By 1640, the errors of reproduction had multiplied so greatly that even Heylyn himself, seeing what had become of 'his' book, 'could no longer call it mine'. In what sense had printing fixed Heylyn's text? Or, to put the question another way, could a copy of it found in a London bookshop be trusted?*

Heylyn spent the years of the civil wars wandering, 'disguis'd both in his Name and Habit', between Cavalier safe-houses, where he hid from Roundhead soldiers in old Jesuit priest-holes. But by the time of Charles's execution in 1649, the memory of that Londoner's gnomic accusation was so insistent that he resolved to settle down and return to his study of the world. The labour was long and difficult. None the less, eventually his new work did appear. The result of all this effort was the transfiguration of the little *Microcosmus* into a massive folio *Cosmographie*. It was published in 1652, and reissued five years later (Figure 7.2)

This intense reading and writing caused Heylyn serious harm. 'His *Brain* was like a *Laboratory*', recalled his contemporaries, 'kept hot with study'. His natural constitution being hot and dry, 'his Brain, heated with immoderate study, burnt up the Christaline humor of his Eyes'. In short, Heylyn went blind.[15] The *Cosmographie* was the last book he himself penned, and he died ten years after its publication.

'Lord, what variety of troubles are incident to the nature of Man!'   *7th Sigh.*——

The Cosmographie *was not only a book presenting natural phenomena. The experience of its own creation could also be understood using notions of nature, and in particular of the human frame. Heylyn's contemporaries habitually employed such notions to understand the experiences of writing – and also of reading – books. That reading and writing could make you blind was not news to them. Physicians had to treat such conditions regularly.*

This should give us pause. For it is routine to site the growth of natural history in the context of a republic of letters characterized by certain new literary and typographic entities: the correspondence network and the learned journal stand out. We assume that the impact of these objects on readers is readily understandable, since reading seems

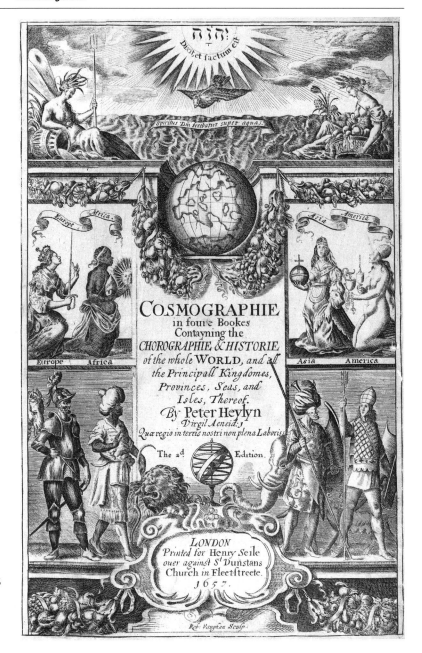

Figure 7.2 Peter Heylyn, *Cosmographie* (London, 1652; second edn., 1657), frontispiece. Cambridge University Library, classmark Adams.3.65.3.

to us a constant, defined as such by the structure of the human body. They therefore seem to constitute a secure foundation for historical understanding. But early modern writers described the body differently. They accounted for the effects of both texts and images in terms of faculty psychology and the physiology of animal spirits, such that they resulted from a complex interaction of imagination, reason and the body. We should take their words seriously, since people made decisions

*about what, when and how to read on the basis of such knowledge.
Those decisions were the foundation of any natural historical community
constituted by print. It matters, then, that reading has a 'natural history' of its own.*[16]

As Heylyn himself had put it, 'Books have an immortality above
their Authors'. Each impression was 'to them another being'. His
Cosmographie was no exception. Both the artefact itself and its readership were being transformed. *Microcosmus* had been a modest
quarto, aimed at a single royal reader. The 1650s *Cosmographie* had
not enjoyed such patronage. Without a dedication, it bore only an
address from Heylyn 'To The Reader' which stipulated how he
was now attempting to delimit readership. 'The greatness of the
bulk, and consequently of the price', apparently made Heylyn confident that 'none but men of judgement and understanding' would
read his work. None the less, he baulked at including more than
the minimum of illustrations. To do otherwise would, he declared,
'increase the Book both in bulk and price, and consequently make
it of less publick use than I did intend'.[17]

The *Cosmographie* suffered more changes over the following
years, being reissued nine times between 1665 and 1682 – a complex history which testifies to intense competition in the ranks of
the book trade.[18] But the greatest transformation was still to come.
In 1686, an impoverished Tory named Edmund Bohun (1645–
1701), shivering in a 'dark, stinking and inconvenient' garret, was
put to work by equally Tory bookseller Charles Brome to produce
for him a new edition. Cosmography was about to enter the empire
of Grub Street.

The *Republica Grubstreetana*, as Jonathan Swift called it, was no
place for an unprotected book. The first injury the *Cosmographie*
suffered was physical dismemberment. Heylyn's frontispiece,
along with significant portions of the book's content, was hijacked
for Bohun's other project, a little octavo *Geographical Dictionary*
(Figure 7.3). Such a small work was sorely needed, Bohun
insisted: 'it is a great mistake [to think] that all useful Books must
be of the largest size, whereas some are the more useful because
cheap and small'. Readily portable, it was designed to be useful
for travellers. But more realistically, it was also ideal for those
whose only expedition was to the nearest coffee-house. There,
Bohun claimed, 'News being one of the most usual Entertain-
ments, the knowledge of places is of absolute necessity'. Explaining
this in terms of the physiology of reading, Bohun alleged that without such knowledge, news impressed only 'faint and confused
Notions on the Minds of the Readers'. But the 'first and Principal'
use of the book, he concluded, was to act as a 'General Index'
to all geographical books. Printing had created such a plethora of

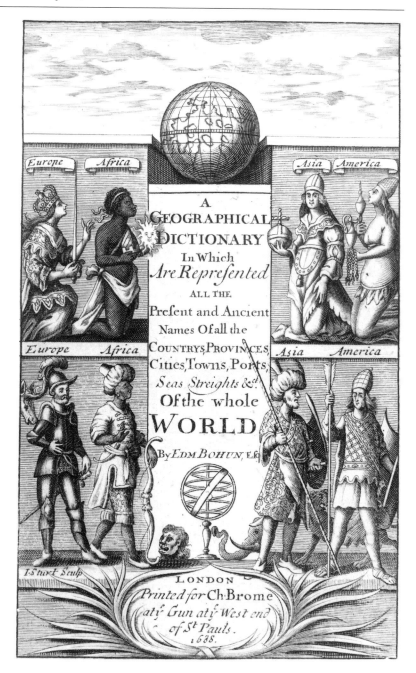

Figure 7.3 Edmund Bohun,
A Geographical Dictionary
(London, 1688), frontispiece.
Cambridge University
Library, classmark N.10.64.

publications that without such a guide it had become 'above the strength of Human Faculties' even to know where to look for knowledge.

Not even Bohun was in control now. The bookseller, Brome, issued later versions of the dictionary without his consent, and although he advertised Bohun's involvement, he in fact employed another writer – one 'said to be a Jacobite' – to amend them. His emendations were controversial; readers especially objected to his entry for the River Boyne, in Ireland, which failed to laud William III's victory there. At length the identity of this second writer was revealed. He was John Augustine Bernard – none other than Peter Heylyn's grandson. A convert to Catholicism, Bernard had been forced on Oxford University by James II as professor of moral philosophy. After the Glorious Revolution he had fled to join James in Ireland, where he wrote propaganda for the Jacobite cause. In late 1690 he had returned to England 'very poor and bare', and in this condition had been discovered and set to work.

Seeing Brome's advertisements, a horrified Bohun attempted to publish his own notices repudiating the work. But no printer would accept them. Bohun predicted he would be 'proved a *Jacobite*', and he was right: thanks to the dictionary, he was saddled with a lasting reputation for Jacobitism. This had dramatic effects, moreover, for in 1692 he became licenser of the press – in effect, government censor. A Whig outcry ensued. Bohun was sacked, and the entire licensing system – a regime of press regulation and literary property which had persevered for almost 150 years – came to an end, never to return. The destruction of Bohun had led to the fundamental restructuring of the cultural politics of print.[19]

'Lord, 'tis good for me that I have been in trouble, that I may learn   8th Sigh.——
thy Statutes'.

Books were subject to the communities which made and used them, and to their practical conventions – the social and cultural 'statutes' which governed the world of printing. Problematic cases such as Bohun's help us see this, for one of the most important statutes was that the conventions themselves be invisible. Controversy brought them to light.

Heylyn's work had now gone through a complex series of different formats, texts, titles and 'authors'; its last edition, attributed to the dead Bohun, was financed by subscription [Figure 7.4]. In this it is representative of a central aspect of print. For natural historians classified not just natural entities, but bibliographic ones too. They needed to master not just natural objects, but the vastly increased number of titles purporting to provide knowledge of them. Konrad Gesner is the best-known example. At the same time as his pioneering natural history work, Gesner was publishing a 'universal library', aspiring to list all the printed books then available. Indeed, he pioneered the enterprise of bibliography. Bohun, too, compiled a Universal Historical Bibliotheque *which incorporated natural history books. Transmuted into a geographical 'index', Heylyn's cosmography had now become a synthesis of the two enterprises: a summary both of the order of books, and of that of the world.*

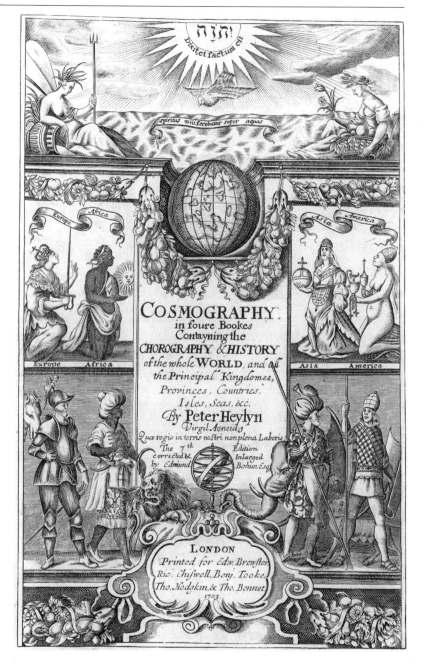

Figure 7.4 Peter Heylyn/ Edmund Bohun, *Cosmography* (London, 1703), frontispiece. Cambridge University Library, classmark N.7.38. Note the multiplication of booksellers' names: from one in 1657 to five in 1703. This represents a change in the social structure of publishing. The figures named here were prominent among London booksellers co-operating in wholesaling alliances called 'congers', to share costs and deter piracy. In another development since Heylyn's early quartos, this edition was produced by subscription.

That world was in transition. The practice of reading was coming to be performed silently and in isolation. New objects facilitated the change: small books, like Brome's Dictionary, *portable by hand and in the pocket. So did indexes – 'index-learning' became the subject not of Jacobean awe but of Augustan disdain. Eventually, a revolution in the practice of natural history would become possible, as workers could take their cross-referenced handbooks into the field.*

That a work from the 1620s was appropriated so drastically in the 1690s is striking, especially when its putative reviser disclaimed responsibility, when he himself was the man entrusted by the state with overseeing the book trade, and when his resulting humiliation played such a large role in the end of the entire regime of press regulation. What replaced licensing, eventually, was the new notion of copyright. It is perhaps the greatest irony of Bohun's story that in the ashes of his own good name lay the origins of a new authorial dispensation.

Bohun's subsequent fate also illustrates the new dispensation in natural history. His ambitions frustrated, he emigrated to Carolina. There he became the colony's foremost natural historian. He spent the following years, not in the library, but in the field. Confessing that it was 'the first time I ever did any thing of that kinde', he travelled the length and breadth of the region, searching out plants, insects and minerals, and procuring correspondents for the London naturalist James Petiver. In 1701 he was about to set sail for England to deliver his collection to Petiver when a ten-month illness suddenly worsened, and Bohun died. The edition of Heylyn's Cosmographie *on which he had laboured finally appeared two years later, financed by subscription and supported by a 'conger' of wholesaling booksellers, joined in a league to combat piracy. It was to be the last of all.*[20]

Books are peculiarly mutable objects. Given its different structure, typography, size, price, audience and interpretation, was Bohun's edition of the *Cosmographie* the same book as Heylyn's? In what sense? It should be clear that our answers to such questions must be historical in character. An alternative may be proposed, then, to our common-sense view of print. This alternative will accept the importance of printing for the history of natural history, and it will acknowledge that that importance derives in large part from fixity and increased rates of production. But it will assert that fixity and production – the very credibility of print, which now seems so obvious – need to be explained. This will allow us to restore to centre-stage the work which Gesner, Ray, Plot and Buffon needed to do to maintain an appearance of constancy and intelligibility for their books. In other words, it will see print as something fully and truly historical.

That said, perhaps it is appropriate to end with one last sigh. This sigh comes not from a seventeenth-century writer, though, but from the ultimate source: Scripture. Since Richard Atkyns concluded his own sighs by remarking on the parallels between his sufferings and those of Job, it is apt to finish with the Book of Job itself. There, persecuted as much by disbelief in his prophecy as by torture of his body, the prophet is eventually reduced to a state of anguished exasperation. What can he do to make his words ring out 'for ever'? With splendid anachronism, James I's translators appropriated this cry, and had Job hit upon their own greatest hope for a *machina ex deo* capable of transcending the strife of their world. 'Oh that my words were now written', they made Job sigh: 'Oh that they were printed in a book!'[21]

Further reading

Blunt, W., *The Art of Botanical Illustration*, rev. W. T. Stearn (London, 1950; 1994 edn.).

Chartier, R., *The Cultural Uses of Print in Early Modern France*, trans. L. G. Cochrane (Princeton, NJ, 1987).

'Texts, printings, readings', in L. Hunt (ed.), *The New Cultural History* (Berkeley, CA, 1989), pp. 154–75.

The Order of Books: Readers, Authors and Libraries in Europe between the Fourteenth and Eighteenth Centuries, trans. L. G. Cochrane (Cambridge, 1994).

(ed.) *The Culture of Print: Power and the Uses of Print in Early Modern Europe*, trans. L. G. Cochrane (Cambridge, 1989).

Dance, S. P., *Classic Natural History Prints*, 5 vols. (London, 1990–1).

Darnton, R., *The Business of Enlightenment: A Publishing History of the 'Encyclopédie' 1775–1800* (Cambridge, MA, 1979).

'History of reading', in P. Burke (ed.), *New Perspectives on Historical Writing* (Cambridge, 1991), pp. 140–67.

Eamon, W., *Science and the Secrets of Nature: Books of Secrets in Medieval and Early Modern Culture* (Princeton, NJ, 1994).

Eisenstein, E. L., *The Printing Press as an Agent of Change: Communications and Cultural Transformations in Early Modern Europe*, 2 vols. (Cambridge, 1979). Abridged as *The Printing Revolution in Early Modern Europe* (Cambridge, 1983).

Febvre, L. and Martin, H.-J., *The Coming of the Book: The Impact of Printing 1450–1800*, trans. D. Gerard (London, 1984).

Fissell, M. E., 'Readers, texts, and contexts: vernacular medical works in early modern England', in R. Porter (ed.), *The Popularization of Medicine, 1650–1850* (London, 1992), pp. 72–96.

Jardine, L., *Erasmus, Man of Letters: The Construction of Charisma in Print* (Princeton, NJ, 1993).

McKenzie, D. F., 'Printers of the mind: some notes on bibliographical theories and printing-house practices', *Studies in Bibliography*, 22 (1969), pp. 1–76.

Bibliography and the Sociology of Texts (London, 1985).

II Virtuosity, improvement and sensibility

8 Natural history in the academies

In seventeenth- and eighteenth-century France, the monarchy established a network of learned societies, starting with Paris, where the Académie Française was one of the great state foundations prior to 1670. Here also, under Louis XIV and his minister Colbert, the Academy of Sciences was established and reorganized between 1666 and 1700, at the same time as the Académie des Inscriptions et Belles-Lettres was being formed. In the capital, the Crown had at its disposal a number of official institutions which it had founded by harnessing or organizing private assemblies. Each, in its own field, served in an advisory capacity and supplied experts in different areas of learning. The Parisian network was completed by the addition of the Academy of Painting, Sculpture and Architecture, that of Music, and that of Surgery between 1720 and 1830. In the 1770s the Royal Society of Medicine joined the number of royal establishments.

Contrastingly, the provincial academic movement occurred in three principal stages. Before 1700 some ten societies were formed, benefiting from the support of local patrons and from royal protection. Between 1715 and 1760 about twenty new societies appeared, fostered by *lettres patentes*, and these were followed by a further ten academies after 1760 (Figure 8.1). The movement as a whole, which coincided with the appearance of other, less formally structured societies throughout France, especially after the 1760s, indicates the development of relations between the intellectual world and the administrative, royal, and regional governing bodies. In Paris the societies welcomed true professionals, who received pensions, but in return often carried out an active role in the service of the Crown, such as in censorship or in royal factories. They gave advice and established desirable prizes. In the provinces these societies formed a meeting place for amateurs recruited from the local social elites, and early on set up competitions and funded savant activities. Through their ceremonies, finally, they came to be integrated in the social and political life of the town.

The encyclopaedic project benefited from the success of the academic movement, which supplied it with a proportion of its Parisian and provincial collaborators.[1] The *Encyclopédie* was

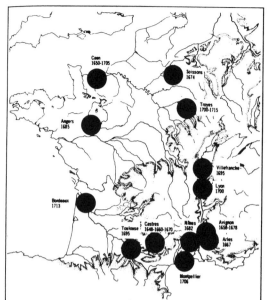

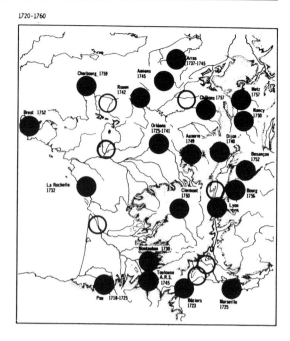

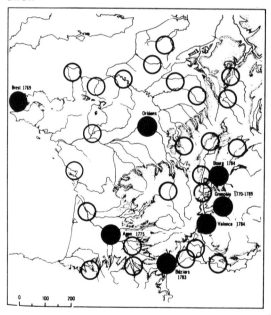

Figure 8.1 Map of France, showing the sites of academies founded 1650–1715, 1720–60, and 1760–89, respectively. From D. Roche, *Le Siècle des lumières en province: Académies et académiciens provinciaux, 1680–1789* (Paris, 1978), 2 vols., vol. II, 'Cartes', figs. 1–3. Empty circles denote the locations of academies pre-dating the period in question.

initially run by Denis Diderot (1713–84) and the great mathematician Jean le Rond d'Alembert (1717–83). Diderot was a *philosophe*, renowned for his savant works, his translations from the English, and soon also for his philosophical audacity, as the observations of the inspector of police reveal. Together they conceived and constructed a general systematic inventory of knowledge, as the *Tableau préliminaire* to their joint work shows. Here the entirety of nature is evoked in words and illustrations to show how it was possible to integrate the sciences and arts within a coherent philosophical system. The great dictionary appeared after more than twenty years of hard work, in spite of censorship and criticism, thanks to the support of a section of the royal administration, in particular that of the director of the Librairie, Chrétien-Guillaume de Lamoignon de Malesherbes (1721–94). Provincial amateurs, professional men, physicians, surgeons, and professors did not by any means accept all the daring philosophical and religious statements contained within the *Encyclopédie*, but they accepted the subject matter, the factual analyses, the inventory, and the technical processes described, which they in turn helped to improve and disseminate.[2]

In its breadth and in the diversity of its articulations, the *Encyclopédie*'s definition of natural history allows us to understand the essential attraction that this branch of science could exercise over the world of provincial amateurs.

The object of Natural History is as extensive as nature; it comprises all the beings which live on the earth, which lift themselves into the air, or which remain in the bosom of the waters, all the beings which cover the surface of the earth, and all those that are hidden within its entrails. Natural history in its widest extent would embrace the entire universe, because the stars, the air and meteors are included within nature; thus one of the greatest philosophers of Antiquity, Pliny, produced a natural history under the title of History of the world, *historia mundi*. But the more we have acquired knowledge, the more we have been led, and even forced, to divide it into different kinds of sciences. This division is not always exact, because the sciences are not so distinct that they do not have relations with one another; that they do not merge into one another and connect at several points, in both their generalities and their particulars.[3]

Diderot, to whom this article has been attributed, continued the description of the tree of knowledge whose three principal branches – animals, plants, and minerals – constitute the diverse subjects of a realm in which 'accurate and complete descriptions', observation as the basis of comparison, and the effort necessary 'to see the progress of nature in her productions', were to reign.

Anatomy, medicine, botany, and agriculture were thus to unite savants of all orders and 'work together for the happiness of men'.

Here we can see a range of the concerns which formed a part of the academicians' patrimony. The link between these concerns was strengthened by the fact that it depended less upon a precise and programmatic definition which operated similarly everywhere, than upon a wider concern to generate order which proved capable of channelling a variety of abilities and different cultural levels into a single programme. Within this programme, 'history' was the study of facts taken as true and no longer that of facts taken as false, and thus enabled the amalgamation of a genuine civic utilitarianism with a concern for the sciences.[4] Natural history reveals the concerns which pulled the academic world in different directions: both curiosity and science based upon precise observations and experiments; both the taste for the extraordinary and the concern to establish true facts; both the application of Kant's 'adventure of reason' (that is, speculation about the origins of living beings) to the real world and the desire to serve the public. The practices of provincial amateurs, and their fundamentally utilitarian goal, brought closer, as much as they distanced, the rational ideal of a science based on the rejection and articulation of facts, rather than on their accumulation.

An understanding of the social conditions of such concerns underlines the real divisions which could exist between the theoretical discourses and scientific practices of Parisian academicians, on the one hand, and those of the motley environment of the provincial pillars of society, on the other. The former were principally those who, at the Royal Academy of Sciences, worked to 'establish science' and to 'experience/experiment on the living world', whilst escaping the pitfalls of animism and reductionism. In the provinces, however, the power of the institutions, as well as the very different relationship between practitioners of the exact sciences, geometers, and naturalists, did not encourage the emergence of the life sciences with the same rapidity and visibility.[5]

Natural history and the 'academic city'

It is hard to fix a precise date for the conversion of the academic city to science, and even more difficult to follow the rapid growth of the multifaceted interest in the sciences of nature with any rigour. The Parisian example did not serve as a model for all the provincial societies, which were originally founded to celebrate language and power, on the lines of the Académie Française. From the first half of the seventeenth century, savant circles in Toulouse (with Fermat), in Clermont (with Pascal), and in Caen (with Huet), had an active and public existence, but their meetings, within which the eighteenth-century academies would perceive a continuous tradition, initially did not follow Parisian programmes.

In the case of Caen, D. S. Lux has shown how its Parisian foundation, under royal patronage, put an end to the activities of local patrons and provoked a grave crisis from which the Caen institution needed several years to recover.[6] It is in Montpellier and Bordeaux that the first naturalists are to be found, and in both cases the new establishments were founded under the watchful eye of Parisian academicians.

In the case of Montpellier, the opportunity for the development of natural history arose as a result of the close ties between the old faculty of medicine and the Court, and from the wish to create a centre of savant and utilitarian culture in the southern provinces which would be strong enough to collaborate with the Royal Academy of Sciences. The Montpellier society thus obtained total equality with the Parisian society, a prerogative that no other society would ever attain. However, in the period 1706–40, it was confronted by the same problems as those facing the Caen Académie de Physique in its members' attempt to organize themselves and collaborate within a centralized and hierarchical scientific life. Physician–naturalists and numerous amateurs engaged in activities which would enable the Montpellier society to be relaunched in the second half of the eighteenth century, but with less Parisian pressure.

By contrast, in Bordeaux, it was the wealth and culture of a group of noble amateurs, dominated by *parlementaires*, and among whom Montesquieu was to be found from an early stage, which controlled the scientific enterprise, captured Parisian patronage, and ensured the continuity of research. But we should note that the Bordeaux project placed natural and literary history on the same footing. From 1715, in a project which would continually reappear on the agenda, the Bordeaux amateurs located themselves in a dual tradition, that of the great provincial political, civil, and textual histories, exemplified by the Benedictine model, and that of the history of nature. This last was guided by the Aristotelian taste for collection and enquiry, being empirical rather than experimental, a genealogy and an investigation of facts simultaneously. As at Montpellier, the mental horizons of Bordeaux academicians were rather those of the curious observation of nature and of the living world than those of natural philosophy or physics.

In the first phase of academic formations, a debate took place over the status of science and its relation to the humanist and rhetorical heritage which formed the common environment of the provincial amateurs. At the start of the eighteenth century the provincial model was still dependent upon an older culture of humanism and curiosity. This was fuelled by the teaching of the *collèges*, Furetière's 'human sciences' (1690), and the Jesuit and Oratorian '*physique*'. The provincials were former pupils of the *collèges*, and for them the power of discourse underpinned their whole cultural project; the triumph of Parisian language was thus that of a particular political perspective. Simultaneously, another controversy was taking place, also recognizable in the creation of the Paris Academy of Sciences. It involved a choice between two structuring principles, that of a Baconian utilitarianism which aimed to help man control his environment and society, and that of a larger vision, defended by Claude Perrault, which aimed to involve all talents in the service of experience and within a general understanding of culture which was associated with the notions of organization and professionalization.

Natural history was situated at the pivot of these two conceptions, as can be seen from the number of physicians who classed themselves amongst the physicists of the Parisian society (five out of seven academicians),[7] and from the numbers in which provincial savant societies recruited members of the faculties or colleges of medicine. Of 3,000 academicians identified over four generations, more than a quarter of the ordinary members (28 per cent) and more than a third of the correspondents and associates belonged to this group.[8] A sociological constant underlay the permanence of this scientific concern: two learned societies were never to elect physicians, the Jeux Floraux of Toulouse and the Academy of Montauban. In Paris, as in the provinces, physicians were not always elected as practitioners, but it was nevertheless as such that they occupied a place in the urban fabric,[9] and as practitioners that they made theoretical and practical claims, which were backed up by their social position, in academic debates.

As it progressively escaped the influence of the *belles-lettres* model, the provincial movement was able to develop an orientation which favoured the pursuit of the sciences. Between 1715 and 1760 the number of academies was increased by the addition of twenty new societies. Their titles indicate the role of the appeal to the sciences: nine included the term *belles-lettres*, faithful to the old practices of the 'human sciences' as defined by the humanities; all the rest placed themselves under the polymathic heading of sciences, arts (meaning the technical and fine arts simultaneously), and letters. The absence of savant specialization meant that everyone could study what he liked: and it was this quality that made the provincial situation unique. The reconciliation of the Baconian ideal with the wide ambitions of Perrault allowed all recognized amateurs to be naturalists, observers, and experimenters, rather than speculators. The rule of objects, rather than words, was what the Parisian specialists – *physiciens*, anatomists, botanists, chemists – and the provincial polymaths had in common. A single hope could unite Parisians in their separation and provincials in their unity: the prospect of success for a community that validated

individual discoveries and guaranteed their use for the transformation and progress of the academic city.[10]

Here we can doubtless detect the effects of physics teaching
within the *collèges*. Between 1700 and 1760 almost all the Oratoire
establishments dispensed mathematical and physical education, as
did the Jesuit *collèges*.[11] The elite was educated by means of a
humanization of geometry. *Collège* professors employed a '*méthode
riante*' in which the phenomena of physics were studied less in
terms of a mathematization of the world than through the sight of
picturesque experiences, and within the happy disorder of an
induction which was not wholly codified. It offered a space for
very diverse subjects, as the public exercises and surviving courses
reveal: the barometer, colours, the air, natural geography, anatomical matters, the elements of botany, and natural observations. At
the École de Sorreze, directed by the Benedictines, natural sciences
entered the curriculum after 1775 and physics exercises gave way
to meteorology and physiology; the programme for natural history
courses in 1784 included botany, pharmacology, medicine, agriculture, and observations on the animal and vegetable kingdoms.
Students began to be taken into the countryside on herborizing
trips.

The conversion of provincial amateurs to natural history not only
presupposed this transformation, but also depended upon it and
accelerated it, for there were profound, old, and durable links
between the world of the teachers and that of the academies.[12]
Thus Amiens, Auxerre, Besançon, Béziers, Brest (which was
affiliated to the Academy of Sciences), Châlons-sur-Marne, Clermont-Ferrand, Dijon, Lyons, Metz, Nancy, Pau, Rouen, and Toulouse opted, with variations, for a title which placed them under
the banner of the sciences. La Rochelle and Montauban remained
faithful to the letters, but none the less opened their doors to the
sciences. A number of the first-founded academies imitated them
after 1750, and the last to be created, Cherbourg, Arras, Bourg-en-
Bresse, Agen, Grenoble, Valence, and Orleans, followed the trend.
The academicians, masters of language, still existed in a universe
of inventories and taxonomies, but they were already inspired by
the organized enquiry which guided both the new norms of experiment and the drive for utility. The academic function of natural
history should be deciphered and understood within this innovative realm lying between the sciences and the arts. Natural history's theoretical framework was of less importance than its applications and its utility: its relationship with the mastery of the real
world.

Growing savant ambitions offered more room for the history of
nature. At Dijon, in the context of a literary society dominated by
the *beaux-esprits* of the parliament, the doyen Pouffier proposed a

vast programme, characteristic of the pre-encyclopaedic spirit, for academic discussions: within this, in effect, physics and medicine addressed every aspect of the knowledge of nature, covering what the university omitted, and they led to a science of *mores*. Let us examine this provincial manifesto, written in the testamentary and notarial style, some time after October 1725.[13]

The modernity and utility of natural history

Choices of subject for discussions . . . It is thus only in relation to the *mores* of the soul, and to the other part of man, which is the human body, that useful observations and discussions can be made: with regard to this second part, man cannot have a perfect knowledge of the parts and properties of his body unless he should also have one of those very simple parts of matter which are the principles and the elements which enter into the composition of all mixed bodies and of his own, and unless he should also know the effects and properties of all the natural bodies which surround him and act upon him by virtue of the relation between them and the human body. Physics, being the knowledge of all natural bodies, must be in part the object of these discussions.

Pouffier's aim is a history (rather than a natural philosophy) which could be useful to the academic city. His physics and medicine reallocate the substance of physics to chemistry, that of anatomy to physiology, that of botany to zoology, that of geology to topography, that of the science of nature to the science of man. The natural history of the provincial academicians cannot be understood if we neglect this reference to its potential for utilitarian synthesis and the opportunities it offered for playing a socially useful role. The inventory of natural and artificial objects under the surveillance of the authorities allowed culture to be reconciled with nature, enabled cultural inequalities to be wiped out, and permitted the conquest of a new modernity. This perspective was at once historical and archaeological, accumulative and interpretative, empirical and theoretical, utilitarian and objective.

We can imagine the distance of this project from Parisian specialization and its regulated 'appeal to experiment', where 'progress is no longer an accumulation but a change in the nature of science, which passes from conjecture to the truth, from letter and rule to calculus, from confrontation to experience',[14] and where the knowledge of living things found its place within the permanent framework of the Baconian model. In the provinces natural history was an element in a cumulative whole which was the expression of a sensitivity towards nature. What concerned amateurs more than savants, even if there were genuine savants amongst the former and amateurs amongst the latter, were the requirements of an age whose attention was focused upon the world and itself, and which was concerned with the discovery of the planet in its entirety, but also with the investigation of its proximate environment, both its surface and its depths. Academic modernity included natural history in the list of its activities because it was simultaneously distracting and useful, indicative of the progress of both knowledge and society. Both experience and invention could contribute to it. It was in this way that it was portrayed in academic discourse, as expressed in the opening speech of M. Lantin, doyen of the parliament, inaugurating the meetings of the academy at Dijon in 1741:

What is there to repel in the sciences? . . . Physics unveils the most hidden secrets of nature, it opens her treasuries to us; nothing is more instructive, more curious, more amusing. Morality teaches us the duties of civil life; can one ignore them without indecency? Medicine frees us from the diseases to which the human condition makes us subject, or at least serves to assuage them; I see nothing more desirable.[15]

Similarly, again, at Rouen in 1745, M. de Cideville, a friend of Voltaire, justified the utility of the academies:

How many treasures this beautiful province encloses in its bosom or displays at its surface, and which are still unknown; how many important facts in its past remain to be elucidated; how many branches of industry to perfect or to make known, how many processes useful to agriculture, to the rearing of domestic animals, to the growth of commerce . . . One day, a majestic edifice will grow from all your useful dissertations on so many subjects which I have only indicated, a complete body of civil, physical and political history of this province. . . .[16]

We would search in vain for a rigorous definition of natural history, Cideville's 'physical history', and the conglomeration of 'faculties' proposed by Pouffier. What we can find bears witness to a faith in progress and on the other hand, as we will see, to a faith in method.

After the mid-century, in Paris as in the provinces, the alliance of science and the authorities did not cease to favour the progress of the disciplines of Nature. In his essay *De l'interpretation de la nature* ('On the Interpretation of Nature'), Diderot foresaw the end of the rule of geometers:

We are reaching the moment of a great Revolution in the Sciences. Given the leaning that the wits have towards morality, towards the Letters, towards the history of nature and experimental physics, I almost dare to predict that before a hundred years have passed, it will be impossible to count more than three geometers in Europe. This Science will stop short.[17]

If the prophecy was not to be fulfilled, at any rate it is evidence of the tension which spanned the savant world at the rise of a taste removed from abstract speculations, and at the new association of natural history with literature, above and beyond the association of the physical and moral which founded a new anthropology.[18] In short, this was the rise of *physique particulière*, which equated with the *sciences naturelles* (or natural sciences) mentioned in the *Encyclopédie's* table of knowledge. Between *physique particulière* and a more global re-evaluation of man's place in nature and in the city was but a short step, which the provincial amateurs took without hesitation.

This tendency developed in the academic world in two forms. The first can be studied through the development of the disciplines composing natural history, whether through internal analysis, through the comparison of important problems and the comparison of questions with discoveries, speculations, successes, and failures; or through an analysis comparing institution and experience. The second can better be studied from the perspective of the growing demand which can be seen to develop in public manifestations of savant and academic practices.

Firstly, we might note the increase in routes for the diffusion of knowledge which could ensure the wide success of new attitudes: literary societies, reading chambers or clubs, and different groups where books, information, and debates could circulate. Among these developments, the royal societies of agriculture merit a special position, for agronomy was a major element of natural history. The agricultural societies, which captured some of the political and intellectual territory of the older societies, were important in diffusing interest in nature.[19] Between 1757 and 1763 fifteen societies were established on the Breton model: Tours, Paris, Limoges, Lyons, Orléans, Riom, Rouen, Soissons, Alençon, Bourges, Auch, La Rochelle, Montauban, Caen, and Valenciennes. Four others completed this network before 1789: Poitiers, Aix, Perpignan, and Moulin. Co-operation or competition yielded alliances or divisions, which affected the same social circles throughout France. Thanks to the network of local administration, such changes filtered down into the geography of planting, appealing to those who shared concerns for a better, more controlled understanding of nature and her kingdoms. To improve cultivation required a better knowledge of soils, of plants and their enemies; the improvement of animal products entailed the study of natural and artificial meadows, and of ways of improving forage, combating epizootics, and multiplying the best species. These societies of notables assembled improvers of all kinds, from the great chemist Lavoisier to the labourer, and encouraged a general faith in the effectiveness of experimental science and the management of the economy. This was the context within which applied natural history could become a means of growth.[20]

In the field of natural history, just as for all areas of science, we need to question the import and sense of academic involvement. We cannot resort to changes in underlying principles as a way to explain how disciplines develop, nor ought we to ignore the fact that, in reality, there was a very slender representation of innovators in provincial societies when compared with the situation in Paris. Here efforts could be boosted from experience to experimentation, from collection to classification. We must recognize the social pressures which served to define the limits of the creative capacities of the provincial environment and to dictate its role in the popular diffusion of knowledge. This process can be measured in terms of the establishment of instrumentation and through the meaning given to the disciplines of natural history in the privileged moments of academic life.

From instrumentation to practice

In the *Encyclopédie* article, Diderot is sensitive to the fact that the savant movement could benefit from specific material conditions. 'The taste' for this science is spreading, 'men of letters make it an object of study or entertainment'. Some observe and perfect the science, doing real work; others admire and collect the objects of their admiration to display them to the ever-growing public gaze. They both gather together

the productions of Nature in these cabinets which increase daily, not only in the capital cities, but in the provinces of all the states of Europe. The great number of these cabinets of 'Natural History' manifestly proves the taste of the public for this science; they can only be formed through painstaking research or through a considerable expenditure, since the price of natural curiosities has currently reached a very high level. Such an expenditure of time and money implies a desire to instruct oneself in 'Natural History', or at least to demonstrate one's taste for that Science which is sustained through example and emulation.[21]

The academic movement as a whole seems to illustrate Diderot's comments and the entry of the world of fashion into the sciences.[22] In Paris, as in the provinces, rich amateurs, great lords, and wealthy townsmen collected (with greater or lesser degrees of discrimination and science) the objects and instruments which populated private laboratories; after 1750, able popularizers such as the abbé Nollet, the abbé Pluche, Sigaud de la Fond, and the abbé Rozier published manuals and popular works. Cabinets were rarely specialized and assembled everything, from instruments to specimens, from natural to artificial objects. Dezallier d'Argenville and Daubenton offered rich amateurs the necessary principles for the up-to-date arrangement of cabinets.[23] The general craze benefited from the work of Parisian academicians and professor–demonstrators (such as those at the Jardin du Roi), from the diffusion of examples through books and journals, and from the development of a market, as recorded by Diderot. There were certainly degrees of specialization between the simple cabinet of curiosities and the cabinet of natural history which was, as Lamarck put it, 'advantageous to the progress of the Sciences';[24] but all attracted the curious just as they engaged savants to experiment on the variety of nature.[25]

The map of provincial and Parisian cabinets indicates Parisian pre-eminence: over 230 existed in the capital, and a good hundred in the provinces with a clear preponderance in important academic towns: 4 in Besançon, 5 in Dijon, 9 in Lyons, 20 in Marseilles, 10 in Montpellier, 9 in Rouen. The map of botanic gardens supports this savant geography, which was very clearly shaped by the action of the authorities in protecting the academic movement and thus permitting certain societies to develop a new pedagogy of science around collections, sites of demonstration, and experimental laboratories. From a very early stage academicians had interested themselves in botanical gardens. Often, particularly in the seventeenth century, they employed these as a means to realize their dream of escaping from the world, of rebuilding the Athenian garden of Academe. During the eighteenth century the taste for agronomic and medical experiments, herb and fruit gardens, and the requirements of professors of medicine developed this need still further, just like the increased demand for cabinets of natural history and for laboratories. However, the lack of resources meant that it was rare for societies to possess their own garden. Most frequently they made use of private gardens placed at their disposal by members. Amiens, Bordeaux, Clermont, Dijon, Montpellier, Rouen, and Toulouse were better endowed, thanks to donations by associates. The garden was a site of instruction for apothecaries, physicians, and surgeons, but it also supplied provincial amateurs of natural history with a place of relaxation and solitude, a way to link the utilitarian dream with the moral utopia.

Scientific practices no longer depended only upon the use of books – half the societies possessed a library – but also upon the use of instruments and the management of specimens. The collections of the Academy of Dijon, its laboratory where Guyton de Morveau worked, Séguier's cabinet and garden at Nîmes, the Lafaille cabinet at La Rochelle which was left to the Academy, the collections of the Bibliothèque Delphinale, those of the Montpellier Royal Society, those of the Academy of Lyons (which, in 1775, bought the famous Pestalozzi cabinet), the cabinet of the Toulouse Academy of Sciences, and that of the Academy of Marseilles, all illustrate by their composition the diversity of these new interests,

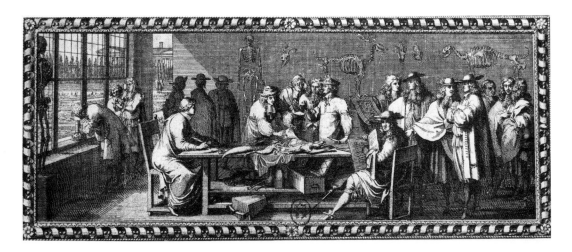

and, more particularly, their switch to the domain of the natural sciences. Societies which did not possess one could benefit from access to collections assembled by one of their members. The history of laboratory practices remains to be written, but we know that their space was not reserved for experimental physics or chemistry. Chemistry was the principal occupant, as one can see at the Paris and Dijon Academies of Sciences, but increasingly, following Nollet's example, the instrumentalization of the sciences of the living world gained ground in the academic framework.[26] In the first years of the Royal Academy of Sciences it is possible to recognize an encounter between geometers, the practitioners of pure, mathematical science, and experimenters who practised natural philosophy. The latter, through chemical analysis, botanical observations, vivisection, experiments of all kinds, and publications such as the *Mémoires pour servir à l'histoire naturelle des animaux* (1671) and *Mémoires pour servir à l'histoire des plantes* (1676), made the experimental enterprise indispensable to the knowledge of natural objects. That relationship enabled physiology to develop within the laboratory, disengaging itself from the wider medical programme and the hospital context. In the eighteenth century the relation between experimental philosophy and natural history developed into experimentation on the living world, overstepping the epistemological boundaries of the older natural history. Simultaneously, the living individuality of plant and animal life was recognized.

The victory of the natural sciences was the result of a shift in the activities of the academies. If one analyses the 30,000 papers presented in ordinary meetings, there is a clear increase in representation of the sciences, even if that figure is purely indicative, given the important lacunae of the sources and their heuristic heterogeneity.

Figure 8.2 Experimentation at the Académie Royale des Sciences at the end of the seventeenth century. Through the window, the academy's garden can be seen, whilst academicians observe with hand lenses and microscopes by the light of the window. On the table, an anatomical dissection is in progress, and skeletons line the walls. Academicians reading anatomical texts look on. From A. Stroup, *A Company of Scientists: Botany, Patronage and Community at the Seventeenth Century Parisian Royal Academy of Sciences* (Berkeley, 1990), plate 3.

This serves to confirm the quantitative analyses undertaken to date on periodicals (*Journal des Savants*, *Mercure de France*, *Correspondance Littéraire*, etc.), registers of publishing privileges and tacit permissions to publish, which reveal the state of the kingdom's book trade. 60 per cent of works can be classified under the rubric of sciences and technics. Within this quantity, it is hard to draw precise boundaries between disciplines, especially between theoretical reflections and practical activities, but it is incontestable that the true passion of enlightened provincial academics was for natural history. This boom reflects two factors. The first, indicated by Diderot, relates to the fact that, in contrast with the abstract sciences where

a single man of superior genius can advance with great strides without any outside help because he can draw facts, principles and their consequences from within himself, in the sciences of Nature facts can only be acquired through long and difficult observations; the number of facts necessary for this Science surpasses the immense number of the productions of nature. A single man is thus incapable of such a vast labour.[27]

The academic space provided the opportunity for that gathering of facts. Secondly, the fundamentally amateur recruitment allowed a plethora of observers to be involved and to contribute to the work of joint elaboration. It is here that the connection between science and utility can be found because, for many provincial academicians, it was less a case of systematic and ordered reflection than one of extending their professional activity or their concerns for the management of men, wealth, and spaces to produce a piece of research which might be densely textured, but which would be more useful for all in the short term. This is why physicians and surgeons of provincial cities were to be found among the first rank of naturalists, alongside great amateurs from the robe nobility or Church, or from the urban bourgeoisie in general.

Three main types of enquiry channelled this enthusiasm: in the first place, medicine, which included anatomical descriptions and physiological reflections, and mixed clinical definitions with pharmaceutical propositions. Globally, it made up almost a quarter of the work surveyed, and sometimes far more: 44 per cent at Amiens, 55 per cent at Auxerre, 36 per cent at Lyons, 32 per cent at Montpellier, and 34 per cent at Toulouse. Botanical, zoological, and geological and mineralogical concerns composed the remainder, in variable proportions. An important interest was also manifest in the agronomic concerns which filled more than a quarter of the scientific and technical papers. Such concerns were scattered and fragmented. They depended upon a double preoccupation: on the one hand, an empiricism of curiosity, linked by an obsession with taxonomy and directed by a desire to make the art of healing less conjectural; and, on the other, a descriptive framework which collected accounts of local features and whose explicit (but rarely attained) goal was to be the writing of a natural history of the provinces. At Bordeaux, where the academy was well supplied with naturalists, one can follow the two sides of this development, which was motivated by the hope of forming a local inventory. Between 1715 and 1720 academicians heard the abbé Olivier, the physicians Cardoze and Douan, and Monsieur de Montesquieu, presenting, in no particular order, papers on 'the prejudices of medicine', 'indigestion', 'vomiting', 'poisonings', 'rabies', 'sudden death', 'the use of tobacco', 'fevers', 'drunkenness', 'wigs', 'diseases due to the wet winter and warm spring', 'the salivary glands', 'hair', 'the appendix', 'the sheep's liver', 'autumn flowering', 'the growth of cress', and 'the flower of the grapevine'.[28] M. de Navarre's speculation in 1717 about the most widespread obstacles to the advance of the sciences and about the means of combating them is understandable. Fifty years later, the link between dissertations remained similarly unclear.

Nevertheless, between 1715 and the Revolution, the route sometimes suggested as a means of integrating the concerns of savant naturalists and others was a return to the idea of undertaking the natural and literary history of Guyenne.[29] This work, 'impossible for an individual', as Navarre, the secretary, noted, was to inspire academicians charged on 17 August 1715 to distribute their tasks: which they did in geographical order. Caupos chose the peculiarities of the Médoc; M. de Vezis, those of the Entre-deux-Mers; Donzan, the diseases of Bordeaux; Bitry chose Garonne; Grégoire and Cardoze, the botany of the Bordeaux region; the abbé Sabathier selected mines; the abbé Descors the vine; and the abbé Bellet, history. It was the latter who would return to the subject with more constancy, although the project was never to be completed. None the less, it suggests the persistence of a concern which attracted scattered but intense research: research which was only outlined, but which aimed to popularize particular habits and practices and to locate the practitioner within a more rational ideal of knowledge.

Other societies, such as Dijon, Montpellier, or Besançon, could offer numerous examples of this revealing distribution and indicate its variations.[30] At Montpellier, seat of the professors of medicine, the entirety of work concerning *materia medica* ran to less than 20 per cent of physiological or anatomical memoirs. The analysis of clinical cases and surgical procedures carried the day. At Dijon medicine occupied 43 per cent of meetings, 632 memoirs in fifty years; botany inspired 135 papers, zoology around twenty, and geology just a dozen.[31] At Besançon, where the Academy gratefully received the collections of curiosities belonging to the marquis de Montrichard in 1762, the presence of salt pans constrained chemists to occupy themselves with medicine and physicians with chemistry. Here again, the idea of a natural history of the province directed energies towards the topographic geography of the Jura, and to the produce of the forest and vineyard.[32]

In the general context of a systematizing project, the success of natural history was clearly linked to the vitality effectively originating within the faculties; natural history expressed the perpetuity of the spirit of curiosity which guided and corrected medical scepticism, and it was suitable for a world of amateurs lacking training or particular specialization. This interpretation enabled new crazes to be assimilated, and benefited from the publicity of public meetings and competitions, although in the latter botany, medicine, and applied natural history occupied less space than the subjects of physics and chemistry. The subjects considered almost always had a utilitarian orientation, and here it was in the provinces, rather than in Paris, that a form of scientific practice was developed from which an improvement in the condition of all was anticipated.[33]

Above all, interest in the natural sciences expressed itself in terms of a general promotion of science. The natural sciences were particularly favoured for their apparent facility, and in the index to Bachelard's *La Formation de l'esprit scientifique* (Paris, 1965, pp. 25–7) one can find a number of those naturalists who were in fact only false savants, lovers of pointless questions and inventories. In the ordinary meetings even more than in the public ones, natural history was represented, theatricalized, far more than it would be in an experimental situation or in reasoned classification. The popularizing savant could conquer his public by showing them objects with a familiar appearance such as plants, sick people, or animals; or by appealing to the vertigo of the marvellous and of the monstrous. Anatomy, zoology, and botany were opportunities to parade monsters and freaks, deformed children and animals, monstrous vegetables, humans, fish, dogs, horses invaded by polyps, headless lambs:[34] in fact an irrational seduction. It was less a case of 'changing the real' than of discovering and exhibiting what everyone could do.[35]

The source of the craze for natural history was to be found in a social imaginary where appearance and analogy still reigned. It was further enriched by the promotion of the senses and the entrenchment of the individual within familiar realities. It is in this way that we can understand the significance of botany and the academic taste for herborization, the collection of plants, and gardens. These practices were favoured by an intellectual and social sensitivity and a new morality. The provinces readily subscribed to Jean-Jacques Rousseau's view of this, and the *Lettres sur la botanique* were read in circles close to the Lyons Academy. 'The epistemological concern set in motion a vital concern',[36] for botany participated in the progress made against the excesses of urban civilization and authorized a relationship with nature and with the world that was richer and more sensitive, an unmediated knowledge whose fervour encompassed the wisdom of the greatest, such as Linnaeus.

Nevertheless, the true science gained. Firstly, because naturalists had in common a set of principles and practice, a method, a privileged role for experience, an antipathy towards systems, and the organizing framework of classification. It was not enough to keep account of the facts, one must be a philosopher. Thus the request of the Chevalier de Vivens to the Bordeaux academicians:

That which is the cause of our having so few good observers, and which, in consequence, greatly hinders the progress of our knowledge, is the neglect of almost all the matters within our grasp, however worthy they may be of admiration and curiosity. Assiduous spectators, yet indifferent to or distracted from the most marvellous effects of nature, we are only surprised by them in reflection, when we perceive that their cause is still hidden. But before reaching that point, one must be a philosopher and have wasted much time explaining matters perhaps too distant from us, or too rarely observed, for us to be able to judge concerning them.[37]

This can explain the apparent paradox of the precocious provincial triumph of Linnaeus, whilst at Paris, Buffon ruled over the botanists. It was at Montpellier, with Gouan and Gerard, at Lyons, with Devillers, Claret de La Tourette (Rousseau's correspondent), Gilbert, and Mouton-Fontenille, at Bordeaux with Latapie, at Agen with Saint-Amans, at Toulouse with Picot de La Peyrouse, at Angers, Rouen, Nancy and everywhere within the academic context that the network of convinced Linnaeans was woven. It would end by imposing itself on Paris with the creation of the Société d'Émulation, the Société Philomatique, and finally, in 1790, the Société Linnéenne.[38]

If from the dawn of the enlightened century, therefore, there was a dialogue between the men of science in Paris and those in the provinces, the amateurism of the provincial historians of nature may have been a hindrance, but the conditions of a new philosophy of nature were equally decisive. The change did not take place in a completely linear manner, but rather through relays of curiosities, waves of interest, and the progressive impulse of marginal disciplines; and within the academic structure in both Paris and the provinces one sees taking shape all that hindered and all that was to come with the triumph of experimentation, 'with the positive revision of the relations admitted by natural history, a revision which was aided, according to subject, by physics, physiology, and even, after Lavoisier, by chemistry, but also by anatomy':[39] the irruption of the problem and the end of the concern with the description of objects. The study of this break with natural history requires a change of framework and other tools of inquiry.

Further reading

Brockliss, L. M. B., *French Higher Education in the Seventeenth and Eighteenth Centuries: A Cultural History* (Oxford, 1987).

Darnton, R., *The Business of Enlightenment: A Publishing History of the 'Encyclopédie', 1775–1800* (Cambridge, MA, 1979).

 The Great Cat Massacre and other Episodes in French Cultural History (London, 1984).

Gillispie, C. C., *Science and Polity in France at the End of the Old Regime* (Princeton, 1980).

Hahn, R., *The Anatomy of a Scientific Institution: The Paris Academy of Sciences, 1666–1803* (Berkeley, 1971).

Keohane, N. O., *Philosophy and the State in France: The Renaissance to the Enlightenment* (Princeton, 1980).

McClellan, J. E., III, *Science Reorganized: Scientific Societies in the Eighteenth Century* (New York, 1985).

Roche, D., *Le Siècle des Lumières en province: Académies et académiciens provinciaux, 1680–1789*, 2 vols. (Paris, 1978).

Roger, J., *Les Sciences de la vie dans la pensée française du XVIIIe siècle: la génération des animaux de Descartes à l'Encyclopédie* (Paris, 1963).

Salomon-Bayet, C., *L'Institution de la science et l'expérience du vivant: Méthode et expérience à l'Académie Royale des Sciences, 1663–1793* (Paris, 1978).

Stroup, A., *Royal Funding of the Parisian Académie Royale des Sciences during the 1690s* (Philadelphia, 1987).

 A Company of Scientists: Botany, Patronage and Community at the Seventeenth Century Parisian Royal Academy of Sciences (Berkeley, 1990).

9 Carl Linnaeus in his time and place

When visitors arrived in Uppsala, Sweden, to meet the Enlightenment's most famous naturalist, Carl Linnaeus (1707–78), they were surprised to encounter 'a somewhat aged man, not tall, with dusty shoes and stockings, markedly unshaven and dressed in an old green coat'.[1] At the Swedish court, Linnaeus's charm was considered to be precisely that his entire person remained that of a provincial parson. This slovenly, argumentative little man even admonished the queen herself. And she was the sister of Frederick the Great of Prussia, and a formidable woman who bottled her own stillborn child for her curiosity cabinet. Yet Linnaeus's surly self-confidence only enhanced the fact that alongside the guenon monkeys, Sami servants, African slaves, and all the other wonders, he himself was part of the royal curiosity collections which he curated.

European visitors, however, were disappointed to meet in Linnaeus a provincial. Apart from a mongrel Latin, he spoke only his own vernacular south Swedish. He read no modern languages, and lacked both general culture and the 'new science'. His foreign students especially disliked the way in which, every Sunday, Linnaeus called in a tenant farmer to play the fiddle, and then watched as they danced the reel with his four unmarried daughters (girls kept semi-literate so as not to compromise Gothic housekeeping with French fashions).

Still, Linnaeus was considered 'the greatest Botanist that the world ever did or probably ever will know'.[2] He was compared to Solomon, Socrates, Galileo, and Newton. Yet Linnaeus had neither mathematized living nature, nor identified general laws explaining life's diversity. He was chiefly a floral classifier, and without a single, towering achievement to his name. His great reputation rested instead in the democratizing accessibility of his achievement. For the value of Linnaeus's classifications lay in their humdrum, everyday usefulness, for casual and serious users alike. In his guides and handbooks, and in the structure of his systems as such, Linnaeus lowered the educational and financial entrance fee to the study of nature.

The sexual system of plant classification

In *Systema naturae* (1735), Linnaeus first presented his global classification of natural productions to the international public. One Dutch friend nicely summarized its significance:

[With Linnaeus's] Tables we can refer any fish, plant, or mineral, to its genus, and, subsequently, to its species, though none of us had seen it before. I think these Tables so eminently useful, that every body ought to have them hanging in his study, like maps.[3]

Linnaeus's *Libellus amicorum* ('Booklet of Friends') (1734–8), the souvenir booklet which he prepared for his 1735–8 travels abroad, similarly planned

all three of nature's kingdoms depicted on maps or paintings printed under the title *Geographia Naturae*.[4]

Linnaeus began his 'geography of nature' as early as 1727, as a twenty-year-old student at Lund University, south Sweden. In the following year he transferred to Uppsala University. Both these centres of learning were sleepy little towns, teaching the rudiments of Lutheran orthodoxy to the future parsons and civil servants of the Swedish state, a Spartan war machine now, after the defeats of the Great Northern Wars (1700–18), without purpose.

Linnaeus was largely self-taught. Between 1728 and 1731, however, he worked as a guide in the Uppsala botanic garden, as a tutor in the home of an Uppsala professor of medicine, and as a plant collector to a professor of theology, who in preparing a 'Biblical *Botanica*' also searched Scandinavia's conifer forests for clues to Sinai's desert shrubs. Linnaeus's own first flora, from 1727, enumerated plants around his childhood home, Stenbrohult parsonage, in Smolandia. Later he inventoried the 'small and swampy Uppsala [botanic] garden by the Black Creek',[5] complaining that it 'daily decays, so that there now are hardly two-hundred [species]'.[6]

Linnaeus wrote three catalogues of plants in the Uppsala garden (each entitled *Hortus Uplandicus*), and one more in 1731, and it was then that he shifted from the systematics of Joseph Pitton de Tournefort (1656–1708) to his own '*methodus propria*' or sexual system of plant classification. In 1728, he had been much taken by a magazine review of Sébastian Vaillant's (1669–1722) work on plant sexuality. He now made it the focus for his classificatory botany. But Linnaeus also drew on other scholars, retaining, for example, most of Tournefort's genera. For the general structure of his systematics he relied on the work of the Italian Andrea Cesalpino (1519–1603), and ultimately on Aristotle.

Linnaeus created one global classificatory tree encompassing all life on earth, and divided into five levels of generality: class, order, genus, species, and variety. He privileged the rank of genus, and sub-divided plant genera according to the number, size, placement, and shape of stamen and pistils. But he grouped animals by broad and variable characteristics, such as their teeth, locomotion, type of blood if any, and habitual home (e.g. land or water). Like Cesalpino, Linnaeus thus affirmed his theoretical allegiance to scholastic logic while preserving common-sense groupings. He also bunched most invertebrates together as *Vermes*, a folk category akin to the English vernacular 'bugs'.

Linnaeus's divisions of fauna (a term which he coined, alongside flora) was influential. He was the first to name us *Homo sapiens*, and to class us as primates. Linnaeus's central innovation, though, was the self-consciously artificial way in which he devised his floral classification. He even compared it to an alphabetical list.

By 1735 Linnaeus had essentially completed his systematic theory. In his own view, it now only remained to slot nature into his classificatory tree. He estimated that the earth housed about 40,000 plant and animal species, and a few hundred minerals. He thus could, and did, position himself as a final arbiter of all natural productions. His first major applications of his system were *Flora Lapponica* (1737) and *Hortus Cliffortianus* ('Clifford's Garden') (1737). Their exoticizing frontispieces nicely parade their Arctic and tropical plentitude. These works drew on Linnaeus's two formative projects outside of a university setting: his 1732 Lapland journey, sponsored by the Societas Regia Literaria et Scientiarum Sueciae (Sweden's first scientific society, founded in Uppsala in 1710 as the Collegium Curiosorum); and his work as a curator of the botanic garden of a Dutch banker, George Clifford, during his 1735–8 stay in Holland.

As he extended his systematizing, Linnaeus scoured earlier literature for species citations. But as a point of method, he privileged first-hand experience. He made no distinction between herbarium studies and collecting and travelling, however. At times he also cribbed unverified data from secondary sources. Yet, because many Linnaean holotypes (the specimens on which he based those of his species descriptions that are now regarded as foundational) have survived, we can evaluate Linneaus's observational style in detail. He was a gifted and rigorous examiner of nature.

Linnaeus emphasized that both learned and lay people could cooperate in his mechanized classificatory work. Whilst involving himself in learned correspondence networks, he also reached out to new audiences and collaborators. His botanical handbooks were brief enough to be read with ease, and small enough to carry into the field. He wrote them in a straightforward, unornamented Latin, and encouraged vernacular translations. In their format – they were each divided into twelve chapters and 365 aphorisms –

they were reminiscent of the Lutheran almanac. Linnaeus probably imagined his followers learning an aphorism by heart every day, just as his father's parishioners rehearsed their daily catechism.

Linnaeus bragged that because the sexual system depended on a few, easily observable features, coloured or engraved images were superfluous. Poor students and the common people could now become proficient botanists, for an expensive library and previous instruction were no longer essential to a botanist's training. 'Yes, even for Women themselves',[7] botany was now a possible science. As further encouragement, Linnaeus gave botanic discoverers honorific floral names. He named plants for women, farmers, and artisans, and once even, to the envy of his 'master', for a Surinam field slave.

Linnaeus also spelled out botanical practices to aid the novice. Thus *Philosophia botanica* (1751), his most important botanical guide, ends with a series of one-page instructions, teaching the reader how to set up a herbarium, organize an excursion, plant a garden, and even embark on a voyage of discovery. He added ten full-page diagrammatic line drawings of plant parts, as well as several indexes.

Philosophia botanica's regime stemmed from Linnaeus's own practices. As early as 1733, Linnaeus guided fellow students on excursions around Uppsala. In 1739 he combined his natural history lectures to the Swedish House of Nobility with floral ramblings on Stockholm's islands. In the 1740s, at Uppsala University, he led day-tours for as many as 300 men and women.

For these hikes Linnaeus regulated reading-lists, departure times, and public 'demonstrations'. He itemized objects to collect, such as 'little birds that are shot', and listed field equipment such as vascula, field microscopes, magnifying glasses, note-papers, butterfly nets, insect-pins, and pocket-knives. He also established labour divisions: 'Sharp-Shooters' killed birds, 'Annotators' protocolled results, and 'Fiscals' guaranteed 'the troops' discipline'.[8]

Linnaeus repeatedly promised to replace his sexual system with a natural system. This he variously likened to a chain, countries nestling on a map, knots in a fishing-net, and a grove of trees. And he believed it was somehow encrypted in the relation between all seven basic parts of fructification (calyx, corolla, pericarp, pistil, seed, stamen, and receptacle). Another clue, he suspected, was to be found in his hypothesis that modern species, while probably fixed in the present, had hybridized from a small number of Edenic life-forms, each representing one of the present-day orders. Yet his efforts, as he lamented, always remained fragmentary.

However, in Linnaeus's view, an artificial and a natural classification were not two distinct schemes between which one was forced to choose. Rather, the one was linked to the other, as a provisional but necessary means to a more perfect but remote end.

Only when his own giant file index of nature was completed, he thought, would the regularities of nature's diversity reveal themselves. In the meantime, Linnaeus initiated an atheoretical systematics, where lay collectors enjoyed the same status as natural philosophers.

Similarly, Linnaeus's prefaces encapsulated his classifications in an Aristotelian physics, with its four elements of Water, Fire, Earth and Air, and a Christian theology, emphasizing the perfection of God's material creation. But this framing effort did not intrude on the systematizing endeavour itself.

Binomial nomenclature

In *Philosophia botanica* (1751), Linnaeus suggested that each life-form should be labelled with a 'trivial name', or a two-word reference, denoting its genus and species. Altogether, Linnaeus thus named around 7,700 plants and 4,400 animals. Today Linnaeus's binomial nomenclature is considered his only lasting contribution to science. The names of plants (1867), animals (1906), bacteria and viruses (1948), and cultivars (1953), are all generated according to taxonomic codes that start in Linnaeus's *Species plantarum* (1753), for flora or *Systema naturae* (10th edn, 1758) for fauna. Historians have argued, variously, that Linnaeus's binomials were inspired by scattered precedents in Renaissance herbals, by folk names (such as 'barn owl'), or by his habit of abbreviating bibliographical references.[9] My own archival research suggests that alongside such general influences, a specific story lies behind Linnaeus's binomials.

Early modern botanists, Linnaeus included, constructed diagnostic phrase names (a brief species description which simultaneously functioned as a proper name) by pairing opposed characteristics, descending from the general to the specific. These phrase names were up to half a page long. They also varied from author to author. Thus some scholars initiated zoology trees with a four-footed–two-footed division. Others began with a blooded–bloodless split. Such trunk choices in turn determined how living beings were grouped (and named) in the branches of the classificatory trees. Thus, when in 1729 Linnaeus was first examined on his knowledge of individual plants by an Uppsala professor, as he later remembered, he 'answered them all with the names after Tournefort's method'.[10]

Diagnostic phrase names also changed over time, even in a single author's *oeuvre*. For they were constructed so as to distinguish the bearer of a name from its congeners (members of the same genus). Each time a new species was added to a system, all its congeners needed (in theory at least) new names.

Thus a young scholar might find himself haplessly reclassifying an entire kingdom. Indeed, when in 1730 Linnaeus set up a botanic system of his own, he simply followed the routine procedure of the botanic novice. In 1729 Linnaeus's best friend (also an Uppsala student) had similarly written a local flora, 'ordered', as he put it, 'after the very simplest and clearest Method', namely his own.[11]

At Uppsala University in the 1740s Linnaeus noticed that his students found diagnostic phrase names difficult to use as proper names in their encounters with both unknown flora and older texts. As was common in the period, they generated their own home-made abbreviations. But together Linnaeus and his students also began experimenting with communal, and more practical, ways of referring to individual species. They used, variously, folk names, bibliographic references, and the consecutive numbers which species were then commonly assigned in classificatory publications (somewhat as we number colour plates in art books today).

However, since in subsequent editions newly discovered species were listed in their appropriate genera, consecutive number sequences were unstable over time. Linnaeus and his students therefore used a set of numbers (assigned in the 1745 edition of Linnaeus's *Flora Svecica*) as stable and independent references. They had invented an arbitrary numerical nomenclature.

Yet around that time (1748–9) Linnaeus and his students also began using what we today know as true binomials. These first appear in printed form in *Pan Svecicus* (1749), a brief tract on cattle fodder directed against the import of cattle feed. Importantly, *Pan Svecicus* was a collaborative work. Linnaeus assigned each student a farm animal (e.g. pig, goat, hen, or cow). Clutching ink-pots, goose feathers, and scrap papers, the students tracked their animal experimenters, noting throughout the day the plants they fed upon. Altogether they listed some 850 plant species under the difficult condition of seeing the specimens disappear down the throats of their test animals at the very moment they needed to identify them.

Linnaeus's binomials thus emerged out of his students' work practices, in the context of his economic botany, and as a stop-gap measure to make his students more efficient support staff and collaborators. Later the binomials spread from Linnaeus's economic pamphlets to his classificatory tomes. By the 1760s, in an idealist reversal of reality, Linnaeus bragged of the way in which he had suddenly thought of (rather than laboriously worked towards) a true binomial nomenclature.

Voyages and collections

Linnaeus was a typical Enlightenment improver. To his Scandinavian patrons and public, he spoke of his science as serving the state's economic needs. Also, his voyages and collecting were undertaken in part for economic reasons. As he put it in 1746:

Nature has arranged itself in such a way, that each country produces something especially useful; the task of economics is to collect [plants] from other places and cultivate such things that don't want to grow [at home] but can grow [there].[12]

As Linnaeus and his students also formulated their botanic acclimatization theory, they hoped to 'fool', 'tempt', 'teach', or 'tame' 'Indian' (i.e. American and Asian) cash crops to grow even on the Arctic tundra. Thus, Linnaeus daringly predicted, Europe could become as rich even as China.

Linnaeus himself passed up offers to explore at the Cape, in Canada, and in Surinam. But he directed and took part in explorations at home, intended to identify indigenous life-forms to substitute for imports. In a series of state-sponsored journeys, he and his students exhaustively researched Sweden's provinces. In this way his students also trained for their own long-distance travels.

From 1745 on, nineteen of Linnaeus's students left on voyages of discovery. Their teacher solicited travel funds from the Levant, Greenland, and East India companies, the Bureau of Manufactures, the Academy of Sciences, Lund and Uppsala universities, the Estates, the cabinet of ministers, and the court, as well as from individual patrons. He even staged public lotteries.

Linnaeus's travelling students included Daniel Solander, Joseph Banks's botanist on Captain Cook's first circumnavigation of the earth (1768–71); Anders Sparrman, Johann Reinhold Forster and Georg Forster's botanist on Captain Cook's second voyage (1772–5); Carl Peter Thunberg, who as a ship's surgeon in the Dutch East India Company botanized at the Cape, Java, Sri Lanka, and Japan in 1770–9, and is a crucial figure in the history of Japanese medicine; Pehr Löfling, who, in the employ of the Spanish Crown, explored the natural productions of Spain and Spanish South America in 1751–6; Pehr Forsskål, who participated in a Danish royal expedition through the Ottoman empire and the Arabian peninsula in 1761–3; Pehr Kalm, who explored north-west Russia in 1744–5 and North America's eastern seaboard in 1748–51; and Johan Petter Falck, who as part of the Russian Orenburg expedition criss-crossed parts of the Caucasus, Kazan, and West Siberia in 1768–74. Other students of Linnaeus travelled to

Lapland, the Arctic Sea, Surinam, coastal India, China, and in Africa the Atlas, mountains, Senegal, Sierra Leone, and the Cape.

Falck slit his throat, a crazed opium addict, in Kazan. Löfling and Forsskål died of tropical fevers. Other now forgotten Linnaean travellers also died during their travels, or returned insane or mortally ill. Still others survived, but lost large parts of their collections (Thunberg), or published their results only tardily (Sparrman), if at all (Solander). Mostly they became country parsons and provincial professors.

The Linnaeans' role in the history of the voyage of discovery is still important, and indeed undervalued in the historiography. But their actual achievements did not match their own and their teacher's expectations. For Linnaeus had great hopes for his travelling 'apostles', as he termed them.

Linnaeus issued these students with 'memorials', or order-lists of plants and animals. There he especially admonished them to study local peoples' knowledges of the natural world and their manufacturing techniques. For Linnaeus believed new natural knowledges could form by way of a cross-cultural mediation between high and folk/tribal knowledges. This syncretic 'new science' he regarded as simultaneously an epistemology and a technology, that is, as both a way to know, and a material tool.

Linnaeus preferred his students to travel the well-worn Cadiz–Ghuangzhou trade route, instead of visiting 'wild deserts' (as he even-handedly designated Pennsylvania and Yemen). For it was there that such prizes as tea-seeds, herbal medicines, and techniques of porcelain manufacture might be found; and thus, Linnaeus held, these voyages' ultimate aim, namely to abolish the conditions that now made them necessary, might be fulfilled.

The Linnaean ideal voyager, then, was an industrial spy in the busy cities of high civilizations, not a lone wanderer in a pristine natural world. Even if Linnaeus faithfully entered into the *Systema* the animals and plants that his travelling students sent to Uppsala, thus adding to his register of earth's life, that was not the only or even the main aim of their travels. For example, only at the very end of his 1745 'memorial' for Kalm's voyage to America does Linnaeus suggest, in a single offhand line, that next to his studies of Amer-indian economies, Kalm make his own 'observations on Birds and Fishes, on Snakes and Insects, on Plants and Trees, on Stones and Minerals'.[13]

Linnaeus carved out for himself a lucrative role as a governmental adviser on voyages, collections, and colonial economies. To take a single example, having read in travel descriptions about Chinese fresh-water pearl plantations, Linnaeus, brandishing a few small pearls which he had managed to grow in the tepid stream that runs through Uppsala, persuaded the Swedish cabinet of ministers to fund a project to inoculate Lapland river mussels. In 1762, the Estates (the Swedish parliament) sold Linnaeus's 'secret', as he styled it, to a Gothenburg whaler, and granted the inventor 6,000 silver thalers. The Estates also granted Linnaeus the right to dispose of his university chair, and ratified the king's recommendation to ennoble Linnaeus. Thus Linnaeus bought a country estate, secured his son's succession, and was dubbed 'von Linné'.

Actual pearls were never produced. This came as no surprise to Sweden's most distinguished economic historian, Eli F. Heckscher: 'naturally, the whole apparatus resulted in nothing.'[14] Yet this is only true from the perspective of the present. From Linnaeus's point of view, his nine small pearls, supposedly demonstrating the possibility of domesticating Lapland's foamy rapids (now long since dammed for hydroelectricity) resulted in a 'national award', a family lineage, and noble status. In 1762 Linnaeus became, as he had hoped as early as 1748, 'the lord of all of Sweden's clams'.[15]

Linnaeus cast himself, then, as a political economist and an acclimatization experimenter. He took little interest in the exact sciences, however, or in technological progress (e.g., ferrous metallurgy and hydrodynamics). Nor did he interest himself in such instruments as diving-bells, steam engines, air pumps, telescopes or even – though he used them – microscopes. Linnaeus's work-spaces more recalled a Renaissance *studiola* or a curiosity cabinet. For he believed that his science reflected nature's harmony, which in turn was analogous to the order of his own study. As he put it in 1754, 'the earth is then nothing else but a museum of the all-wise Creator's masterpieces, divided into three chambers'.[16]

From his student days onwards, Linnaeus arranged around himself a home which was a microcosm of that 'world museum'. In this emporium of art and organic nature, parrots and squirrels, and even a young orang-utan, played among potted plants, insect specimens, mineral samples, scientific instruments, and herbarium sheets. Some 3,000 plant species grew in the botanic garden in which the house was placed. Over thirty species of songbirds nested in Linnaeus's chambers (he provided them with huge tangled branches). Botanic prints served as wallpaper, and were covered in turn by portrait engravings of botanists, paper sheets with handwritten botanical annotations or pressed plants, and shells and conches dangling on iron nails.

Next to family portraits and plaster medallions of royalty, Linnaeus arranged likenesses of his guenon monkeys, his tame raccoon named 'Sjubb', and a whale captured off the coast of Norway in 1719. Latin mottoes adorned doorways. And on top of cabinets, he balanced pieces of china decorated by his own heraldic flower,

Linnaea borealis, as well as Chinese shell arrangements and Spanish cork statuettes of a type sold to sailors, and depicting Africans covered by mussel-shells. Over the sanded, broad-planked floors, Linnaeus scattered his botanical manuscripts, where blinded nightingales splattered them with droppings and raccoons played and clawed among them. He dressed the ceilings in birdskins. And 'together with other curiosities' he hung his 'Lapp' costume (a trophy from the 1732 Lapland voyage) on the wall.

International fame

Philosophia botanica and the binomial nomenclature it recommended were received with enthusiasm. Reprinted ten times in Latin between 1755 and 1824, the book was also translated into English, Dutch, Spanish, German, French, and Russian. An abundance of Linnaean primers also appeared, often written by gardeners and 'empirics'. One prototype of the genre, James Lee's *Introduction to Botany* (1760), went through eight editions. A competitor, Philip Miller's *Short Introduction . . . to Botany* (1760), even reached fifteen editions. (Indeed, England especially celebrated Linnaeus.)

Numerous botanic dictionaries were also published. They were modelled on Linnaeus's own plant vocabulary, *Termini botanici* (1762), which was itself reprinted twenty-two times before 1811. By the 1760s and across Europe, local floras, botanic plate publications, natural histories of foreign countries, and even species monographs and children's books, now all used a Linnaean vocabulary. Linnaeus was a household word among the educated public, and a fashion spread across Europe for the botany he had made accessible to lay people.

In the 1730s and 1740s certain naturalists had suggested that Linnaeus's sexual system was immoral (J. Siegesbeck, St Petersburg), that he should return to Theophrastus (J. J. Dillenius, Oxford), or that he should instead develop an alphabetical taxonomy (Sir Hans Sloane, London). More importantly, at Göttingen, the eminent Albrecht von Haller (1708–77) advocated biogeographical criteria for classification. At Edinburgh the anti-sexualist school cast doubt even on the notion that plants had two sexes. And in Paris, Michel Adanson (1727–1806), counting in the preface to *Familles des plantes* (1763–4) over sixty natural botanic systems so far, went on to demonstrate (contra Linnaeus's sexual system) that no known single characteristic could satisfactorily divide plant groups.[17]

The century's most famous naturalist next to Linnaeus himself, Georges-Louis Leclerc de Buffon (1707–88), vigorously mocked his competitor in the 'Initial Discourse' (1749) of his celebrated *Histoire naturelle*. 'This large tree which you see is perhaps only a bloodwort. It is necessary to count its stamens in order to know what it is.'[18]

Through Denis Diderot and L. J. M. Daubenton, Buffon's condemnation became a commonplace in salons and in the *Encyclopédie*.[19] Julien Offray de la Mettrie even wrote a pornographic *L'homme plante* (1748), which was dedicated to Linnaeus and diagnosed a 'plant-woman' according to the sexual system.

Linnaeus in turn read his critics through 2 Samuel 7:9: 'And I . . . have cut off all thine enemies out of thy sight'. This biblical quotation he regarded as prophesying his eventual victory over that 'Frenchman named Buffon, who' (Linnaeus always dismissed the *comte* as a kind of gardener) 'lived in the Botanical Garden in Paris, as Inspector, and always wrote against Linnaeus'.

Linnaeus also became famous for wider cultural reasons. In his handbooks he condemned rhetoric, and he attempted to banish from his science the use of language as a means of persuasion or for emotional effect. *Philosophia botanica* explicitly bans tropes such as synecdoche, metaphor, and irony. This preference for a plain style is of course itself a rhetoric, related to Linnaeus's cultivation of himself as a 'Gothic' moralist opposed to all things courtly and French. Yet the neo-classicists, primitivists, and Romantics of the later eighteenth century were themselves at once nostalgic and stern, and in pursuit of an unmediated language of authenticity. They thus could claim Linnaeus as their precursor, even if his attacks on civilization derived from very different sources, namely his Lutheran and Carolingian childhood.

Thus it was for the moral qualities believed to be intrinsic to his botany that Linnaeus was so admired by Jean-Jacques 'Russau' (as Linnaeus, who was hazy on who he was, spelled his name). Rousseau's hugely popular *Lettres élémentaires sur la botanique*, written between 1771 and 1773 to educate a four-year-old girl, in turn molded the female fad for Linnaean botany in the later eighteenth century. It also inspired the Parisian Société d'Histoire Naturelle (founded by 'lovers of freedom' in 1790) to erect a statue of Linnaeus in the Jardin des Plantes, and thus symbolically to close off Buffon's reign.

In England, Erasmus Darwin's *The Botanic Garden, Part II: Containing the Loves of the Plants* (1789), dedicated to 'ladies and other unemploy'd scholars', was also important. And in Germany, Johann Wolfgang von Goethe, who himself was much taken by Rousseau's *Letters*, made debating Linnaeus a pastime in fashionable Romantic circles. Like Rousseau, Goethe carried *Philosophia botanica* along on his country walks. Goethe also claimed to have had only three teachers in life – Shakespeare, Spinoza, and Linnaeus – adding pointedly for the last, 'not however in botany!'[20] It was the provincial naturalist's anti-rhetorical stance, and his appeals to virtue, that appealed to Europe's most celebrated man of letters.

Religious beliefs

In 1739 Linnaeus gave the inaugural speech at the Swedish Academy of Science which he had just co-founded. It discussed the economy of nature by means of 'curiosities among insects', and on the model of the Lutheran sermon. Linnaeus subscribed to the natural theology of William Derham's *Physico-Theology or, Demonstration of the Being and Attributes of God from His Works of Creation* (1713) and John Ray's *Wisdom of God Manifested in His Works of Creation* (1686). Nature, he believed, was one single, self-regulating global entity. Since unchecked population growth would outstrip nature's sustaining capacity, the regulatory mechanism of natural equilibria was thus what he called 'a war of all against all' and regarded as embodied both in predator–prey relations and in general competition for resources. Linnaeus understood competition as taking place between, and not within, species. He also saw it as a principle of continuity, not of change.

At other times Linnaeus considered nature as only the animated setting for salvation history. Domesticating the wilderness, he believed, meant restoring it to an Edenic state. He could not grasp the real possibility of extinction of life-forms by human agency. From his doorstep, he placidly rejoiced in the wild fauna's self-evident and continuing presence as storks waded in his backyard ditch, and eagles (hunting hens) dove into his garden.

Yet in Linnaeus's days the North Atlantic heather moors were still spreading. He himself describes how, in a rainy Swedish countryside which goat-grazing was turning into a desert, he walked on sand dunes so large and so fast-moving that along the dune ridges there protruded from the sand the tops of still green and leafy tree canopies.

Still Linnaeus assumed nature to be indestructible. One of his students, describing Surinam's rain forests in 1755, agreed. 'I am sure, that this huge Hothouse will stand un-mutilated, as long as the earth [itself].'[21] In Linnaeus's eyes, too, 'the most wonderful [region] in the world' was Scania (south Sweden). For this grain-field desert, shaded only by a few coppiced willows and horn-beams, was, as Linnaeus fondly recalled, 'a plain without mountains, hills, stones, rivers, lakes, trees or bushes'.[22]

Linnaeus thus viewed nature as a prelapsarian paradise, containing within each nation all the natural productions necessary for a complete and complex economy. Yet why then did people suffer and even starve in the midst of this posited material plenty? In his spiritual diary *Nemesis Divina* (late 1750s), a testimonial to God's working in history, Linnaeus elaborated a bleak theodicy. Discounting miracles or indeed any benign divine intervention, he collected some 200 case histories of divine retribution. Some stories linger over how unwed mothers, or 'whores', die by scaldings and burns, as if that was the appropriate end to their inflamed passions. Others chronicle how officers, themselves mutilated in battle, had mangled hapless civilians in drinking bouts. Yet others explained the sufferings of moral innocents (such as infants) by ancestral sins. Linnaeus also noted that a colleague who dared to argue with him at a faculty meeting, at that instant fell down in a swoon, to be carried home, a cripple from that day on.

With such punishments, Linnaeus additionally suggested, God's interest in humankind ceased. For, while idealizing nature, Linnaeus was less sure of man's role in it. He repeatedly asked whether he should 'call man ape or vice versa?'[23] (As a counterpart to *Homo sapiens*, he described a *Simia sapiens* that played backgammon.) He argued that animals possessed souls and that, conversely, humans were mortal. Immortality belonged solely to a personal entity he called 'the eternal force', 'the world soul', or 'nature', and understood as animated creation, put in place by a distant sky-god he called 'the wisdom' or 'the intention'.

Redemption, then, had no place in Linnaeus's faith. Linnaeus even suspected that God made nature for his own enjoyment, as a kind of toy, since otherwise he would have created the globe as 'an [edible] mass, wherein we would have ate and slept like the worm in the cheese'.[24] He bitterly noted that 'when [God] doesn't want to keep / us, / we are removed. / [He] let others be put in our place. / Thus nature dooms us, contra Theology'.[25] In his diaries the grief-stricken father also likened his dying children to 'snow crystals' and 'candle ends'.

On a more optimistic note, Linnaeus positioned his own natural science as a material theodicy, enabling humankind to harvest and prepare the natural productions of their homelands. Here he explained suffering, and especially Scandinavia's many famines, by ignorance. 'The sciences are thus the light that will lead the people who wander in darkness.'[26]

A parson's son in the fifth generation, Linnaeus analogized himself to a Luther of science. In a manuscript classification of naturalists, he entered next to the category '*Reformatio*' only two words – '*ego*' and '*mihi*'. In the same spirit, he divided botanists into 'heterodox' (non-Linnaeans) and 'orthodox' (Linnaeans). And he took 1 Kings 17:8, 'and the word of the LORD came unto him', to be a Biblical prophecy of his binomial nomenclature.[27] Linnaeus also likened himself to Moses on the mountain, and even, as on the frontispiece of the 1760 Lange edition of *Systema naturae*, to Adam. Collapsing time, he at once names the animals and writes the *Systema*.

Medicine and anthropology

Linnaeus's doctoral dissertation (Harderwijk, Holland, 1735), was on a medical topic, the cause of malaria. Between 1738, when he returned from Holland, and 1741, when he was appointed professor of medicine at Uppsala University, Linnaeus practised medicine in Stockholm, and was made the chief physician to the Swedish navy. He typically sought environmental roots of diseases, arguing for example that epilepsy was caused by washing one's hair, ergotism by eating radish seeds, and malaria by drinking muddy water.

Linnaeus was also deeply involved in questions of child-rearing. Together with his Uppsala colleague Nils Rosén von Rosenstein, a founder of paediatric medicine, he battled wet-nursing, baby farms, swaddling, and the doping of infants with gin. He described, with a furious sorrow, how he encountered small victims of physical abuse, 'lame, hunchbacked, or covered with runny wounds'.[28]

Later in life, Linnaeus felt that nature's division into male and female principles provided an overarching 'double key to medicine'. Earlier, however, he had argued that medicine should be 'taken from the principles of zoology . . . man is an animal and ought to live like an animal'.[29] Linnaeus rewrote Sweden's pharmacopoeia for mostly indigenous herbal simples, and he and his students carefully studied folk medicine, of both 'wild nations' (tribal people) and Scandinavian peasants. Thus, on his 1732 Lapland voyage, Linnaeus emphasized the indigenous remedies of Arctic people, and ignored the Sami's use of common Scandinavian medicinals such as tobacco, vodka, beaver glands, and a crude form of moxibustion. He noted instead their use of reindeer cheese for frost damage, birch bark for wound dressing, and milfoil for intestinal parasites.

Linneaus also metaphorized his zoological medicine as a 'return' to older and more 'natural' mores. This, he held, had the power of eradicating what he saw as a single whole: poverty, disease, ignorance, and sin. It was thus a materio-moral enterprise, conflating mind, body, and spirit. Broadly speaking, he envisioned a great chain, or universal scale, of health, reaching from 'wild nations' at the top, through European farmers and townsmen, to the lowest of all, French courtiers with their vile diseases.

Linnaeus's primary example of a 'wild nation' was the Sami, the indigenous people of Scandinavia's Arctic regions. They in fact suffered from colonial diseases such as alcohol addiction, and severe forms of measles and influenza. Linnaeus thus spent his 1732 Lapland voyage convincing himself, in spite of overwhelming evidence

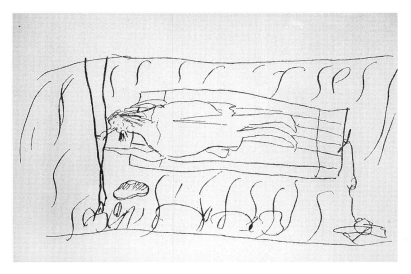

Figure 9.1 Here we see Linnaeus's clumsy attempt to portray a pearlfisher's craft, as he observed it at Purkijaur, Swedish Lapland, in 1732. Note the long wooden tongs with which the pearlfisher, lying on his craft, picks mussels from the bottom of the river. Ink on paper drawing in his Lapland diary, *Iter Lapponicum* (1732), in E. Ährling (ed.), *Carl von Linnés ungdomsskrifter* (Stockholm, 1888), p. 163.

to the contrary, that he had 'discovered' noble savages living in a natural state and a belated Ovidian Golden Age.

Since the Sami were Edenic beings, Linnaeus reasoned in turn, it followed that original sin was not all-pervasive. Neither Lutheranism (salvation by faith), nor Calvinism (salvation by grace), nor Catholicism (salvation by good works) applied. Instead, empirical field studies of 'the Lapps' customs, economy, diet, etc.'[30] could reverse both sin and disease among Europeans. In Linnaeus's hands, then, the noble savage became a token of a proximate salvation. Or, as he put it, 'The Lapps are our teachers'.[31]

Yet at the same time, Linnaeus's Lapland voyage aimed to eradicate the very possibility of this postulated new knowledge formation by cross-cultural mediation. He and his sponsors regarded the journey as an exercise in '*oeconomia*', or part of the ongoing exploitation of 'our West-Indies' (as Swedes called their Arctic frontier). And this colonial venture in turn was predicated on erasing indigenous culture, as the 'wild' Sami and their herds were chained to the engines of industry.

Reception history and conclusion

Linnaeus was given a grand funeral, on a dark winter's evening in 1778. 'His farmers, dressing in mourning, followed the carriage with torches.'[32] All over Europe memorial addresses were delivered at scientific academies. (They were composed, as was customary in the period, by Linnaeus himself.) Eulogies were published in both the scientific and the popular press.

Yet in his homeland, Linnaeus was largely forgotten after his death. Indeed, natural history itself declined. Linnaeus the

Figure 9.2 This illustrates Linnaeus's opposition of metaphor to plain description: one of the only two erotic drawings in his hand, and a plant description. Ink on paper drawing in his Lapland diary, in E. Ährling (ed.), *Carl von Linnés ungdomsskrifter* (Stockholm, 1888), p. 68.

Younger failed entirely at Uppsala University. Other ageing students turned into pedantic and provincial professors of 'practical economics' (as Scandinavian chairs of Linnaean natural history were named). The study of nature only rejuvenated a generation later, with the Scandinavian reception of German Romantic morphology.

Linnaeus's first public monument in Sweden was thus a modest funeral plaque placed in Uppsala Cathedral in 1798, twenty years after his death and eight years after Paris acquired a public statue. It remained his sole official commemoration until 1811. Linnaeus also fared badly within the historiography of his discipline. In 1875 the German plant physiologist Julius Sachs's famous *History of Botany* categorized Linnaeus as an Aristotelian essentialist. For the next hundred years, most historians of science agreed. In 1965 Linnaeus was even given a prominent role in 'two thousand years of [scientific] stasis'.[33]

Between the 1870s and the 1930s, however, Linnaeus's fame underwent a renaissance in Scandinavia. His place of birth was made into a museum in 1866, as were his country seat in 1879, his childhood home in 1935, and his town-house in 1937. 'Linnaeus liqueur' and 'Linnaeus cakes' were sold in cafés. Temperance lodges, cultural associations, and baby girls were all named after the great Swede, and poems, songs, and even a ballet were composed in his honour. His birthday, May 23rd, was pronounced 'Linnaeus Day' and sponsored by various state agencies as a national competitor to May 1st, International Workers' Day. Flags were flown, schools closed, parades arranged, and keepsakes manufactured. In a time of rapid industrialization, mass emigration, and widespread

Figure 9.3 Linnaeus's ethnography illustrated by a rare drawing of a Same man carrying a canoe: note that 'Same' is the singular form of 'Sami'. Ink on paper drawing in his Lapland diary, in E. Ährling (ed.), *Carl von Linnés ungdomsskrifter* (Stockholm, 1888), p. 38.

social unrest, Sweden's conservative elites thus launched Linnaeus as a 'flower king' who recaptured in his science the victories won by Gustavus Adolphus in the Thirty Years War and Charles XII in the Great Northern Wars. Over time, such sentiments grew increasingly racialist, and the small, dark naturalist transmogrified into a blond and blue-eyed giant.

In the 1930s, and with the advent of a Social Democrat government, this cult of Linnaeus faded away. When a small town in his home region staged a 'flower march' for the 250th anniversary of his birth in 1957, Linnaeus was played by a schoolboy and placed on a tractor-drawn 'flower-float' next to a medieval saint, a Viking Amazon, and a magic midget. Further crowding the float were two 'Lapps' and 'a great many small school-children more or less disguised as flowers'.[34] Such marches hardly rekindled interest in Linnaeus. Nor was a second Linnaean renaissance inaugurated by the breathless tabloid rubrics of the 1960s, such as 'Linnaeus TV hero – if TV had existed',[35] 'Linnaeus – our greatest PR-man?',[36] and 'Not just flowers for the sex radical Linnaeus'.[37]

Linnaeus thus dwindled into a local hero whom (as a 1957 questionnaire revealed) people dimly recalled as a 'famous tee-totaller', and who was listed in tourist brochures alongside midgets and Amazons. This trajectory of reception may seem like a funnel, where the magisterial Enlightenment scientist diminishes into a small bronze statue of a pre-pubescent boy, gazing at a flower by the grass-roofed cottage where he was born. Yet as it grew ever more localized and localizing, Linnaeus's image also reflected a profound truth about him: he was from the start a quintessentially local man.

Further reading

Atran, Scott, *Cognitive Foundations of Natural History. Towards an Anthropology of Science* (Cambridge, 1990).

Blunt, Wilfrid, *The Compleat Naturalist. A Life of Linnaeus* (London, 1971).

Broberg, Gunnar (ed.), *Linnaeus: Progress and Prospects in Linnaean Research* (Pittsburgh, 1980).

Frängsmyr, Tore (ed.), *Linnaeus, The Man and His Work* (Berkeley, 1983; rev. ed., Canton, MA, 1994).

Heller, John Lewis, *Studies in Linnaean Method and Nomenclature.* Marburger Schriften zur Medizingeschichte, Band 7 (Frankfurt am Main, 1983).

Larson, James L., *Reason and Experience. The Representation of Natural Order in the Work of Carl von Linné* (Berkeley, 1971).

Linnaeus, Carl, *A Tour in Lapland* . . . ed. James Edward Smith, 2 vols. (London, 1811; fac. repr. New York, 1971).

Miscellaneous Tracts relating to Natural History, Husbandry, and Physick. To which is added the Calendar of Flora, rev., ed. and trans. B. Stillingfleet (3rd ed., London, 1775; fac. repr. New York, 1977).

Smith, Sir James Edward, *A Selection of the Correspondence of Linnaeus, and Other Naturalists, from the Original Manuscripts* (London, 1821).

Stafleu, Frans Antonie, *Linnaeus and the Linnaeans. The Spreading of their Ideas in Systematic Botany, 1735–1789* (Utrecht, 1971).

Stevens, P. F. and Cullen, S. P., 'Linnaeus, the cortex-medulla theory, and the key to his understanding of plant form and natural relationships', *Journal of the Arnold Arboretum*, 71 (April 1990), pp. 179–220.

10 Gender and natural history

> The *stylus* of the female [plant] is the vagina while the vulva
> and the mons Venus . . . correspond to the *stigma*. Thus the
> uterus, vagina, and vulva make up the *pistil* – the name that
> modern botanists give to all the female parts of plants. . . . As
> far as we men are concerned, a quick look is enough: sons of
> Priapus,[1] spermatic animals, our stamen is rolled as in a cylindri-
> cal tube. The *stamen* is the penis, and the sperm is our fecundat-
> ing powder.
>
> Julien Offray de La Mettrie,
> *L'Homme plante* (1748)

From Aristotle through Darwin, Freud and beyond, nature has
been infused with sexuality and gender. Carl Linnaeus, the greatest
taxonomist of his age, imagined that plants have vaginas and
penises and reproduce on marriage beds. Gender traits ascribed to
animals changed with shifting notions of masculinity and femininity in Western culture. For Aristotle, writing long before the historical rise of the passionless female, mares were thought sexually
wanton, said to 'go a-horsing' to satisfy their unbridled appetites.
If not impregnated by a stallion, these dissolute females would be
fertilized by the wind. By the late eighteenth century, females
throughout nature – with the exception of Erasmus Darwin's lusty
flowers – were said to evince a patient modesty. Even among
insects, females were thought to 'repel the first [sexual] attacks of
the males' and in so doing win the respect of their paramours.[2]

My purpose in this chapter is to explore how gender – both the
real relations between the sexes and ideological renderings of those
relations – shaped European science in the eighteenth century, and
botany in particular. Crucial for our story is the fact that, in the
seventeenth and eighteenth centuries, Europeans who described
nature were almost exclusively male, though for a while botany
was considered a science particularly suited to women.[3] While it
is true that ladies of the middle and upper classes botanized
actively, they primarily collected and dried plants, perhaps corresponded with leading botanical figures and prepared illustrations
for publication. They were not taxonomists, nor among those

shaping the future course of the science. Even Maria Sibylla Merian, the adventurous German botanical illustrator and entomologist who travelled to Surinam in search of exotic caterpillars in the early eighteenth century, presented only her observations, leaving classification to her male colleagues. In later years the traditional informal routes into science taken by Merian and others (the noble networks and craft guilds) were closed to women.[4] Ann Shteir has shown that looking at women's work in early nineteenth-century botany takes us into domestic spaces; women botanized in their kitchens and around their breakfast tables.[5] When prescribed for women, botany was to provide pleasure and instil virtue; for botany in this era, especially in England, was considered a branch of natural religion leading to an appreciation of God and his universe.

In this chapter I explore not women's role in natural history, but how gender moulded certain aspects of early modern botany. We are aware of the consequences of exclusion for women: women have long lived at the margins of intellectual life. But what have been the consequences for science and human knowledge more generally? As we shall see, gender became one potent principle organizing eighteenth-century science, a matter of consequence in an age that looked to nature as the guiding light for social reform.

Plant (hetero)sexuality

Today it is recognized that many plants reproduce sexually. Plants (and plant parts) are also typed as male and female. But what does it mean to call a plant female or male? How did botanists in early modern Europe address this question?

As extraordinary as it seems today, it was not until the late seventeenth century that European naturalists began recognizing that plants reproduce sexually. The ancients, it is true, had some knowledge of sexual distinctions in plants. Theophrastus knew the age-old practice of fertilizing date-palms by bringing male flowers to the female tree. Peasants working the land also recognized sexual distinctions in trees such as the pistachio. Plant sexuality, however, was not the focus of interest in the ancient world. An eighteenth-century observer charged that the ancients were ignorant of what he considered the essential nature of sexuality in plants: they sometimes called the seed-bearing plant 'the male', and the barren plant 'the female'.[6]

As late as the Renaissance, botanists gave names to what we today call the sexual parts of flowers that were not associated in any way with reproduction. The male organ was called the *stamen*, a Latin word denoting the warp thread of a fabric. The female organ was called *pistil*, a term suggesting the resemblance of those flower parts to a pestle.[7] Even the sixteenth-century revival of botany brought no immediate interest in the sexual nature of plants. Until the seventeenth century most botanists discounted the entire notion of plant sexuality as just another fable.

Plant sexuality exploded onto the scene in the seventeenth and eighteenth centuries. Everyone wanted to claim the honour of having discovered the sexuality in plants. In France, Sébastien Vaillant and Claude Geoffroy tussled over priority. In England, Robert Thornton complained that the honour of this discovery was always given to the French, though properly it belonged to the English. Linnaeus, always keen to reap his due reward for scientific innovation (and not, in fact, the first to describe sexual reproduction in plants), claimed that it would be difficult and of no utility to decide who first discovered the sexes of plants.

Systematic investigations into the sexuality of plants became a priority for naturalists in the late seventeenth century. Interest in assigning sex to plants ran ahead of any real understanding of fertilization, or the 'coitus of vegetables', as it was sometimes called.[8] Botanists distinguished certain parts of plants as male and female, Claude Geoffroy reported, 'without knowing well the reason'.[9] Botanists such as Nehemiah Grew, who was the first to identify the stamen as the male part in flowers, developed their notion of plant sexuality from their knowledge of animals. In *The Anatomy of Plants* (1682) Grew reported that Sir Thomas Millington, a distinguished physician, had suggested to him that 'the attire' (Grew's term for the stamen) performs the function of the male in reproduction. In quiet prose Grew worked out his analogies between plant and animal parts:

The blade (or stamen) does not unaptly resemble a small penis, with the sheath upon it, as its praeputium [prepuce]. And the . . . several thecae, are like so many little testicles. And the globulets [pollen] and other small particles upon the blade or penis . . . are as the vegetable sperme. Which as soon as the penis is errected, falls down upon the seed-case or womb, and so touches it with a prolific virtue.

Thus Grew ascribed maleness to the 'blade' of plants because it looked and functioned like the penis of animals. In a moment of unbridled enthusiasm, Grew declared that sexual dimorphism pervaded the vegetable kingdom, that 'every plant is . . . male or female'.[10]

By the early part of the eighteenth century, the analogy between animal and plant sexuality was fully developed. Linnaeus, in his *Praeludia sponsaliorum plantarum* ('Preludes to the Betrothals of Plants'), related the terms of comparison: in the male, the filaments of the stamens are the *vas deferens*, the anthers are the testes, the pollen that falls from them when they are ripe is the seminal fluid;

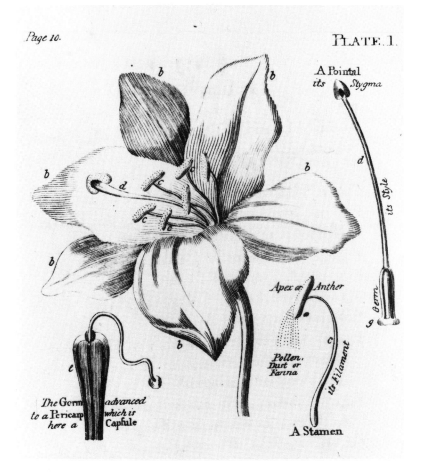

Figure 10.1 The English woman of letters, Priscilla Wakefield, nicely illustrated a hermaphroditic flower (a flower with both male and female parts) in her *Introduction to Botany* (London, 1796), plate 1. Letter 'b' shows the petals. Letter 'c' refers to the male stamen with its pollen-producing 'anther'. Letter 'd' refers to the female pistil (she calls it a 'pointal') with its stigma, style, and germ (ovary).

in the female, the stigma is the vulva, the style becomes the vagina, the tube running the length of the pistil is the Fallopian tube, the pericarp is the impregnated ovary, and the seeds are the eggs. Julien Offray de La Mettrie, along with other naturalists, even claimed that the honey reservoir found in the nectary is equivalent to mother's milk in humans.

Most flowers, however, are hermaphroditic with both male and female organs in the same individual. As one eighteenth-century botanist put it, there are two sexes (male and female) but three kinds of flowers: male, female, and hermaphrodites or, as they were sometimes called, androgynes. While most eighteenth-century botanists enthusiastically embraced sexual dimorphism, conceiving of plants as hermaphroditic was more difficult: they could not or would not recognize an unfamiliar sexual type. William Smellie, who rejected the whole notion of sexuality in plants, distanced himself from the term hermaphrodite, noting when using the word that he merely spoke 'the language of the system'.

Though eighteenth-century botanists were correct to recognize that many plants do reproduce sexually, they gave undue primacy to sexual reproduction and heterosexuality. Linnaeus was so taken with heterosexual coupling that he attributed this form of reproduction to his *Cryptogamia* ('plants that marry secretly' – by which he meant ferns, mosses, algae, and fungi) – organisms that display little fixed sexuality let alone long-term relationships. Sexuality characterizes reproduction among the higher organisms but not the earth's majority of organisms. The very fact that non-sexual reproduction is called *a*sexual (a term of nineteenth-century origin expressing the absence of sex, though parthenogenesis had been recognized since the 1740s) reveals the normative preference given to sexual reproduction.

Linnaeus did not, however, stop with simple definitions of maleness and femaleness. Not only were his plants sexed, but they actually became humans; more specifically, they became husbands and wives. When Linnaeus introduced new terminology to describe the sexual relations of plants, he did not use the terms stamen and pistil but *andria* and *gynia*, which he derived from the Greek for husband (*aner*) and wife (*gyne*). The names of his classes of plants end in '*andria*' (*monandria, diandria, triandria* and so on); his orders end in '*gynia*' (*monogynia, digynia, trigynia* and so forth). One of the most striking elements in Linnaeus's system is that plant sexuality takes place almost exclusively within the bonds of marriage. His text is filled with tender embraces of duly wedded couples:

The flowers' leaves . . . serve as bridal beds which the Creator has so gloriously arranged, adorned with such noble bed curtains, and perfumed with so many soft scents that the bridegroom with his bride might there celebrate their nuptials with so much the greater solemnity. When now the bed is so prepared, it is time for the bridegroom to embrace his beloved bride and offer her his gifts.[11]

His renowned 'Key to the Sexual System' is founded on the *nuptiae plantarum* (the marriages of plants); the plant world is divided into major groups according to the type of marriage each plant has contracted – whether, for example, they have been wed 'publicly' or 'clandestinely'. These two types of marriage, in fact, characterized custom in much of Europe; only in 1753 did Lord Hardwicke's Marriage Act do away with clandestine marriages by requiring a public proclamation of banns.

It is significant that Linnaeus focused on marriage when he thought of sexuality. As Lawrence Stone has shown, marriage underwent rapid change throughout the late seventeenth and eighteenth centuries. Upper-class parents and even wealthy peasants less often arranged marriages for their children out of property

Figure 10.2 'Love attacks the herbs themselves': sexual relations between a female (left) and a male (right) plant as portrayed in Linnaeus's *Praeludia sponsaliorum plantarum* (1729), in N. H. Lärjungar and T. Fries (eds.), *Smärre Skrifter af Carl von Linné* (Uppsala, 1908).

considerations alone. Increasingly love and affection became legitimate reasons for marriages. Men and women fell in love and created for themselves elaborate rituals of courtship as the middle-class romance was born.[12] Linnaeus's own marriage followed this pattern. He courted with tender expressions of love the daughter of a small-town physician (who also brought a substantial dowry to the marriage). As he wrote some years later, he left the running of his house entirely to his wife, and concerned himself with the works of nature.[13]

The notion that plants and animals reproduce within marital relations persisted into the nineteenth century. The term 'gamete' – adopted by biologists in the 1860s to refer to a germ cell capable of fusing with another cell to form a new individual – derives from the Greek *gamein*, meaning to marry.[14]

The eighteenth century also saw the rise of modern pornography with its emphasis on explicit description of genitalia and romantic encounters. The same forces – the loosening of traditional social controls – leading to Stone's 'companionate marriages' also drove the relationship Margaret Jacob has described between the rise of the pornographic novel and mechanistic natural philosophy.[15] Though Linnaeus's plants celebrate nuptials, the majority do not engage in lawful marriage. Only one class of plants – Linnaeus's *monandria* – practises monogamy. Plants in other classes join in marriages consisting of two, three, twenty or more 'husbands' who share their marriage bed (that is, the petals of the same flower) with one wife. Plant husbands of his 'class xxiii' – *polygamia* – live with their wives and harlots, later called concubines, in distinct marriage beds. Each of these 'marriages' signifies a particular arrangement of stamens and pistils on the flower. *Monandria* have but one stamen or husband on a hermaphroditic flower. *Diandria* have two stamens (or husbands) on a flower with one pistil, and so on.

There is no evidence, however, that Linnaeus, the Swedish country parson's son, consciously wielded a pornographic pen, though he was sometimes accused of doing so. Raised in an upright, thrifty, Protestant family in rural Sweden, he was conservative in his religious views (all of nature celebrating the glory of its Creator) and in his attitudes toward women. He would not allow his four daughters to learn French for fear that with the language they would adopt the liberties of French custom. When his wife placed their daughter Sophia in school, Linnaeus immediately took her out again, stopping what he considered 'nonsensical' education. He also refused Queen Louisa Ulrika's offer to receive one of his daughters at court, thinking the court environment apt to corrupt morals.

Though we cannot be sure, it seems likely that Linnaeus did not introduce his explicitly sexual imagery as an affront to social custom. He simply tended to see anything female as a wife. He considered 'Dame Nature' as his other wife and true helpmeet. The celebrated botanical illustrator, Madeleine Basseporte, who worked at the Jardin du Roi in Paris, he called his second wife. Linnaeus called his own wife 'my monandrian lily'; the lily signifying virginity and *monandrian* meaning 'having only one man.'[16]

Plants were not the only organisms to suffer misplaced anthropomorphism. The enormous influence that gender exerted on definitions of sex in this period can also be seen in the sexing of bees. Jeffrey Merrick has shown that from the time of Aristotle until the mid-eighteenth century naturalists spoke of the ruling bee as 'the king bee', despite the fact that these 'kings' gave birth. Even after Swammerdam correctly identified the genitalia of the queen in the

1670s, naturalists persisted in their belief that the ruler of a hive must be a king. In this instance social function – the act of wielding sovereignty – held greater sway when assigning sex than did the biological act of giving birth. When, in the eighteenth century, entomologists finally dethroned the king bee, they domesticated the newly crowned queen, emphasizing her maternal role in order to accommodate their notions about the natural destiny of women.[17]

Gendering taxonomy

It is possible to distinguish two levels in the sexual politics of early modern botany: the *explicit* use of human sexual metaphors to introduce notions of plant reproduction into botanical literature, and the *implicit* use of gender to structure botanical taxonomy. In the section above we saw how botanists introduced plant sexuality through explicitly anthropomorphic thinking, ascribing to plants human form, function and even emotion. In this way botany contributed to the great interest in sexual difference characteristic of the eighteenth century.[18] Sexual difference weighed heavily on the minds of Europeans as the Enlightenment issued its challenge that 'all men [often interpreted as including women] are by nature equal.' If women were not to be given rights in the newly imagined democratic orders, the nature of the female had to be investigated and shown to be unworthy. Debate over these issues, referred to in this era as the 'woman question', took place in a variety of texts and contexts, including medicine, literature, theology, coffee-houses, literary salons and political assemblies. The new botanical sciences participated in the making and remaking of sexuality in the Enlightenment.

There is a second level in the sexual politics of botany in this period, namely, the way that implicit notions of gender structured Linnaean taxonomy, to which we now turn. In the uproar that surrounded the introduction of notions of sexuality into botany, no one noticed that Linnaeus's taxonomy – built as it was on sexual difference – imported into botany traditional notions about sexual hierarchy.

The ardent sexualization of plants coincided with the 'scientization' of botany. Within medieval cosmology, plant classification generally emphasized the usefulness of plants to human beings as foods and medicines. By the seventeenth century, botany still remained closely allied with medicine. Herbal texts, often arranged alphabetically, classified plants according to their use; each entry included a description of a plant's appearance and its varieties, the season and place it could be collected, the parts to be used and methods for preserving it. Also included were a plant's degrees of heat and moisture, its powers against particular ailments, dosages, and methods of preparation and administration.[19] Knowledge of plants at this time was local and particular, derived from direct experience with plants in agriculture, gardening or medicine, or from knowledge handed down based upon that experience.

In the seventeenth century, academic botanists began to break their ties with medical practitioners. New plant materials from the voyages of discovery and the new colonies flooded Europe (the number of known plants quadrupled between 1550 and 1700) at the same time that an emphasis on observation increased discord between ancient texts and modern knowledge. The proliferation of knowledge required new methods of organization. Emphasis in classification turned from medical applications to more general and theoretical issues of pure taxonomy, as botanists sought simple principles for classification that would hold universally.

Linnaeus based his system on sexual difference because he recognized the importance of preservation of kind. For this reason, he considered the generative parts of plants, the stamens and pistils, 'the very essence of the flower'. But the success of Linnaeus's system did not rest on the fact that it was 'natural', that it captured true affinities between organisms. Determined to find *the* natural system, Linnaeus acknowledged none the less that his sexual system was highly artificial.[20] Though focused on the most important parts of the plant, his system did not capture fundamental sexual functions. Rather it concentrated on purely morphological features (i.e. the number and mode of unions) – exactly those characteristics of the male and female organs *least* important to reproduction.

Linnaeus devised his system in such a way that the number of a plant's stamens (or male parts) determined the *class* to which it was assigned, while the number of its pistils (the female parts) determined its *order* (Figure 10.3). In the taxonomic tree, class stands above order. In other words, Linnaeus gave male parts priority in determining the status of the organism in the plant kingdom. There is no empirical justification for this outcome; rather Linnaeus brought traditional notions of gender hierarchy wholecloth into science. He read nature through the lens of social relations in such a way that the new language of botany incorporated fundamental aspects of the social world where women were legally subordinate to fathers and husbands. Although today Linnaeus's classification of groups above the rank of genus has been abandoned, his binomial system of nomenclature remains, together with many of his genera and species.

It was not by chance or genius, then, that Linnaeus's system gained such currency in the eighteenth century, amidst upheavals surrounding the nature and definition of sexuality and sexual roles in that century. Linnaeus focused on sexuality as one of his principal taxonomic divisions because he saw the sex organs as the most important organs of the plant. He saw plants in this way because he viewed them through an eighteenth-century lens. Sexual images were prominent in botanists' language at the same time that botanical taxonomy recapitulated the most prominent and contested aspects of European sexual hierarchy.

Plant sexuality, sensibility and the 'woman question'

Linnaeus's sexual system of classification was one among many proposed in the eighteenth century. By 1799, in his popular version of the Linnaean system, Robert Thornton counted fifty-two different systems of botany.[21] Botanists based their systems on different parts of the plant, such as the flower, calyx, seed, and seatcoat or the corolla and the fruit. Others, including Albrecht von Haller, continued to argue that geography was crucial to an understanding of plant life and that development as well as appearance should be represented in a system of classification.

Despite the number and variety of systems, Linnaeus's sexual system was widely adopted across Europe after the publication of the second edition of his *Systema naturae* in 1737, and until the first decades of the nineteenth century was generally considered the most convenient system of classification. Only in France did Linnaeus face strong opposition. For one thing, French natural history remained vibrant throughout the eighteenth century. The Parisian dynasty of naturalists – de Jussieu – took little interest in Linnaeus's artificial system of classification, preferring instead to continue their countryman Tournefort's efforts to develop what they considered a 'natural' system. Linnaeus also encountered a formidable opponent in Georges-Louis Leclerc, comte de Buffon, director of the Jardin du Roi (now the Jardin des Plantes). Buffon, born the same year (1707) as Linnaeus and his principal rival, opposed system building generally and ridiculed Linnaeus's system in particular for being too abstract and artificial, and for depending on characteristics so minute and inconsequential that a naturalist had to carry a microscope into the field in order to recognize a plant.[22]

Linnaeus took England by storm in the 1750s and 1760s. His sexual system gained easy acceptance in Britain because academic natural history had been in decline since the 1720s; the classificatory advances of John Ray (one of the first to develop a system based on natural affinities) had persisted since the late seventeenth century but without further interest or development. During this same period, however, natural history – especially entomology, conchology and, eventually, botany – became popular among the

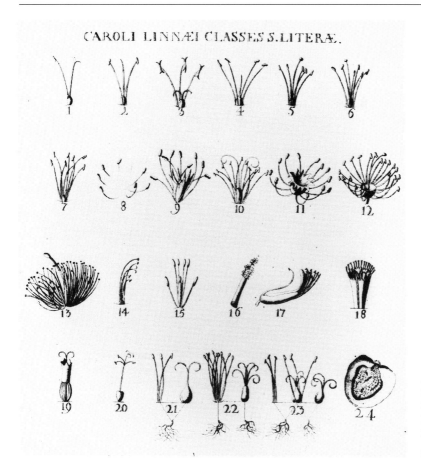

CAROLI LINNÆI CLASSES S.LITERÆ.

Figure 10.3 'Carolus Linnaeus's Classes or Letters'. Printed with Linnaeus's *Systema naturae* beginning with the second edition (1737).

fashionable. Well-born ladies, including the Duchess of Beaufort, the Duchess of Portland and Mrs. Eleanor Glanville, led the way, collecting rare and exotic plants from all over the world.[23] The royal family (George III, Queen Charlotte and his mother Augusta – all botanical enthusiasts) further enhanced the popularity of botany by serving as influential patrons and enlarging the Royal Botanical Gardens at Kew.

But it was also in Britain that bloody and protracted battles erupted almost immediately over the scientific and moral implications of Linnaeus's sexual system. 'Anti-sexualists' (those opposing Linnaeus's system) attacked Linnaeus's work primarily on empirical grounds. William Smellie, chief compiler of the first edition of the *Encyclopaedia Britannica* (1771), blasted the 'alluring seductions' of the analogical reasoning upon which the sexualist hypothesis was founded and argued that it did not stand up to facts of experience. Many animals (he mentioned polyps and millipedes) reproduce without sexual embraces, and if many species of animals are destitute of 'all the endearments of love', what, he asked, should induce us to fancy that the oak or mushroom enjoy these distinguished privileges?[24]

In addition to his ontological qualms, Smellie denounced Linnaeus for taking his analogy 'far beyond all decent limits', claiming that Linnaeus's metaphors were so indelicate as to exceed the most 'obscene romance-writer'.[25] Smellie's sentiments were shared by others. The Revd Samuel Goodenough, later Bishop of Carlisle, wrote to the founder of the Linnean Society in 1808, 'a literal translation of the first principles of Linnaean botany is enough to shock female modesty. It is possible that many virtuous students might not be able to make out the similitude of *Clitoria*.' Even Goethe thought the innocence of the young, particularly girls, should not be exposed to works setting out the 'dogma of sexuality'.[26] Faced with such opposition, the authors who popularized Linnaeus's system in England – John Miller, James Smith and John Rotheram – made little use of his sexual imagery.

Erasmus Darwin (Charles Darwin's grandfather), however, writing during the upheavals surrounding the French Revolution brought the Linnaean sexual system to full bloom in his *The Loves of the Plants*, elaborating Linnaeus's ideas in a way that might well have shocked Linnaeus himself. Unlike Linnaeus's, Darwin's plants did not limit sexual relations to the bonds of holy matrimony. Rather, they freely expressed every imaginable form of heterosexual union. The fair *Collinsonia* – sighing with sweet concern – satisfied the love of two brothers by turns. The *Meadia* (an ordinary cowslip) bowed with 'wanton air', rolled her dark eyes, and waved her golden hair as she gratified each of her five beaux. Three youthful swains succumbed to the riper years of the *Gloriosa*.[27]

Darwin was not the conservative that Linnaeus was. He was an atheist, an advocate of liberty, equality (for middle-class men) and abolition of the slave trade, and he welcomed the French Revolution. Darwin, under the cover of poetic licence, may well have been advocating the free love that he himself practised after the death of his first wife. Though Darwin was a democrat and materialist, it should be pointed out that his radicalism with respect to women was measured. His *Plan for the Conduct of Female Education*, written for the school set up by his illegitimate daughters at Ashbourne, was in step with the new middle-class prescriptions in advocating distinct roles for men and women in society.[28]

Remarkably, neither Darwin's style nor his liberal politics disturbed the public when his *The Loves of the Plants* first appeared as part II of his *The Botanic Garden* in 1789. The initial indifference to Darwin's unorthodox opinions has been explained as a result of the relative social stability that England had enjoyed since the 1750s.[29] In this atmosphere mild expression of unorthodoxy by men of the gentry and professional classes could safely be tolerated. As

Roy Porter has argued, the Enlightenment had also ushered in more tolerant views of human sexuality. Sex was no longer seen as a sin or vice, but as part of the economy of nature – a natural impulse that should find free expression. Free love was not only discussed among elites, it was practised: pornographic journals began appearing from the 1770s; erotic novels proliferated; men of substance walked in public with their mistresses; and bastards grew up as accepted members of the family (though without inheriting the family name or property). Sexuality expressed within the bounds of upper-class sensibility and decorum could be tolerated because it did not pose a serious threat to social order.[30]

The French Revolution shattered this calm. The Rev. Richard Polwhele, writing shortly afterwards, asserted that the open teaching of the sexual system in botany encouraged unauthorized sexual unions. For him, democratic tendencies, liberated and irreligious women and free love all threatened to undermine English society. In *The Unsex'd Females*, Polwhele attacked that Amazonian band of 'female Quixotes of the new philosophy' for adopting the sentiments and manners of republican France and singled out Mary Wollstonecraft, author of *A Vindication of the Rights of Woman*, as the prophetess of the movement. In a striking passage Polwhele exploited the full potential of botanical allegory, which he so despised, in order to paint for the reader a vivid picture of Wollstonecraft's 'disgraceful' life:

> But hark! lascivious murmurs melt around;
> And pleasure trembles in each dying sound.
> A myrtle bower, in fairest bloom array'd
> To laughing Venus streams the silver shade . . .
> Bath'd in new bliss, the Fair-one [Wollstonecraft] greets the
>   bower
> And ravishes a flame from every flower;
> Low at her feet inhales the master's sighs,
> and darts voluptuous poison from her eyes.
> Yet, while each heart pulses, in the Paphian grove,
> Beats quick to Imlay and licentious love.[31]

Polwhele is wilfully ambiguous in the poem; the reader is left uncertain whether Wollstonecraft actually becomes one of Darwin's 'adulterous' plants or if her libertine relations simply take place in the heaving floral bower.

In any case the message was clear: association with plants leads to licentious love. In his notes Polwhele explained how Wollstonecraft's liaisons in England with Henry Fuseli, the well-known painter who provided several illustrations for Darwin's *The Botanic Garden*, and in France with the American writer Gilbert Imlay, led her to attempt suicide from which she was rescued by Godwin only to die soon thereafter in childbirth. An early death, in

Polwhele's view, was a just end to a dissolute life. Our botanizing girls, he wrote, are worthy disciples of Miss Wollstonecraft. These 'unsex'd females', sworn enemies to blushes, he lamented, throw aside their modesty – that 'brilliant ornament' of their sex.[32]

On the heels of the Revolution, when women failed to receive political rights and professional privileges, separation of spheres set specific limits to women's ambitions and opportunities. In this pre-Victorian era plants were often stripped of sexuality, especially for women of the middle classes. Priscilla Wakefield's *Introduction to Botany* (1796) is interesting in this regard. In these letters on botany written to her sister she traced the canonical analogy between plants and animals from the bark serving as 'skin' to sap as the plant's 'blood', but completely desexualized the plant. The anther (the male part) does not resemble a penis but a 'kind of a box' that opens when it is ripe. Fertilization between plants was reduced to 'communication'. Wakefield allowed only that the seed resembles the eggs of animals. In another passage Wakefield remarked that male and female orchids are distinguished from one another 'but without any reason for that distinction'.[33]

Gender in science

The question of women engaging in science is not just a question of equality—whether all people should have an equal opportunity to pursue careers of their choosing. Nor is it a question of 'man-power', having enough scientists to sharpen a chosen nation's competitive edge, as is often discussed today. It is a question of knowledge. Only recently have we begun to appreciate that who does science affects the kind of science that gets done.

Had women been among eighteenth-century taxonomists would debates about plant sexuality and the place of female organisms in nature have been different? It is difficult or even impossible to say how things might have been. The point is to appreciate how knowledge has been moulded historically. What I have tried to show is how knowledge is shaped by who is included in science and who is excluded, which projects are pursued and which ignored, whose experiences are validated and whose not, and who stands to gain in terms of wealth or well-being and who does not.[34] That Europeans modelled plant sexuality on culturally sanctioned heterosexual unions cannot be blamed on the short-sightedness of a few individuals alone, but can be traced to broader social trends of which science was a part.

The sex of participants in science should not necessarily be important to the results of science. We must recognize, however, that Western culture is highly gendered. Scientists added their increasingly authoritative voices to the explosive debates about women and their role in society. Yet scientists did not work from a privileged vantage-point above the rough and tumble of political struggle – science itself was part of the terrain that divided the sexes.[35] Often scientists' fundamental understandings of males and females, plants and animals were informed by and shaped emerging ideals of modern middle-class masculinity and femininity, ideals which reinforced women's professional and political disenfranchisement. It is in this context that it makes a difference who does science today.

We are just beginning to unravel how deeply gender has been worked into nature's body. Historical exposé, of course, is not enough, for like a variant on Penelope's shroud, what we unravel by night is often rewoven by day in ongoing workaday institutions of science. Science – its methods, priorities and institutions – must be recast to allow women and their concerns to fit comfortably within it. Scientists need to become aware not just of how history shapes the present, but also how what is studied – and what has been neglected – grows out of who is doing the studying, and for what ends.

Further reading

Abir–Am, Pnina and Outram, Dorinda (eds.), *Uneasy Careers and Intimate Lives: Women in Science, 1789–1979* (New Brunswick, 1987).

Benjamin, Marina (ed.), *A Question of Identity: Women, Science, and Literature* (New Brunswick, 1993).

Bennett, Jennifer, *Lilies of the Hearth: The Historical Relation Between Women and Plants* (Camden, Ont., 1991).

Broberg, Gunnar (ed.), *Linnaeus: Progress and Prospects in Linnaean Research* (Stockholm, 1980).

Delaporte, François, *Nature's Second Kingdom: Explorations of Vegetality in Eighteenth Century*, trans. Arthur Goldhammer (Cambridge, MA, 1982).

Frängsmyr, Tore (ed.), *Linnaeus: The Man and His Work* (Berkeley, 1983).

Harding, Sandra, *Whose Science? Whose Knowledge? Thinking from Women's Lives* (Ithaca, 1991).

Harding, Sandra and O'Barr, Jean (eds.), *Sex and Scientific Inquiry* (Chicago, 1987).

Keller, Evelyn Fox, *Secrets of Life, Secrets of Death, Essays on Language, Gender, and Science* (New York, 1992).

Kevles, Bettyann, *Females of the Species: Sex and Survival in the Animal Kingdom* (Cambridge, MA, 1986).

Schiebinger, Londa, *The Mind Has No Sex? Women in the Origins of Modern Science* (Cambridge, MA, 1989).

Nature's Body: Gender in the Making of Modern Science (Boston, 1993).

Stroup, Alice, *A Company of Scientists: Botany, Patronage, and Community at the Seventeenth-Century Parisian Royal Academy of Sciences* (Berkeley, 1990).

Tuana, Nancy (ed.), *Feminism and Science* (Bloomington, 1989).

11 Political, natural and bodily economies

The word 'economy' suggests to us today a financial system which structures our individual lives, our society and the hierarchy of political power. That notion of a global economy of money, 'the economy', did not exist in late eighteenth-century Europe; Enlightenment writers were engaged in defining what an economy was, and indeed what wealth was, during precisely this period. We still retain many of the expressions they employed, however. Today we often discuss the economy in natural terms – we speak of the 'health' of the economy or about its 'recovery'. It has been argued that, in the use of such bodily terms to describe State operations, one can trace moves from mechanical to organic analogies of the State.[1] But eighteenth-century writers did not treat the State as a fixed and disembodied entity, something that late twentieth-century society takes for granted. In examining how the age of improvement constructed the relationship between State and individual, between political and natural economies, it is useful to invert the argument: to look at the ways in which nature and natural bodies were discussed in economic terms in the eighteenth century.

Nowadays it is a commonplace among members of Western industrial societies to be concerned with the 'balance' of the economy. Eighteenth-century naturalists, too, were deeply concerned to interpret the balance of nature for their audience. Typical of the human interpretations given to such views was Carl Linnaeus's argument in *De politia naturae* ('On the Polity of Nature') (1760) that nature made men have wars in order to reduce their numbers. For Linnaeus (1707–78), as for many of his contemporaries, both the functions within the individual and the balance of species in the natural world as a whole formed an integrated 'economy of nature': a universal plan which was open to man's understanding. By studying the laws of nature, naturalists could come to understand how God managed the natural world through the placing of powers within it. In turn, man was privileged to make use of that economy for his own ends; and, providing he followed the laws that naturalists could reveal, man could follow the Divine example and manage nature.

In eighteenth-century France, domestic animals were 'governed', but so were the poor; natural metaphors abounded in discussions on government, policing, management, and ruling. In English, French, and German discourses at this time, children, in being educated, were being 'cultivated' just like plants.[2] Thus, the link that naturalists like Linnaeus repeatedly affirmed, between natural historical language, Divine action, and naturalists' role in government related the natural historical enterprise closely to the new forms of government of the lower orders that were being invented in the second half of the eighteenth century.

Improvement and management

For Linnaeus, natural productions had been placed in the world to serve man's purposes. Like many of his contemporaries, the Swedish naturalist foresaw great benefits for the human species accruing from increasing knowledge of the relationships between natural productions. Related plants, for example, might share similar medicinal or nutritional virtues. But he maintained that it would be necessary to know all natural beings before the universal natural order could be found.

Man's task, in Linnaeus's economy, was to study the natural history of his country and the world in order to perfect the self-sufficiency of the nation, and the physical and social state of its inhabitants. Such an improving perspective fuelled many natural historical writings and practices in this period, but it was also, often in close connection with natural history, being elaborated in the writings of eighteenth-century political economists such as Adam Smith and the physiocrat François Quesnay. Improvement became immensely popular in the latter half of the eighteenth century, as Europe's monarchs and ministers came to see natural history and the introduction of new species of plants and animals as a certain way to increase national revenues and private wealth.[3] Enlightened proprietors began to ameliorate their own lands; implementing the new agricultural and industrial practices was seen as the route to moral self-improvement, which simultaneously served the nation. The 1760s were the decade of agromania, as one contemporary commentator noted.[4]

Naturalists connected to State establishments of natural history also found improving policies to their liking, since they offered naturalists the opportunity to fill their gardens with new plants, whilst at the same time becoming the most patriotic of scientific practitioners. Linnaeus's success in introducing the banana at the Hartekamp garden in the 1730s became legendary, and was used by other naturalists to encourage sceptical patrons to fund other attempts to introduce new plants across Europe.

Notions of natural and national economy were closely linked for

Figure 11.1 The first banana plant to flower in Europe. From C. Linnaeus, *Musa Cliffortiana, quo in horto Hartecampiano floruit* (1736).

managers of big natural historical centres like Linnaeus. Good management of institutions won patrons' respect and increasing freedom of action for enlightened naturalists such as André Thouin (1747–1826) and Louis-Jean-Marie Daubenton (1716–99) at the Jardin du Roi in Paris, and Joseph Banks (1743–1820) and William Aiton at Kew. These naturalists regarded good management of the natural economy as the way for nations in particular, and the human race in general, to improve and to achieve happiness. Establishments like Kew and the Jardin du Roi were, in themselves, sites at which models of good government could be tested by enlightened managers like Thouin, who was responsible for the good behaviour of large numbers of specimens, subordinates, and visitors every day.

The moral economy of the body

Besides the wider notion of an economy of nature, naturalists also employed a more local notion of economy in talking about the body. Animals and plants had individual economies which were open to investigation by naturalists and others, including medical men, philosophers, and theologians.[5] Medical accounts of the soul as the controller of bodily function derived from the classical medical writings: the Hippocratic corpus, Aristotle, and Galen. It is to the Latin for 'soul' (*animus*) that we owe the term 'animal'; but classical accounts also conferred a vegetative soul, controlling growth, upon plants. The animal soul controlled functions which plants did not possess, such as movement and feeling. Man, at the head of the scale of being, possessed a rational soul which enabled thought and judgement. Over several centuries, the classical medical soul had been assimilated into the tradition of Western Christianity, and physicians debated its relations with the Christian immortal soul, which was able to receive revealed knowledge. Thus, well before the eighteenth century, discourses about the soul had not only acquired natural historical or medical meanings but also implications for religion and government.

In 1740 a young man, Abraham Trembley, working as a tutor in Normandy, wrote to his patron in Paris, René Ferchault de Réaumur (1683–1757), to tell him of a new insect which he had found and which appeared to have startling powers. When Trembley had dissected it, he left the remains in his collecting jar for a few days, and found to his astonishment that, instead of putrefying and dying, each half of the little polyp had regenerated to produce two new individuals. This faculty for regeneration from parts produced a shock-wave which reverberated through the Parisian scientific world and beyond. The Paris Academy of Sciences had waged a long campaign over previous decades to defeat a view of the animal body as a purely material mechanism, a view derived from the writings of Descartes. Since the 1660s Cartesians had used this materialistic model of the animal body to argue for an atheist account of the natural world, one in which God's intervention had no place.[6]

By contrast, the Paris Academy of Sciences supported a worldview that explicitly sought to place God at the centre of the operations of the natural world; and it was as a part of this enterprise that they adopted a Newtonian programme of investigation into

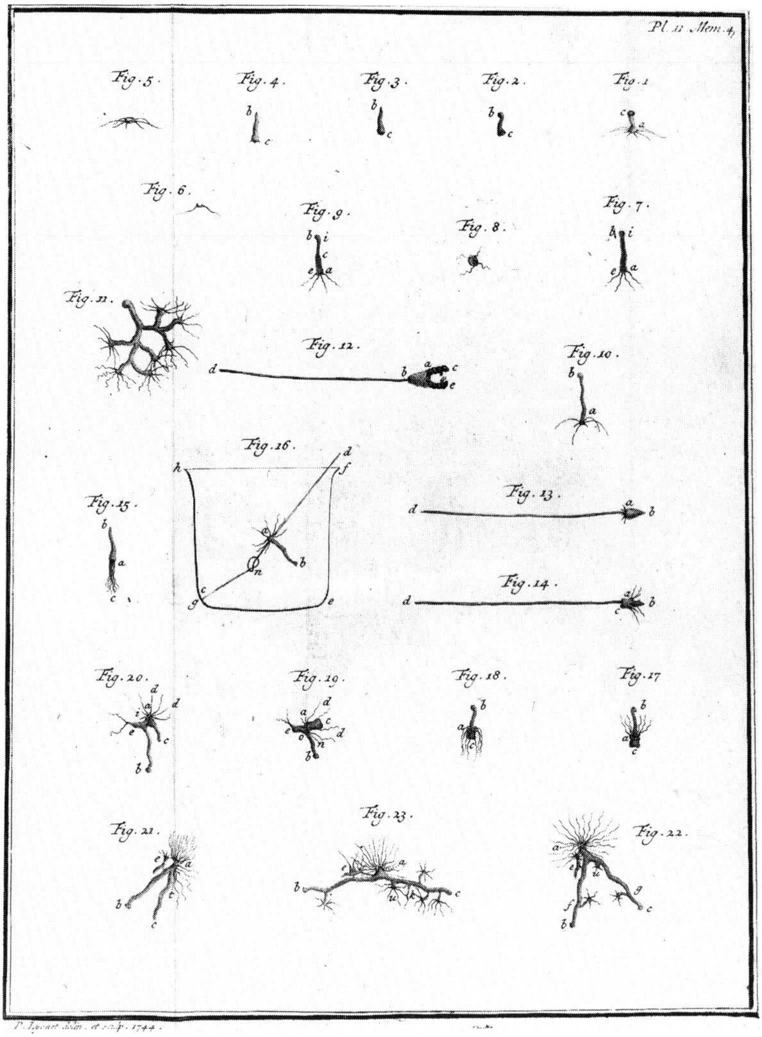

Figure 11.2 Different experiments on the body of the polyp. From A. Trembley, *Mémoires pour servir à l'histoire d'un nouveau genre de polypes* . . . (Leiden, 1770), plate 11. Figures 2 to 4 show the polyp regenerating from a cut section.

divinely-implanted natural forces like electricity, gravity and magnetism.[7] Experiments could display the powers of the soul, which were likewise of Divine origin. In the early eighteenth century, French academicians gathered evidence proving that animals had souls, in order to combat the Cartesian notion of the beast-machine; the beaver's ability to build dams, for example, was highly prized evidence for the existence of its soul. Naturalists wrote at length on animals' power to feel and think. In the assimilation of the medical and theological soul to the Paris Academy's natural philosophical programme of the early eighteenth century, the animal soul had itself acquired the characteristics of unique individuality characteristic of Christian (human) souls.

By the 1730s the debate had largely been resolved in favour of the souled animal, rather than the beast-machine – at least, it no longer appeared in the press. The new discovery of the polyp seemed to shatter that position, however. If new individuals could be made simply by the mechanical process of cutting, the implication was that even the powers of the soul with which anti-Cartesians endeavoured to discredit the mechanistic world-view were, in fact, no more than purely material – inherent in the fabric of the body itself. Such problems were made explicit in the work of the radical medical reformer Julien Offray de La Mettrie (1709–51), who initially portrayed himself as an ardent Newtonian. After 1740, however, his writings frequently referred both to the polyp and to the other animals which a frenzy of experimentation showed had similar 'budding' properties.[8] In *L'Animal machine* ('The Animal a Machine') (1746), *L'Homme plante* ('Man a Plant') (1748) and finally *L'Homme machine* ('Man a Machine') (1748), La Mettrie's view of how the world worked provided progressively more mechanical and materialist explanations for the functioning of animals, plants and man. Over the same period he shifted from praising Newton to praising Descartes and attacking the Lockean theory of the animal soul; eventually he concluded that the soul was absolutely indistinguishable from matter. His writings were violently attacked and banned by the royal censors, which meant that copies were confiscated and publicly mutilated. In an atheist, non-Providential world, neither the divine right of kings nor the stability of society could be guaranteed; La Mettrie's world-view threatened all aspects of the careful structures for the maintenance of public order that existed in the old regime in France, and contemporaries regarded him as mad.

In his *Contemplation de la nature* of 1764, which was widely read, Trembley's relative Charles Bonnet (1720–94), the Genevan natural philosopher, attempted to combat materialist accounts of generation with his own version: all living things pre-existed as invisible germs, encapsulated one within the other *ad infinitum*. Nature's plan was discernible in the unfolding through time of pre-formed creatures. This preformationist doctrine derived from accounts of animal generation advanced by seventeenth-century natural philosophers attacking Cartesian atheism and the notion of spontaneous generation (the inexplicable appearance of worms in cheese, for example). Many natural philosophers used Bonnet's pre-existing germs to account for peculiar facts of generation such as regeneration of parts and the division of the polyp. A lost part in lower animals could reform from a germ for that part contained in the body without the need to appeal to self-active forces, which risked ascribing too much power to Nature and not enough to the Creator.

Perplexity over the polyp thus spread to many parts of educated

European society. Natural historical phenomena were of intense interest to the wide and varied literate public. The *philosophe* Diderot produced a materialistic account of nature in his *Pensées sur l'interpretation de la nature* ('Thoughts on the Interpretation of Nature') of 1754, and suggested that human beings were themselves composed of germs that acted together like a swarm of bees. In his *Système de la nature* (1756), the naturalist Pierre-Louis Moreau de Maupertuis (1698–1759) criticized Bonnet's generation and Newton's physics of attraction for failing to account for the *purposeful* arrangement of matter into living forms. Questions about generation and the nature of the soul were theologically problematic in the mid-eighteenth century, and simultaneously served to negotiate the validity of the Newtonian world-view.

Birth and breeding

The problem of inheritance had particular relevance for the Enlightened elite of French society. Many naturalists worked for or were themselves improving landowners; thus they had an acute interest in the formation and disappearance of races, the role played by different sexes in passing on traits, and the effects of environment upon offspring. It was still quite common for physicians to claim that monstrous children resulted from some shock or spectacle experienced by the mother before or after conception.[9] The results of breeding experiments were widely disseminated during this period. Maupertuis, for example, carried out many years' work on the inheritance of polydactyly; what interested him was not only the question of 'heredity', but the regularity in what had previously been explained as an accident of nature, something that occurred by chance and which therefore should not obey regular laws. The enlightened elite's claims to social and intellectual superiority were increasingly related to the debate over whether such hierarchies were located inalterably in ancestry and anatomy or were local accidents of circumstance, and whether such accidents could be reversed or inherited.

Georges-Louis Leclerc, later comte de Buffon (1707–88) was also a committed Newtonian and a member of the Academy of Sciences when the polyp storm broke. During the late 1740s Buffon and the English naturalist John Turberville Needham observed the decomposition of seminal fluids into minute moving particles, concluding that all living beings were made of specific fundamental particles, the 'organic molecules'.[10] Inanimate and living beings did not merely possess different properties; the distinction between the living and non-living realms of nature commenced with the matter of which they were composed. Buffon and Daubenton's widely read, multivolume *Histoire naturelle* (Paris, 1749–67) treated life as the chief discontinuity in the natural order, a new view which came to be widely adopted.

For Buffon, problems of conformation and parental inheritance were explained by means of an internal mould, the *moule intérieur*. Attractive forces operated to move organic molecules to the *moule intérieur* from food ingested by the parent, thus generating new individuals. The cycle of life of a living being consisted of a series of stages; birth, growth, maturity, decline and death were all part of nature's plan for the perpetuation of the species. This, Buffon argued, should be regarded as a succession of individuals descended from one another – in other words, as a succession of generative acts. Only thus could the species constitute a real entity in nature.[11] Such concerns led to his programme of breeding experiments, designed to test where the limits of different species lay. He found, for example, that the wolf and the dog were one species.

Political economists and naturalists made explicit comparisons between the productivity of the earth and the productivity of the human race: both revealed the unfolding of Nature's plan. European natural history and debates about generation were not merely concerned with polyps; the naturalists were simultaneously writing natural histories of society. During the eighteenth century, a new field of social mathematics was developed to investigate changes in the population of nations, in order to determine how healthy – in moral and physical terms – that nation was. The foundation of this new mathematics of productivity of the population and of the soil was the calculus, the same mathematical instrument used by Buffon as the basis for his definition of the species. Correct use of the calculus allowed one to form predictions about future events through knowledge of a very large number of past events; thus medical problems such as the mortality rate of a population, and political problems such as that of fair voting procedure in elections, could all be studied by recourse to social mathematics. Buffon was among the first to utilize social mathematics as an instrument to generate inductive knowledge about the French population.

Generation, degeneration and regeneration

Following the discovery of the polyp, the Swiss naturalist and polymath Albrecht von Haller (1708–77) explicitly rejected his earlier views of generation, and began to explain natural phenomena in terms of forces similar to Newton's; he had written poetry in praise of the English natural philosopher. Above all, he was concerned to underline the primacy of God in controlling the formation of new individuals. Haller's Pietist search for ways of proving God's activity in the world led him to regard accounts of self-active forces operating in the body as tending towards self-active matter, whereas activity in the natural world should be ascribed to God. In 1752, in a preface to the second volume of his translation of the *Histoire naturelle*, Haller attacked Buffon's claim that the internal mould could account for organization; the penetrating force which Buffon supposed to be organizing the organic molecules would have to be purposive to get right all the millions of linkages in the body.[12] In the 1750s Haller experimented on chicken eggs, sending his results to Bonnet in Geneva. His work, published in *Sur la formation du coeur dans le poulet* ('On the Formation of the Heart in the Chicken') (1757) showed that the structures of the unborn chick were pre-existent in the membranes of the egg, before the approach of the male: organizing forces were not necessary.

In 1776 Buffon attacked Haller and Bonnet for attempting to prove the existence of God by means of arguments drawn from the design of living beings. 'I only accuse them', he said, 'of prejudice for their absurd system of pre-existing germs, a system which, despite its absurdity, may still last for a long time in the heads of those who imagine that it is linked with religion.'[13] Buffon's criticism reflected his opposition to a tradition of natural theology, most popular in England, by which the functions of the living body and of the economy of nature supplied direct evidence of Divine craftsmanship. Accounts of the natural world, for the French, must be based only upon the knowledge which was directly accessible to the senses; accounts of physical nature must be given in terms of physical laws. However, this meant that it was hard for French naturalists to ensure that their accounts of the world were not perceived as materialist or atheist; as Wood's essay shows, Buffon was not immune to such charges. Peers regarded such writers as failing to uphold their responsibility to improve public morals, for naturalists' accounts of the operation of the animal world were simultaneously claims about the best course of action for particular societies and for the human race as a whole.

The expert knowledge that naturalists were perceived to have about the living world made them valuable public servants. The most renowned European naturalists – Banks, Buffon, Haller, Linnaeus – served as consultants to the Crown in agricultural and medical matters. Vast sums of money were involved in colonial trade, for individual and State alike; thus naturalists' discussions about economies were highly political. They could contribute to the growing imperialism of European nations through controlling the movement of new species into and out of centres of natural history. They could, through their expertise in the internal economies of animals and plants, work to ensure the success of new endeavours at cultivation and breeding in the colonies and in the central gardens, parks and farms of Europe; this was a process which continued well into the nineteenth century and reached its peak during that time.[14] However, there was another sense in which knowledge of the animal economy had political implications. Naturalists accounted for the stability of the natural world – the ability of species to perpetuate themselves indefinitely through successive generations. They accounted for the maintenance of the balance of nature by treating the natural economy in the same way as the political one; but they were also faced with the problem of accounting for the changes that they saw in the world.

This was a problem common to natural and civil historians alike. Civil historians such as Edward Gibbon (1737–94) and political economists such as Jean-Jacques Rousseau (1712–72) – and the two roles were not easily distinguishable – were deeply concerned with the ways in which nations became civilized and enlightened, or fell into decline and degradation.[15] Until the mid-century many historians, natural and civil, regarded history as a balanced, but cyclical, process: nations would rise from barbarism to a state of enlightenment and the pursuit of the sciences and arts, but inevitably, with increasing civilization, there would be increasing corruption of morals, leading to an inescapable decline and fall.

The concerns of naturalists with the physical history of animals and man reflected this concern with degeneration. Buffon regarded non-whites, for example, as having degraded from an original perfection – the European type. This was the consequence of generations of people having been exposed to the deleterious effects of a hot climate. Local climatic conditions and food were the two principal causes of degeneration. Inorganic particles were assimilated into living bodies from their surroundings, destroyed the perfection of the form that the species had received at origin, thereby affecting its moral and physical nature, sometimes irreversibly.[16] In the context of the expansion of European colonial holdings after 1750, this fear of changing climate had an almost mythical resonance in the travel accounts consumed by an eager readership during this period.

Following the writings of the *agronomes*, in France and to a lesser extent in England, several political commentators (including the permanent secretary of the Paris Academy of Sciences, the marquis de Condorcet) began to suggest that society might be infinitely perfectible, and that the necessary decline into barbarism might be avoidable.[17] This was possible for man because, unlike the animals, he possessed the power of reason – the power to develop the arts and sciences – by means of which he might improve himself both physically and morally. Practitioners of the sciences endeavoured to make themselves into expert dispensers of human regeneration, fighting against the gradual decline of nature. The growing interest of naturalists towards the end of the century in studying other human races was also symptomatic of a view of the history of the human species as a process of perfection in permanent tension with a natural tendency to degenerate.

For nearly half a century Buffon experimented on the limits of degeneration through bad food, heat and other natural influences, such as electricity. Meanwhile, trials by his colleague Daubenton during the 1760s suggested that cross-breeding could improve degenerate animals within the space of one or two generations. The extent to which animals and, by implication, man, could be changed through different breeding choices excited considerable interest, particularly in France, at the very end of the eighteenth century. Apparently Buffon carried out breeding experiments among the peasants on his estate by arranging marriages between them, to see if he could regenerate them: this practice was not popular with the peasants!

Scales of being

Concerns about orders of perfection were characteristically expressed in classificatory schemes. In 1774 Buffon gave permission for a young botanist, Antoine-Laurent de Jussieu (1748–1836), to reorder the plants in the Jardin du Roi according to his uncle Bernard's natural method of classification. Jussieu based his new order on the external marks of internal function – the indicators of the vegetable economy. He ordered genera and species in a hierarchy of physical complexity and perfection: the most perfect species, such as trees, came at the head of the plant kingdom and the least perfect ones, such as fungi, at the bottom of the series. 'Perfection' was thus defined on the basis of the complexity of the functions exhibited by the body of the being. This notion of a scale of being was derived from the classics; in different versions, it was to be found in the writings of many eighteenth-century naturalists. Jussieu's method, however, published in *Genera plantarum* (Paris, 1789), revealed a comparatively new concern for classifying by function; similar views were shared by zoologists like Félix Vicq d'Azyr (1748–94), Daubenton's student. For these naturalists, a 'natural' classification was one that expressed the structural and functional relationships which bodies exhibited as a consequence of possessing life.

Anatomy had not achieved the same recognition as botany in eighteenth-century Europe, in part because of its lack of appeal for an aristocratic audience compared to the widely popularized charm of flowers. Linnaeus had classified animals by using outer parts such as the teeth and feet. From the 1740s, however,

Daubenton, Haller and others stressed the value of internal parts as the basis for investigation of the true nature of animals and man. Daubenton's programme of comparative anatomy allowed the comparison of animal species with the well-known anatomy of man to reveal the degree of perfection of each species. Thus it would be possible to establish a degenerative series of animal species based on complexity of function.

Vicq d'Azyr's *Traité d'anatomie et de physiologie* (1786) utilized the faculties of the animal economy to demarcate the boundaries of animal classificatory groups, as had Daubenton. Vicq also indicated his debt to a new science of the animal economy developed by Haller in the 1740s and 1750s – a science Haller called physiology, literally 'laws of nature'. Its object was to study the processes by which the presence of life in animals and plants was indicated; everything, in fact, that set the living world apart from the non-living – locomotion, sensation, irritability, respiration, nutrition and generation. By the 1780s those functions were of central interest to naturalists across Europe in their attempts to order the natural world, although there was very little agreement among naturalists as to the nature of the principles acting in animals and plants. Investigating the bodily economy, therefore, raised questions about how far living beings were subject to mechanical laws, and how specific forces of life, if they existed, operated within the body. Studies of these principles contributed to the ordering of living beings in a hierarchy of complexity. But they also contributed to a wider account of the relationship between man and the world, and of the degree of autonomy possessed by nature. The location and nature of the will in voluntary movements was central to such debates, since the nature of the will defined the limits of bodily autonomy. Contemporary discourses about the location of political authority similarly revolved around the question of whether political will and authority *naturally* came from above (the monarch, the parliament, God, the church, the ruling educated elites) or from below (the people). In particular, the debate over the materialism of the soul, which divided natural philosophers and natural historians throughout eighteenth-century Europe, brought the question of the relationship between mind and body to the forefront of concern, since, as La Mettrie put it, 'the different states of the soul are . . . always co-relative to the states of the body'.[18]

Sensibility and structure

The eighteenth-century term 'sensibility' was used in several contexts to describe the ability to feel. In natural history, sensibility was a characteristic of more organized animals. In 'On the Sensible

Figure 11.3 'Changing a quadruped into man'. From P. Camper, *Oeuvres* (Paris, 1803), *Atlas*, plate 33: accompanying the second 'Discours sur l'analogie qu'il y a entre la structure du corps humain et celle des quadrupèdes, des oiseaux et des poissons'.

and Irritable Parts of the Human Body' (1752), Haller distinguished between the qualities of irritability and sensibility. Irritability was an innate property in muscles, independent of the will, which was experimentally demonstrable. Sensibility, on the other hand, related to properties of animal bodies that were controlled by the will; but it, too, resided in the fabric of the animal body. These views were partly a response to the work of the Edinburgh physician Robert Whytt (1714–66), whose *An Essay on the*

Vital and Other Involuntary Motions of Animals (1751) had suggested that the soul, coextensive with the fabric of the body, was
the direct source of voluntary movement. Haller sought a physical
cause for these natural phenomena; but he also objected to the
materialist positions that he saw Whytt, La Mettrie and others
adopting. God, the prime mover of matter, had placed attractive
forces such as irritability and gravity in matter and none of these
were self-directing. Their acerbic debate prompted an anonymous
Edinburgh student to comment that apparently 'neither Baron
Haller nor Dr. Whytt were deficient in irritability'.[19]

Vicq d'Azyr, like many others, drew a close connection between
function, organization and degree of perfection. In a series of
detailed dissections depicted in the *Traité*, Vicq showed that the
degree of sensibility an animal possessed was related to its brain
structure and to the degree of complexity of the nervous system.[20]
Unlike Vicq, the Dutch anatomist Petrus Camper (1722–89) made
explicit connections between the anatomical structure of different
species of animal or human races and their moral qualities, a standard concern in this period.[21] The relation of physical structure
to moral qualities was not merely of interest to naturalists. Travel
accounts, medical writings, discussions of gender differences and
books on etiquette all evidenced similar concerns. The revival of
an ancient art, physiognomy, by a Swiss pastor, Caspar Friedrich
Lavater, in the 1760s and 1770s marked the start of a craze for
physiognomical portraits which swept the fashionable streets of
Europe. Physiognomists could predict moral character from an
observation of the facial structure; the examples given in physiognomical guides, however, clearly show that class, gender and context were highly important elements in assessing an individual's
virtues and vices.

The term 'sensibility' was also increasingly being used to
describe human feelings and responses in a more narrowly defined
way – people with a greater sensibility were those who had stronger
feelings, particularly in love, friendship or when reading the
romantic novels which became popular at the end of the century,
such as Rousseau's *La Nouvelle Héloïse* (1758).[22] Simultaneously,
the sensitive plant, *Mimosa pudica*, attracted much interest from
European naturalists seeking to establish the site of feeling, intellect and imagination. It also fuelled the accounts of Romantic poets
like Coleridge, Shelley and Wordsworth in their writing on sensibility. In his *Botanic Garden* (1794), Erasmus Darwin ascribed perception, sensation and volition to plants, for how else, he argued,
could plant sex organs know how and where to find their mates?
What directed them must be a 'sensation of love'.[23]

The language of sensibility in the late eighteenth century –
especially when using animals and plants as an example – was used

Figure 11.4 Frontispiece to J. de Vaucanson, *An Account of the Mechanism of an Automaton. Translated out of the French Original, By J. T. Desaguliers* (London, 1742). These automata were first exhibited in Paris in the 1730s, as part of Vaucanson's hugely popular touring show. They could well have served as models for La Mettrie's notion of the 'man-machine'. Automata, popular throughout the eighteenth century, performed human acts such as speaking and writing, and some were made to pronounce political statements. Revolutionary political writers warned their audiences of the dangers of becoming automata through sacrificing their autonomy to despotic rulers.

to describe, discuss and legislate about human sexual behaviour in a society for which the role of the Church was increasingly uncertain. But studying the moral qualities of man and animals was more than just a way of discussing individual human relationships. It could also provide naturalists with a means to comment upon the processes underlying the 'proper' operation of human society as a whole. Like other naturalists, Daubenton discussed animal behaviour in human terms: 'The hamster is of a singular ferocity and an astonishing courage for its size . . . It seems to have no other passions than that of anger which leads it to attack everything in its path, without paying attention to the superiority of the

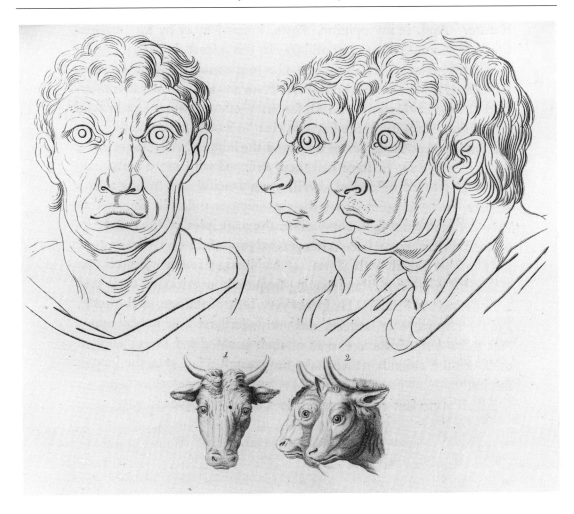

enemy's forces'.[24] Such descriptions reveal the great political power of natural history in the eighteenth century: since it was so widely accepted that animals could serve as an example of what 'natural man' would be like, claims about natural government made by expert naturalists were important claims that could not be dismissed lightly. The structure of eighteenth-century government was frequently compared by historians, rulers and political economists to that of the family; and naturalists also couched normative claims about family life in discussions of how animals conducted their domestic existence. Many political commentators argued that exhibiting sensibility was a perfect way for statesmen and public figures to prove that they were fit to govern, by demonstrating in public that they were not so corrupt as to have lost touch with their natural feelings. Those concerned with the problems of ordering people in society thus drew upon natural historical and medical discourses about the body. Natural history, in particular, depended for its

Figure 11.5 'Gross brutality, rudeness, force, stupidity, inflexible obstinacy, with a total want of tenderness and sensibility'. From J.-C. Lavater, *Essays on Physiognomy*, trans. Henry Hunter, 3 vols. (London, 1789–98), Vol. II, p. 108. The great fear of the eighteenth-century elite was that of degenerating from humanity to bestiality, as this quote accompanying the 'caricatures of men forced into a resemblance to the ox' reveals.

success on its romantic appeal to a public which did not draw boundaries between nature and human experience of it.

Social, sexual and racial differences were mediated in natural historical and medical texts by recourse to anatomical accounts of structural difference. Writers such as La Mettrie stressed the relative strength of the nerve fibres of men and women as the cause of their differences. The swifter motion of nervous fluid in women's bodies rendered their sensibility greater, their sensory impressions more vivid and their thoughts more fleeting and inconsequential. Thus they were regarded as less intelligent and fitted only for certain sorts of activity, usually domestic and family-based.[25] By the 1780s and 1790s most physicians and anatomists gave accounts of the operations of the bodily economy in terms of the circulation of subtle fluids identical to, or analogous to, electricity, magnetism and gravity. For the botanist Antoine-Laurent de Jussieu, the circulating principle was electricity: it formed the link between the natural economy, the productivity of the population of France and of French soil, and the individual moral and physical lives of French men and women. This fluid was also the cause of all sensation, hence the ultimate cause of character formation and of the customs and degree of civilization of different nations.

By the time of the French Revolution, concerns about de- and regeneration were associated with a view of the living body as extremely malleable in its functions: as subject to change in response to external conditions and internal will. In England, the Dissenter and Royal Society chemist Joseph Priestley showed through a series of experiments in the 1770s that plants had the power to purify the air of the poisons with which animals corrupted it.[26] This, he said, was proof of the fact that God had designed even plants to have an important role in the natural economy. The economy of the living body thus formed a small part of a greater cycle of generation, degeneration and regeneration through which nature's economy was balanced.[27]

Conclusion

By the end of the eighteenth century, naturalists managing institutions throughout Europe had established themselves as the experts in collection, classification, cultivation and breeding, catering to a wealthy and enthusiastic public and to the increasingly imperialist aims of European governments. Many of them, such as Buffon and Banks, had risen to the upper echelons of European elite society through their efforts in these areas. Their increasing ability to present themselves as expert mediators between nature and society allowed them to work on the boundary between the social and the natural worlds. Besides their power to comment upon the political state of nations, naturalists in late eighteenth-century Europe began to make increasingly rigorous claims about the ways in which women and other social groups ought to behave, based on the observation of their structure – claims which are still widespread in our society today.

Observing structure and controlling economies conferred a great deal of political and social power, allowing naturalists to make order in the social, as well as the natural, world. Increasingly, towards the end of the century, justifications for social, racial and gender hierarchies were located within the fabric of the body itself. The debate over natural and bodily economies was political, because it was a debate over the location of proper government: how far natural and bodily change could occur from within, and how far such changes represented a divergence from an original or potential perfection. Late eighteenth-century French natural history, by enclosing the bodily, political and moral economies within the economy of nature, would become central to Revolutionary policies of universal regeneration.

Further reading

Bowler, Peter, *The Fontana History of the Environmental Sciences* (London, 1992).

Chisick, Harvey, *The Limits of Reform in the Enlightenment: Attitudes Towards the Education of the Lower Classes in Eighteenth-Century France* (Princeton, 1984).

Cross, Stephen J., 'John Hunter, the animal oeconomy, and late eighteenth-century physiological discourse', *Studies in the History of Biology*, 5 (1981), pp. 1–110.

Darnton, Robert, *Mesmerism and the End of the Enlightenment in France* (Cambridge, MA, 1968).

Gasking, Elizabeth, *Investigations into Generation, 1651–1828* (London, 1967).

Heckmann, Herbert, *Die andere Schöpfung. Geschichte der frühen Automaten in Wirklichkeit und Dichtung* (Frankfurt am Main, 1982).

Jordanova, Ludmilla and Porter, Roy (eds.), *Images of the Earth: Essays in the History of the Environmental Sciences* (Chalfont St Giles, 1979).

Lenhoff, Sylvia G., *Hydra and the Birth of Experimental Biology, 1744: Abraham Trembley's* Mémoires *Concerning the Polyps* (Pacific Grove, CA, 1986).

Porter, Roy, 'Making faces: physiognomy and fashion in eighteenth-century England', *Études Anglaises*, 4 (1985), pp. 385–96.

Spary, Emma, 'Making the natural order: The Paris Jardin du Roi, 1750–1795', Ph.D. dissertation (Cambridge, 1993), ch. 2.

Walker, W. Cameron, 'Animal electricity before Galvani', *Annals of Science*, 2 (1937), pp. 84–113.

Wellman, Katherine, *La Mettrie: Medicine, Philosophy and Enlightenment* (Durham, NC, 1992).

Weullersse, Georges, *Le Mouvement physiocratique en France de 1756 à 1770* (Paris, 1910).

Williams, Elizabeth A., *The Physical and the Moral: Anthropology, Physiology, and Philosophical Medicine in France, 1750–1850* (Cambridge, 1994).

12 The science of man

If nothing else, the spate of publications marking the quincentenary of Columbus's landfall in the Americas reminds us that the intellectual revolution which occurred in early modern Europe involved more than just a change in cosmology. For too long, historians of science have focused their attention almost exclusively on the 'birth of a new physics', and have largely ignored the equally profound transformation of European conceptions of human nature prompted by the discovery of hitherto unknown lands and peoples across the seas. Visions of humankind derived from the Bible and from antiquity were gradually destablized by a growing literature which chronicled the natural and civil histories of Asia and the Americas. Through these texts, Europeans learned of peoples whose appearance, customs, and beliefs were radically different from their own, and the information thus compiled challenged not only traditional views concerning man's physical nature, but also widely held assumptions about morality, politics, and religion. Arguably, therefore, a revolution in the human sciences took place during the 'long eighteenth century' in Europe which rivalled that which shook the physical sciences in the period. In what follows, I shall trace some of the consequences of this momentous refashioning of European images of humankind through a discussion of the relationships between natural history and the science of man in the Scottish Enlightenment.[1]

The Baconian legacy

Given the almost Biblical status of Sir Francis Bacon's writings in eighteenth-century Scotland, it is significant that in his *Advancement of Learning* (1605) and *De augmentis scientiarum* (1623) Bacon (1561–1626) incorporated the classification and description of human character types in the branch of learning he christened 'the culture of the mind'. Furthermore, in the brief tract *Parasceve*, appended to the *Novum organum* (1620), Bacon catalogued the various natural histories which he thought were requisite for a firm empirical foundation for his inductive philosophy, and he called for histories of the individual human senses, the passions, and the faculties of the mind. In Bacon's works the science of the human mind thus takes on an important natural historical dimension, and his methodological message was taken up by a number of leading Scots.

During his student days, the physician John Gregory (1724–73) outlined a plan for the founding of a society dedicated to the pursuit of medical enquiries according to 'the strict method of naturall history & accurate induction from it' set out by Bacon. Among the subjects to which he thought the proposed society should devote itself, Gregory included natural histories 'of the Senses & the method of preserving & improving them' and of the faculties of the human mind. Significantly, he stressed the value of comparative evidence drawn from the animal kingdom for such enquiries, as he called for 'A Naturall history of such Animals & such parts of nature as by their anatomy, constitutions, diseases & instincts can throw any light on' humankind. But one of the 'desiderata' of medicine he listed stands out, not least because it attests to his indebtedness to Bacon's *De augmentis scientiarum*. Under the heading 'Desiderat[um] 4' Gregory wrote 'History of the connexion between the mind & the body & their mutuall effects in producing & curing one anothers diseases. This comprehends the effects of imagination, music &c in the production & cure of diseases & the manner of operating upon & changing the faculties of the mind by medicines'.[2] In Book IV of the *De augmentis*, Bacon too recommended the study of 'the bond and connexion between the mind and body', which for him encompassed physiognomy, the interpretations of dreams, the discovery of the effects of the body on the mind, and the tracing of the influence of the passions and imagination on the body. There can be little doubt that Gregory took his inspiration from Bacon's remarks, especially as it was in Book IV of the *De augmentis* that Bacon discussed improvements to the art of medicine. Gregory therefore drew on Bacon's methodological legacy to develop a natural historical and comparative method which incorporated the investigation of both body and mind, along with their interconnections.

Gregory's early interests and approach later bore fruit in his *A Comparative View of the State and Faculties of Man with those of the Animal World* (1765), wherein he championed the comparative approach to the study of the human mind. Gregory contended that one of the major obstacles to progress in this branch of knowledge had been the mistaken belief that there was no analogy between the constitution of man and that of animals; because nature was one continuous chain, he argued, 'no one link of the great chain can be perfectly understood, without the knowledge, at least, of the links that are nearest to it'. Consequently, he urged the legitimacy of juxtaposing not only the physiology of man and animals, but also their 'states and manner of life'.[3] For Gregory, the main principle of action common to man and animals was instinct. He maintained that within their proper limits man's instincts were 'a sure and infallible guide' (p. 15), but he thought that life in civilized societies had rendered them corrupt. Hence he believed that in order to determine the natural operations of instinct, man in his savage state had to be studied, along with wild and domesticated animals. What differentiated humankind from the beasts, according to Gregory, was the superiority of human reason, our moral sense, and our enjoyment of the various pleasures of the mind derived from curiosity, the pursuit of the sciences, and the appreciation of the fine arts. While each of these was liable to corruption, Gregory nevertheless affirmed that these faculties of the mind gave humankind its greatest advantages over animals, namely religion, taste, and life in society.

Gregory also proposed a radical widening of the scope of physiology, which reflected both his belief in a unified science of man and his debt to Bacon. In his *Lectures on the Duties and Qualifications of a Physician* (1772), he stated that it 'belong[ed] to physiology to enquire into the laws of the union between the mind and the body; into the effects of culture and education upon the constitution [and] into the power of habit, the effects of enthusiasm, and [the] force of imagination' (p. 79). For Gregory, physiology was thus not concerned simply with the animal oeconomy of the body treated as a self-contained, fixed system; rather, it focused on the dynamic interactions of body, mind, society, and environment. Gregory also specified in some detail what he thought was encompassed by the study of the mutual influence of mind and body. His list of topics included natural histories of the power of imagination, enthusiasm, dreams, custom and habit, the effects of the fine arts on the mind and body, language, and our capacity to imitate others. This indicates that he regarded the study of the relations between mind and body to be essentially a natural historical one, and that Bacon was the primary inspiration for his approach to this subject. Gregory thus propagated a Baconian vision of the science of the mind, and in so doing he helped to popularize the natural history of man, which he regarded as 'a more interesting subject . . . afford[ing] a larger scope for the display of genius, than any other branch of natural history' (p. 218).

One the most interesting Scottish examples of the use of the Baconian natural historical method in the science of man is a pair of discourses delivered before the Aberdeen Philosophical Society in 1758 and 1760 by Gregory's associate, the physician and naturalist David Skene. Skene took as his starting point Bacon's discussion in the *De augmentis* of the proper subject matter of natural history and the philosophy of mind, but argued that Bacon had mistakenly excluded the human mind from the works of nature, and hence from the province of natural history.[4] Skene maintained that the mind was 'as truely a Natural Production as the Human Body', having 'its own particular Constitution . . . compounded of distinct principles united in various degrees, which have their origin, progress & decline'. Consequently, he urged that the human mind was a legitimate object of natural historical enquiry. Skene proposed that the 'first grand division of Natural History should be into that of Mind & that of Matter', and he recommended that the natural history of the mind should encompass 'all the facts relative to the Structure or Anatomy of the Mind & the Combinations of its several Principles'.[5]

Skene then divided the natural history of the human mind into two main branches: the 'Abstract Philosophy of the Mind' or metaphysics, and the 'Philosophy of Life & Manners'. In the first, the mind was conceived of 'as an extensive Genus of Beings endued with certain Powers & faculties, which are in some degree common to every one of the Species' and, according to Skene, the aim of metaphysics was to 'investigate these powers & their laws of Operation' by means of introspection. In the second branch, Skene stipulated that the faculties of the mind 'are considered as they are found existing in individuals . . . [and] Man is contemplated not only as a rational, but a sociall active being'. Here the aim was to discover 'a just & accurate picture of Human Nature', and to delineate 'a set of Characters engag'd in various important transactions, which are faithfully recorded & their Causes & Consequences justly deduc'd'. Echoing Bacon, Skene argued that the natural historian can identify and classify fixed human character types by turning to history, biography, plays, novels, and the pages of the periodical press for his evidence.[6] Skene's application of the natural historical method to the science of man was thus both highly innovative and thoroughgoing, and his scheme for the study of the mind clearly reflects his preoccupations as a practical naturalist, for he was acutely aware of the problems of classification and systematic arrangement.[7]

In the Scottish Enlightenment Bacon's ideas served to legitimize academic as well as methodological reforms, as can be seen in Alexander Gerard's *[A] Plan of Education in the Marischal College and University* (1755), which served as an apologia for the restructuring of the arts curriculum carried out by the college in 1753. Gerard (1728–95) here cited Bacon's requirement that the study of natural histories should precede the use of the inductive method to justify the relegation of logic from the first to the third and final year of the new philosophy course. Gerard insisted that logic was 'one of the most abstruse and difficult branches of Philosophy, and therefore quite improper to begin with'. Instead, students had to be acquainted first with the natural history of the mind; Gerard wrote that '[t]he natural history of the human understanding must be known, and its phaenomena discovered; for without this, the exertions of the intellectual faculties, and their application to the various subjects of science will be unintelligible' (p. 9). In his defence of the reformed curriculum, Gerard thus followed Bacon in calling for a natural historical foundation for the science of the mind, and he expounded a similar methodological message in the moral philosophy lectures he delivered at Marischal during his tenure of the Chair from 1752 until 1759.[8] Indeed, for much of the eighteenth century it was something of a methodological commonplace at Marischal to conceive of moral philosophy as having a natural historical dimension, and we shall see that Gerard's views were echoed by other Scottish moralists.

Another feature of the Baconian legacy which had important consequences for the natural history of man in the Scottish Enlightenment was Bacon's treatment of the various branches of history in his scheme of human knowledge. In the sixteenth and seventeenth centuries, the boundaries between civil and natural history were redrawn in such works as the Spanish Jesuit Josè de Acosta's *The Naturall and Morall History of the East and West Indies* (orig. 1590; 1604). Bacon evidently endorsed Acosta's practice, for in his *De augmentis scientiarum* he classed together natural and civil history on the grounds that these two branches of learning were ultimately rooted in the faculty of memory. Well into the eighteenth century, the concerns of British virtuosi reflected the close connections envisaged by Bacon between these two branches of history, insofar as they commonly engaged in the collection of both antiquities and specimens from the three kingdoms of nature.[9] Within the Scottish universities, Bacon's map of the sciences was embodied institutionally at Aberdeen, where civil and natural history were taught together in the first year of the philosophy course following the curriculum reforms of 1753 at King's and Marischal Colleges.[10] Moreover, the various museums set up in the Scottish universities during the eighteenth century resembled the cabinets of the virtuosi in their juxtaposition of natural historical specimens with ethnographic materials, coins, and antiquities.[11] Thus enlightened Scots saw a fundamental continuity between civil and natural history, and it is arguable that it was their Baconian vision of the structure of human knowledge which in part led writers like Henry Home, Lord Kames, and Adam Ferguson to regard their histories of society as being natural histories of man in his social state.

John Locke and the natural history of the mind

In his *An Essay concerning Human Understanding* (1690), John Locke (1632–1704) self-consciously adopted an 'Historical plain Method' in dealing with 'the discerning Faculties of . . . Man', and in Book II of the *Essay* he claimed to have 'given a short, and . . . true *History of the first beginnings of Humane Knowledge*'.[12] Hence Locke was primarily concerned with the classification and description of our ideas, of the powers of the mind, and of the epistemic status of our beliefs, though it should be recognized that he also chronicled the temporal history of the human mind, insofar as he reconstructed the genesis of our ideas and the progressive development of our intellectual faculties. With this shift to the natural historical mode came a more systematic appeal to comparative evidence. Because he rejected the Cartesian view that beasts were mere machines, Locke compared the operations of the human mind with those of the higher animals in order to ascertain the precise differences between them. He also considered the mental states of idiots and madmen in an effort to illuminate the normal functioning of the mind. Finally, Locke cited the evidence of the 'whole course of Men in their several Ages, Countries, and Educations' in support of his 'history' of the human mind, and he invoked the evidence of language to illustrate how the mind operates.[13] Locke thereby offered a means of transcending the limitations of the introspective method, and opened up new perspectives on history, anthropology, and the comparative study of languages which were to be explored by enlightened savants in Scotland and elsewhere during the eighteenth century.

Locke's natural history of the mind proved to be controversial in the decades following the publication of the *Essay*. For example, in his *Lettres philosophiques* (1732), François-Marie Arouet de Voltaire remarked that 'after so many deep thinkers had fashioned the romance of the soul, there came a wise man who modestly recounted its history: Locke has unfolded to man the nature of human reason as a fine anatomist explains the powers of the body'.[14] D'Alembert later echoed Voltaire's enthusiasm, but readers such as the Oxford logician Edward Bentham criticized Locke (and others) for devoting too much attention to the analysis of the 'nature of our Souls, our Sensations, our Passions and Prejudices', which Bentham claimed was more properly 'a part of the natural History of Man, rather than a part of Logick'.[15]

Scottish thinkers, however, tended to endorse Voltaire's assessment, and emulated Locke's methodological example. For example, William Duncan (who taught at Marischal College from 1752 until his death in 1760) wrote in his highly influential

Elements of Logick (1748) that the science of logic 'may be justly stiled the History of the human Mind' (p. 4), and his fellow Aberdonian Thomas Reid (1710–96) likewise assumed the Lockean role of the natural historian of the mind in his classes at King's College, Aberdeen. In an outline of his philosophy course dating from 1752, he described his pneumatology lectures as constituting 'the History of the Human Mind and its Operations & Powers', and Reid later discussed the methodological problems involved in attempting to chart such a history in the introduction to his *An Inquiry into the Human Mind, on the Principles of Common Sense* (1764). Reid urged the need for an accurate 'anatomy of the mind', which would exhibit the original powers, perceptions, and notions of the 'thinking principle' within us. Yet he recognized that mere introspection could not reveal these powers and perceptions, since 'before we are capable of reflection, they are so mixed, compounded and decompounded, by habits, associations and abstractions, that it is hard to know what they were originally' (p. 10). Consequently, he maintained that we must analyse the mind, in order to enumerate 'the original powers and laws of our constitution' and to explain 'the various phenomena of human nature' (p. 12). Thus Reid's projected anatomy of the mind, as sketched in the *Inquiry*, can be seen as a fusion of the description and classification of the 'furniture of the human understanding' (p. 528) with Newton's method of analysis.

It is arguable too that Reid deployed a natural historical approach in the lengthy third essay of his *Essays on the Active Powers of Man* (1785), wherein he catalogued the different 'principles of action' of the human mind. Earlier in the century, British philosophers like George Turnbull (who taught Reid at Marischal College in the 1720s) and Bishop Joseph Butler (whom Reid read closely at a formative stage in his intellectual development), routinely combined the perspectives of the moralist and the natural historian.[16] Reid's classification of the various principles of action and his use of evidence drawn from the animal kingdom to illustrate the workings of those principles shared by man and beast was similar to that of his predecessors, although it must be said that Reid's utilization of such evidence was much more systematic.

Outside of Aberdeen, one Scot who might profitably be seen in this Lockean context is David Hume (1711–76), who, like Reid, conceived of himself as an 'anatomist of the mind'. While most scholars tend to regard Hume's remarks in the introduction to the *Treatise of Human Nature* (3 vols., 1739–40) concerning the use of the experimental method in the science of man as *the* statement of his methodological principles, Hume's comments on the analysis of the mind in the *Enquiry concerning Human Understanding* (1748) are no less revealing. He argued that, because of the limits of introspection, it is very difficult to distinguish between the various operations and powers of the human mind, and consequently recommended that we should endeavour to classify accurately our mental faculties. It may be that Hume's methodological statements in the *Enquiry* reflect his adoption of the natural historical approach exemplified in Locke's *Essay*; if so, his *The Natural History of Religion* (1757) can be seen (in part) as a product of this approach.[17]

The French connection

Like the works of other naturalists, Georges-Louis Leclerc, comte de Buffon's *Histoire naturelle* (1749–1804) served as a fund of factual information about the various modes of life of the human species in different parts of the globe and about mankind's place in the animal kingdom. But Buffon's theoretical interpretations of the facts found less favour. One particularly contentious aspect of the *Histoire* was his account of racial differences. Echoing the Baron Montesquieu's famous dictum that the 'empire of climate is the first of all empires', Buffon maintained that climatic influences were responsible for the production of racial types and for what he considered to be the degenerate state of the flora and fauna of the New World.[18] Following Hume's critique of Montesquieu in 'Of National Characters' (1748), Scottish men of letters sought to counter Buffon's theory of race by appealing to 'moral' causes of variation such as the state of society. Thus Lord Kames argued that differences in national or racial temperaments were attributable primarily to social factors. Yet Kames also contended that God was the only cause powerful enough to produce the physical differences among human races, and hence asserted each race had been specially created. However, Kames's idiosyncratic theory was no more acceptable to his contemporaries than than that of Buffon, and his speculations sparked off a heated debate which saw his opponents trying to steer a safe passage between the Scylla of environmental determinism and the Charybdis of polygenism, not least because they saw Kames's ideas as subverting the foundations of religion and morality.[19]

The Scots were equally critical of other parts of the *Histoire*. Writing to the fledgling *Edinburgh Review*, Adam Smith (1723–90) praised Buffon's style but dismissed his explanations of 'the formation of plants, the generation of animals, the formation of the foetus, [and] the development of the senses' as being 'almost entirely hypothetical' and occasionally unintelligible.[20] Although Smith seems not to have been apprehensive about the implications of Buffon's hypotheses, the Aberdonians Thomas Reid, Robert Eden Scott (1770–1811), and the natural historian James Beattie were highly critical of his ideas, for they perceived in the *Histoire* hints of atheism and materialism.[21] Such concerns notwithstanding, Reid, Scott, and Beattie were in different ways indebted to the *Histoire* for empirical information about humankind and the animal kingdom, as were many of their Scottish contemporaries. In a similar vein, Adam Smith incorporated data from Linnaeus' *Systema naturae* (1735), as well as his own field observations, in his essay on the human senses published posthumously in 1795.[22]

But the Scottish moralists were indebted to natural historians for more than just their 'facts'. Adam Ferguson's whole approach to the study of man and society was grounded in the methods of the natural historians. In his moral philosophy lectures, Ferguson (1723–1816) insisted that 'before we can ascertain rules of morality for mankind, the history of man's nature, his dispositions, his specific enjoyments, and sufferings, his condition and future prospects, should be known'.[23] Consequently, he believed that moral philosophy should be based on 'Pneumatics, or the physical history of mind', which he divided into the history of the species and the history of the individual, the former encompassing the description of man's physical nature as well as the modes of man's social and political existence, and the latter the description of the faculties and powers of the mind (pp. 15–16). Ferguson likewise identified himself as a natural historian of man in his *An Essay on the History of Civil Society* (1767), a work which was clearly intended as a contribution to the history of the human species. Ferguson invoked the methodological practices of natural history, and criticized the theories of the state of nature advanced by Thomas Hobbes and Jean-Jacques Rousseau on the grounds that they had substituted hypotheses for historical facts. Ferguson insisted that their methods of reasoning were illicit, and argued that our notions of human nature should be based solely on the historical record which, he claimed, demonstrated that man is naturally a sociable animal.

Ferguson's natural historical view of man received its fullest exposition in the revised and enlarged version of his lectures published in 1792. Even though Ferguson distinguished between the history of man and the science of morals, he nonetheless reiterated his conviction that the former served as the foundation for the latter.[24] Consequently, as a basis for his review of the principles of politics and morals he outlined fairly detailed histories of both the human species and the human mind, in which he traced analogies between man and the rest of the animal kingdom, while at the same time emphasizing man's uniqueness as a self-conscious, purposive being. Ferguson also underlined the necessity of compiling such histories, for he maintained that introspection had to be coupled with the history of the species 'as it may be observed by

Figure 12.1 John Kay's caricature of Lord Kames, the advocate and historian Hugo Arnot, and Lord Monboddo.

any indifferent spectator' in order to reveal the secrets of human nature (vol. I, p. 49). Moreover, he explicitly recommended the use of the descriptive and classificatory methods of the natural historian in the science of the mind. According to Ferguson, the mind was no different from any other constituent of the creation when considered 'historically', and its operations could be classified in much the same way as any other natural object. Thus, while Ferguson may have been a moral preacher in the classroom, we should not neglect the fact that in both his lectures and his published works he also regarded himself as a natural historian of human nature, and saw moral philosophy as being rooted in the natural history of man.

Like Ferguson, John Millar (1735–1801), Lord Kames (1696–1782) and James Dunbar (d.1798) all saw themselves as contributing to the 'history of the species'. Millar stated that his work on the origin of rank in civil society was 'intended to illustrate the natural history of mankind in several important articles', and he made adroit use of the four stages theory of history to classify the types of human societies discernible in the annals of history and the accounts of travellers.[25] Kames's *Sketches of the History of Man* (1774) grew out of his desire in the 1740s to write 'a natural history of man' in which he hoped to chart the 'History of the Species, in its progress from the savage state to its highest civilization and improvement'.[26] Consequently, in the *Sketches* Kames described the similarities and differences between man and the rest of the animal kingdom, and narrated the progress of human society, social institutions such as property, government, manners, the useful and fine arts, the moral sciences, and, significantly, the status of women. In effect, Kames collapsed Bacon's weak distinction between natural and civil history, and treated the whole of man's past as the subject of the natural history of humankind. Furthermore, Kames's *Sketches* demonstrate that for most enlightened Scots the history of the species illustrated the workings of Divine Providence. Moral concerns likewise shaped Dunbar's *Essays on the History of Mankind in Rude and Cultivated Ages* (1780), which focused on the improvement of the human species following the formation of the first primitive societies in order to 'vindicate the character of the species from vulgar prejudices, and those of philosophic theory' (p.i).

The paradoxical Rousseau

Another formative influence on the natural history of man was Jean-Jacques Rousseau (1712–72), whose works were quickly taken up and debated by the Scottish literati.[27] As Adam Ferguson's remarks (referred to above) on Rousseau's concept of the state of nature suggest, at least some Scottish readers regarded the *Discours sur l'origine et les fondemens de l'inégalité* ('Discourse on . . . Inequality') (1755) as a natural history of man, charting the 'progress' of our species from innocent rudeness to corrupt refinement. Given the complexities of the *Discours* such a reading may seem problematic, but Rousseau's unequivocal assertion that he was writing the true history of the species, his use of evidence drawn from travel accounts and from natural historians such as Buffon, and his speculations on man's relations with apes and our earliest physical attributes all serve to underline the natural historical dimension of the *Discours*. Moreover, the significance of the *Discours* cannot be underestimated, for Rousseau showed how the evolution of mind and the successive stages of society were interconnected, and thereby indicated that the study of the history of our species provided the key to unlock the secrets of human nature.

The most sympathetic and systematic Scottish response to Rousseau's second discourse came in Lord Monboddo's massive work on the development of language. Like Rousseau, Monboddo maintained that both society and language were human artefacts, and that life in society was a prerequisite for the invention of the spoken (and written) word. But whereas Rousseau could not see how there could be society without language, Monboddo appealed to the evidence of natural history, which he thought showed that animals formed primitive societies even though they lacked speech. To account for the origins of language, Monboddo was obliged to reconstruct the history of man from his largely solitary beginnings to his life in civil society, 'collected from facts, in the same manner as we collect the history of any other animal'.[28] Moreover, the history of our acquisition of language was for Monboddo the history of human nature, insofar as he held that speech was the distinguishing characteristic of the human species. He said that his work traced the progress of humankind from the '*birth* of human nature . . . to its state of *maturity*' (vol. I, p. 2), and, following Rousseau, drew attention to the beneficial or harmful effects of the different modes of man's social existence on his intellectual and physical character.

Rousseau's vision of the corruption wrought by the civilizing process was also taken up by John Gregory. Gregory affirmed that civilization created a 'softness and effeminacy' of manners and a 'debility and morbid sensibility of the nervous System, which lays the foundation of most of our diseases, and deprives us at the same time of the spirit and resolution to support them'.[29] In particular, Gregory targeted the harmful effects of 'modern luxury' on the health of children, and urged his readers to follow the dictates of nature and common sense when caring for the young. For Gregory, as for Rousseau, rearing children in a 'natural' way was of paramount importance, because it promised to be one of the most efficacious antidotes for the ills of civilized society.

Gregory went on to issue a more extreme attack on the evils of commercial societies in a preface which he added to later editions of the *Comparative View*. He held up the stage of society depicted in MacPherson's *Works of Ossian* (1765) as the happiest, and lashed out at the moral malaise engendered by the pursuit of wealth. According to Gregory, humankind cannot be truly happy in the savage state primarily because 'the nobler and more distinguishing principles of human Nature lie in a great measure dormant'. Consequently, the savage is like 'a beast of prey [who] passes his time generally in quest of food, or in supine sloth' (p.iv). But it is in the next transitional stage of society, as described in *Ossian*, that humankind is truly happy, for Gregory claims that man here retains the full vigour of his body and his animal functions, possesses simple manners and warm social affections, acts as a disinterested patriot, cultivates the expressive arts as opposed to the rational sciences, and has few wants. However, this stage 'seldom lasts long'; the few who hold power soon abuse it, thereby unleashing the dangerous passion of ambition, and the progress of the mind coupled with foreign trade leads to moral corruption through the creation of new pleasures and wants. Gregory believed that in commercial polities money becomes 'the universal idol to which every knee bows', and to which men sacrifice their virtue, religion, patriotic feelings, and health. He emphasized that the pursuit of wealth placed human nature in 'the most unhappy state in which it can ever be held', because 'the constitution both of body and mind becomes sickly and feeble, unable to sustain the common vicissitudes of life without sinking under them, and equally unable to enjoy its natural pleasures, because the sources of them are cut off or perverted' (pp.v–xiii).

Following his move to the Glasgow Chair of Moral Philosophy in 1764, Thomas Reid explicitly addressed the argument of Rousseau's *Discours sur . . . l'inégalité* in his lectures on 'the culture of the human mind'. Given Reid's emphasis on the crucial role of education in the formation of polite, virtuous, and godly citizens, he necessarily had to refute Rousseau's depiction of the state of nature and of the earliest forms of human society. To do so, Reid turned Rousseau's principles and methods of analysis against him. Starting from Rousseau's notion that human nature is perfectible, Reid argued that it is a law of human nature that the mind can be improved, and that this improvement can only be brought about by culture. Reid then sketched out a history of the development of our mental powers, modelled on the *Discours*, in which he argued that 'wild men' and savages living in primitive societies are unable to benefit fully from humankind's natural capacity to learn because their largely solitary habits prevent them from acquiring the use of artificial, as opposed to natural, language. Without an artificial language, Reid maintained, savages cannot employ the rational or moral faculties of the mind, and hence the cognitive and emotional horizons of the primitive state were extremely limited. By contrast, Reid believed our life in society promoted the acquisition of the primary vehicle of human improvement, namely artificial language. He affirmed that the higher powers of the human mind only begin to evolve once we live in a sufficiently advanced social state, wherein we finally start to form ethical concepts, refine our knowledge of nature, and experience true happiness. Because of the manifest disparity between the mental worlds of the savage and civilized states, Reid dismissed Rousseau's condemnation of the civilizing process as an empty paradox, and insisted that we must look to the advance of society to nurture our intellectual and active powers.[30]

Conclusion

Reid's response to Rousseau well illustrates a number of key features of the natural history of man in the Scottish Enlightenment. First, Reid mobilized natural historical evidence drawn from the three kingdoms of nature and information drawn from travel accounts to substantiate his view of human nature. Secondly, Reid was attentive to the comparative anatomies of the minds of animals and of man, in order to establish their distinguishing characteristics. Thirdly, Reid provided an account similar to Rousseau's of the effects of different stages of society on the development of human nature. Reid was interested in the ways in which the various stations in life produced markedly different manners and moral attributes, and thus shared the common Scottish preoccupations with the social formation of the human mind, and the issue of moral culture. Rousseau's historical diagnosis of the moral ills arising from the advent of civil society meant that the cultivation of virtue required an understanding of the natural history of man and society. Scottish moralists like Reid responded by developing moral systems which rested on natural histories of the human species, modelled on the practices of the natural historians of the vegetable and animal kingdoms. Reid's natural history of man was, therefore, a characteristic Scottish blend distilled from the writings of assorted naturalists and voyagers, as well as the texts of Bacon, Rousseau, Locke, and Buffon.

Further reading

Bryson, Gladys, *Man and Society: The Scottish Enquiry of the Eighteenth Century* (Princeton, 1945).

Duchet, Michèle, *Anthropologie et histoire au siècle des lumières: Buffon, Voltaire, Rousseau, Helvétius, Diderot* (Paris, 1971).

Glacken, Clarence J., *Traces on the Rhodian Shore: Nature and Culture in Western Thought from Ancient Times to the End of the Eighteenth Century* (Berkeley, 1976).

Greene, John C., *The Death of Adam: Evolution and its Impact on Western Thought* (Ames, IA, 1959).

Meek, Ronald L., *Social Science and the Ignoble Savage* (Cambridge, 1976).

Pagden, Anthony, *European Encounters with the New World: From Renaissance to Romanticism* (New Haven, 1993).

Rousseau, G. S. and Porter, Roy (eds.), *Exoticism in the Enlightenment* (Manchester, 1990).

Schiebinger, Londa, *Nature's Body: Gender in the Making of Modern Science* (Boston, MA, 1993).

Vyverberg, Henry, *Human Nature, Cultural Diversity, and the French Enlightenment* (Oxford, 1989).

I3 The natural history of the earth

In the natural history museums of London, Berlin, Vienna, New York and many other cities, there are always substantial displays of minerals, rocks and fossils. These public exhibitions often date back to the eighteenth century and derive from still older collections of curiosities. Such cabinets were a common feature of courtly furnishings; they served the combined functions of display, entertainment and improvement. Gemstones and other visually appealing minerals, rocks and striking natural productions – together with countless works of art – fascinated the wealthy and powerful.

Following the great spiritual awakening of the Renaissance, an interest in minerals and other productions of the earth's crust began to grow in educated circles. It was not only collections of rare specimens and curiosities that gave rise to diverse interpretations. Conceptions of the nature of the earth, its origin and destiny, were charged with theological meaning. Moreover, the increasing commercial importance of mining made the possession of accurate knowledge about ores and their natural distribution essential. Even the highest levels of society took a great interest in natural history, as a comment published in Ulm in 1760 reveals:

Natural history, as is well known these days, is the particular science which is both universally beloved and almost universally practiced . . . Nowadays even emperors, kings and princes consider it so little opposed to their supreme and high dignity to occupy themselves with natural history, that they number it amongst their important sovereign duties to be connoisseurs, protectors and encouragers for those who devote themselves to the knowledge of nature and who make new and generally useful discoveries therein.[1]

This was especially true in the case of minerals, rocks and other natural products of the earth, as well as for mining.

Attention was paid not just to attractive, curious or useful qualities. Scholars also began to collect natural productions in a systematic manner: to study, compare and describe, and, on this basis, to classify. They created a multiplicity of systems that would encompass nature into an order which could be surveyed.

The mineral kingdom and mining in natural history

Reporting and recording observations provided the empirical basis for the systems that were elaborated for inorganic nature. Those features which could be experienced directly through the senses (colour, transparency, lustre, form, weight and smell) were of principal significance in grasping the distinguishing characters of minerals.

The Swedish scholar Carl Linnaeus (1707–78) was celebrated for his classificatory systems. As early as 1735, the year of publication of his *Systema naturae*, Linnaeus concerned himself with the mineral kingdom, dividing it into *Petrae* (rocks), *Minerae* (minerals) and *Fossilia* (petrifactions), which were then further subdivided. In the many later editions of the *Systema*, he expanded the treatment of minerals. Linnaeus attempted to show that minerals were natural kinds akin to those found in the animal and plant kingdoms. However, approaches to classification derived from chemistry and mining proved more successful, not least because the Linnaean sexual system, so fruitful in botany, could not be applied to minerals.[2]

Johann Gottschalk Wallerius (1709–85), professor of chemistry and pharmacy at the University of Uppsala, also produced a system of minerals in 1747. In a subsequent elaboration of this system in 1768 the author presented his reflections on the distinctive characters and foundation for such classifications. He distinguished external from internal characters, and, in natural historical style, he took external qualities such as colour, form, taste, smell, uses and occurrence as criteria for classification. If these features still yielded an uncertain and incomplete picture, reactions to fire or chemical reagents could also be tested, thereby revealing internal characters which could then be used to classify the mineral. Wallerius's subsequent revision of his system (1778) not only expanded his earlier work, but made wider use of chemical criteria for division. Swedish chemists such as Axel Cronstedt (1722–65) and Torbern Bergman (1735–84), who succeeded Wallerius at Uppsala in 1767, extended this approach. Chemical methods were beginning to penetrate regions of natural history.[3]

Against the background of these interpretations by Swedish scholars, Abraham Gottlob Werner, in Saxony, developed an approach which he applied first to minerals, then to geological questions. He reached a broad audience and proved extraordinarily influential. Werner was born in 1749 to a family which had been connected with metallurgy for generations. Even as a child he had been interested in the nature of the earth, and, having studied

ABRAHAM GOTTLOB
WERNER.

Figure 13.1 Abraham Gottlob Werner developed a consistent method of mineral determination based on natural history criteria; for decades he taught geology according to the tenets of Neptunism at the Mining Academy at Freiberg.

mining and metallurgy at the Mining Academy of Freiberg in Saxony, he went on to follow a course in law, languages and natural history – especially mineralogy – at the University of Leipzig. In 1775 he obtained the Chair of Mining and Mineralogy at Freiberg, where he worked productively until the end of his life in 1817.[4]

Werner's book *Von den äusserlichen Kennzeichen der Fossilien* ('On the External Characters of Minerals', Leipzig, 1774) brought him widespread recognition. In this work he presented a technique for identifying minerals through the human senses. These were characters such as crystal form, external surface, external lustre, internal lustre, fracture, form of fragments, transparency, streak, colour, hardness, flexibility, adhesion to the tongue, and sound. He described individual characters in a minutely detailed manner, and scrupulously subdivided them in such a way as to maximize their utility for mineral identification. For the colour red alone, Werner distinguished thirteen different varieties: morning or aurora red, hyacinth red, brick red, scarlet red, cochineal red, blood red, copper red, carmine red, crimson red, peach-blossom red, flesh red, cherry red and brownish red. His student Dietrich Ludwig Gustav Karsten (1768–1810) even went so far as to distinguish twenty-nine different variations of red.

Through the diagnosis of specific combinations of qualities, based on external characters, mineral type could be recognized rapidly and by relatively simple means. Werner became very famous, and was hailed as the supreme master of a method of identification unmatched in mineralogy for some time. With this work, he provided a perfected version of the natural historical method of mineral identification, and, simultaneously, a methodology for mineralogy as a discipline, which began to emerge as a science distinct from natural history.

In the course of more than forty years as a mineralogist, Werner published several versions of his classification, which increasingly came to be based upon chemical criteria. By 1777 Johann Friedrich Gmelin (1748–1804) could list more than twenty-five mineral systems, produced between 1650 and 1775 by different European authors. One might add numerous accounts of crystals, including several by the French mineralogist Jean-Baptiste Romé de l'Isle (1736–90), who compiled them while cataloguing the cabinets of wealthy patrons. Each of these accounts, in their own way, indicated the nature of natural history, which aimed primarily to record and systematize natural objects, initially without a historical and chronological perspective.

These major developments in the natural history of the earth drew on centuries of experience derived from intensive mining practices. The mining centres in Sweden, the Tyrol, Transylvania and Thuringia – and notably those in Bohemia and Saxony – were particularly significant. As early as the sixteenth century, a variety of small books on mining and assaying were published in central Europe. They supplied the economic, technological and natural knowledge necessary to obtain minerals. In his writings, the Saxon physician Georgius Agricola (1494–1555) depicted minerals and phenomena such as the erosive effects of water, volcanoes and earthquakes; his *De re metallica* (1556) provided encyclopaedic descriptions of metallurgy and mining technology with impressive illustrations. The book remained a classic resource for mining in

Europe, and even in China and South America, for two centuries.

Many authors prepared works on mining and metallurgy, which dealt with prospecting for valuable mineral deposits, mining boundaries, extraction procedures, water management, weather machines, ore mining, smelting and metallurgy. Some of the leading scholars of the era concerned themselves with similar questions, since mining and metallurgy were the most important areas of commercial production throughout the period. At the end of the seventeenth century Gottfried Wilhelm Leibniz (1646–1716) drew up extensive plans for wind-driven mechanisms and the management of water in the mines of the Harz mountains. During the Baroque period full descriptions of mountain sites and mining technology were produced, such as *Magnalia dei in locis subterraneis* (1727) by the physician Franz Ernst Bruckmann (1697–1753) of Brunswick, which described over 1,600 mines throughout the world. In the same period the practical engineer Jacob Leupold's multivolume *Theatrum machinarum* (1725) embodied the results of his extensive experience drawn from mining production. Mining provided the foundations for much contemporary technological knowledge.

Mining practices also suppled vital information concerning the properties and distribution of minerals, which needed to be recognized quickly and unmistakably. For miners, knowledge about ores and the rocks in which they occurred was very valuable – especially knowledge about lodes from which precious and semi-precious metals could be extracted. Systematic texts described the different kinds of lodes; their horizontal and vertical distribution; their relations with the surrounding rocks; and ways of recognizing them in nature.

Knowledge about the earth drawn from mining had thus achieved considerable breadth by the eighteenth century. In most countries it had become essential to the discovery of ores and other valuable minerals. The centralized state bureaucracies of the Enlightenment recognized that mining officials required not only legal and engineering knowledge, but an understanding of the natural history of the earth. The universities offered only a very incomplete training in this regard, and one, moreover, that was wholly theoretical.

The growing need for training was filled by mining academies founded in the last third of the eighteenth century. On the heels of the success of institutions offering instruction in the German lands, the Austrian empire, Scandinavia, Russia and South America, the first academic chair of mining was founded in 1762 at the University of Prague. Mining academies followed in rapid succession in many different countries: the Freiberg school in 1765; Schemnitz, Austro-Hungary in 1770; Berlin, Prussia in 1770; St

RERVM METALLICARVM SVBSIDIVM

ACADEMIA
FREIBERGENSIS
D. XIII NOV MDCCLXV FVNDATA.

Figure 13.2 The mineral collection of the Freiberg Mining Academy. The study of mineral specimens was a major part of the teaching programme of the technical institutes founded in the second half of the eighteenth century. Title-page from *Bericht vom Bergbau* (Bergakademie zu Freyberg, 1769).

Petersburg, Russia in 1773; Almaden, Spain in 1777; Paris, France in 1783; and Mexico City in 1792. This was the first generation of technical institutes of higher education to coexist with the universities, and in which the curriculum centred on the natural history of the earth. It was in this way, through the burgeoning demand of governments for trained officials, that disciplines such as mineralogy and geognosy acquired an institutional basis for the first time. But there were also other sites within which knowledge about the earth's nature was being generated.

Physicotheology, mystery and natural history

Natural history permeated the natural theology or physicotheology of the eighteenth century. The study of nature was designed to render the scale of God's natural creation comprehensible and to affirm Christian faith. Religious tracts, treatises and sermons featured countless discussions of animals and plants, water and rocks, thunder and lightning, the universe and the earth, treated in the strictly natural historical sense. Such phenomena could, for the most part, be experienced through the senses. Authors described them in great detail, drawing together many earlier facts. Particularly in the first half of the eighteenth century, publication and wide distribution of books and pamphlets made it possible for knowledge about nature to spread to many parts of society, even though natural history played a role subordinate to medicine or philosophy in university teaching. In a period when 'enlightened' thinking was becoming widespread, natural history (especially in the Protestant countries of northern and central Europe) could be used to provide proof for Biblical claims, and thus confute atheists and sceptics.

Thomas Burnet's *Telluris theoria sacra* (1681) was a key publication, and after translation into English and German it remained widely read throughout the following century. With great conviction and a captivating style, Burnet (1636–1715) explained the formation of the earth, including its chief alterations and current circumstances, from the perspective of contemporary natural knowledge. In presenting a physical interpretation of Biblical claims, however, he laid himself open to criticism from other scholars and theologians. Within a short space of time the English natural philosophers William Whiston (1667–1752) and John Woodward (1665–1728) produced analogous theories of the earth, which also achieved a wide distribution in Europe. In his *Essay toward a Natural History of the Earth* (1695) Woodward explained that the universal Deluge described in the Bible had produced the layers of the earth's crust and the fossilization of plants and animals.[5] In doing so, he drew on earlier accounts of the cause of these natural phenomena, and his work gave them great currency during the eighteenth century. Such 'theories of the earth' demanded a historical and genetic perspective, which began to take shape in the decades around 1700.

As early as 1669 the Danish scholar Niels Stensen (1638–86) had suggested that mountains had not always existed in their present form.[6] From the ordering of strata in Tuscany, he reconstructed the sequence of their formation and, in this way, developed a historical perspective with regard to changes in the earth's nature. At the same time, the London natural philosopher Robert Hooke (1635–1703) found that different animals and plants seemed to be present in different parts of the strata sequence. What he believed to be fossilized organisms could potentially be used to produce a chronological order of rocks in the same way that coins established a chronology of human history. Previous accounts of natural objects had been essentially in the form of stories, in which their position in a chronological sequence had no part.

In 1700 naturalists continued a fierce debate over the nature of puzzling objects found in the earth, especially those with apparently organic forms. Some interpreters sought signs in natural productions, such as handwriting in 'graphic granite', 'thunderstones' shot from the heavens and *Ruinenmarmor* (marble with pictures of ruined cities in it). Unusually shaped rocks were supposed to result from a specific natural force or *vis plastica*, which had produced 'sports of nature' (*lusus naturae*), a concept dating from the Middle Ages that remained widespread among scholars. In these circumstances, forgeries of 'figured stones', which repeatedly appeared in prints of the first half of the eighteenth century, made consensus about a correct interpretation hard to achieve.

A striking example was the 'Würzburger Lügensteine' (lying stones of Würzburg), which appeared in Ebelstadt, to the southeast of Würzburg in Germany, in 1725 (Figure 13.3). These figured stones, resembling animals, plants, letters, stars and so on, were described in Johann Bartholomeus Beringer's *Lithographiae Wirceburgensis* (1726).[7] The stones, however, were fakes constructed by academic colleagues to ridicule Beringer (*c.*1667–1738), who was known to be fascinated by fossils. Beringer did not believe that all these unusual figured stones were necessarily the remains of once-living creatures, and thought the whole phenomenon unclear, doubtful and inconsistent. By publishing his work on the stones, Beringer hoped to encourage other scholars to investigate and comment upon them. The nature of different figured stones was at the forefront of debates about the natural history of the earth.

The Zürich physician and natural philosopher Johann Jakob Scheuchzer (1672–1733) had studied a tremendous variety of different phenomena relating to the earth, and described their origins in line with the physicotheological tradition. Originally, he too had interpreted fossils as the results of an unusual power which had shaped them into figured stones by chance. But after reading Woodward's *Essay*, he began to explain them as remains of living beings which had perished in the Deluge. In a short treatise of 1708, he portrayed different fish lamenting their death in the Deluge (Figure 13.4).[8] Petrifactions could thereby be interpreted as impressions of once-living animals and plants, even if they were engendered within a brief period by a single event. Scheuchzer also explained many other changes on the surface of the globe as products of the Biblical flood.

It was vital for natural history that Scheuchzer's new interpretation stimulated him to collect plant and animal fossils, and to have descriptions and engravings of them printed. His *Herbarium diluvianum* first appeared in 1709, and by 1723 it was already into a third edition. The discovery of a fossil skeleton in a chalk quarry near Oeningen, in 1726, was an important event for Scheuchzer. He held the remains to be the leg bones of a wicked human, whose

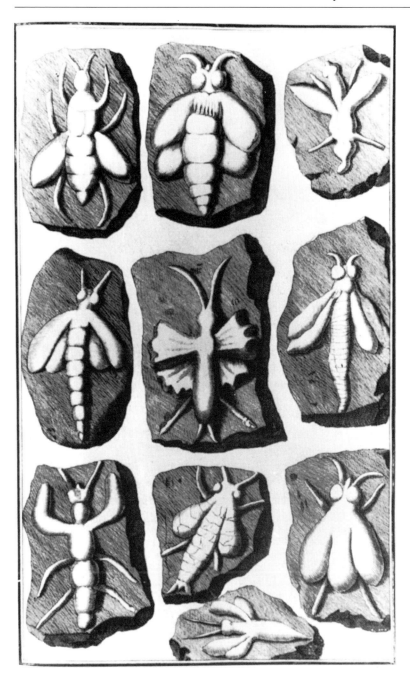

Figure 13.3 Some of the strange fossil specimens illustrated by J. B. A. Beringer in his *Lithographiae Wirceburgensis* (Würzburg, 1726). These proved to be fakes planted by jealous colleagues at the University of Würzburg.

sins had been punished by the coming of the Deluge. Scheuchzer portrayed the fossil in natural historical terms, but gave its creation a moralizing Scriptural interpretation. It was not until a century later that Georges Cuvier reclassified this fossil as the remains of a giant salamander, which he named, in honour of its discoverer, *Andrias Scheuchzeri*.[9]

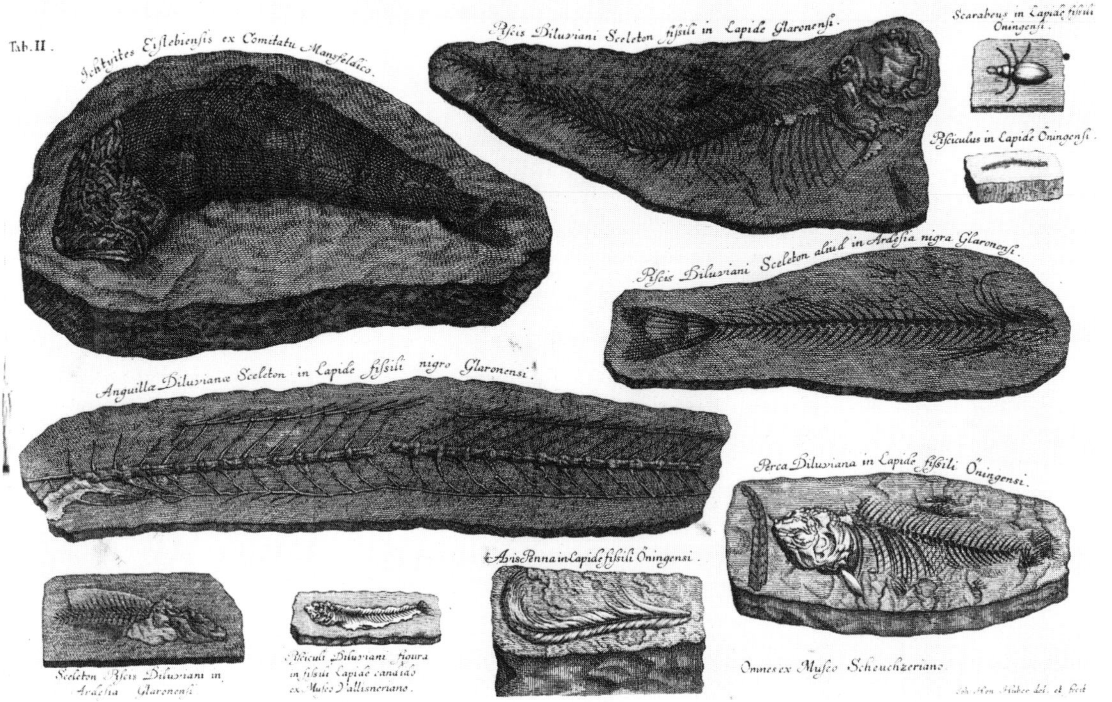

Figure 13.4 Portraits of fossils by Johann Jakob Scheuchzer in his work on 'The Different Fish that Died in the Great Flood'. From Scheuchzer, *Bildnissen verschiedener Fischen und dero Theilen, welche in der Sündfluth zu Grund gegangen* (Zürich, 1708), Table II.

The wider religious culture of this period contributed substantially to knowledge of nature. It was for religious reasons that Scheuchzer encouraged an interest in fossils, developed the depiction of strata sections, carried out regular measurements of air pressure with a barometer, repeatedly observed the glaciers of his native country, refuted the view that mountain crystals had formed from snow, and, as early as 1705, climbed to the summit of St Gotthard, which at that time some regarded as the highest mountain in Europe. In all, he earned himself a name for the study of the natural history of Switzerland, and furthermore provided an impulse for similar studies in other countries.

Scheuchzer published works which he conceived as natural historical explications of Biblical texts. He died before completing his comprehensive *Physica sacra*, which was published in four folio volumes in Augsburg and Ulm from 1731 to 1735. In 750 copper-plate engravings and more than 2,000 pages of text, he presented a wealth of observations and ideas which give an impressive conception of the permeation of natural history by religion.[10]

Further works in which the affirmation of religious faith

provided a goal for describing natural objects and circumstances were also produced during this period. In the German-speaking lands, Friedrich Christian Lesser (1692–1754), the evangelical pastor of Nordhausen, Thuringia, had a great interest in natural history. In 1732 he published his work on lithotheology, which he wished to be read as a natural history and religious study of stones. He wished to demonstrate the omnipotence, goodness and justice of the great Creator, to explain the claims of the Scriptures, and to rouse men to the admiration, praise and service of the supreme Deity. In this light he described minerals, rocks and fossils systematically from a natural historical standpoint, including their uses and misuses. In later editions of his lithology he claimed to have recognized a representation of Christ's crucifixion in an agate, and depicted it in his book as a miraculous natural form (Figure 13.5). This picture, evidently drawn by a human hand, was intended to confirm religious belief, although by the mid-eighteenth century such portrayals were increasingly criticized.

Figure 13.5 The 'image of Christ crucified on the cross'. As the lower scroll exclaims, 'such a wonderful vision has Nature made in an agate'. From F. C. Lesser, *Lithotheologie* (Hamburg, 1732), no. II. S. 294.

From the very first volume of his *Histoire naturelle* (1749) Georges-Louis Leclerc, comte de Buffon (1707–88), attacked the Christian conceptions of Burnet, Whiston and Woodward. His own deistic claims about the earth's age, based on experiments with cooling globes, soon came under fire from the Sorbonne's theology faculty, since they contradicted expert chronological studies of the Bible showing that the earth was 6,000 years old.[11] Soon after, the German philosopher Immanuel Kant (1724–1804) published his *Allgemeine Naturgeschichte und Theorie des Himmels* (1755), in which neither God nor the Bible directly underpinned earth history. In 1763 the Russian Mikhail Vasilievich Lomonosov (1711–65) also objected to the claim that the Deluge had led to petrified shells on the tops of mountains. Such views combated physicotheological ideas and treated human understanding as the only valid basis for ascribing meanings to natural phenomena. The methods that scholars adopted in their natural historical work increasingly embodied the aims of the Enlightenment.

Rocks, water and fire

Around the mid-eighteenth century new ways of thought began to become established, which, in particular, adopted the notion that there had been a genuine alteration in nature in the past. Although natural history had initially only consisted of stories in which static accounts of natural objects were given, observations involving a historical perspective could now also be made, so that a true history (in the modern sense) of nature could be produced. The view that the earth had undergone multiple transformations since the Creation or the Deluge, and thus had a history, became more common. Rocks and fossils could be repeatedly produced, and did not result from a single great flood. The age of the earth might be considered in terms of millions of years, and need not accord with the 6,000 years calculated from the Bible. This was a deep-rooted transformation in the conception of the natural history of the earth.

Extensive observations encouraged new possibilities for explanation. Debate about the essential nature of fossil objects became less important than their potential use for understanding the earth's past. Scholars of different countries made observations on mountain formation which yielded a global picture. Starting from accounts of different European regions, they began to develop a common language for representing the basic divisions of the upper parts of the earth's crust.

In 1786 Werner compiled the findings of many scholars into a single classification of rocks and mountains. He demonstrated a succession of formations, each consisting of beds or layers stacked on top of each other. By 'formations' (*Gebirge*), Werner meant rock complexes of the same substantial composition and mode of origin. These changed in character from upper layers (4) to lower ones (1); together they constituted the structure of mountains:

(4) *Alluvial rocks*: sand, gravel, clay, loam, peat, bituminous earth, etc.;

(3) *Volcanic rocks*: lava, ashes, pumice, tufa, sulphur, hot springs, etc.;

(2) *Floetz rocks*: limestone, sandstone, coal, grauwacke, chalk, salt, etc.;

(1) *Primitive rocks*: granite, gneiss, mica schist, clay slate, porphyry, basalt, serpentine, etc.

Werner related these formations to directed changes in conditions during the earth's past.

Similar but less elaborate conceptions of the structure and history of mountains were widespread in the mid-eighteenth century; the French diplomat Benoît de Maillet (1656–1738) suggested aspects of it in his famous *Telliamed* (1748). In a work published in Berlin in 1756 the mining administrator and academician Johann Gottlob Lehmann (1719–69) distinguished between the Primitive mountains, which he claimed had originated at the Creation, and Secondary mountains, which he ascribed to the Mosaic Deluge[12] (Figure 13.6). Above these lay a third class of rocks, which had been formed by volcanoes or massive flooding. The court physician at Rudolstadt, Georg Christian Füchsel (1722–73), described a succession of strata he had observed in Thuringia in analogous terms. At around the same time in Italy, the mining expert Giovanni Arduino (1714–95) drew upon the treatises of the physician Antonio

Vallisnieri (1715) and the cleric Lazzaro Moro (1740) in his researches into the composition, structure and development of mountains near Padua, Vicenza and Verona. Arduino divided them into Primitive, Secondary, Tertiary and Volcanic rocks. In Uppsala, Torbern Bergman's physical description of the earth in 1771 had produced a threefold division of the formations on the earth's crust. Peter Simon Pallas (1741–1811) came to broadly similar conclusions as a result of extensive travels in Russia.

These observations began to be presented not just verbally, but in strikingly visual terms. In 1756 Lehmann published a horizontal section for a sequence in Thuringia, and Füchsel brought together several of his conceptions in 1761 in a distribution map (Figure 13.7). And in the manuscripts of Arduino, sketches of sections (1758) represent phenomena from the natural history of the earth; these were a form of illustration that Moro had used in 1740 for describing strata. These pictorial genres later became preconditions for the development of stratigraphy as a part of historical geology, based upon knowledge about the character and layering of rock masses.

Many hypotheses were put forward concerning the history of the earth. Joining in the anti-clerical campaigns of the Enlightenment, many authors abandoned the aims and interpretative styles characteristic of the older forms of physicotheology. In the final two decades of the eighteenth century a pair of opposing theoretical concepts emerged as fundamental. Both 'Neptunism' and 'Plutonism',

Figure 13.6 Section of stratification at the Mansfeld Kupferschiefer, one of the first sections to be based on a specific locality. From Johann Gottlob Lehmann, *Versuch einer Geschichte von Flötz-Gebürgen* (Berlin, 1756).

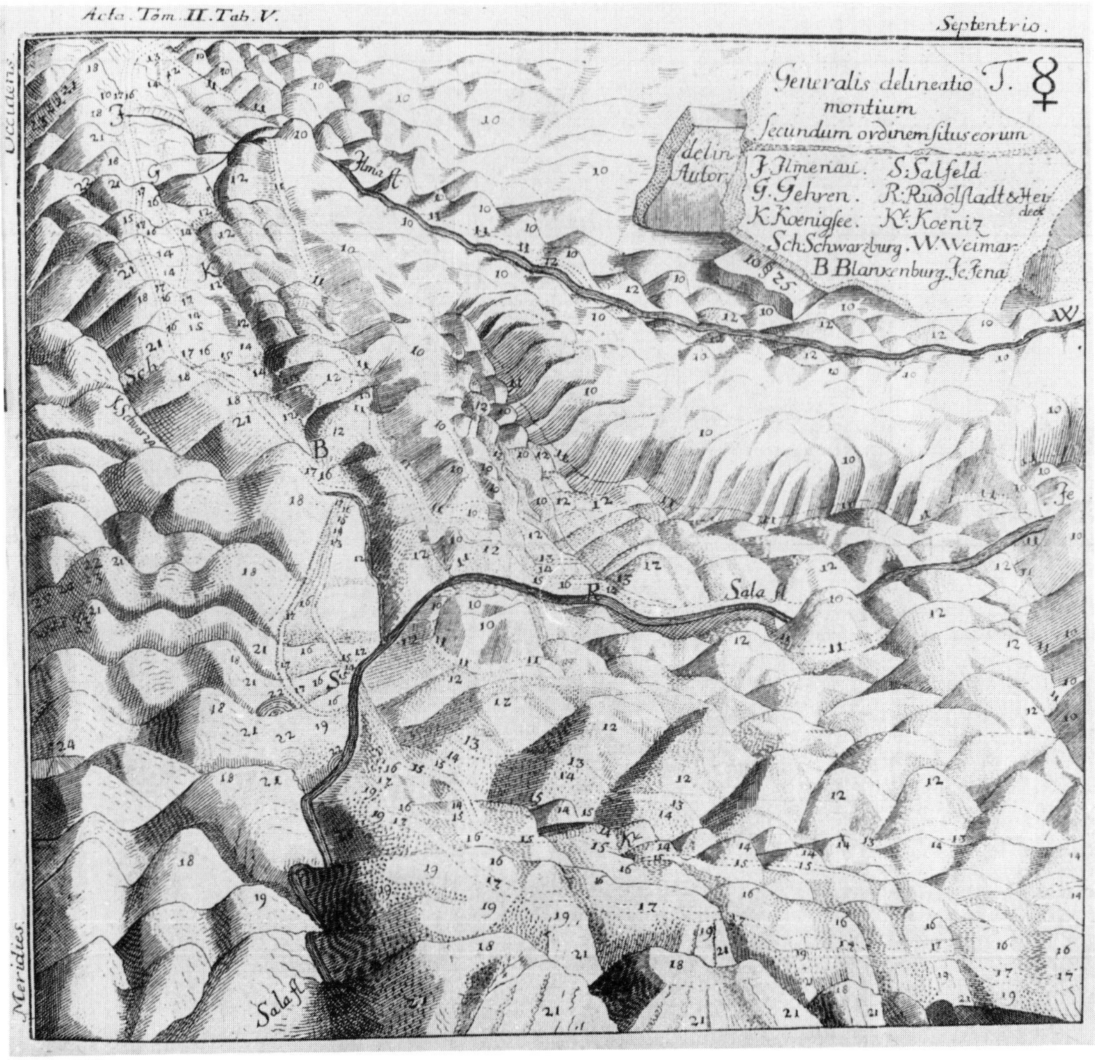

Figure 13.7 Map of Thuringia, published in 1761 by Georg Christian Füchsel. The distribution of rocks is indicated by a system of numerals. From Füchsel, 'Historia terrae et maris, ex historia thuringiae, per montium descriptionem', *Actorum Academiae electoralis moguntinae scientiarum utilium quae Erfordiae est, Erfordiae II* (1761), pp. 44–254.

as they came to be called, had roots in the traditional natural history of the earth, but also offered new perspectives. In both, the representation of the character of the earth's past was at stake, especially the causes of change.

Werner was one of the most influential supporters of Neptunism. He believed the whole surface of the earth to have been covered originally by a primitive ocean, and assumed that nearly all formations had developed in water or had been shaped by water. He proposed: 'The solid globe, insofar as we know it, was originally formed entirely from water . . . the work that has been done on it by fire is insignificant in comparison with the whole.'[13] Werner, a deist, did not explicitly invoke the Deluge, the Bible or God in his explanations.

Werner held that approximately one million years ago conditions were very different from the present. A great fog floated above the vast water surface, so that little sun reached it. Since there were no terrestrial masses, land plants and animals were absent. The water of this great ocean contained many chemical components and thus few or no organisms. These were the conditions for the development of the Primitive rocks, which were seen primarily as precipitates from solution, largely without mechanical sediments or fossil remains of plants or animals.

As the water level gradually began to descend, dry land emerged. On the dry surface processes of destruction occurred, producing debris and mechanical sediments. Salts were also deposited. Plants began to cover the land, and animals multiplied in the sea and on the continents. Under these conditions limestones, sandstones, bituminous sediments and coal were produced, and the formation of fossils increased. These circumstances were preconditions for the development of the Secondary or Floetz rocks in a second great stage of the earth's history.

Werner demarcated a further chapter in the earth's past by an increase in the mechanical sediments, to which the new phenomenon of volcanoes was added. Neptunists rejected deep-lying heat sources in the earth's crust as causes for volcanoes, and explained them by means of burning bituminous layers in the younger and higher layers near the earth's surface. These were the conditions for the formation of Volcanic rocks. The Alluvial rocks were the youngest, formed mostly from sand, lime and clay. This phase has lasted until the present, with its characteristic forces of terrestrial change.

Although many Neptunist views seem speculative, they helped to produce a picture of earth history with successive phases. Individual phases possessed particular characters, but there were gradations from one to another, so that each new phase built upon the preceding. Werner himself did not enter into detailed explanations based on chemistry, preferring to see processes of precipitation and consolidation in mechanical terms. His many followers, including the philosopher Henrich Steffens (1773–1845), the traveller Alexander von Humboldt (1769–1859), and the Scottish mineralogist Robert Jameson (1774–1854), greatly increased the role of chemistry. All the Wernerians based the development of the earth within the history of nature upon unchanging physical laws.

Whilst water and its operations were fundamental for the Neptunists, the Plutonists saw formations as caused above all by heat and fusion. James Hutton (1726–97), an independent gentleman in Scotland, was the leading supporter of Plutonism, and had worked out a detailed theory of the earth. He had studied medicine in Edinburgh, Paris and Leiden, but also had an interest in chemistry, philosophy, rural economy and, above all, geological issues. After travelling in the Netherlands, Belgium and France, he settled in Edinburgh and enthusiastically undertook excursions in Scotland. A man of the Enlightenment, Hutton founded a small society, the Oyster Club, together with the economist Adam Smith (1723–90), the chemist Joseph Black (1728–1819) and the mathematician John Playfair (1748–1819) and others. Over communal meals they discussed problems in natural philosophy. Unlike Werner, Hutton never taught, and his views spread primarily through printed works published between 1785 and 1795.[14]

In his *Theory of the Earth* (1795, vol. II, p. 355), Hutton claimed 'that subterraneous fire and heat had been employed in the consolidation of our earth, and in the erection of that consolidated body into the place of land'. High temperatures and fire were the decisive powers in the concretion of loose sedimentary masses into rocks, and the motor for the alterations in the earth's nature; the role of water was rejected. On this basis Hutton also explained the underlying processes in the earth's crust, processes which required vast spans of time (Figure 13.8). He started from the position that the earth possesses a solidified crust with enclosing shells of water and air above it. In the past the interchanges between these regions had produced a multitude of changes. There was a constant exchange of land and sea, which manifested itself in the erosion of continents and the formation of rock on the sea bed.

Hutton distinguished three stages in the terrestrial cycle. The first was typified by the habitable continents of the present surface of the earth. A second stage could be observed in the eroded masses of older parts of the crust, which had been washed into the ocean and there collected on the ocean floor. In the third stage, heat fused these sediments to form new rocks. Raised out of the water, these rocks replaced older parts of the crust which had been destroyed – only to be eroded themselves in due course. In Hutton's theory one continent followed another in an eternal circulation of nature. Earth history was thus an indefinitely continuing series of cycles, with heat and fusion as the essential causes of the genesis of new rocks and their uplift. Hutton described this system as part of a divinely instituted natural order.

Followers of Werner and Hutton clashed throughout Europe and America, particularly over the question of the means and manner of the origin of basalt. Whilst Neptunists saw basalt as a rock formed in water, Plutonists claimed to recognize its subterranean genesis through observations of its frequent association with volcanoes. The basalt controversy attracted intense public interest, and the great German poet Johann Wolfgang von Goethe (1749–1832) even portrayed it in the second part of his *Faust*.

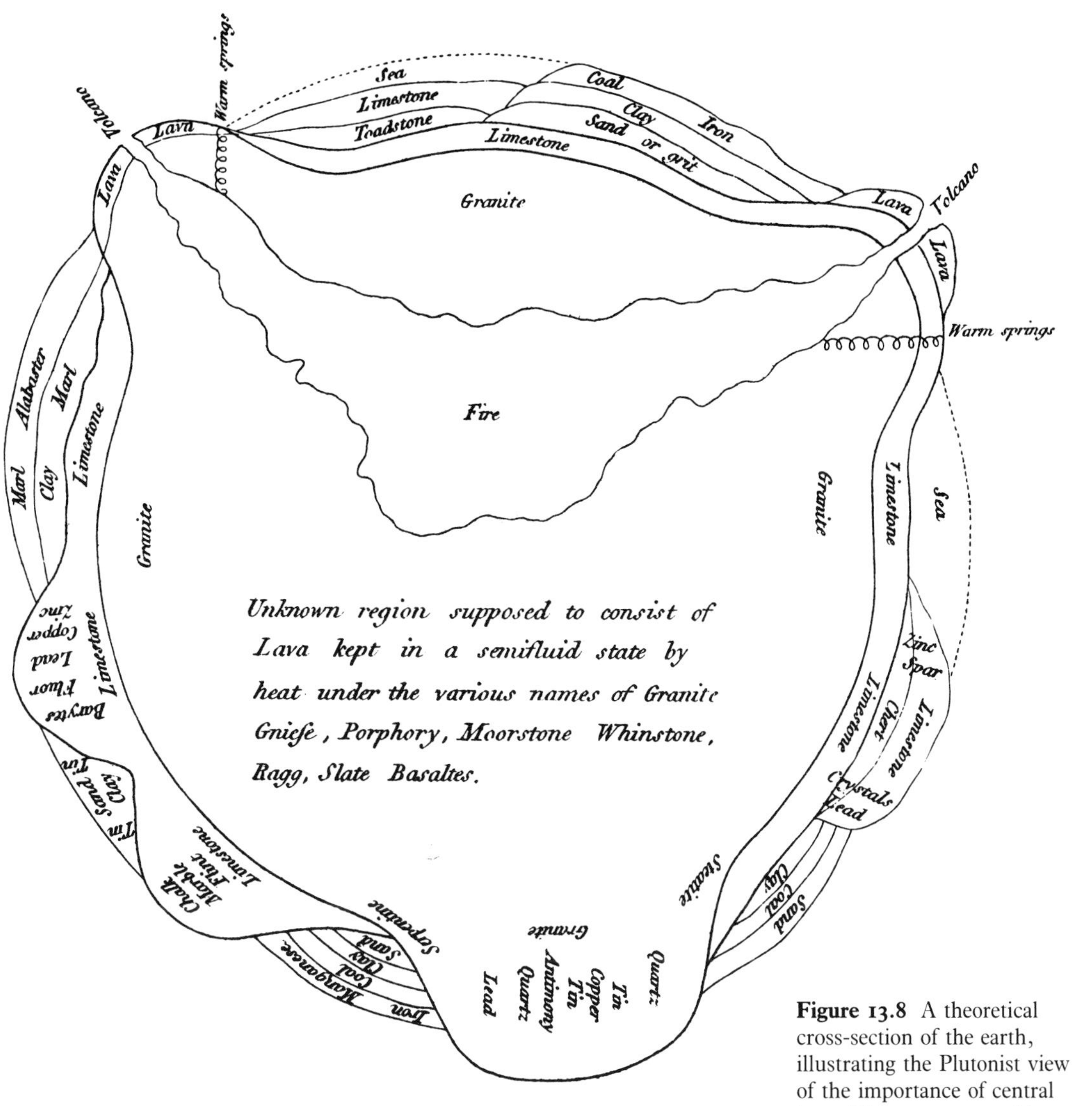
Volcano
Warm springs
Lava
Sea
Limestone
Toadstone
Coal
Clay
Iron
Sand or grit
Limestone
Granite
Lava
Volcano
Lava
Warm springs
Fire
Alabaster
Marl
Marl
Clay
Limestone
Granite
Granite
Limestone
Sea
Unknown region supposed to consist of
Lava kept in a semifluid state by
heat under the various names of Granite
Gniese, Porphory, Moorstone Whinstone,
Ragg, Slate Basaltes.
Zinc
Copper
Lead
Fluor
Barytes
Limestone
Zinc
Spar
Chert
Limestone
Limestone
Crystals
Lead
Tin
Sand
Clay
Tin
Chalk
Marble
Flint
Limestone
Steatite
Serpentine
Sand
Clay
Coal
Manganese
Iron
Granite
Quartz
Tin
Copper
Tin
Antimony
Quartz
Lead
Clay
Coal
Sand

Figure 13.8 A theoretical cross-section of the earth, illustrating the Plutonist view of the importance of central heat. This was prepared by James Hutton's friend, the poet Erasmus Darwin. The original caption reads: 'Section of the Earth. A sketch of a supposed Section of the Earth in respect to the disposition of the Strata over each other without regard to their proportions or number'. From E. Darwin, *The Botanic Garden: A Poem, in Two Parts* (London, 1791), 'Additional Notes,' facing p. 67.

Conclusion

The natural history of the earth had become, next to botany, a preferred field of study for scholars, administrative officials and the educated laity during the eighteenth century. Many eagerly occupied themselves with minerals and fossils, and keenly followed debates about the formation and development of the earth, so that the latest publications were widely read. The host of books, pamphlets and journal articles on geological or mineralogical themes could scarcely be assimilated. This prompted Johann Ehrenreich Fichtel (1732–95), of Transylvania to sarcastic exaggeration in 1792:

I was told that in the last but one decade of this century, more had been written on minerals than on theology, philosophy and jurisprudence put together in half a century – in fact, that mineralogical papers were as common as hay and straw in a good year nowadays.[15]

Mineralogy had become fashionable, and the natural history of the earth was a vital aspect of polite learning.

Natural history had changed its nature. The study of the three realms of nature that it had contained – minerals, plants and animals – developed into mineralogy, botany and zoology. The recognition that nature altered with time was important in determining this transformation: it marked a passage from representations in the form of stories and descriptions of nature to a history of nature with variable circumstances and qualities. This was especially true of the natural history of the earth, which became the basis for mineralogy. Eighteenth-century mineralogy included geognosy, from which geology would be developed, and oryctognosy, from which modern mineral science was created.

In wider historical terms, curiosities collected to elucidate the natural history of the earth became essential elements in the system of the modern sciences. It was on these foundations that biostratigraphy, historical geology, physical geology and other new earth sciences would be constructed in the nineteenth century.

Further reading

Faul, H. and C. Faul, *It Began with a Stone: A History of Geology from the Stone Age to the Age of Plate Tectonics* (New York, 1983).

Frängsmyr, T., 'Linnaeus as geologist', in T. Frängsmyr (ed.), *Linnaeus: The Man and his Work* (Berkeley and Los Angeles, 1983), pp. 110–55.

Guntau, M., 'The emergence of geology as scientific discipline', *History of Science*, 16 (1978), pp. 280–90.

Hutton, J., *Theory of the Earth with Proofs and Illustrations*, 2 vols. (Edinburgh, 1795).

Laudan, R., *From Mineralogy to Geology: The Foundations of a Science, 1650–1830* (Chicago, 1987).

Porter, R., *The Making of Geology: Earth Science in Britain, 1660–1815* (Cambridge, 1977).

'The terraqueous globe', in G. S. Rousseau and R. Porter (eds.), *The Ferment of Knowledge: Studies in the Historiography of Eighteenth-century Science* (Cambridge, 1980), pp. 285–324.

The Earth Sciences: An Annotated Bibliography (New York, 1983).

Rossi, P., *The Dark Abyss of Time: The History of the Earth and the History of Nations from Hooke to Vico* (Chicago, 1984).

Rudwick, M. J. S., 'The shape and meaning of earth history', in D. C. Lindberg and R. L. Numbers (eds.), *God and Nature: Historical Essays on the Encounter between Christianity and Science* (Berkeley and Los Angeles, 1986), pp. 296–321.

Werner, A. G., *On the External Characters of Minerals*, trans. A. Carozzi of *Von den äusserlichen Kennzeichen der Fossilien* (orig. 1774; Urbana, IL, 1962).

Short Classification and Description of the Various Rocks, trans. with introd. by A. Ospovat of *Kurze Klassifikation und Beschreibung der verschiedenen Gebirgsarten* (orig. 1786; New York, 1971).

14 *Naturphilosophie* and the kingdoms of nature

In the section 'Histoire naturelle' of his report of 1810 to Napoleon's Council of State, Georges Cuvier, acting in his capacity as a Secretary to the Institut National, offers a summary of the new 'German philosophy of nature', condemning it for its confusion of the moral with the physical and the metaphorical with the logical, whilst conceding that it includes men of real talent who have enriched natural history with 'precious facts'.[1] Not all received the new German *Naturphilosophie* so critically. The Germanophile Madame Germaine de Staël wrote in 1810 with approval of the works of Friedrich Schelling, Franz Xaver von Baader, and Gotthilf Heinrich von Schubert 'in which the sciences are presented from a point of view which captivates reflection and imagination'.[2] A few years earlier her friend, the 'Grand Tourist' Henry Crabb Robinson, attending Schelling's lectures at Jena, had vaunted the new German mystical philosophy over the cold rational quibbling of the English and French.[3] And a few years later Samuel Taylor Coleridge, embarking on his reading of the geological works of Schelling's disciple, the *Naturphilosoph* Henrich Steffens, enthusiastically endorsed the project for a developmental history of the earth and its productions; and, though appalled at Steffens's 'pantheistic blasphemies', he even dreamed of studying under him.[4]

The natural historians with whose works we shall be concerned were all closely involved in the culture of German Romanticism. I shall, accordingly, first indicate some key points in the Romantic vision of nature and history. There follows a brief account of the *Naturphilosophie* of Schelling and his circle. The ambitions of the *Naturphilosophen* were boundless: nothing less than a re-enactment of the creation, a reintegration of spirit and nature, and success (where the French Revolutionaries had failed) in initiating the Millennium. But in academic and disciplinary terms their attainments were more limited. As Cuvier noted, it was only in the German lands that they achieved a general recognition, and only in the fields of medicine and natural history that they secured a measure of control. In the third section of the chapter I indicate the range of the natural historical interests of the *Naturphilosophen*, examining the major works of three of them: the geologist Henrich Steffens, the botanist Nees von Esenbeck, and the comparative anatomist Lorenz Oken. I shall conclude with some remarks on the origins and cultural significance of this remarkable brand of natural history.

German Romanticism and *Naturphilosophie*

If there is a single mood characteristic of Romantic writing, it is *Sehnsucht* – longing or nostalgia – for the morning of the world when mankind was at one with itself and nature, for homeland and childhood, for past experiences, past loves, past intimations of immortality.[5] How can mankind recover the lost solidarity with nature? There was a fair measure of agreement on how *not* to proceed. Imposition on nature of a static, mechanical philosophy is no route to understanding. Along with bureaucratic despotism, codified law, pragmatic history, mimetic poetry, and all the other apparatus of 'that absence of ideas that dares to call itself Enlightenment' (Schelling) it is symptomatic of mankind's alienation from nature.

As for positive prescriptions for human redemption, it is harder to find common ground. In Novalis's (Friedrich von Hardenberg) *Die Lehrlinge zu Sais* ('The Apprentices at Sais'), a paradigm of High Romanticism, the central theme is precisely the multiplicity and divergence of the roads to reunion with nature. And Romantic artistic and literary activity is indeed marked by an extraordinary proliferation of new styles, genres, and philosophies. If there is a single value characteristic of German Romanticism it is the barely translatable *Eigentümlichkeit*, the singularity, individuality, distinction, and groundedness of a person, work of art, scene, or object as constituted by local history and setting – their resistance to generalization, translocation, or representation. It is in their *Eigentümlichkeit* that subjects possess their freedom and their moral and aesthetic character. A pervasive theme in Romantic writing about mankind and nature is fragmentation, the dark side, so to speak, of *Eigentümlichkeit*. The lost unity is not to be pieced together by reason, but glimpsed by intuition in traces, relics, particular viewpoints. Understanding of nature is not to be set out once and for all in treatises, theories, and allegories, but rather in productive forms, susceptible like nature itself of multiple interpretations: fragments, aphorisms, symbols, sketches.

Naturphilosophie and natural history are, for the Romantics, by no means the most direct ways back to nature. The most authentic reunion with nature requires not the discursive exercise of the mind, but immediate engagement: the innocent gaze of the child (Novalis), the sensitivity of the nervously disordered (Schubert), the absorption of the artist–genius lost in the work of creation (Schelling). It is in aesthetics rather than philosophy that the Romantics theorize the reunion with nature. And it is, of course, in the arts – in expressive poetry, in sublime landscape painting, above all, in absolute music – that we find the great monuments to their enterprise. For all that, *Naturphilosophie* and natural history have a part in the culture of Romanticism: they *are* roads to redemption, though minor and devious ones.

There is no canonical work of the new German *Naturphilosophie*. Its most famous exponent, F. W. A. Schelling (1775–1854), produced not one, but half a dozen systems and sketches of systems in as many years. In the same years Karl Eschenmayer and Franz Xaver von Baader sketched divergent systems;[6] and even among Schelling's declared followers some (Johann Wilhelm Ritter, Steffens, and Oken, for example) departed widely from his teaching. There are, however, certain features of *Naturphilosophie* that clearly set it apart from other schools in the natural philosophy and sciences of the period.

First, the *Naturphilosophen* were committed to a very strong form of vitalism. Like many medics and physiologists of the period, they postulated vital forces to explain the development and activities of living beings. And like many historians, both natural and civil, they used organic terms – 'growth', 'development', 'maturity', 'decay' – to describe the history of the universe, of the earth and its rocks, of the cultural and political fortunes of mankind. But they went much further, treating the cosmos itself as a living being, the source of all particular lives.[7] As the anatomist, gynaecologist, and landscape painter Carl Gustav Carus declared: 'If once we have recognised nature as being in the process of endless inner linkage, then we must at the same time consider it as the absolute living thing, from whose primordial life (*Urleben*) are derived the appearances of life of each particular living thing'.[8]

The *Naturphilosophen* were further committed to a thoroughgoing dynamism. With Kant, J. H. Lambert, and many others of the final decades of the eighteenth century they sought a developmental history of the heavens, the earth, and the earth's inhabitants. But where these earlier programmes set aside the question of the ultimate origins of the universe, of its inhabitants, and of the human spirit, the *Naturphilosophen* had no such inhibitions. They aimed at a total history, one that would encompass the entire differentiation of the cosmos from the original oneness, through the formation of the solar system and the earth, the proliferation of the three kingdoms of nature (minerals, plants, and animals), to the culmination of the universe in humankind. At least in the case of Schelling and his disciples, such a total history was envisaged dialectically, as a drama of successive partial resolutions and renewed manifestations of the primordial conflict of forces that sprang from the original unity. Nature considered dynamically, *natura naturans*, is this play of forces. The individuals of the visible world, stones, plants, animals, and people – *natura naturata* – are its by-products, each kind representing a particular and temporary balance of forces. That all beings are thus derived from and expressive of an original ideal unity is evident in the network of correspondences that pervades the cosmos – between the macrocosm and the earth, between the earth and the human microcosm, between plants and animals, between lower animals and higher animals. Indeed, there is nothing beyond or prior to *natura naturans*: properly understood the developmental history of nature *is* but the outward manifestation of the history of spirit. As Schelling remarked: 'What we call nature is a poem encoded in secret and mysterious signs, but if the riddle could be solved, we would recognise in nature the Odyssey of the spirit'.[9] Or, as Steffens enthused: 'Do you want to know nature? Turn your glance inwards and you will be granted the privilege of beholding nature's stages of development in the stages of your spiritual education. Do you want to know yourself? Seek in nature. Her works are those of the selfsame spirit'.[10]

There is considerable variety in the methods used by the *Naturphilosophen* to justify their claims and order their writings. First and foremost, there is the method of a priori construction. This is the dialectical procedure whereby the *Naturphilosoph* 're-creates' or 're-produces' the universe, recapitulating the process whereby successively higher and more specialized natural products arise as the successive partial resolutions of the primordial strife. Then there is the 'magic wand of analogy' (Novalis), the working out of the correspondences of structure and function that testify to the unity of plan underlying the development of the cosmos. What of observation and experiment? The *Naturphilosophen* were certainly opposed to sciences 'stuck in the rubbish dump of sensory reflexion' (Steffens) – chemistry based on analytical experiment in the manner of Lavoisier, botany based on standardized descriptions in the manner of Linnaeus. But they certainly did not, as is often alleged, repudiate observation and experiment outright. Rather they valued *Eigentümlichkeit* in the realm of phenomena, seeking experiences that bring into play the aesthetic and introspective faculties of the observer, allowing nature to speak directly to us. This led to an extraordinary emphasis on autoexperimentation, typified by the Galvanic experiments performed by Alexander von Humboldt and Johann Ritter.[11] And in the domain of natural historical observation it led to a quest for the primordial or ideal types from which the diversity of natural beings can be derived. Exemplary for this approach were Goethe's widely emulated morphological studies, in particular his presentation of plant organs as successive transformations of the primordial leaf, and of the skull and vertebrae as successive modifications and fusions of the primordial vertebra.[12]

Kingdoms of nature

Let us start with Schelling himself.[13] In 1799, the year after his call to a professorship of philosophy at Jena from Leipzig (where he had studied mathematics, physics, and medicine), Schelling set out his programme for a 'wholly new natural history' in his *Erster Entwurf eines Systems der Naturphilosophie* ('First Sketch of a System of Nature Philosophy'). He first notes that some have interpreted the succession of organizations as evidence of a genealogy of types, even going so far as to suppose that all types of living beings may be the progeny of a single ancestral type. (Schelling evidently has in mind Kant's speculations, in *Critique of Judgement*, sect. 80, on the derivation of all living beings from a single original organization, 'the womb of mother Earth'.[14]) This is impossible, Schelling claims: 'The distinctness of the stages at which we now see the organisations fixed evidently presupposes a ratio of the original forces peculiar to each one; whence it follows that nature must have initiated anew each product that appears fixed to us.' When properly viewed, comparative anatomy and physiology testify not to a genealogy of species, but to a development which realizes an original ideal. Given that the various types of organization are determined by and expressive of ratios of organic forces, it should be possible in principle to construct a priori the entire sequence of types of organization. In a striking passage Schelling goes on to contrast such a 'history of nature herself' both with the standard descriptive natural history and with the genealogical natural history that Kant had proposed.

Natural history has up to now been only the description of nature, as Kant has very rightly remarked. He himself suggests the name 'natural history' for a special branch of the science of nature, namely knowledge of the gradual alterations that the various organisations of the Earth have undergone through the influence of external nature, migrations from one climate to another, etc. If only the idea just set out were practicable, the name 'natural history' would assume a much higher import, for it would then actually convey a history of nature herself, namely of how through continual deviations from a common ideal she gradually brings forth the whole multiplicity of her products and thus realises that ideal, not indeed in individual products, but in the whole.[15]

Here we have a proposal for the a priori derivation of the entire natural system, a system that is conceived not as the plan of a transcendent creator, but as the realization of an ideal immanent in nature. However, for sustained attempts at such a derivation we have to turn from Schelling to his disciples.

Let us start with the study of minerals, the first kingdom of nature. Henrich Steffens (1773–1845), Norwegian by birth, studied at Copenhagen and Kiel, where he obtained his doctorate in mineralogy.[16] In 1798 he met Schelling, becoming his ardent follower, and embarked on further studies of mineralogy under Abraham Gottlob Werner at the famous Freiberg *Bergakademie*. He then taught geology, mineralogy, and *Naturphilosophie* at Halle, Breslau, and Berlin, of which he became Rector Magnificus in 1833.

Steffens's first major work, *Beyträge zur inneren Naturgeschichte der Erde* ('Contributions to the Inner Natural History of the Earth') of 1801, dedicated to Goethe, is by *naturphilosophische* standards fairly accessible. The work opens sedately proposing to combine chemistry with geognosy in the study of minerals and rocks. Werner defined geognosy as 'that part of mineralogy which acquaints us systematically and thoroughly with the solid earth, that is, with its relationship to those natural bodies that surround it and which are familiar to us, and also, especially, with the circumstances of its external and internal formation and the minerals of which it consists according to their differences and modes of formation'.[17] Steffens follows Werner's geognostic theory of formation of rocks by deposition from a primal ocean, though unlike Werner he insists on the prevalence of chemical rather than mechanical causes. The chemical processes of rock formation, Steffens claims, express the two fundamental vital powers, a carbon-based power of vegetation and a nitrogen-based power of animation. The vegetative power has given rise to the primitive siliceous rocks in which vegetable fossil remains predominate. The animating power has given rise to the more advanced calcareous rocks, in which animal fossils predominate. Having elaborated these general correspondences between vital, chemical, and geognostic processes, Steffens turns to a consideration of the origin of the individual inhabitants of the world: minerals, plants, and animals. The perfect natural history would derive the manifest diversity of beings a priori, step by step from the original ideal: but Steffens reluctantly settles for a more modest 'reductive' approach, one by which laws are conjectured on the basis of comparative observations and then, with luck, shown to be in agreement with higher laws derived a priori. As a specimen of this approach he presents a classification of the metals into two series. Allied to Nitrogen there is the 'fluid', vegetative series, from Arsenic to Mercury; allied to Carbon is the 'coherent', animated series, from Antimony to Gold. After a long account of the chemistry of the metals and their ores Steffens sketches an a priori derivation of the two metal series and their earthly distribution. Building on Schelling's *Erster Entwurf eines Systems der Naturphilosophie*, he derives a 'double polarity' of the earth, North–South embodying the duality of magnetism and its polar representatives Nitrogen and Carbon, East–West embodying the duality of electricity and its polar representatives Hydrogen and Oxygen. He infers from this that the more coherent metals must be more abundant near the Poles, the more fluid ones more abundant in the equatorial regions. The final section deals with plants and animals. His account is based on the 'Law of Succession of Organic Forces' that Carl Friedrich Kielmeyer (1765–1844), teacher of zoology at the Hohen Karlsschule at Stuttgart, had presented in his address *Über die Verhältnisse der organischen Kräfte* ('On the Relations between Organic Forces') of 1793.[18] In plants and the lower animals the reproductive force predominates, in insects the force of irritability, and in the higher animals the force of sensibility. Within each of these major groups there is a development from generalized to more specialized forms, the structure of each form being expressive of a particular ratio of the organic forces. The entire sequence culminates in man, 'the most individual of all forms'.

Steffens's later natural historical works build on the themes of this one. In his *Geognostisch-geologische Aufsätze* ('Geognostic-Geological Essays') of 1810, he greatly expands his account of the chemical genesis of rocks and speculatively relates the distribution of fossils to past climatic changes. In his *Vollständiges Handbuch der Oryctognosie* ('Complete Handbook of Oryctognosy') of 1811–24, devoted, as the term 'oryctognosy' implies, to the classification of minerals, he proposes a scheme which combines the Wernerian genetic method with his own chemical approach. His *Grundzüge der philosophischen Naturwissenschaft* ('Foundations of Philosophical Natural Science') of 1806, opens with an impassioned attack on the Enlightenment 'science of appearance' and a defence of the new 'science of the inner life of nature' inaugurated by Schelling. Through the new science the history of the universe will be fulfilled, mankind achieving at the level of reflection the lost union with nature, nature being completed by her mirroring in science. In the body of this aphoristically presented work Steffens expands his earlier Schellingian cosmological speculations; he amplifies his treatment of the carbonic and nitrogenous series of metals; and he recasts his account of the development of living beings as a story of the progressive realization of spirit in matter.

As protagonist for plants, the second kingdom of nature, I have chosen Christian Gottfried Daniel Nees von Esenbeck (1776–1858), radical democrat, Catholic social reformer, protégé of Goethe, Professor of botany and Director of the botanic garden first at

Erlangen then at Breslau, and from 1818 to his death President of the Leopoldina (Leopoldinisch-Carolinische Akademie der Naturforscher).[19] In fact, on his own reckoning Esenbeck was an expert on two kingdoms of nature, for as a mycologist he argued heatedly and at length for the recognition of fungi as an independent realm.

Esenbeck's major work of *naturphilosophische* botany, the two-volume *Handbuch der Botanik* of 1821–2, dedicated to Goethe, was published in the massive series of textbooks through which Gotthilf Heinrich von Schubert (1780–1860), Professor of Natural History at Erlangen, aimed to disseminate the new 'scientific' natural history. Esenbeck's book is full of bizarre, to us almost surreal, analogies and aesthetic speculations; and it is hard to envisage its use as a textbook. (He is, however, outclassed in oddity by his friend Georg August Goldfuß, Professor of Zoology and Mineralogy at Bonn: see Figure 14.1.) In the opening sections of the work Esenbeck sketches a priori constructions of the kingdoms of nature and of the basic structures and functions of plant anatomy and physiology. Thus, using Steffens's, 'double polarity' of the Earth, Esenbeck associates fungi with the primordial polarity, the North Pole, and Earth: 'Mushrooms are the expression of the Earth, being-for-itself, founded on itself, reverting to itself (the first simple polarity, ±). So mushrooms belong to the North Pole, are northern plants and ever seek their way back into rest, sleep, and death.' Higher plants represent the second polarity, $±/-$, the South Pole, and the Sun. The animals represent the West, the third polarity and midnight; man represents the East, the fourth polarity and midday. From the formula $±/-$ Esenbeck derives the principal types and dispositions of organs, tissues, cells and fibres in the higher plants (Table 14.1). The body of the work is concerned with 'organography', providing for each plant organ – root, stem, leaf, flower, seed, etc. – expositions of anatomy and physiology followed by extensive accounts of their metamorphoses, that is, their life-histories, their transformations in the ideal sequences of plant types, and their modifications caused by sickness. Here Esenbeck pays repeated tribute to Goethe's *Die Metamorphose der Pflanzen* ('The Metamorphosis of Plants') of 1790, in which the series of plant appendages – from seed-leaves to organs of fructification – were derived through processes of expansion, contraction, and perfection from the primordial leaf (*Urblatt*).[20]

Esenbeck's *Handbuch* was not a success, and his later natural historical works are more orthodox in content and presentation. In particular, his masterpiece, *Naturgeschichte der europäischen Lebermoose* ('Natural History of the European Liverworts') of 1833–8, though introduced in the Romantic manner as the first of a

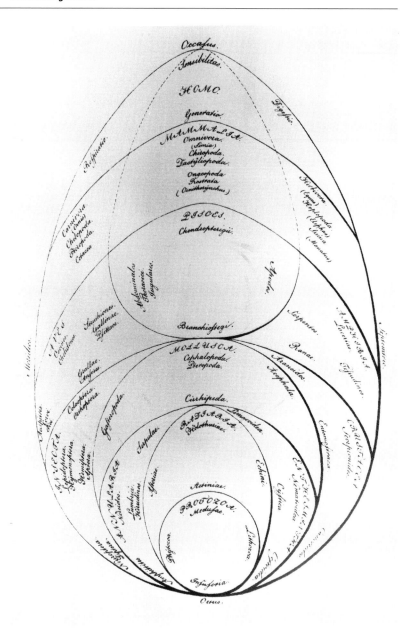

Figure 14.1 The ordering of the animal kingdom in conformity with the 'great egg of nature'; from G. A. Goldfuß, *Ueber die Entwicklungsstufen des Thieres* (Nuremberg, 1817); on Goldfuß see H. von Querner's introduction to the facsimile reprint (Marburg, 1979).

projected series of 'memories of the Riesengebirge', is devoid of *Naturphilosophie*, and his substantial treatise on *Naturphilosophie* of 1841 is almost devoid of natural history.

As historian of the animal kingdom I have chosen another political radical, Lorenz Oken (1779–1851), student of J. F. Blumenbach at Göttingen, disciple then rival of Schelling, friend then enemy of Goethe, holder of Chairs in Medicine and Natural History at Jena, Munich, and Zurich.[21] As a prolific and innovative

Table 14.1. *Esenbeck's derivation of the fundamentals of plant anatomy from the double polarity of the Earth:* Handbuch der Botanik, *vol. I, p. 40. Note that the growth of the flower is described as 'regressive' because it ceases after producing a definite number of organs, by contrast with the open-ended 'progressive' growth of the stem.*

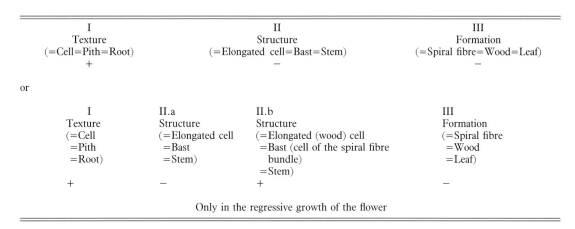

I	II	III
Texture	Structure	Formation
(=Cell=Pith=Root)	(=Elongated cell=Bast=Stem)	(=Spiral fibre=Wood=Leaf)
+	−	−

or

I	II.a	II.b	III
Texture	Structure	Structure	Formation
(=Cell	(=Elongated cell	(=Elongated (wood) cell	(=Spiral fibre
=Pith	=Bast	=Bast (cell of the spiral fibre	=Wood
=Root)	=Stem)	bundle)	=Leaf)
		=Stem)	
+	−	+	−
		Only in the regressive growth of the flower	

comparative anatomist, founder of *Isis* (a major European forum for natural history), and mastermind of the Gesellschaft deutscher Naturforscher und Aerzte (the model for the British Association for the Advancement of Science), Oken was by far the best known of the *naturphilosophischen* natural historians.

In a speech of 1809, Oken protests that the cultivation of natural history only for its practical and commercial fruits in medicine and agriculture leads to a 'senseless enumeration, description and naming of animals'.[22] In place of such an ignoble, 'profiteering' natural history, he pleads for a natural history integrated into the new *Naturphilosophie*. This noble natural history will unify the German people with themselves and the world; it will give them an understanding of their own nature and of their relations to plants and animals; and it will imbue them with manly resignation when their power falls short of their understanding. Such is the natural history set out by Oken in his *Lehrbuch der Naturphilosophie* ('Textbook of Nature Philosophy') of 1809–11 and its companion *Lehrbuch der Naturgeschichte* ('Textbook of Natural History') of 1813–26. The first of these works offers a derivation from God, the 'primordial zero', of a comparative anatomy and natural classification of living beings. The second uses the anatomy and classification as the basis for a comprehensive descriptive natural history.

The 'Textbook of Nature Philosophy', dedicated to Schelling and Goethe, is an extraordinary document. It consists of numbered paragraphs (3,562 in the first edition) and combines an elaborate dialectical construction in the manner of Schelling and Steffens with a plethora of often apparently weird analogies. Oken's friend

Alexander Ecker aptly remarked that 'the language seems to come to us out a remote past as though wafted from the tongues of Egyptian priests'.[23] The work opens with a 'mathesis' in which Gravity, Light, Heat, and Fire are derived as direct manifestations of God, and an 'ontology' which combines elements of Schelling's cosmogony and Werner's geognosy in an account of the formation of the solar system, the Earth, and the principal rock formations and mineral types. The account of living beings, 'biology', starts with the primordial units of life, formed by the action of the air on the primordial sea-slime (*Urschleim*). By construction Oken argues that these units must be vesicles (*Bläschen*): 'The organic must become a vesicle, since it is a galvanic process which can take place only between the elements. The action of air is necessarily an external one, so it divides the slime inwards into the earthy and the watery, cell-wall and cell-content.' For good measure, he adds a brisk argument by analogy: 'The organic must be a vesicle because it is the image of the planet.' In isolation the vesicles occur in water as *Infusoria*; variously combined they constitute other types of organism. The second stage, rather perfunctorily executed, is the construction of the plant kingdom. The third stage, which synthesizes the other two, is the construction of the animal kingdom, culminating in man, the complete and perfect realization of God. (In fact, following Goethe and Petrus Camper, Oken believed that the full perfection of man, though realized in ancient Greek statuary, was yet to be attained: see Figure 14.2.) The basis of this construction is provided by a ranking of the organic processes associated with the four elements. From this is derived a partition of the animal body into tissues, organ systems, and organs. The series of types of animals is then built up by addition and reduplication of successively higher ranking organs, culminating in man, who possesses all organs in their highest form – hence Oken's pronouncement: 'The animal kingdom is but a dismemberment of the highest animal, man.'

The principal ranking of animal types is the sequence of seventeen classes, from infusorians to mammals, shown in Table 14.2.[24] In the course of its development from fertilized egg to adult, an animal of a given class passes in turn through stages representative of each of the classes that rank below it.

The foetus is a representation of all animal classes in time. At first it is a simple vesicle, stomach, or vitellus, as in the Infusoria. Then the vesicle is doubled through the albumen and shell, and obtains an intestine as in Corals . . . With the appearance of the osseous system, into the class of Fishes. With the evolution of muscles, into the class of Reptiles. With the ingress of respiration through the lungs into the class of Birds.[25]

The criteria used to demarcate and rank the classes of animals are iteratively applied within each class to provide a demarcation and

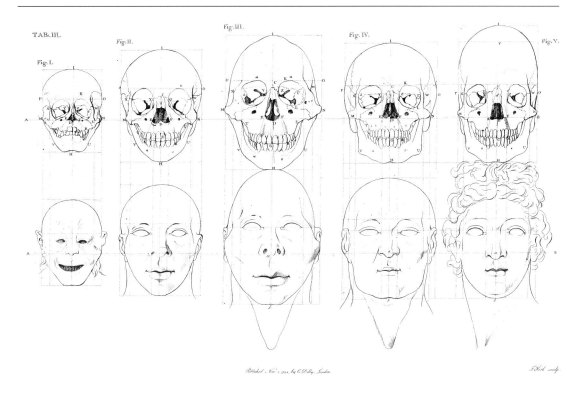

Figure 14.2 The perfection of the skull, culminating in the European and the Antique Ideal; from *The Works of the Late Professor Camper, on the Connexion between the Science of Anatomy and the Arts of Drawing, Painting, Statuary*, trans. T. Cogan (London, 1794).

ranking of orders, families, and genera.[26] In the resultant scheme there are correspondences between types of different categories in the hierarchy (between classes and families, between families and genera) and also between lower and higher types of the same categories (between fishes and birds, between birds and mammals, for example). All these correspondences are reflected in 'analogies' of anatomical structure. Some of these analogies involve transformations of a structural type similar to those postulated by Goethe. Indeed, Oken and Goethe became involved in a bitter priority dispute over the 'discovery' that the skull is derived from a transformation and fusion of a series of primordial vertebrae.[27] Others of Oken's analogies, however, invoke transformations stranger and more extreme than anything countenanced by Goethe: 'the nose is the thorax repeated in the head'; 'the limbs of insects are the ribs of mammals'; 'the fish is a mussel from between whose shells a monstrous abdomen has grown'.

Oken's system is an extraordinary feat of synthesis. It takes a decisive stand on every one of the major controversial issues in the natural history of the period – the basis of the process of generation, the form of the natural system, the relation between form and function, the role of God in the natural world. Moreover, it is a system which tightly integrates the description, classification, anatomy, physiology, and chemistry of living beings.

Table 14.2. *Oken's construction of the series of animal classes. From the 3rd edn. of his* Lehrbuch der Naturphilosophie *(Zurich, 1843).*

Dominant element	Dominant sense	Dominant organ-system	Circles	Classes
Earth		Alimentary Gastric Intestinal Absorbent	Protozoa	Infusorians Polyps Jellyfish
Water	Tactile	Vascular Venous Arterial Cardiac	Conchozoa	Shellfish Snails, slugs Squids
Air		Respiratory Cutaneous Branchial Tracheal	Ancyliozoa	Worms Crustaceans Insects
Fire	Taste Smell Hearing	Osseous Muscular Nervous	Sarcozoa	Fish Reptiles Birds
	Sight	Sensory	Aesthesiozoa	Mammals

Concluding suggestions

Given the dearth of critical studies of early nineteenth-century German natural history, I shall conclude only with some questions and tentative suggestions.

To start with, there is a question of scope. In the opening decades of the nineteenth century in many of the universites of the German lands – Jena, Heidelberg, Munich, Erlangen, Giessen, Leipzig, Breslau, Bonn, Berlin – the study of natural history was dominated by *Naturphilosophie*. Is this extraordinary development best considered on the local German scale, or as an aspect of more general European changes? There are grounds for taking the latter view. In the first half of the nineteenth century we find a substantial body of natural historical writings outside Germany which show affinities with the publications of Steffens, Oken, Esenbeck, and their circle. Obvious examples are the works of Etienne Geoffroy Saint-Hilaire and Etienne Serres in France; of Robert Knox, John Goodsir, Richard Owen, William Sharp Macleay, and William Swainson in Britain; and of Louis Agassiz in the USA. These are marked by a fairly well defined set of commitments: to interpretation of the diversity of living beings as an unfolding or enactment of original ideas and forces; to the specification of morphological types and morphological laws; and to the tracing of parallels between individual development and the ideal succession of living beings.[28] It is tempting to relate such 'transcendental' natural histories to more general processes of secularization; for in them

God's plan becomes a purposiveness immanent in nature, sacred history is transposed into history of the cosmos, and theology is absorbed into natural philosophy. Alternatively, one may seek to relate these transcendental natural histories to the global change of *episteme* around 1800 which Michel Foucault saw as leading natural history from a static concern with classification and external characteristics to a dynamic concern with inner development, function, and structure.[29]

Any firm answer to the question of scope must await much further investigation of the sources for and the reception of the natural historical works of the *Naturphilosophen*.[30] It is, however, my suspicion that many of the developments that I have discussed may profitably be viewed as local responses to specifically German predicaments of natural history. One such predicament has to do with the programme for the pursuit of natural history established by Blumenbach at Göttingen, the Mecca for German natural historians, and theorized by Kant in his *Critique of Judgement* of 1790. In this 'teleomechanical' approach vital forces were postulated to explain both the development of individual living beings and the derivation of races and species from ancestral types in response to migration and climatic change.[31] These teleological vital forces, whilst inscrutable in nature, were supposed to act through discoverable material and chemical causes. This was the framework for a considerable body of empirical research in the decades around 1800. However, it placed severe restrictions on enquiry, all questions about the nature of the vital forces and the ultimate origins of life and organization being declared 'unscientific' and 'beyond the bounds of sense'. The *naturphilosophischen* natural historians were generally explicit about their indebtedness to this programme, and in particular to the work of Blumenbach's famous pupil Kielmeyer. And they presented their techniques of a priori construction as ways of getting to grips with all the fascinating questions that Blumenbach and Kant had deemed illegitimate.

A further, and, I believe, crucial local predicament of natural history has to do with university reform. In the period between the French Revolution and the opening of the new University of Berlin in 1810 there was agitation throughout the German universities for promotion of philosophy from its traditional role as a 'lower' preparatory faculty to that of a higher 'scientific' faculty on a par with law, medicine, and theology.[32] (Kant's *The Conflict of the Faculties* of 1798 was an important early contribution to this debate.) It is significant that in the writings of Oken, Steffens, and Carus *Naturphilosophie* is presented as the means whereby natural history can cease to be a mere utilitarian *Brotstudium* and become a science.[33] *Naturphilosophie*, in other words, was perceived as offering natural history a rise in status from a mere appendage of the medical faculty to full membership alongside mathematics, philology, and physics in a higher philosophical faculty. Of course, this provides only a partial explanation of the spread of *naturphilosophische* natural history. Almost all the universities of the German lands were under direct State control, and a fuller explanation would have to look in detail at the negotiations through which the Ministers of Culture and their bureaucrats were recruited to the *naturphilosophische* cause.[34]

Historians of science have offered sharply contrasting assessments of the impact of *Naturphilosophie* on medicine, natural history, and the sciences. The majority have followed the lead of Liebig, Du Bois-Reymond, and other luminaries of the new empirical natural sciences in dismissing it as an aberration – speculative, ill-disciplined, and irrational. And certainly the writings of the *Naturphilosophen* I have considered contain much to encourage such a view. Esenbeck and Goldfuß, for instance, often appear wilfully arbitrary, fantastic, even frivolous. As already noted, many of them reject outright the Enlightenment ideals of cosmopolitan, polite learning in favour of esoteric, privately communicated, local knowledge; moreover, their attempts to enter the public domain were often dismal failures – societies that collapsed after a couple of meetings, journals that produced but a single issue with only a couple of contributors, etc.

However, this cannot be the whole story. Some at least of the institutions of the *Naturphilosophen* were highly effective: the *Journal für die Chemie und Physik*, edited by J. S. C. Schweigger, and Oken's Gesellschaft deutscher Naturforscher und Aerzte, for example. Indeed, in recent years a number of historians have taken a very different view, arguing that displacement of the Enlightenment ideal of encyclopaedic learning by the Romantic ideology of genius fostered the research ethic by allowing that the individual *Forscher* could make substantial contributions to the progress of the sciences.[35] They have insisted that the defeat of *Naturphilosophie* by the empirical natural sciences in the 1840s and 1850s is a polemical construct rather than a finding of historical research; and that despite their undoubted rejection of much *naturphilosophische* theory, the new natural scientists retained much of the agenda of *Naturphilosophie*.[36] To this we may add the suggestion that, for all their denunciations of the 'hare-brained aesthetic blathering' (Du Bois-Reymond) of the *Naturphilosophen*, the experimental practices of the new scientists owed as much to German Romantic *Eigentümlichkeit* as to French analysis.[37] And we may add the more general suggestion that the Romantic subversion of the Enlightenment commonwealth of polite learning paved the way for the empire of natural science.

At present, I fear, we lack both the historical data and the historical categories needed to resolve these issues. I have already noted the scarcity of studies of reception of the works of the *naturphilosophischen* natural historians; to this we may add that, despite the fact that they were university teachers, field observers, experimenters, directors of gardens, curators of cabinets, suppliers of *materia medica*, we know virtually nothing about the impact of *Naturphilosophie* on the practices of natural history.[38] Finally, we must remember that the works of Steffens, Esenbeck, Oken, and their circle embody Romantic categories – of poetic science, of genius as re-creation of the world, of the encyclopaedia in fragments – categories that subvert our essentially scientistic dichotomies of science/art, discipline/anarchy, reason/unreason.

Further reading

Amrine, F., Zucker, F. J., and Wheeler, H., *Goethe and the Sciences: A Reappraisal* (Dordrecht, 1987).

Bowie, A., *Schelling* (London, 1993).

Cohen, R. S. and Wartofsky, M. W. (eds.), *Hegel and the Sciences* (Dordrecht, 1984).

Cunningham, A. R. and Jardine, N., *Romanticism and the Sciences* (Cambridge, 1990).

Engelhardt, D. von, 'Bibliographie der Sekundarliteratur zur romantischen Naturforschung und Medizin 1950–1975', in R. Brinkmann (ed.), *Romantik in Deutschland* (Stuttgart, 1978), pp. 306–30.

Gode-von Aesch, A., *Natural Science in German Romanticism* (New York, 1941; repr. New York, 1966).

Gusdorf, G., *Le Savoir Romantique de la nature* (Paris, 1987).

Jardine, N., *The Scenes of Inquiry: On the Reality of Questions in the Sciences* (Oxford, 1991), ch. 2.

Lenoir, T., *The Strategy of Life: Teleology and Mechanics in Nineteenth-Century German Biology* (Dordrecht, 1982).

Oken, L., *Elements of Physiophilosophy*, trans. A. Tulk (London, 1849).

Rehbock, P. F., *The Philosophical Naturalists* (Madison, WI, 1983).

Russell, E. S., *Form and Function: A Contribution to the History of Animal Morphology* (London, 1916, repr. Chicago, 1982).

Schelling, F. W. J., *Ideas for a Philosophy of Nature*, trans. E. E. Harris and P. Heath (Cambridge, 1988).

Stallo, J. B., *General Principles of the Philosophy of Nature: With an Outline of Some of Its Recent Developments among the Germans, Embracing the Philosophical Systems of Schelling and Hegel, and Oken's System of Nature* (Boston, MA., 1848).

Ziolkowski, T., *German Romanticism and Its Institutions* (Princeton, 1990).

III Discipline, discovery and display

15 New spaces in natural history

It is really only in one's study that one can roam freely throughout the universe.

Georges Cuvier, 'Analyse d'un ouvrage de M. Humboldt intitulé: *Tableaux de la nature* . . .' (1807), p. 6.[1]

I have become deeply aware of the double aspect of space, this outer one that surrounds me, this room in which I am writing . . . and this inner one which is the space my body takes up, and is not lit at all, it's dark and incommunicable in words, indescribable – but not empty, it's warm and rich, full of an odd sort of joy, though a profane kind . . . To perceive the created world one does have to relate to the inner private sea, just as Traherne said.

Marion Milner, *Eternity's Sunrise: A Way of Keeping a Diary* (London, 1987), pp. 30, 113.

Natural history in transition, 1780–1830

Between the eighteenth and nineteenth centuries natural history underwent profound transformations. An overwhelming interest in evolving classification systems for specimens of plants and animals was slowly edged out, though never completely replaced, by investigation into aspects of the inward functioning of their physiological systems. 'Natural history' itself slowly separated into separate sub-disciplines such as physiology or palaeontology, each with their own methods, agendas, and subject-matter. At the same time, natural history began to separate from theology, especially in continental Europe, though at a slower pace than in Britain. By the early nineteenth century active men of science began to see natural history as distinct from attempts to argue from the nature of the created world to belief in and knowledge of a benevolent deity.

During the same period natural history as a profession also underwent great changes. Much new knowledge about nature continued to be produced by gentlemanly and gentlewomanly amateurs working in largely domestic environments, supported by an extensive network of clubs and societies. But at the same time, especially in Europe, natural history was increasingly given the backing of major new state-funded and state-controlled institutions staffed by paid full-time expert researchers. Such, for example, were the Natural History Department of the British Museum in London, the Zoological Gardens in London, natural history museums such as that founded by Charles Willson Peale in Philadelphia, or the geological and zoological museums of the new University of Berlin and, above all, the Muséum National d'Histoire Naturelle in Paris, refounded in 1793 on the basis of the former royal botanical gardens or Jardin des Plantes. It is upon this key institution that much of this essay will focus.

The Muséum National d'Histoire Naturelle

The Muséum was undoubtedly the best-known institution in this field in at least the first half of the nineteenth century, attracting many researchers from outside France, and being the home (literally as well as professionally) of many of the leading names in natural history in this period, of men like Georges Cuvier (1769–1832), Jean-Baptiste Lamarck (1744–1829), and Etienne and Isidore Geoffroy St.-Hilaire (1772–1844 and 1805–61 respectively). The Muséum, whose grounds also contained a zoo, a publicly accessible botanical garden, lecture theatres, display galleries, a library, and preparation and dissection rooms, lay on a twenty-acre site on the south bank of the Seine. Throughout this period the site expanded, until it reached almost the same size as the institution occupies today. Unlike today's Muséum, however, the Muséum in the early nineteenth century was not simply a site of scientific labour and of public resort: it was also a domestic space which housed the professors and their families, as well as those of many of the technical assistants and the non-tenured professional staff. Often, their living space was shared between several individuals, and there was little separation between domestic space and working space. Cuvier, for example, reconstructed his own living space so that, by opening a door, he could walk straight from his apartment into the anatomy galleries.

The boundaries of the Muséum enclosed a highly complex space which contained many ambiguous and contested elements. It was a public space which also contained domestic space. It was an open, park-like space in the middle of one of Europe's largest urban centres. It was a space ostensibly dedicated to professional natural history at a time when such paid careers were monopolized by men, and yet was also a space lived in by both genders and all age groups. It was a 'professional space' which was also open to the amateur or even merely slightly interested public. It was a place which focused professional labour on one site, and yet from that site, field naturalists were sent out to wild and inaccessible parts of the world to collect otherwise unknown flora and fauna. The 'open' spaces of the gardens and zoo were flanked by the 'closed' spaces of the galleries and lecture theatres; and each sort of space carried with it a different facet of natural history practice and organization. The Muséum's space was a microcosm which mirrored the debates over the different uses of space which were crucial, in this period of transition, for the future direction of the discipline. Cuvier, for example, expanded the space allotted to his collections in the anatomy galleries, at the expense of other professors, notably Lamarck, with whose classification systems he disagreed. He and his brother Frédéric, who directed the zoo in the Muséum, clashed over allocation of extra space between the living animals and the anatomy galleries, a clash which mirrored their debate over the importance of work on the living animal, or the dead specimen divorced from habitat.

The Muséum was a space where not only contestations about classification could take place, but also one where entirely new approaches to nature could be worked out. Naturalists like Cuvier were creating the new science of comparative anatomy, by which it was hoped to generate a new way of looking at the relationships between living beings. Cuvier tried to relate living beings to each other by comparing their internal structures rather than their external characteristics. For museum naturalists often working from imperfectly preserved specimens, this could have two consequences. Firstly, an emphasis on detailed anatomical exploration of internal physiological structures, to discover the particular ways in which systems related, and to use those relationships as the basis for grouping living beings together. This way of classifying tended to privilege function over form. Secondly, the expansion of comparative anatomy also enabled much exploration in palaeontology to take place. Fossil remains were being discovered in Paris in great numbers from the 1790s, due to excavations for the expanding city planned by Napoleon. Naturalists like Cuvier also devoted much time and energy to obtaining careful copies or drawings of specimens of fossils known elsewhere. As French conquests in Europe proceeded, newly conquered territories contributed to the collections of the Muséum. It was in this way that complete skeletons, such as the Madrid *Megatherium* (Figure 15.1), became known in Paris. All this meant a change in the nature of the Muséum space. In the later eighteenth century the Muséum had been largely a botanical *garden*; after 1793 it began to be an institution which remained an outdoor setting for the display of nature but also began to house increasing numbers of built spaces like galleries and dissection rooms to house an increasingly indoor science. This also put great pressure on the vocational ideals of naturalists. Where was their science located? Indoors or out? Were the

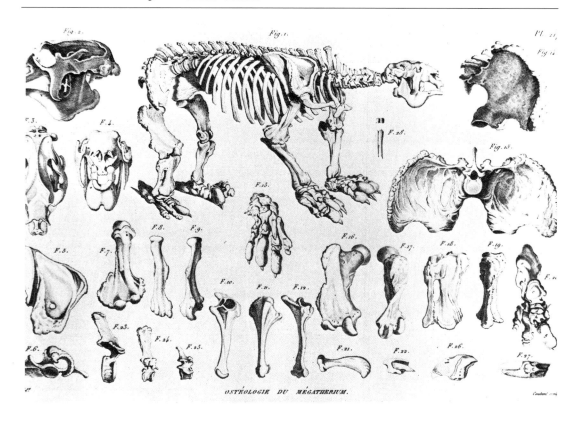

OSTÉOLOGIE DU MÉGATHERIUM.

Figure 15.1 Skeleton and bones of the Madrid *Megatherium*, which became well known to French naturalists after the French conquest of Spain in 1808. This illustration, from Cuvier's *Recherches sur les ossemens fossiles* (1836 edition), also well illustrates the complexity of reassembling extinct creatures from many separate fragments of skeleton, work on which Cuvier's indoor version of comparative anatomy mixed with palaeontology was increasingly to focus. Reproduced in Alan Moorehead, *Darwin and the Beagle* (London, 1969), p. 84.

systems of explanation created by the work of indoor anatomists superior to the intimate knowledge of living creatures in their habitats which was traditional field natural history?

The meaning of space

Recently, the relation between spatial disposition and intellectual authority has become a new focus of study in the history of science and beyond. Spatial metaphors abound in recent philosophical works. Charles Taylor, for example, in an important recent work, *Sources of the Self* (1989) posits that the basis of the self is a 'moral space', from which the self speaks. Alasdair MacIntyre in *After Virtue* (1985) argues that rational enquiry has intentions that are only intelligible in virtue of their settings. Both would agree that what passes as rational belief has not merely a history, but a geography as well. Spatial metaphors in studies of the distribution of scientific influence, in the transmission of scientific instrumentation, and the disposition of space in 'laboratory life' have been investigated in well-known works.[2] The 'built spaces' of science, especially the anatomy theatres and public lecture rooms, as well as institutions of popular science such as mechanics' institutes,

Figure 15.2 The zoology galleries of the Muséum in 1815, in a print by Courvoisier in the Library of the Muséum. This clearly shows the dominance of the galleries of prepared specimens over the layout of the Muséum. The many figures in the foreground also indicate the Muséum's parallel importance as a place of public resort. Reproduced in Jean Théodoridès, 'Quelques documents inédits ou peu connus relatifs à l'oeuvre et les relations de Georges Cuvier', *Biologie médicale* (numéro hors série), (1961), pp. 21–50.

have been investigated extensively by architectural historians. Such buildings, as has been shown, both manifest and impose structures of authority between teachers and taught, expert and public. The design of buildings can also both pressure and structure the access to knowledge of different social classes. Historians have long known that, in a broader sense, science varied with location. The styles of science are regionally differentiated, as is the reception of scientific themes. Sociologists like Erving Goffman have pointed out that the spaces of scientific knowledge have often determined the degree of credence given to claims to expert knowledge. This is a point also made in Steven Shapin and Simon Schaffer, *Leviathan and the Air Pump* (1985).

On the other hand, little attention has been paid to the 'spaces' of science which lie outside the built environment, spaces like the public botanical and zoological gardens of the Muséum, or indeed the whole area of 'wild nature' into which its 'field' naturalists ventured to find specimens for examination by the experts of Paris. How did they see 'the wild' or 'the field'? Also, little attention has been paid to the use of 'domestic' space in science. What were the interactions *between* these very different sorts of spaces? In the confines of this chapter we cannot hope to go more than a little way towards resolving these questions, but we can at least open them up. We can begin to ask more fundamental questions about how contemporaries categorized and experienced different forms of space: we can also explore our own presuppositions about what space is.

'Space', after all, is a profoundly ambiguous concept. The word itself is unclear. It can mean either 'the area contained within a given boundary', or it can mean the opposite, unbounded area, as in the expression 'outer space'. Phrases like 'space, the final

Figure 15.3 The greenhouse at the Muséum, devoted to the acclimatization of exotic plants. The greenhouses reinforced the character of the Muséum as a place where all nature, however exotic, could be gathered on one spot. Reproduced from P. Bernard and L. Couailhac, *Le Jardin des plantes* (Paris, 1842).

frontier', resonate as they do precisely because they encapsulate the contradiction inherent in our idea of space. Secondly, it has also been the case that spatial perception has been treated very largely as equivalent to *visual* experience alone. However, much recent discussion has been redefining spatial perception as an experience in which the *whole* body is implicated.[3] In other words, to reconstruct the spatial experience of the natural historians of this period we have to think about not just what they might have seen, but also about what sort of psychological structures mediated their response to space. It is not possible to explore space as if it were a simple or autonomous entity, without questioning the inner world of its human perceivers. To find out what space perception was, we have to come to grips with the whole-body experience of the men of the past. Their 'inner-space' also structured what spaces were seen, and how.

'Space', therefore, can mean many different things. It can mean the bounded, built spaces in which natural history located itself. It could also, especially in natural history 'fieldwork', be experienced as 'movement over' terrain, which allowed the naturalist to observe a rapidly changing population of living beings in their

Figure 15.4 Cuvier's house at the Muséum. Cuvier lived here with his family of children and step-children, his household of assistants, his library and work-room, which had immediate access to the anatomy galleries, and was also in the middle of the public gardens of the Muséum, with little separating the house from the Muséum public. It thus shows the characteristic Muséum combination of family space, working space, and public space, in close proximity – a mixture which was very characteristic of spatial arrangement in most cities before industrialization instituted more rigid divisions between home and work, as did increasing bureaucratization of labour for the upper classes and for intellectual workers such as indoor naturalists. Reproduced from L. Boitard, *Le Jardin des plantes* (Paris, 1842), in Jean Théodoridès, 'Quelques documents inédits ou peu connus relatifs à l'oeuvre et les relations de Georges Cuvier', *Biologie médicale* (numéro hors série), (1961), pp. 21–50.

habitat. In this kind of natural history, knowledge-gathering was inseparable from movement *through* space, inseparable therefore from bodily involvement. In this way, natural history fieldwork was squarely associated with a tradition which linked curiosity to movement.[4] Conversely, other kinds of natural history which concentrated on the dissection of specimens collected by field natural historians, carried out in bounded, built environments such as anatomy rooms, carried quite a different epistemological tradition.

Different spaces too had different social and political connotations. The Muséum was also, as we have already noted, a public space, a place to which the ordinary population of Paris was able to resort freely, where it could wander among the gardens and admire the live animals in the zoological gardens, or, indoors, take in the wonders displayed in the galleries devoted to displays of the dead specimen. It was a place for gentle exercise and refined entertainment in the heart of the metropolis. In 1800, the Muséum's professor of chemistry, Antoine Fourcroy, described the

Figure 15.5 This famous animal was a gift to the French government by Ali Pasha, Viceroy of Egypt, and was walked by its keepers, pictured here in oriental costume, from Marseilles to Paris, where it was housed in the Muséum, after a journey by barge up the Seine. The giraffe, then an extreme rarity in Europe, caused massive enthusiasm, and tickets for admission to the Muséum to see it changed hands at staggering prices. The picture shows the crowds of enthusiastic spectators, and illustrates the function of the Muséum as a showground of public entertainment based on natural history. Reproduced in Jean Théodoridès, 'Quelques documents inédits ou peu connus relatifs à l'oeuvre et les relations de Georges Cuvier', *Biologie médicale* (numéro hors série), (1961), pp. 21–50.

gardens of the Muséum as an 'Elysium', or abode of the gods.[5] Fourcroy's description should alert us to the extent to which the Muséum's character as a 'public' space was also dependent on its definition as a space which could be assimilated to the mythological structures of classical antiquity, and could thereby be conceptualized as a space outside space: a heavenly place off the grid of real-world maps, and hence outside the focus of social and political production. This way of conceptualizing the Muséum was reinforced by the way in which its publicly accessible collections brought together in one place the whole range of the natural order, which, in the 'real world', would never be found together in one space. The rapidly growing palaeontological galleries also meant that the visitor to the Muséum could see not only the denizens of many different parts of the earth's surface together in one spot, but also the products of many different *eras* of the earth's history. In thus making visible nature from all times and from all spaces, the Muséum intensified its character as an ideal place, a place where all other times and places were gathered, a claim which remained extremely powerful because it was supported by the spatial layout of the Muséum itself. Thus the Muséum could remain unaffected, for the ordinary public, by the quarrels about the order of nature which caused such conflict amongst its professional staff.[6]

This 'ideal' character of Muséum space was heightened by its location in the heart of a great capital city. The Muséum was resolutely divorced from its congested, built-up setting, in a part of Paris which was undergoing considerable 'development' in the early nineteenth century. Yet paradoxically this very setting acted to emphasize, by its contrast, the Muséum's character as an ideal space, not generated by other surrounding spaces. This ideal character at once obscured, and at a profound level mirrored the political conflict of the French Revolution in which the Muséum itself lay embedded. The idea of a botanic garden as an Eden or an earthly Utopia long antedated the foundation of the Muséum in its modern form in 1793. But the cultural impact of the idea was to become far greater, when the Muséum became accessible to the general public, and was notionally owned by the nation which claimed to represent that general public. In the case of the Muséum, this change in relation to the larger public was particularly dramatic, and shows the interaction between the growth of representative politics, and the increase in public consumption of natural history.

At the height of the French Revolution, in 1793, the Muséum was given its own constitution, which entailed the transformation of the former royal Jardin des Plantes into a national possession, designated as a place of public resort and instruction in natural history. At a profound level, its transformation was in line with the violent transformation of France from monarchy to republic, from dynastic state to representative republican government. Its transformation was also congruent with the Jacobin rhetoric which erected 'nature' and the 'natural' almost into ethical norms. The Muséum, in this revolutionary mode, was supposed to contribute to the moral, and hence political, transformation of Frenchmen into citizens of the Republic of virtue, by rendering the order of nature visible, accessible, and transparent to the sovereign people, who had replaced the sovereign monarch.

In reality, the Muséum owed much to violence, and was characterized by internal strife. Its collections were hugely amplified by the assault of Revolutionary terror on the aristocracy, and later by the victories of Revolutionary armies over the monarchs of Europe, which allowed their 'liberated' collections to be transported to the Paris Elysium. Muséum staff frequently quarrelled over what the order of nature actually was, resulting in the display of mutually incompatible classification systems in the public anatomy galleries. The 'space out of space' of the Muséum also often had to be brutally constructed, put together with a sledge-hammer *bricolage*. Cuvier, for example, described in his autobiography the typically direct way in which he extended his own display space, by breaking through from the Muséum buildings into another adjacent building left vacant by the demise of an old-regime administrative body.

But it was the same Cuvier who in 1819 was to describe the

Muséum public gardens as 'a temple', as 'the largest and most beautiful ever consecrated'.[7] Why was there such investment in the 'ideal' of the gardens, in their image as an earthly paradise, where the public could find moral uplift and civic regeneration by contact with the ordered display of nature, even by men who well knew the conflict and chaos which had gone into the making of those particular Elysian fields?

Partly this was because an image of a public space outside social production, fitted in so well with the new ideas of the collectivity, the 'nation', which was the political innovation of the Revolution itself. In 1793 military crisis on the frontier, and civil war within France led the Jacobin politicians in power to put even greater stress on the collectivity which they called the 'nation'. The 'nation' was supposed to replace the old-regime social order, which divided groups and regions from each other by privilege. The 'nation' on the other hand, was supposed in theory to be a collectivity of individuals equal before the law, without class, gender, or political conflict. The political realities of the Revolution and succeeding periods were far from reflecting this ideal. Just as the social fabric was now to be conceived as seamless, without social forces or conflicts, so too the Muséum was to be conceived as a 'Utopian' space equally removed from conflict, whose visitors would not see reflected in front of them the fragmented world of actuality, but a seamless order of nature.

This idea of the gardens of the Muséum as an Elysium also fits in with the famous analysis by the social theorist Walter Benjamin, in which he likened such Utopian spaces as parks to a collective dream:

In the dream in which every epoch sees in images the epoch which is to succeed it, the latter appears coupled with elements of pre-history, that is to say, a classless society . . . to give birth to the utopias which leave their traces in a thousand configurations of life from permanent buildings to ephemeral fashions.[8]

In exactly the same way, the Muséum offered the Paris public access to a space outside space, a space which refused to mirror the social and political conflicts so harshly mapped on to the surrounding social world. The Utopian space of the Muséum also corresponded to the essential characteristics of the political categories such as 'the nation' or 'the general will' or 'the sovereign people' which had arisen from the Revolution. This is an important point to make, because Benjamin associates the rise of Utopian public space with the onset of industrialization. The example of the Muséum seems to show, rather, that Utopian public space is generated by the enormous shift in political categories which accompanied the French Revolution, and thus long preceded large-scale industrialization in France.

Field natural history and wide open spaces

The public spaces inside the Muséum site thus associated it, and thereby natural history, with a Utopian and panoramic view of nature. The Muséum's connection with 'field' natural history was also associated with the production of images of the practice of natural history which were heroic and exemplary. Field naturalists, whether dominant and heroic figures like Alexander von Humboldt, or the less well-known figures sent out by the Muséum as '*naturalistes-voyageurs*', were imagined by a growing public for popular science as struggling over remote and dangerous terrains in dedicated pursuit of new and strange plants and animals. Public veneration of such figures, who came closer than any other men of science to emulating the heroic men of action so central to nineteenth-century European imperialist mythology, however, masked an increasing struggle within natural history over the value of field natural history.

The concept of 'the field' itself is a complex one, and we must await its definitive exploration by historians. But it is nonetheless important for our purposes, because the idea of 'the field' is pivotal in its union of spatial metaphor and epistemological assumptions.

Field natural history was closely associated with a particular approach to nature, as well as with ideals of heroic, manly endeavour. Increasingly, however, the field naturalists, and the naturalists who theorized about the order of nature, or who conducted wide-ranging comparative research in taxonomy, physiology, or anatomy, usually within built environments, ceased to be the same people. Charles Darwin, who published his *Origin of Species* in 1859, the year of Alexander von Humboldt's death, was probably the last man of science who combined all these different aspects of natural history in his own life.

Earlier in the century the battlelines had already been drawn. In 1807 the Muséum professor Georges Cuvier reviewed a report of field research by the scientific hero Humboldt. He commented:

Usually, there is as much difference between the style and ideas of the field naturalist ('*naturaliste-voyageur*'), and those of the sedentary naturalist, as there is between their talents and qualities. The field naturalist passes through, at greater or lesser speed, a great number of different areas, and is struck, one after the other, by a great number of interesting objects and living things. He observes them in their natural surroundings, in relationship to their environment, and in the full vigour of life and activity. But he can only give a few instants of time to each of them,

Figure 15.6 A naturalists' expedition into the Brazilian forest, illustrating the luxuriance of exotic surroundings, distracting the eye, the rapid passage over terrain, in this case by boat, and the frequently perilous character of such journeys. From Prince Maximilian of Wied-Neuwied, *Travels in Brazil in the years 1815, 1816 and 1817* (1820).

time which he often cannot prolong as long as he would like. He is thus deprived of the possibility of comparing each being with those like it, of rigorously describing its characteristics, and is often deprived even of books which would tell him who had seen the same thing before him. Thus his observations are broken and fleeting, even if he possesses not only the courage and energy which are necessary for this kind of life, but also the most reliable memory, as well as the high intelligence necessary rapidly to grasp the relationships between apparently distant things. The sedentary naturalist, it is true, only knows living beings from distant countries through reported information subject to greater or lesser degrees of error, and through samples which have suffered greater or lesser degrees of damage. The great scenery of nature cannot be experienced by him with the same vivid intensity as it can by those who witness it at first hand. A thousand little things escape him about the habits and customs of living things which would have struck him if he had been on the spot. Yet these drawbacks have also their corresponding compensations. If the sedentary naturalist does not see nature in action, he can yet survey all her products spread before him. He can compare them with each other as often as is necessary to reach reliable conclusions. He chooses and defines his own problems; he can examine them at his leisure. He can bring together the relevant facts from anywhere he needs

together under a single lens, both information and perception. Field naturalists were validated by their heroism in physically encountering and overcoming distance from metropolitan centres. Conversely, sedentary naturalists were forced to argue that their *psychic* distance from the object of their study guaranteed the superior truth-value of their brand of natural history. It was not a big step from the establishment of distance as a cultural value over which it made sense to struggle, to the production of the idea of objectivity, meaning precisely the placing of 'distance' between the observer and the observed, between the knower and his own responses. For the eye of the field naturalist, seduced by the dazzle of passing events, Cuvier's sedentary naturalist substitutes an observation which is distanced, and thereby dominating in its control over the whole range of the natural order. It was in this way that a previous era had often conceived of the eyes of princes moving over the conspectus of their own territories, just as the sedentary naturalist passed all nature 'in review' before him, and through this gaze controlling it, rather than running the risk of being overwhelmed by it.[12]

The naturalist and inner spaces

The 'inner space' of the sedentary naturalist, with its high degree of compartmentalization between observation and response, inner and outer worlds, comes very close to exemplifying what Norbert Elias, in his groundbreaking study of the history of emotional structure, has called '*homo clausus*' (enclosed man). By this, he refers to a human type which he sees as emerging from courtly society between the Renaissance and the end of the eighteenth century, one whose identity was based precisely on the maintenance of his own 'distance', emotional and physical, from other human beings, with a strong consciousness of himself as separate, walled off from others by invisible barriers. With the emergence of this human model came restrictions on the exploration of the world through the whole body. Conduct books, for example, recommended children to understand new things by looking but not by touching. *Homo clausus*, just like Cuvier's sedentary naturalist, was supposed to refuse access to his closely guarded inner space to the disorderly rush of sensation, the bodily effort and response, which accompanied field natural history. In contrast, Muséum work in natural history needed close physical control involved in the preparation of delicate specimens. Correct assessment of specimens depended on a highly controlled orderly environment which screened out the inessential or the background of detail.

I have argued elsewhere that *homo clausus* is also the human type on which the new political world of the French Revolution was

Figure 15.8 An allegory of natural history, end-piece to Georges-Louis Leclerc de Buffon's best-selling *Histoire naturelle, générale et particulière* (1749–67). The muse of natural history is shown in a setting which is neither indoors nor outdoors, but contains elements of both. The field beckons over the balustrade, and contains both familiar and exotic beasts. The interior is full of preserved specimens.

to be based, the model for the sort of human being who was an autonomous, and thereby legitimate, political actor, and hence was also the privileged location of 'rights'.[13] It is no accident, therefore, that natural history should also begin to construct its own 'ideal person' in a different way from the past, and that that construction should involve the same conflict between inner and outer spaces and be as deeply influenced by the impact of the new political values. Just as the Utopian 'public spaces' of the Muséum were constructed at the height of Revolutionary terror, and were congruent with the Revolution's Utopian, conflict-free idea of 'the nation' so too was the replacement of the responsive inner space of the roving field naturalist by the controlling eye of the confined sedentary enquirer.

'Curiosity' had indeed been replaced by a crisis and conflict over definitions of both inner and outer spaces in natural history, a crisis that was to provide obstacles to the maintenance of natural history as a unified discipline.

Further reading

Appel, T., *The Cuvier–Geoffroy Debate* (Oxford, 1987).

Atran, S., *Cognitive Foundations of Natural History: Towards an Anthropology of Science* (Cambridge and Paris, 1990).

Corsi, P., *The Age of Lamarck* (Berkeley and Los Angeles, 1988).

Foucault, M., 'Knowledge, space, power', in P. Rabinow (ed.), *The Foucault Reader* (London, 1986), pp. 239–56.

Lefebvre, H., *The Production of Space* (Oxford, 1981).

Limoges, C., 'The development of the Muséum d'Histoire Naturelle of Paris, 1800–1914', in R. Fox and G. Weisz (eds.), *The Organisation of Science and Technology in France, 1808–1914* (Cambridge and Paris, 1980).

Ophir, A. and Shapin, S., 'The place of knowledge: a methodological survey', *Science in Context*, 4 (1991), pp. 3–21.

Outram, D., *Georges Cuvier: Vocation, Science and Authority in Post-Revolutionary France* (Manchester, 1984).

'Uncertain legislator: Georges Cuvier in his intellectual context', *Journal of the History of Biology*, 19 (1986), pp. 328–68.

The Body and the French Revolution (New Haven, 1989).

16 Minerals, strata and fossils

Animal, vegetable or mineral? The opening move of the traditional guessing game preserves part of what was the taken-for-granted structure of the sciences, at least until the end of the eighteenth century. 'Natural history' was at that time still a highly esteemed branch of human knowledge, and no merely amateurish pursuit. It was not an archaic synonym for what would now be called biology, for it ignored the boundary between the living and the non-living: it included mineralogy as one of three divisions or 'kingdoms' of equal importance (the others were, of course, zoology and botany). But 'mineralogy' was much wider in meaning than the modern science of the same name; it was roughly the equivalent of 'earth sciences' today. The term 'geology' had indeed been proposed, but it was a neologism that was neglected or even rejected, for reasons that will become clear later in this chapter. In fact the shift from 'mineralogy' to 'geology', as the most usual term for what would now be called the earth sciences, encapsulates the dramatic changes in the culture of inorganic natural history that occurred between the late-eighteenth and the mid-nineteenth centuries.

A science of specimens

In the late eighteenth century, throughout Europe and wherever European culture extended, mineralogy was first and foremost a matter of mineral specimens: specimens collected, sorted, named and classified. Specimens were extracted from mines and quarries, hammered out of coastal cliffs or mountain crags, picked out of stream-beds or off the surface of fields, and assembled indoors in museums or private 'cabinets'. Those who collected these specimens called themselves 'mineralogists' or, more broadly, just 'naturalists'. Some, for example the owners or managers of mines, made their collections for strictly practical reasons; but most did so as a socially acceptable part of polite culture, valuing above all the unusual and the spectacular, with motives that might be at the same time aesthetic, scientific and monetary. Rare or valued specimens were exchanged between enthusiasts, purchased when a

HORACE BENEDICT DE SAUSSURE.

Figure 16.1 A highly emblematic portrait of Horace Bénédict de Saussure (1740–99), one of the most distinguished of the late eighteenth-century naturalists who studied the mineral kingdom. He is dressed to match his social status in the polite society of Geneva, but he is portrayed outdoors as if doing fieldwork. He has in his hands a miner's hammer – the badge of the mineralogist – and a rock specimen obtained by its use; by his side are mineral specimens and the bag in which he has collected them, and instruments for surveying the topography and measuring the inclination of rock strata; and he looks up – in a gaze recalling the pious poses of saints in an earlier iconography – towards the Alpine peaks in the background. By 1796, when Saint-Ours painted the portrait on which this engraving is based, the fifty-six-year-old Saussure had in fact suffered a paralysing stroke and his fieldwork career was over. From a print in the author's collection.

deceased or bankrupt naturalist's collection was put up for sale, and – not least – bought from the miners, quarrymen and peasants whose daily toil enabled them to find what these noblemen and gentlemen (and a few ladies) were prepared to pay for.

At least in the more serious collections, specimens were then compared with those of other naturalists, identified and named. As standards of comparison, collections of specimens that had been named authoritatively were particularly valued, and were exchanged between individuals or institutions. Comparisons were often made, however, not with other real specimens, but with what were in effect the *proxy* specimens pictured in publications (Figure 16.2). These were usually engravings, which were sometimes

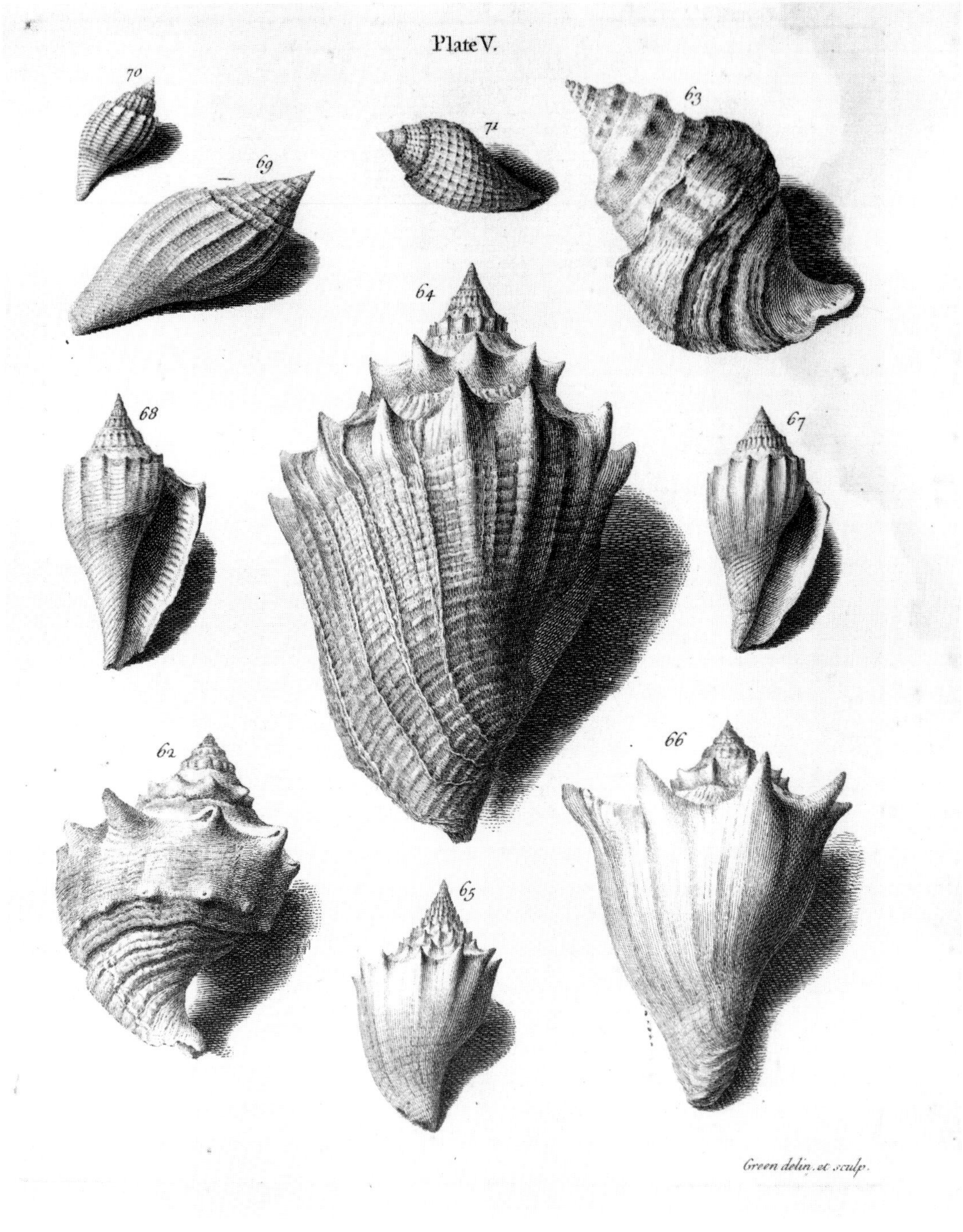
Plate V.
70
71
63
69
64
68
67
66
62
65
Green delin. et sculp.

Figure 16.2 A typical set of eighteenth-century illustrations of fossils: these are of well-preserved mollusc shells from Secondary (in modern terms, Cenozoic) strata in the south of England. Such engraved representations – minerals were depicted in a similar style – formed highly effective proxies for the real specimens. From Gustav Brander's *Fossilia Hantoniensia collecta* (London, 1766), illustrating specimens preserved in the British Museum in London.

hand-coloured with astonishing *trompe l'oeil* realism. Books with illustrations of mineral specimens, often recording a celebrated collection or material from some famous locality, were in effect proxy museums, and they spread their authors' descriptions and identifications as widely as the volumes were bought and sold.[1]

In all this, mineralogy differed little from the other branches of natural history. As in botany and zoology, the fundamental scientific goal was simply to describe, name and classify the diverse riches of nature. Minerals, no less than plants and animals, were to be described in terms of their natural *species*: species such as quartz and felspar, no less than species of daisies and deer. But most mineralogists, like other naturalists, were not content merely to identify and name their specimens. They wanted to construct a classification that would assemble similar minerals into a nesting set of groups, and so reveal the hierarchical structure of the diversity of the whole mineral kingdom.

In this task of identification and classification, it was increasingly regarded as imperative to examine the interior of minerals, as it were, as much as that of plants and animals. While the botanist dissected the intimate sexual parts of the flower, and the zoologist the literally internal anatomy of the animal body, the mineralogist resorted to the laboratory, and performed chemical analysis on his specimens in order to discover their true nature. In this way mineralogy had developed some of its strongest links with chemistry. The emergence of what became known as crystallography, at the end of the eighteenth century, provided a further set of characters for the same task of constructing a truly natural classification of minerals; but it also brought to mineralogy the prestige of being geometrical and quantitative.

To ask about the *origins* of natural species, however, seemed as meaningless in mineralogy as in botany and zoology; or at least, such questions were often regarded as abandoning natural history for the speculative realm of metaphysics. Classifications were intended to reflect the diversity of the world; how its natural kinds had come into being was generally considered to lie beyond scientific investigation, simply because it belonged, in effect, outside time. However, it was in mineralogy that this static conception of the natural world first began to be undermined, as a result of the emergence of problems for which questions of origin seemed both appropriate and soluble.[2]

A science of fieldwork

One of the distinctions that was clarified in the eighteenth century was that between *minerals* and *rocks*. The former term took on a more restricted – and modern – meaning; rocks were interpreted as aggregrates of, usually, more than one kind of mineral. Thus granite was understood as a rock composed of crystals of minerals such as quartz, felspar and mica, and limestone as a rock composed mainly of grains of calcite. Even if the origins of mineral species were considered to be beyond the realm of natural science, the origins of composite entities such as rocks were clearly not. Many rocks, notably 'pudding-stones' (in modern terms, conglomerates or consolidated gravels) and sandstones, were said to be of 'mechanical' origin, being evidently composed of the debris of pre-existing rocks; but many others, such as granite and marble, were composed of crystalline minerals and were considered to be of 'chemical' origin. Whether a chemical origin implied crystallization from an aqueous solution or from a true melt was hotly debated: the contemporary state of chemistry made the former, stressing the chemically active role of water, seem generally the more plausible.

Questions of origin remained problematic, however, for many rocks, particularly for fine-grained ones such as basalt.[3] Significantly, in such cases evidence had to be sought outside the laboratory, in the *field* relations of the rocks. Using fieldwork, the French naturalist Nicolas Desmarest (1725–1815) demonstrated, for example, that at least some basalts – including some with the spectacular hexagonal jointing that made them look like gigantic crystals – were connected to present or former volcanoes, and must originally have been molten lavas.[4] But the field relations of other basalts, found far from any volcanoes and sandwiched between sandstones or other rocks that had clearly been sediments, later suggested to other mineralogists that basalt was a rock of sedimentary origin: this view was propounded forcefully by Abraham Werner (1749–1817), who taught at the great mining school at Freiberg in Saxony. The argument that followed, peaking in the 1790s, pitched the proponents of heat against those of water, or 'Vulcanists' against 'Neptunists'. On the specific issue of the origin of basalt, the Vulcanists eventually won the argument, mainly on the strength of the field evidence. However, most mineralogists – even the Vulcanists – considered that *most* rocks were probably of aqueous origin (though they thought the water might have been very hot and chemically active in some cases) and that volcanoes were relatively minor agents in the earth's economy.[5]

The basalt controversy was important in the long run, less because it settled the origin and classificatory position of one kind of rock, than because its resolution entailed *fieldwork* as an essential part of scientific practice. Until quite late in the eighteenth century, all three branches of natural history were still mainly indoor sciences. Travel and fieldwork were indeed considered essential, but they were undertaken primarily to collect specimens, which were then gathered indoors (or at least into a botanic garden) for

the closer work that made their study truly scientific. It was in mineralogy that this predominantly indoor culture first began to be seriously challenged.

A science of mineral distributions

By the late eighteenth century, mineralogy was already far wider than the modern science of the same name, because it encompassed the geographical dimension of the science of the earth. Some of its most prominent practitioners, such as the Genevan naturalist Horace Bénédict de Saussure, insisted that fieldwork was indispensable, not just for collecting specimens – a task that had often been delegated to assistants or employees – but for seeing with one's own eyes how the various minerals and rock masses were spatially related to one another and to the physical topography of the areas in which they were found.[6] Added to that was the importance of witnessing for oneself the more spectacular features of the mineral world, such as erupting volcanoes and high mountains and their glaciers. Published descriptions of travels could convey only a pale intimation of the grandeur of these phenomena. Even the pictures that increasingly accompanied such texts were no more

Figure 16.3 A view of an extinct volcano in central France, with a solidified lava flow revealed at the river's edge to be basalt with prismatic or columnar jointing. The carriage indicates not only the scale, but also the means by which some gentlemanly naturalists did much of their fieldwork. From an engraving in Barthélemy Faujas de Saint-Fond, *Recherches sur les volcans éteints du Vivarais et du Velay* (Grenoble, 1778).

Figure 16.4 A view by Pietro Fabris of the 1767 eruption of Vesuvius, from a hand-coloured engraving in *Campi Phlegraei* (Naples, 1776–9), Sir William Hamilton's monograph on the volcanic region around Naples. Landscapes such as this were effective proxies for the first-hand experience of the more spectacular features of the mineral world. Hamilton (now perhaps best known as the husband of Admiral Nelson's mistress Emma) was the British ambassador to the court in Naples, and used his residence there to become an outstanding expert not only on volcanoes but also on the antiquities of the region.

than proxies for the first-hand visual experience of remote or distant places; at their expensive best, however, the proxies could be remarkably vivid (Figure 16.4).

'Physical geography' or 'mineral geography' therefore became for many mineralogists the preferred name for their scientific activity. Topographical maps became indispensable tools, with which the distributions of minerals and rocks could be plotted and their spatial regularities perceived.[7] Topographical maps drew attention to river patterns and drainage basins, the location and direction of mountain ranges, the form of coastlines and the distribution of more striking features such as volcanoes; they enabled generalizations about the form of the earth's surface to be perceived and expressed. The occurrence of distinctive or useful rocks and minerals could then be plotted on a map, using conventions adapted from standard cartographic practice: either as scattered symbols, denoting outcrops or quarries, or more boldly – by extrapolation – as a patchwork of colour washes (Figure 16.5).

No mineral geographer, however, could be blind to the third dimension that – at least potentially – converted distributions at the earth's surface into structures in the earth's interior. The relative abundance of rock outcrops and other natural sections in hilly and mountainous regions, and the concentration of useful mineral resources there, focused mineralogists' attention on the hard rocks they termed 'Primary', in preference to the generally softer 'Secondary' rocks of the lower-lying regions.[8] 'Primary' and 'Secondary' denoted the relative structural position of rocks, and only sub-

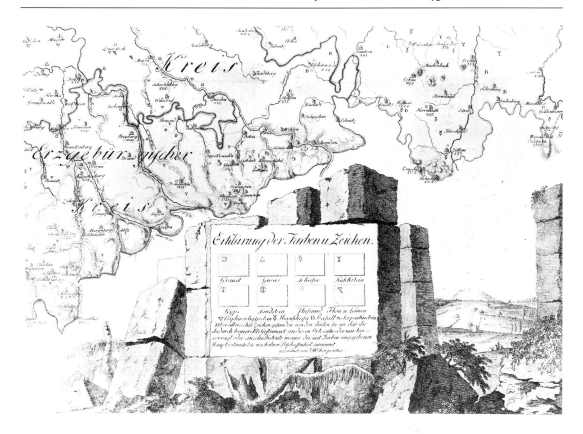

Figure 16.5 The key to a
late eighteenth-century
mineral map: the
distributions of eight kinds
of rock (granite, limestone,
sandstone, etc.) are
represented both by spot
symbols and (in the original)
by colour washes; but they
are not arranged in any
particular order, and the
map represents a pattern of
areal distribution rather than
a three-dimensional
structure. From the
Mineralogische Geographie
(Leipzig, 1778) of part of
Saxony, by Johann
Charpentier.

ordinately their presumed relative age (Primary rocks were
sometimes termed 'Primitive'). The hard rocks of upland regions
were 'Primary' because they appeared to constitute the foundations
of the earth's crust; the softer rocks of lowland regions were 'Secondary' because they manifestly overlay the others, and were at
least partly composed of their debris (although often lower in topographical position, Secondary rocks could be seen to overlie or lap
against Primary ones, wherever the junction was exposed).

The distinction between Primary and Secondary was taken for
granted in the eighteenth century, just for practical convenience
of description. Volcanic rocks of any age were generally treated as
another category on the same level; and 'Alluvial' was used for
superficial deposits of sand and gravel (not rocks at all, in the
everyday sense).

A science of rock formations

These four broad categories – Primary, Secondary, Volcanic,
Alluvial – were much too general to do justice to the diversity of
rocks found in many regions. On the other hand, the individual
layers, beds or 'strata' (for example, specific coal seams) that were

distinguished by miners and quarrymen were often not recognizable beyond a single mine or quarry, or at most some small local area. What came into use in the late eighteenth century, as a category of intermediate generality, was the *formation*.[9] The formation was a concept of immense practical value, despite the impossibility of defining it precisely. A formation was an assemblage of broadly similar rocks, separated more or less sharply from the adjacent formations; the equivalent German term *Gebirge* (literally, 'mountain range') and French term *terrain* both indicate its geographical connotations. A formation might, for example, be termed a sandstone, even if it included some intercalated strata of limestone or shale, provided it had some distinctive overall character and was clearly separated from (say) a limestone formation on one side, or above, and a shale formation on the other side, or beneath. Formations, unlike most of their constituent strata, could often be traced across country throughout some wide region, varying perhaps in thickness and detailed composition, but retaining the same position relative to other formations.

The use of the formation concept made it apparent that minerals required two distinct and complementary kinds of classification: one appropriate to specimens as analysed in the laboratory and stored in the museum, the other to the larger spatial relations of rocks observed in the field.[10] The basic and continuing work of defining, naming and classifying minerals and rocks was work centred on the examination of specimens, and it aimed to display and order the diversity of mineral 'species' and of the rocks that were their aggregates. In contrast, a classification centred on fieldwork included such categories as bed or stratum, Primary and Secondary, and – now – formation; it aimed to display the three-dimensional spatial relations of 'mineral bodies' or rock masses.

The branch of mineralogy that dealt with the classification of rock masses and their spatial relations became known as *geognosy* (literally, knowledge of the earth). The formation concept was central to its practice. Its usual form of publication was a sequential description of the formations found in some specific region. This was often accompanied by a map showing the areal extent of their 'outcrops' at the surface, and one or more sections showing their inferred relations below the surface: together, these allowed the reader to imagine the structure of the area in three dimensions (see Figures 16.6, 16.7). But 'geognosts' (as they called themselves) aimed to define and describe formations that would be recognizable beyond a single region, and ideally even on a global scale. That required a corresponding concept of *correlation*, by which a given formation was identified with its equivalents in other regions or even on other continents, even if it did not have exactly the same character everywhere.

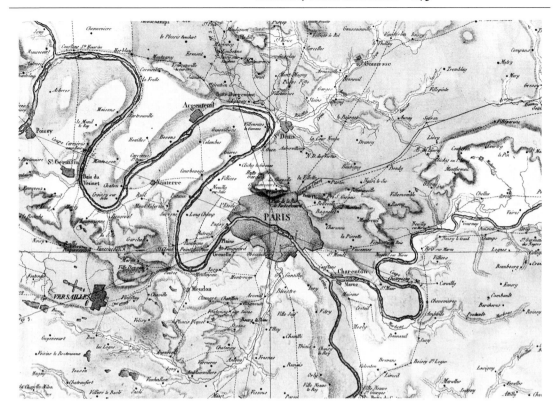

In that task of recognizing formations in different regions, and thereby making the classification as widely applicable as possible, many different criteria were tried out empirically. The kinds of rock were always basic, but many of the same rock types – for example, sandstone or limestone – characterized more than one distinct formation. That criterion was therefore supplemented by others: for example, the altitude at which a formation was usually found, or the degree to which its constituent layers were usually tilted out of the horizontal. However, those criteria proved fallible in practice; what seemed to be the same formation might be found high on mountains in one region and at low elevations in another, or highly tilted in one region and almost horizontal in another. The criterion of 'superposition' proved more reliable: true formations, whatever their altitude or degree of tilt or folding, seemed always to retain the same relative position in the three-dimensional stack of rocks revealed in natural or man-made sections.

Geognosy embodied a primarily structural conception of mineral science. Formations were typically described as 'above' or 'below' others; it was their structural order, as three-dimensional rock masses, that seemed to be reliably invariable, even when in a given region certain formations were missing. The Prussian geognost Leopold von Buch (1774–1853), in a public lecture in 1809,

Figure 16.6 Part of the engraved 'mineralogical map' (hand-coloured in the original) illustrating the monograph by Georges Cuvier and Alexandre Brongniart on the *Géographie minéralogique des environs de Paris* (Paris, 1811). This was based on fieldwork in which the standard procedures of geognosy were supplemented by study of the abundant invertebrate fossils in some of the formations; it allowed relative ages to be assigned to Cuvier's much rarer but spectacular vertebrate fossils. The continuous lines radiating from the centre of Paris indicate the positions of a series of sections; the combination of map and sections enabled the region to be envisaged as a three-dimensional structure.

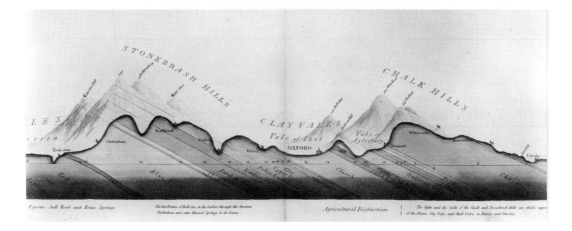

Figure 16.7 Part of William Smith's *Geological Section from London to Snowdon* (London, 1817), showing the succession of formations (in modern terms, mostly of Jurassic and Cretaceous age) in southern England. The section was intended to be 'read' in conjunction with Smith's great geological map of England and Wales (1815), to give a sense of the three-dimensional structure of the country (the 'Stonebrash' and 'Chalk' hills are, respectively, the Cotswolds and Chilterns). The vertical scale – and hence also the 'dip' of the strata – is exaggerated, in order to clarify the relations between the formations. The boundaries between them are drawn boldly with ruled lines, in a style reflecting Smith's work as a civil engineer, although this entailed major extrapolation from the evidence observable at the surface in outcrops and quarries.

explained the concept of formations by using the homely analogy of a row of houses, in which the identity and relative positions would remain unaltered even if some houses were demolished; his fellow Prussian, Alexander von Humboldt (1769–1859), later proposed an elaborate algebraic notation ('pasigraphy') to express the physical place of any formation in a putatively universal order (*Essai géognostique*, 1823). All geognosts were well aware that this structural order of position also represented a temporal order of origin, since it was axiomatic that a structurally lower formation must have preceded in origin any formations that lay above it. But this temporal element was always subordinate to the structural; geognosy was essentially a spatial science, a three-dimensional extension of mineral geography.

A science of characteristic fossils

Around the end of the eighteenth century, yet another criterion – fossils – began to be added to the practice of correlation in geognosy. The mineral specimens that eighteenth-century naturalists collected and classified included many that they considered to be of plant or animal origin. Seventeenth-century debates about the nature of distinctive mineral objects ('fossils' in the original sense) had long been resolved by settling the criteria by which those that were truly the remains of once-living beings could be distinguished from those of inorganic origin. Phrases such as 'extraneous fossils' denoted those of organic origin, but the adjectives were slowly dropping out, leaving the noun with its modern meaning.

Fossils were collected assiduously from Secondary strata, but their perceived significance was limited. The conception of them as 'extraneous' to the rocks in which they were found subtly discouraged any use of them as potential criteria for defining or identifying formations: on the ancient philosophical distinction, they seemed merely 'accidental' characters, not 'essential' ones. Fossil shells were recognized as analogous to those of living marine molluscs (see Figure 16.2); but that simply confirmed that most Secondary rocks had been deposited in the sea and that the sea must formerly have covered the present continents, which was no news to any naturalist. Above all, however, fossils were neglected because scientific attention was focused mainly on the Primary rocks, with their valuable mineral veins and spectacular mineral specimens, and they, by definition, lacked any trace of fossils.

This relative neglect of fossils in geognostic practice ended dramatically in the early nineteenth century when, from two different directions, a new attention was given to the soft and richly fossil-bearing Secondary rocks of some low-lying areas. The English mineral surveyor William Smith (1769–1839) found empirically that fossils were a highly effective means of distinguishing between otherwise similar formations, across wide tracts of the English countryside, where there were only scattered rock outcrops or quarries: specific fossils, he claimed, were 'characteristic' of specific formations. At about the same time, the French comparative anatomist Georges Cuvier (1769–1832), having been attracted to the study of *fossil* bones, realized that their relative ages could be clarified by following geognostic procedures. He and his mineralogist colleague Alexandre Brongniart (1770–1847) augmented that practice, however, by giving close attention to the fossil shells found abundantly in some of the formations around Paris: in their work the fossils became 'essential' features of the formations. Maps and sections, and lists or pictures of the relevant fossils, were published both by Cuvier and Brongniart (1808–11) and by Smith (1815–19).

The priority dispute that ensued had nationalistic overtones – not surprisingly, since France and Britain were at war until 1815 – but the end result was simply to equip geognostic practice internationally with a powerful new tool for correlation. Fossils proved to be generally – though not invariably – reliable indicators of equivalent formations, not only within a given region but also internationally and even globally. The 1820s and 1830s saw the widespread application of the new fossil-based methods to Secondary rocks in many parts of the world; by about 1840 geognosy had been transformed by the empirical success of the fossil criterion.

A standard sequence of formations, now assembled into still larger groupings or 'systems' (for example, 'Carboniferous' and 'Cretaceous'), had been accepted consensually as being valid throughout Western Europe, the most thoroughly explored part of the earth's surface. Its limits were being extended to the Russian empire and to North America, and tested in still more remote

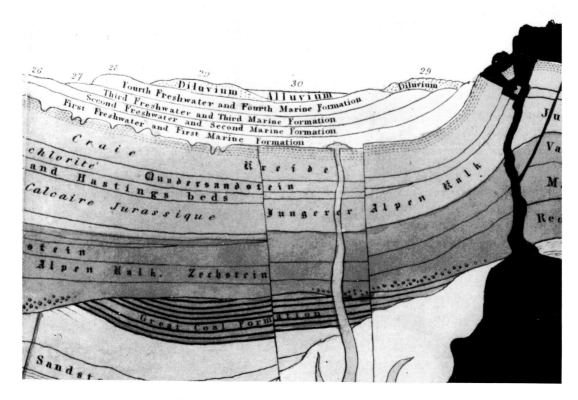

Figure 16.8 A portion of the large generalized or 'ideal section' that illustrated the popular *Geology and Mineralogy* (London, 1836) by the English geologist William Buckland, showing part of what was by then an internationally accepted sequence of major formations of sedimentary rocks (note the French names in italics and German ones in Gothic); igneous or volcanic rocks erupt from the depths. The 'Diluvium' was the peculiar 'boulder clay' or 'till' that was generally attributed to the most recent 'catastrophe'; the 'Alluvium' was the still more recent material (e.g. river gravels) from the human epoch; note the minor role of both in the whole sequence of formations, and hence implicitly the vast scale of *pre-human* earth history. Far thicker formations (not shown in this part of the section) underlay the 'Great Coal Formation' and represented even earlier periods of the earth's history.

regions throughout the world. Even the rocks that Werner had called 'Transition', in which fossils were usually rare or poorly preserved, were yielding to the same treatment (giving rise, for example, to 'Silurian' and 'Devonian'); this pushed the sequence of systems down towards the Primary rocks.

Such descriptive work – later termed 'stratigraphy' – became the foundational practice of what was now almost universally called 'geology'. It was carried out both by gentlemanly 'amateurs' – whose work was anything but amateurish – and by a new and growing breed of professional geologists. The latter were now to be found not only in the management and administration of mines, but also in the new 'geological surveys' instituted and financed by governments in many parts of the world. The first state-supported survey was in France: a team of three geologists began work, significantly, by visiting England in 1823 to study the methods that by then were standard among the members of the Geological Society of London.[11] By 1834 their geological map of France was virtually complete. In the British political climate, less hospitable to state intervention of any kind, an analogous survey started in a precarious and *ad hoc* manner in 1832, but was not established on a permanent basis until the end of the decade. By that time some of the states of the USA had also founded surveys, spreading the model beyond Europe.

Stratigraphical geology remained as structural in orientation as the geognosy from which it had developed. Of course, it provided a basis for a historical understanding of the earth and of life at its surface, but it was not itself primarily historical. Formations continued to be described as 'above' or 'below' one another far more often than they were said to be 'younger' or 'older', and the focus continued to be on their three-dimensional relations as rock masses. Likewise, the study of formations remained as thoroughly descriptive in character as the natural history that had been its origin; it provided materials for causal inferences about the earth and its life; but it was not itself primarily a causal science.

A 'theory of the earth'

Historical and causal analyses of the earth belonged to a different intellectual tradition, which in the late eighteenth century was regarded as distinct, even by those who aimed to contribute to both; only gradually, in the early nineteenth century, did it merge with the descriptive tradition. Mineralogy, mineral geography and geognosy were all regarded as branches of descriptive natural history; theorizing about the history of the earth and its causes, on the other hand, belonged to 'natural philosophy' or, in the old broad sense of the word, to 'physics', the science of natural causes.

Ever since the seventeenth century, causal and historical interpretations of terrestrial phenomena had been termed 'theory of the earth', and in the eighteenth century many important works bore that title. The phrase denoted not so much any particular theory, but rather a genre in which a set of initial conditions (for example, the earth as a molten globe) was coupled with a set of physical principles (for example, the laws of cooling bodies), and used to generate a hypothetical sequence of events or stages through which the earth might have passed in reaching its present state.[12] The *Theory of the Earth* by the Scottish 'natural philosopher' James Hutton (first published in 1788, enlarged into book form in 1795) was a late example of the genre; by the time it appeared, the sheer proliferation of such theories, with each author proudly expounding his own and emphasizing its originality, was leading to a reaction against such unconstrained causal speculation. Saussure, for example, was prominent among those who argued that the variety of these theories proved that all were premature, because they were too little constrained by observational evidence. So when in 1779 Jean-André Deluc (1727–1817), a Genevan resident in England, proposed the word 'geology' (in a mere footnote!) as the terrestrial counterpart of cosmology, it was correctly taken to be a synonym for 'theory of the earth', and was therefore treated with caution or even rejected outright.

Only in the early nineteenth century did the word 'geology' begin to lose its speculative connotations, as 'geologists' – as they then began to call themselves – recognized that more restricted kinds of causal interpretation might be legitimate. The changing status of the word is signalled by its adoption in 1807 by the first scientific society specifically for the study of the earth (the Geological Society of London), notwithstanding its founders' explicit rejection of 'theory of the earth' and strong emphasis on the value of collecting 'facts'. By 1830, when the similar Geological Society of France was founded in Paris, the word had completely lost its earlier dubious reputation, and was used in its modern sense.[13]

One of the earlier overarching theories, however, gave the science a conceptual legacy that transcended its genre. In *Les Epoques de la nature* (1778), the great French naturalist Georges Leclerc, comte de Buffon (1707–88), postulated an initially molten globe that had gradually cooled to its present state. The theory itself was based on little fieldwork, and was widely regarded as outmoded even when published; but the elegant text embodied metaphors that were powerfully influential, although not original to Buffon. His hypothetical story was divided into the six 'epochs' of his title, and he referred to features such as extinct volcanoes and the marine fossils found high above sea-level as the 'archives' or 'monuments' of nature, because they could be regarded as relics surviving from some former state of things. The language of 'epochs' was quickly adopted by others such as Desmarest, but in the service of more modest and local interpretations.[14] Likewise, natural features were used increasingly as evidence for reconstructing an earth history which – because it was ultimately contingent – could *not* be predicted in advance on the basis of any overarching theory. It is no coincidence that this matched the contemporary use of human archives and monuments by historians and antiquarians, in the service of a new historicism, in place of an earlier and deductive style of 'conjectural history'. As with the human world, the diversity of the natural world began to be historicized: a static 'natural history' of the earth began to turn into a truly temporal *history* of the earth.

A science of the history of nature

This newly historical element was evident (as in Desmarest's work) even before geognosy was transformed by the addition of the fossil criterion; but the new attention given to fossils from around the turn of the century greatly accelerated the change. Large bones, apparently those of elephants and other tropical mammals, had long been found in relatively cool climates and high latitudes in both Old and New Worlds. But only in the 1790s did Cuvier's careful study of these bones decisively confirm earlier suspicions that they belonged to species that were distinct from any living mammals and probably extinct. That in turn made it seem more plausible that many fossil shells – far more common than any bones – also belonged to truly extinct species, and not, as had until then seemed possible or even probable, to species still lurking alive in the unexplored depths of the ocean.

With that growing belief in the generality of extinction, it made sense to treat the 'characteristic' fossils of the various formations as being indeed 'essential' to them; for they now became indices of unrepeatable *historical* change, as well as of repeatable environmental conditions. Cuvier's and Brongniart's joint study of the formations around Paris (1808–11) was accepted immediately as a decisive exemplar of a new practice that combined the older geognostic framework with a newly historical and causal dimension. Unlike Smith's rival work on the English formations, with its purely empirical use of the fossil criterion, the French study treated both formations and fossils as evidence for a truly historical interpretation of the Paris region: in terms both of a changing environment – an alternation of marine and freshwater conditions – and of a unique and irreversible history of life.

At much the same time, Cuvier transformed the concept of the earth's 'revolutions' – until then simply a vague notion of major changes in the past – into a much more concrete argument: that *some* (not all) such changes had been 'catastrophes', or sudden changes in environment, which could have caused the extinctions he claimed. Like most of his contemporaries, Cuvier rejected the older style of 'theory of the earth' as premature, and prudently abstained from suggesting what might have caused the sudden catastrophes. But he also rejected the widespread view that the 'actual causes' currently at work in the world were adequate to have produced all the effects observed.[15] The subsequent controversy led geologists in the 1820s to examine much more closely – either by direct field observation or by analysing

Figure 16.9 The frontispiece to Charles Lyell's *Principles of Geology* (London, 1830). This engraving of the surviving columns of the 'Temple of Serapis' near Naples deliberately symbolized Lyell's claim that all the natural 'monuments' of past events can be attributed to the action of processes no more catastrophic than those that are directly observable or recorded in human history. Although the temple (probably in fact a market) was built on dry land in Roman times, a later submergence was recorded by the borings of marine molluscs part way up each column; yet at a still later time the ruin had been elevated back to its modern position at sea-level. This epitomized Lyell's theory of non-directional or steady-state earth history; it suggested how a long succession of similar small-scale changes could have produced even the elevation of mountain ranges and the subsidence of continents, given enough time; and it neatly integrated geological with human history, and past with present, by using a human monument as a witness to geological change.

historical records – just what those agents *were* capable of effecting.

Developing that tradition, the London geologist Charles Lyell (1797–1875) argued persuasively in his *Principles of Geology* (1830–3) that the power of 'actual causes' had indeed been underestimated, and that much geological change had been very gradual, or at least, no more violent than the natural events and processes recorded in human history (Figure 16.9). 'Catastrophists' and 'uniformitarians', as they were called in the 1830s, were never sharply separated parties, and by around 1840 they had in practice reached a kind of compromise: most geological features were agreed to be the work of agencies still observable in the present world, such as sedimentation, erosion, vulcanism and crustal elevation; but it was widely believed that these processes might have slowly declined in their intensity, and might have been much more powerful in the distant past.

A compromise seemed inescapable, because some phenomena continued to resist Lyell's kind of explanation. The peculiar features (erratic blocks, till or boulder clay, etc.) that had been attributed to the most recent catastrophe were particularly puzzling, because no 'actual cause' seemed adequate to account for them. Cuvier had equated the most recent catastrophe with the flood recorded obscurely in the ancient records of many human cultures. His English follower William Buckland (*Reliquiae diluvianae*, 'Relics of the Deluge', 1823) accentuated its identification with one such record – the 'Deluge' of Genesis – and termed the deposits 'Diluvium' (see Figure 16.8). But in practice the 'geological deluge' was conceived as an event very different from any literal reading of Genesis. Its later transformation into an 'Ice Age', due mainly to the Swiss naturalist Louis Agassiz's *Études sur les glaciers* (1840), finally severed any such connection, for all but the scientifically marginal 'scriptural geologists' (mainly in Britain and North America). Above all, however, the acceptance of some kind of glacial period in the geologically recent past served to confirm that the earth had had an unexpectedly eventful history.[16]

A history of earth, life and man

By around 1840 most geologists conceded in practice that the course of earth history could not be predicted in advance by any grand theory – neither by Lyell's steady-state theory nor by the more generally favoured theory of a gradual cooling – but only reconstructed by detailed analysis of the organic and inorganic 'archives of nature'. That conclusion was embodied most strikingly in the first tentative attempts to represent in pictorial form what the world might have looked like in that remote pre-human past (Figure 16.10). It was embodied more formally in the proposal in 1840, by the English geologist John Phillips (1800–74; the nephew of William Smith), that the whole history of life could be summarized in three great eras, the Palaeozoic, Mesozoic and Cenozoic (the eras of ancient, intermediate and recent kinds of animals). With such terms, the historicization of the older geognostic classification of formations and systems was in principle complete.

That this immensely long and complex history was almost entirely *pre-human* was a conclusion that was transformed in these same decades from conjecture to consensus. Back in the late

Figure 16.10 'A more ancient Dorset': a view of life at a remote pre-human period (in modern terms, Jurassic), as drawn by the English geologist Henry De la Beche and lithographed by the artist George Scharf (1830). This was one of the first true pictorial reconstructions of extinct animals, based on a detailed analysis of well-preserved fossils and showing them in their inferred habitat. Large-jawed ichthyosaurs, thin-necked plesiosaurs and flying pterodactyls (in modern terms, pterosaurs) are the most prominent animals. Such imaginary landscapes were quickly adopted for popular books, and served to make the new earth history vividly real to a wide public.

eighteenth century, Buffon's seventh and human epoch (added only shortly before publication) made explicit a speculation already widespread among naturalists: that the whole of human history was but a brief final chapter in a far longer story, recorded in the thick sequences of fossil-bearing Secondary rocks. Buffon's total time-scale of tens of thousands of years – based on scaling up the results of experiments with cooling model globes – was modest by later standards; but it was quite vast enough to be, in a literal sense, almost unimaginable. In the subsequent decades, quantitative estimates of geological time were rarely made explicit, simply because there was little concrete evidence to base them on, and they were widely regarded as merely speculative. But the practice of geologists leaves no doubt that by around 1840 they had surmounted the imaginative hurdle of thinking about vast expanses of time, as successfully as astronomers had become accustomed to thinking about the vast expanses of space.

Geologists' estimates of time were, of course, far too large to be compatible with the traditional chronologies derived from a literal reading of Genesis. However, such literalism had already been discarded by most savants or 'men of science', whether they were personally religious or not, chiefly as a result of the development of scholarly biblical criticism. After the mid-eighteenth century, no major naturalists were seriously hampered by ecclesiastical criticism, still less by persecution, on account of the time-scales they proposed. Buffon, for example, received only the most perfunctory criticism in 1778 for his *Epoques*, in contrast to that which had greeted his earlier theory of the earth in 1749. After the turn of the century the relation between geology and Genesis became a marginal issue for geologists, except for the public relations of their science, and even then only locally (mainly in Britain and North America).

What was more problematic was the relation between the complex history of the earth and its life, as it was reconstructed with increasing confidence and precision by geologists in the early nineteenth century, and the far shorter span of human history, as it was extended backwards by analysis of ancient documents (of which Genesis was only one). Both natural and textual sources continued to be deployed in conjunction with one another, as in Cuvier's and Buckland's work, because both seemed relevant to what was now perceived as a single *history*. Until after mid-century, however, the two sources proved extremely difficult to integrate: the fossil traces of early human life remained sparse and highly problematic, and the concept of a long human 'pre-history' preceding any literate civilization was slow to gain acceptance.[17]

Conclusion

The practice of mineralogy first began to diverge from that of botany and zoology when its problems demanded a geographical, distributional or spatial dimension, and therefore a heightened emphasis on fieldwork. That practice then became increasingly three-dimensional or structural in character, and developed into 'geognosy', the study of rock masses or 'formations' and their world-wide correlation. Geognosy in turn was transformed (into what was later termed stratigraphy) by the striking empirical success of the new criterion of fossils. Initially distinct from all such descriptive practices was the causal project of 'theory of the earth', which aimed to model the likely course and causes of the earth's temporal development. This did not become truly historical, however, until its deductive style was abandoned: its concepts of nature's 'epochs', and of specific features as nature's 'archives', were absorbed into a more inductive and contingent style, by being combined with the newer fossil-based geognosy. By about 1840, a long and complex earth history, dwarfing the whole of subsequent human history, had become a consensual feature of the scientific view of the world.

Further Reading

Ellenberger, François, *Histoire de la géologie*, vol. II, *La Grande éclosion et ses prémices, 1660–1810* (Paris, 1994).

Gohau, Gabriel, *A History of Geology* (New Brunswick, NJ, 1990).

Les Sciences de la terre aux XVIIe et XVIIIe siècles: Naissance de la géologie (Paris, 1990).

Gould, Stephen Jay, *Time's Arrow, Time's Cycle: Myth and Metaphor in the Discovery of Geological Time* (Cambridge, MA, 1987).

Grayson, Donald, *The Establishment of Human Antiquity* (New York, 1983).

Greene, Mott, *Geology in the Nineteenth Century: Changing Views of a Changing World* (Ithaca and London, 1982).

Laudan, Rachel, *From Mineralogy to Geology: The Foundations of a Science, 1650–1830* (Chicago, 1987).

Porter, Roy, *The Making of Geology: Earth Science in Britain, 1660–1815* (Cambridge, 1977).

Rudwick, Martin, *The Meaning of Fossils: Episodes in the History of Palaeontology* (London and New York, 1972; Chicago, 1984).

The Great Devonian Controversy: The Shaping of Scientific Knowledge among Gentlemanly Specialists (Chicago and London, 1985).

Scenes from Deep Time: Early Pictorial Representations of the Prehistoric World (Chicago and London, 1992).

Rupke, Nicolaas, *The Great Chain of History: William Buckland and the English School of Geology, 1814–1849* (Oxford, 1983).

Secord, James, *Controversy in Victorian Geology: The Cambrian–Silurian Dispute* (Princeton, 1986).

17 Humboldtian science

About twenty-five years ago, the American historian of science Susan Faye Cannon began using the term 'Humboldtian science' to characterize a major current in early nineteenth-century natural philosophy, the 'avant-garde', 'the great new thing in professional science in the first half of the nineteenth century'. She used 'Humboldtian' to replace the usual description of the period as 'Baconian', a label which condemned the period to a naive, encyclopaedic empiricism relying entirely on the collection and collation of facts, a fascination with the particular, and a rejection of theory. The vague 'Baconianism' of nineteenth-century scientists, she pointed out, was actually (1) a disciplined practice of observing and measuring a number of physical variables over continental expanses of territory, using the latest advances in instrumentation and attending to all possible sources of error; and (2) a sophisticated conception of the relationship between accurate measurement, sources of error, and mathematical laws. It *was* encyclopaedic, embracing botany, mineralogy, zoology, geology, meteorology, terrestrial magnetism, atmospheric chemistry and tides, and topography, but only insofar as these were capable of numerical expression, arrangement, and comparison. This, Cannon remarked, well describes the enterprise of the Prussian traveller Alexander von Humboldt (1769–1859), and rather than repeat the unhelpful characterization of early nineteenth-century science as 'Baconian', historians would be better off realizing that it is 'Humboldtian'.[1]

The effect of her essay on 'Humboldtian science', published in 1978, was immediate, especially among historians of British and American science, and Humboldtian science quickly found applications in the founding of the British Association for the Advancement of Science (BAAS), the 'Magnetic Crusade', and the professionalization of science in general; it even helped illuminate important aspects of the thought and practice of Charles Darwin, who carried Humboldt's *Personal Narrative of Travels to the Equatorial Regions of the New Continent* (trans. H. M. Williams, 1814–29) with him aboard HMS *Beagle*. The concept clearly 'explained' important aspects of the activities of a broad set of natural philosophers, that had only loosely and uneasily filled the rubric of 'Baconianism', especially the concern for precise measurement and mathematical law.[2]

But while Cannon's invention clarified and deepened our picture of early nineteenth-century science, the label also prematurely 'black-boxed' a complex of concerns and practices, relying on Humboldt's evident stature in the period to define a 'style' or 'complex' and give the package explanatory force. However, the Humboldt which so appealed to Cannon's naturalists was a translated Humboldt, and so a transmuted Humboldt. The evidently useful concept of Humboldtian science requires dismantling; Humboldt needs to be differentiated from the Humboldtians, to expose the illuminating process of translation, the use of Humboldt and his project as a resource in different national and social settings, which the very usefulness of Cannon's concept to historians indicates. To start, I propose to open the black box of Humboldtian science by examining the notion of natural order operative in Humboldt's enterprise.

Born in Berlin in 1769, Alexander von Humboldt began his career as a Prussian mining official in the 1790s, reorganizing the mines and industries of the small, outlying principalities of Ansbach–Bayreuth. He resigned his position in 1797, and after spending two years' searching out instruments, skills, and a boat, Humboldt went to South America in 1799 under the protection of the Spanish king, Charles VII, as part of a planned voyage around the world. Already known for his chemical, geological, and medical investigations, Humboldt returned from the Americas a public hero. He spent the next twenty-two years living and working in Paris, publishing the results of his American travels, which issued in a series of over thirty volumes and exhausted the resources of three publishers by the time the final monograph (on New World grasses) was published in 1834. King Friedrich Wilhelm III of Prussia called Humboldt back to Berlin in 1826 as royal chamberlain, and after a brief expedition to Siberia and the eastern reaches of the Russian steppes in 1829 Humboldt divided the rest of his life between Potsdam and Paris. His intellectual wanderings matched his physical ones. Humboldt's last work, *Kosmos. Entwurf einer physischen Weltbeschreibung* ('Cosmos. Sketch of a Physical Description of the Universe') (5 vols., 1845–62), began its 'physical description' of the universe with the condensation of stars and planets out of nebular clouds and ended with the geography of plants. He was busy with geological stratification when he expired over the preface to the fifth volume in 1859, a few months short of his ninetieth birthday.

By the time he resigned his commission in the mines, Humboldt had committed himself to the pursuit of what he called 'terrestrial physics' (*physique du monde*). He often contrasted the new science with 'descriptive' natural history, as in an 1805 paper read before the Berlin Academy of Sciences:

Until now, little has been achieved by travelling naturalists for . . . terrestrial physics [*Physik der Erde*] (*physique du monde*), because they are almost all concerned exclusively with the descriptive [*naturbeschreibenden*] sciences and collecting, and have neglected to track the great and constant laws of nature manifested in the rapid flux of phenomena, and to trace the reciprocal interaction, the struggle, as it were, of the divided physical forces.

What kinds of phenomena led to 'great and constant laws', manifested the 'interaction of forces', and so attracted the attentions of the terrestrial physicist? Humboldt lists them:

the intensity of magnetism, or the strength of the charge of the globe in different regions and elevations, measured by the oscillations of a magnetised needle; hourly changes of the magnetic meridian; general meteorological phenomena; yearly, monthly and hourly mean decrease in temperature in the upper layers of the atmosphere . . . regular ebb and flood of the aerial ocean, indicated by hourly variations of the barometer . . . the chemical and hygroscopic composition of the atmosphere; the effect of the elevation of the sun and mountainous regions on the electrical charge of the air.[3]

The terrestrial physicist had to carry and care for a remarkable array of precise instruments to measure everything from the passage of time (chronometers) and the angular distances of stars, satellites, and landmarks (telescopes, quadrants, sextants, a repeating circle, a theodolite), the direction and intensity of magnetism (dip needles and magnetic compasses), to the temperature (thermometers), humidity (hygrometers), pressure (barometers), electrical tension (electrometers) and chemical composition (eudiometers) of the atmosphere. Nor would just one of each instrument suffice: chronometers, barometers, eudiometers etc. by different makers or built on different principles should be used together, their errors and constancy compared. In terrestrial physics these instruments now crowded out the specimen cases and collection jars of the voyaging naturalist.

In short, Humboldt's enterprise demanded a new type of naturalist: the naturalist as 'physicist'. In his 1805 *Essai sur la géographie des plantes* (Paris, 1807), the introductory volume to the results of the entire American voyage, Humboldt contrasted the *botaniste nomenclateur*, interested only in the individual structures which distinguish species and genera from one another, with the 'higher, philosophical aims' of the *botaniste physicien*, concerned with the geographical relations of plants to one another and to the geographical variation of other physical parameters. The former concerns were essential, but necessarily preliminary to the latter. Admiration of the ingenuity and variety of the organic structures of an individual or species was to be replaced with admiration for the great laws evidenced in their distribution over the earth. 'We have botanists, we have mineralogists, but we have no *physicists*, of the type demanded by [Francis Bacon's] *Sylva Sylvarum*', he wrote to his banker, explaining his imminent departure for Spanish America in June 1799 by invoking Bacon's exemplary natural history, just as his Humboldtian successors would do. 'My single true purpose is to investigate the confluence and interweaving of all physical forces'.[4] He was pressing the collecting and classifying activities of the field naturalist into the service of terrestrial physics, a master-science, the object of which was the 'mutual interaction of physical forces'.[5]

This new framework for the sciences is nowhere more graphically clear than in Humboldt's *Tableau physique des Andes* ('Physical Profile of the Andes') (Figure 17.1). Complementing his manifesto for a plant geography, with which it was published in 1807, the *Physical Profile* was Humboldt's attempt to summarize the results of his American expedition in a single table, six times the size of the quarto volume within which it was folded. Twenty-one scales framed a schematic portrait of the Andes near the equator, represented by Chimborazo, in present-day Ecuador, then thought to be the highest peak in the world; four of the scales counted elevation in meters and Parisian *toises*, while the other seventeen recorded parameters like the electrical tension and chemical composition of the atmosphere, the nature of rock formations and type of soil, humidity, the refractive power of the air, the intensity of light, the force of gravity, and more, all as they varied with elevation. On the portrait itself were arranged the names of plant species and 'vegetational regions' as they occurred along the flank of the mountain, and for an extra ten francs the hand-coloured edition would illustrate the gradual change in the green of the vegetation and the blue of the sky.

What justified this extraordinary compilation was Humboldt's belief in a 'general equilibrium' of forces:

The general equilibrium which reigns amongst disturbances and apparent turmoil, is the result of an infinity of mechanical forces and chemical attractions balancing each other out. Even if each series of facts must be considered separately to identify a particular law, the study of nature, which is the greatest problem of general physics [*la physique générale*], requires the congress of all the forms of knowledge which deal with the modifications of matter.[6]

Out of the juxtaposition of so many measurements, 'Nature' itself was supposed to emerge, as a dynamic equilibrium of forces. This equilibrium derived not from a mathematically demonstrable balance of forces, like Laplace's self-regulating world system, but from the sheer number of forces acting simultaneously and varying globally; the lawfulness of Humboldt's 'Nature' was born of 'infinity' and complexity, not of mechanism. The greater the number of forces observed and accurately measured, and the greater the extent of the earth's surface such measurements covered, the more fully would the order of Nature emerge.[7]

'Modern physics', balances, and the history of the earth

Humboldt's insistence on the 'co-operation' or 'equilibrium' born of the 'mutual struggle' of physical forces, which formed the chief object of a 'congress' of all the natural sciences, had two, interrelated sources. The first was a lifelong devotion to a liberal, secular ideal of progress and the progressiveness of history, the constructive 'ebb and flood' of peoples and cultures as well as forces and fluids. The second, which gave the first concrete and practical form, was his encounter with the anti-phlogistic chemistry of Antoine-Laurent Lavoisier and the geognosy of Abraham Gottlob Werner at the Royal Saxon School of Mines in Freiberg, where during the summer and winter of 1791–2 Humboldt prepared himself for his commission in Prussia's Department of Mines.

What so excited Humboldt about Lavoisier's *Traité élémentaire de chimie* (1789), which he read thrice over in 1791, was its analytic 'chain of reasoning': in the very language Lavoisier proposed, one could *see* elementary substances condensing and evaporating as they entered into various combinations, determined by the play of chemical affinities and the temperature of the system, as detected by the precision balance.[8] German debates over the status and existence of Lavoisier's new substances led Humboldt to suggest to G. C. Lichtenberg, Professor of Experimental Physics at Göttingen, that natural science as a whole required a general, metaphysical critique, one that would establish Lavoisier's 'reasoning' – the use of the balance to detect subtle matters which enter into manifold combinations with one another, are fixed and unfixed, in large part due to their greater or lesser admixture with caloric fluid [*Wärmestoff*], or heat – as the general philosophical foundation of physical science. 'In any science unable to reach pure knowledge, we can seek certainty to the extent that its concepts can be constructed. Any science to which quantitative reasoning cannot be applied fluctuates . . . *Ought* a chemistry ever go beyond the ponderable?'[9]

Humboldt eagerly applied the principles of what he saw as 'modern physics' to improving Prussian industry. A new

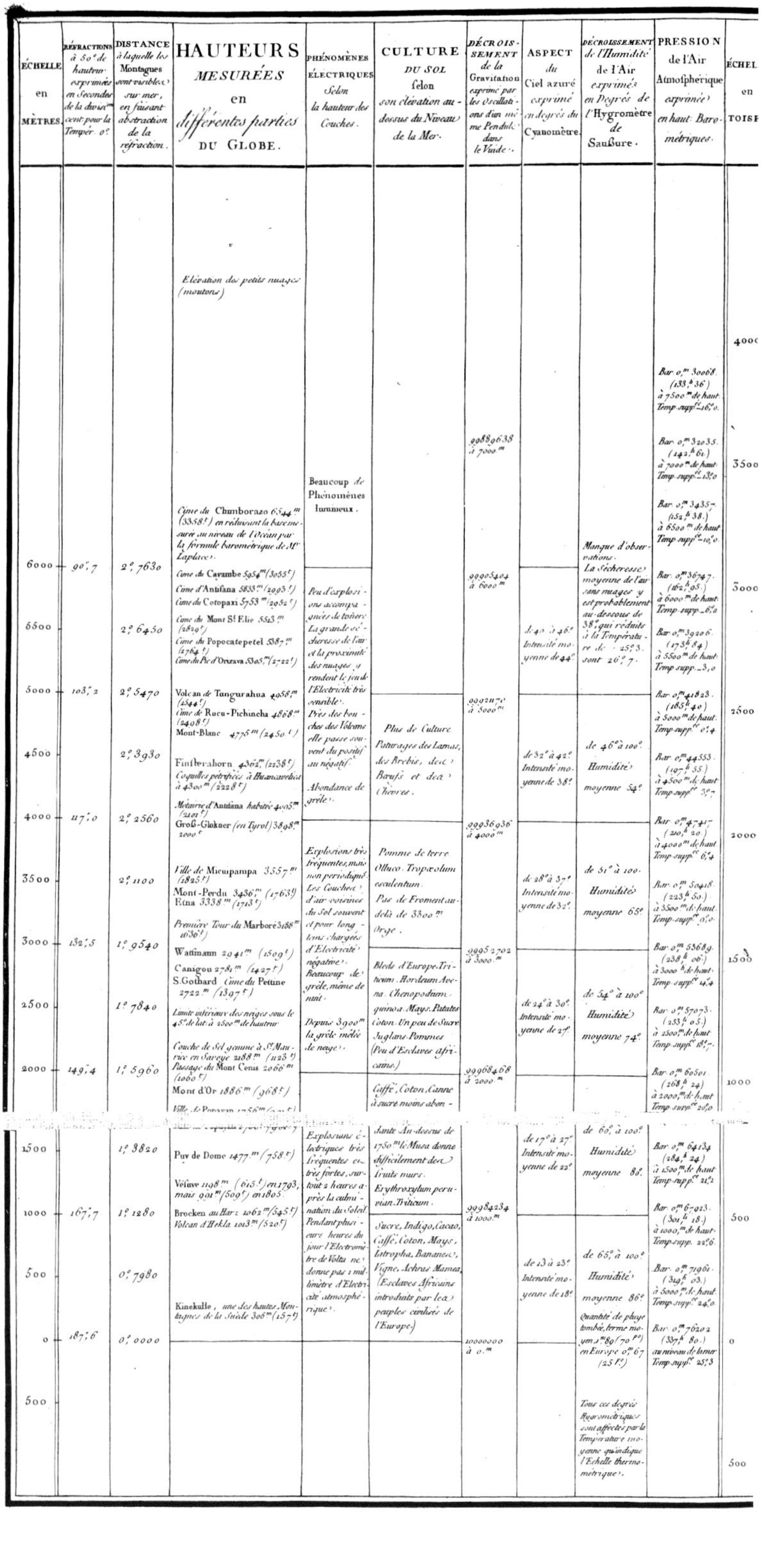

Figure 17.1 Measurements and observations from the 'Physical Profile of the Andes and Surrounding Country' (1807), the engraving intended to accompany Humboldt's 'Essay on the Geography of Plants, with a Physical Profile of the Tropics' (1805). The 'Physical Profile' summarized the results of Humboldt's 1799–1804 expedition from the perspective of a terrestrial physics. The columns, which frame the profile of the Andes, exemplify the collaboration of mathematical theory and precise observation of the widest possible array of physical variables, which characterized Humboldtian science. Humboldt's measurements of atmospheric refraction were corrected by J.-B.-J. Delambre, chief astronomer at the Bureau of Longitudes in Paris; precise elevations were calculated from the barometric measurements of Humboldt and others by Marie Riche de Prony, geometer and engineer at the Paris School of Roads and Bridges, using a formula from Laplace; Humboldt's observations of the hygrometer were converted into humidities by J.-L. Gay-Lussac, a young chemist at the Polytechnic; J.-B. Biot reduced Humboldt's observations of solar intensity to a uniform scale, using a formula from Laplace.

ÉCHELLE en MÈTRES.	TEMPÉRATURE de l'Air à diverses hauteurs, exprimée en maximum et minimum du Thermomètre centigrade.	COMPOSITION CHIMIQUE de l'Air atmosphérique.	HAUTEUR de la limite inférieure de la Neige perpétuelle sous différentes latitudes.	ÉCHELLE des Animaux selon la hauteur du sol qu'ils habitent.	DEGRÉS de l'eau bouillante à différentes hauteurs. Thermomètre centigrade.	VUES Géologiques	INTENSITÉ de la Lumière dans l'air à diverses hauteurs mesurant pour unité son intensité dans le soleil.	ÉCHELLE en TOISES.
						La nature des Roches paraît en général indépendante des différences de latitude et de hauteur. Mais en ne considérant qu'une petite partie du Globe, on découvre que dans chaque région l'ordre de superposition des roches, l'inclinaison et la direction de leurs couches ont été déterminés par un système de forces particulier. On reconnaît qu'il existe de certaines lois locales selon lesquelles s'élèvent les différentes formations au dessus du niveau de la mer.		
					Eau bouillante à 77°,0 (61°,6 R) Bar. 0″,320.	Les Régions équatoriales présentent à la fois les cimes les plus élevées et les plaines les plus étendues du globe depuis 0° à 45° et nulle part ailleurs sur la terre, les montagnes excèdent la hauteur de 5850″. Leur abaissement vers les pôles n'est cependant pas très considérable; car sous les 19°, 45° et les 60° de latitude de bor. on a trouvé des Cimes de 4700″ et même de 5500″ d'élévation.	0,9164	3500
6500			Pas d'Êtres organisés fixés au Sol.			Les Plaines équatoriales mêmes, celles contenues entre la pente orientale des Andes et les côtes du Brésil, sur 700 lieues de long, n'ont pas 70″ à 100″ de hauteur au dessus du niveau de l'Océan.		
6000	Régions trop peu fréquentées pour en connaître la température moyenne, qui cependant y paraît au dessous de zéro. A 5403″ le Thermomètre monte quelquefois à 7°,8		L'air retiré de l'eau de Neige contient 0,287 d'Oxigène.	Le Condor des Andes, quelques mouches ou Sphinx voltigeant dans les airs, poussés élevés en ou Régions par les courans ascendans.	Eau bouillante à 81°,0 (64°,8 R) Bar. 0″,367.	Toutes les formations que l'on a décrites sur le reste du Globe, se trouvent réunies vers l'Équateur. Leur ancienneté relative, qui se manifeste dans l'ordre de leur superposition, y paraît, en général la même que dans les Zones tempérées. Les granites, qui servent de base au Gneiss, au Syénite, au Schiste micacé et au Schiste primitif, les formations secondaires, dans la grès, dans de Gypse et trois de Roches volcaniques, offrent des exemples frappants de l'identité de structure qui règne dans les parties les plus élevées du Globe. La formation problématique des Basaltes, des Amygdaloïdes, des Roches amphiboliques et des Porphyres à la hauteur de 4500″ "Le Charbon de terre et l'Abondance de Porphyres parfois (Porphyre) à une sparse à la haute crête des Andes, comme celle l'est sur celle des hautes chaines de l'Europe.	0,9047	3000
5500		La quantité d'oxigène atmosphérique paraît la même dans les hautes régions et dans les plaines. Mais la proximité des Volcans peut quelquefois, sur les hautes cimes des Andes, modifier la composition de l'air.	Neige perpétuelle sous l'Équateur et sous les 3° lat. bor. à 33° lat. austral 4800″/4464°. Pas de variations de 30″ Neige perpétuelle sous les 20° lat. bor. à 4800″/2361°/maxte elle y descend en hiver à 3800″	Des Vigognes des Guanaco, des Alpaca en bandes nombreuses. Quelques Ours. Con et Eucone Capromal que. Plus de poissons dans les lacs.	Eau bouillante à 84°,7 (67°,7 R) Bar. 0″,418.		0,8922	2500
5000	de -7°,5 à 18°,7 Température moyenne -3°,7 (3°,R.) Il tombe de la neige jusqu'à 4100″							
4500								
4000					Eau bouillante à 88°,0 (70°,8 R) Bar. 0″,474.	Parmi les phénomènes géologiques qui sont particuliers aux Régions équatoriales du nouveau Continent, on doit citer sur tout l'épaisseur des couches et la grande hauteur à laquelle on découvre les matières postérieures au Granite. En Europe le Granite n'est pas couvert par d'autres Roches depuis 3500″ à 4700″. Aux Andes on ne le voit pas au dessus de 3500″. Les Cimes les plus élevées du Globe sont des Porphyres dans lequel l'Amphibole abonde, quelquefois du Quarz et de quelques Minéralogie. On les regarde comme produit d'autre. Les minérales activé par le feu volcanique. Des formations de Grès se trouvent à Huancaveltica à 4500″ "Le Charbon de terre se découvre près de Huaraca à 4400″ de hauteur. Les plaines de Bogota à 2705″ et 2900″ sont couvertes de Grès de Pierre calcaires secondaires, de Gypse et de Sel gemme. Des Coquilles pétrifiées se trouvent aux Andes au dessus de 4500″	0,8787	2000
3500	de 0°,0 à 20° Température moyenne 9° (7°,2 R.) Abondance de grêle, même quelquefois de nuit.	La quantité d'hydrogène contenue dans l'air atmosphérique est moindre de deux millièmes d'une troupe que plus d'hydrogène à 7000″ d'élévation qu'au niveau de la mer.	Neige perpétuelle sous les 35° de latitude à 3500″/1800°/ de hauteur. Neige perpétuelle sous les 40° de latitude à 3100″/1600°/de hauteur.	Des Lama devenus sauvages à la pente occidentale du Chimborazo. Le petit Ours à front blanc. Granda Cerfs. Le petit Lion. Quelques Colibri. Plus de poissons dans les lacs.	Eau bouillante à 91°,5 (73°,2 R) Bar. 0″,536.		0,8640	1500
3000	de 1°,2 à 23°,7 Température moyenne 18°,7 (15°,R.) Grêle très abondante. Brume fréquente et peu élevée.		Neige perpétuelle sous les 45° de latitude bor. à 2500″/1282°/ Aux Pyrénées à 2448″ En Suisse à 2700″ sur les zones isolées; des Montagnes il passe 3100″ Phénomène peu constant.	Vivera maputo Félis tigrina. Grands Cerfs. Palamedos Epinoza. Abondance de l'Aznar de et de Plongeurs. Beaucoup de poux. (Ped. Hum)	Eau bouillante à 94°,3 (75°,4 R) Bar. 0″,600		0,8478	1000
2500	de 12°,5 à 30°		dans les Zones variables.	Petits Cerfs (Cervmeau) Tapir Sus Tai		(en Europe on ne les a pas vu au dessus de 3500″)		
2000				yroso Felis pardalis. Quelques Singes Alouate. Troupial (Oriolus) Coluber corax. Pas de Boa, pas de Crocodile, Beaucoup de Chiques (Ped. penç.		Le Sol du Royaume de Quito contient à 1500″ de hauteur d'énormes ossemens d'Éléphants dont l'espèce paraît détruite.		
1500	Température moyenne 21°,2 (17°,R.) Grêle assez rare. Ciel souvent brumeux.	L'air atmosphérique contient à 200 d'oxigène, 0,28. A toute et environ 0,005 d'acide carbonique. Le maximum de ces variations ne peut en outre recéder un millième d'oxigène.	Sous l'Équateur on voit tomber de la neige à 4800″/2000°/ de hauteur. Au Mexique sous le 19° de latitude elle tombe jusqu'à 1800″ de hauteur.	Singes Sapajou et Alouate. Jaguar (Felis onca) Tigre noir. Lion (Felis onça) laura capibara. Paceroma. Fourmilier. Cervmesse. Armadill. Aptene Aites. Crax Ampeho Boa. Crocodile. Lamentin. Ela mental. Mosquito (Oestr. Human)	Eau bouillante à 97°,1 (77°,7) Bar. 0″,679.	Les Grès de Cuenca où 1560″ l'épaisseur, une formation de Quarz à l'Ouest désinsaturé en à 2400″ La Cordillère des Andes présente plus de 80 Volcans enflammés dont quelques uns sont éloignés de la mer de 37 à 40 lieues marines et dont les plus élevés et les plus rensforcés par les plaines ne viennent pas de laves coulantes, mais des Pierres pumicées. Obsidiennes, des Porphyres et des Basaltes écorchés et sur tout de l'Eau et cette terre subarec dans laquelle est souvent enveloppé un poisson (le Pimelodus Cyclopum).	0,8309	500
1000	de 18°,5 à 38°,4 Température moyenne 25,3 (20°,2 R.) Pas de Grêle. Le Sable souvent à 52°		Neige perpétuelle sous les 70° de latitude bor.	Dans l'intérieur du Globe de mouches ou pince de thermometre rougent les plantes au terrasso.	Eau bouillante à 100°,0 (80 R) Bar. 0″,762			
500		L'Eau de mer à la surface près de l'Équateur hors des courans a 28° mais à la profondeur de 1200″ la mer est 5°,0. La Température de l'intérieur du Globe paraît sous l'Équat. de 27°,3			Eau bouillante à 105°(84 R) Bar. 0″,862.		0,8123	0
0								500

understanding of the role of 'caloric' in evaporation and dissolution enabled Humboldt to classify substances as insulators or conductors, and then reform the construction of salt pans. In the Freiberg *Bergmännisches Journal* ('Miner's Journal'), Humboldt published tables of 'heat-conducting force' for quick, practical reference and published a simplified, Lavoisian account of combustion, 'sufficient for the technician'.[10] Steel production might be rationalized: 'Had the weight [of furnace products] been measured from the beginning of experiments with iron . . . pig iron would have been distinguished from carbonized steel, as amalgam and silver are distinguished, and fewer confusions in iron manufacture would have occurred.'[11]

Humboldt also developed several machines of his own to track the combinations of subtle matters, in effect balances, like his eudiometer, which precisely indicated the evolution and absorption of gases and their more or less intimate combination through the displacement of water. By demonstrating that 'the irritability or tone of the [living] fibre only depends upon *the reciprocal balance between all the elements of the fibre*, azote, hydrogen, carbon, oxygen, sulphur, phosphorus etc.', Humboldt made muscle tissue serve as a chemical balance, its sensitivity varying with the 'irritability' of the fibre.[12] It was in this context that he first staked his claim to a 'terrestrial physics', in a 1796 letter 'on the chemical process of vitality' read before the National Institute in Paris and soon published in the Paris weeklies.[13]

The generality of Lavoisian analysis made it not only the general principle of reform, but the general, rational principle for constructing an earth history. Even before his encounter with Freiberg and Lavoisier, Humboldt had come to the conclusion that the ultimate task of natural philosophy was to develop a history of nature, a secular cosmogony. The speculative cosmology he published in 1799 was advertised as 'an attempt to apply the principles of modern physics to geognosy'. For a long period, everything conduced to raise the temperature at the earth's surface: out of an original gaseous or aqueous solution the primitive rocks began to crystallize, giving up their heat to the surrounding medium, producing evaporation and further crystallization, and so on. Some of this heat went towards developing and expanding the atmosphere, which itself added to the temperature of the earth through its pressure, and the resulting climate made it possible for life to develop everywhere in tropical form and abundance. Eventually, the expansion of the atmosphere and the evolution of heat from crystallization 'reached an equilibrium, towards which they had long striven in vain'. As the atmosphere and the earth cooled, life retreated to the tropics, maintained by the sun's rather than the earth's own warmth. The principles of 'modern physics',

Humboldt emphasized, could thereby account for the dramatic change in climate indicated by fossil remains without resorting to extraordinary hypotheses, like the impact of an errant comet or a calamitous shift in the earth's axis. The equilibria of caloric and fluids that grounded the history of the earth were philosophically an extension of the grand celestial equilibrium proved by 'the most profound analysts of our age, La Place and La Grange, [who] have calculated that the changes in the angle of the ecliptic maintain a cycle which never exceeds $1°21'$ (as a result of the composite gravity of the planets)'.[14]

Lines, types, and means as forms of equilibrium

The 'general equilibrium' studied by terrestrial physics was just the present-day manifestation of the conflict and equilibrium of forces which generated the history of the planet. That history left its traces, Humboldt thought, in continuous global distributions of heat, magnetism, rock formations, vegetational forms, even the elevation of the earth's surface itself. Humboldt developed and promoted a technique of realizing that physical continuity graphically, with isometric lines. Perhaps the most renowned of these lines is the isothermal line: the annual isotherm connects points at sea-level which experience the same mean annual temperature (Figure 17.2). The sinuous pattern of isothermal lines on the map was the result of the simultaneous action of global causes acting locally.

Temperature and magnetism are not like those phenomena which, derived from a single cause or a central force, can be freed of the influence of disturbing causes by restricting attention to the mean results of a great number of observations, in which these foreign effects reciprocally counteract and destroy one another. The distribution of heat, like the declination and inclination of the compass needle or the intensity of magnetic force, is essentially conditioned by location, composition of the soil, by the proper ability of the earth's surface to radiate heat.[15]

Isothermal lines therefore preserved local peculiarities within a general regularity, which might one day supply the 'numerical laws' of meteorology. 'It will become clear', Humboldt wrote in a manuscript note, 'that the state of heat is subject to inequalities contained within narrow limits and that, like the planetary system, it "only oscillates around a mean state, from which it never deviates by more than very slight quantities" (Laplace, *Expos.* p. 20).'[16] As in Humboldt's speculative earth history, Laplace's classic demonstration of celestial equilibration, his *Exposition du Système du monde* (1796), supplied a model of the way in which global order emerged from local conflict. *Physique du monde* was Humboldt's version of the *système du monde*.

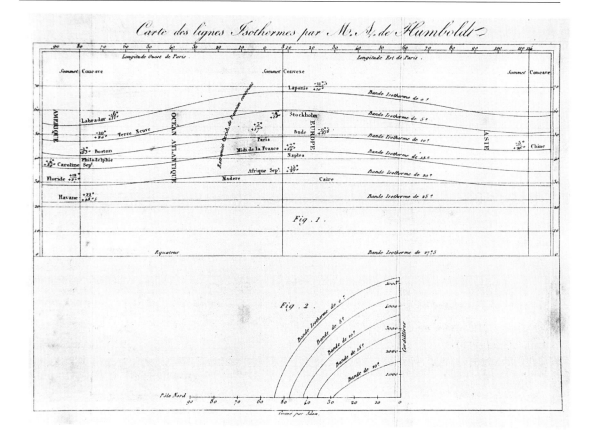

Figure 17.2 'Map of isothermal lines', from *Annales de chimie et de physique*, 5 (1817). On the upper map, lines trace equal annual mean temperatures (isotherms) at the earth's surface, between Asia and the east coast of America; mean winter and summer temperatures for specific locations are placed along the lines, illustrating the variation in annual temperature distributions that can exist along a single isothermal line, and indicating the effect of the shapes of landmasses and coastlines. Despite the variations a lawful and regular sinuosity is observed. The lower graph represents isothermal lines as a function of elevation above sea-level at different latitudes.

The very surface of the earth's crust could be made to yield order in the same fashion. Humboldt promoted the use of barometric surveys (*nivellements barometriques*) and the construction of topographic maps. With these tools the 'physicist' could calculate 'the mean elevation of continents' and draw 'normal surfaces' – horizontal planes that defined, for a given extent of territory, the surface of mean elevation. Using such calculations and surfaces Humboldt showed graphically that mountain chains, for all that their elevated summits attracted 'the common curiosity', actually had little influence on the mean elevation of a large landmass; that rather it is 'those gently undulating plains with alternating slopes' which most determined a territory's normal surface. (Calculations under-taken with the French geologist Elie de Beaumont, for instance, showed that the Pyrenees contributed at most 108 German feet to the mean elevation of France, 816 feet.)[17] Such measurements and maps enabled the terrestrial physicist to reveal natural order in the earth's topography where before only disorder was apparent, per-petuated by the mystifications of geographers who employed the 'hieroglyphic method', representing mountains symbolically and placing little hills around every watershed like 'superstitious Mon-golian priests who erect cairns at every parting of the waters'.

We always begin with what most strikes the imagination, and before we realized that climatology required the exact knowledge of mean tempera-tures, we preferred knowledge of the extremes which heat attains acciden-tally at several points on the globe. The polyhedral form of the surface of our planet offers striking analogies over vast areas: it is the undulations of the plains and plateaux that principally influence the state of agriculture among a people, the heat and humidity of the air; it is the task of the physicist [*physicien*] to contemplate whatever is constant among the physiognomical traits of each continent.[18]

Before the concepts and techniques of isometric mapping became widespread, and before attention was drawn away from accidental extremes to the mean and the normal, fundamental aspects of the natural and social order were obscured. By revealing analogies and typical physiognomies of climate and landscape, the 'physicist' dis-pelled the confusion.

The same concept of how order emerged from the conflict of forces and from a mass of data informed Humboldt's plans for a plant geography.

Plant geography can be considered part of terrestrial physics. If the laws which nature has followed in the distribution of vegetable forms are more complicated than they first appeared, we are no less obligated to submit them to exact investigations. We did not stop making maps when we ran up against the sinuousness of rivers and the irregularity of coastlines. The laws of magnetism manifested themselves as soon as we began to trace lines of equal declination and inclination and to compare a great number of observations that at first appeared contradictory. We would be forgetting the path by which the physical sciences were progressively raised to certain results, if we believed that it is not yet time to seek the numerical elements of the geography of plants.[19]

Humboldt had been writing programmes for a plant geography since the early 1790s as a path to a history of nature, and had used the new science ('which exists in name only') to introduce terres-trial physics as a whole (1805), but only in the 'prolegomena' to the botanical results of his American voyage (1817) did he begin to subject plant geography to quantitative, isometric treatment.[20]

By tabulating the fraction of the total number of phanerogamic plant species present in a given climatic and geographical region comprised by the species of each of the seventeen natural families ('groups based on analogy of forms'), lines could be traced that defined 'maxima of concentration' for each family; Humboldt predicted that such lines would approximate the earth's isothermal patterns. In the absence of sufficiently complete catalogues of the flora of specific regions, which recorded not just the names of specimens collected, but their position and elevation and the geology of their habitat – catalogues compiled by *botanistes physiciens* – the order manifest in the distribution of vegetational forms would remain sketchy and provisional. None the less,

as in all other phenomena of the physical universe, so in the distribution of organic beings: amidst the apparent disorder which seems to result from the influence of a multitude of local causes, the unchanging laws of nature become evident as soon as one surveys an extensive territory, or uses a mass of facts in which the partial disturbances compensate one another.[21]

These were not orders and equilibria that depended on theoretical discoveries or mathematical laws. For Humboldt, nature's lawfulness emerged gradually and progressively from laborious observing, averaging, and mapping over increasingly extended areas. Indeed, for Humboldt, averaging *itself* was the paradigm of lawful behaviour, and mean values were 'the final object, the expression of physical laws; they show us the constant amidst the flux and flight of phenomena'.[22] Isometric lines were a product of averaging and interpolation, connecting well-observed points literally by drawing lines between them on a map. That a continuous line connecting mean values on the earth's surface could be drawn was as little obvious as the very existence of mean values, as Humboldt's own struggles to define and group 'mean daily temperatures' show. Drawing those lines constituted an act of faith in both the physical 'co-operation of forces' and in the emergence of global order out of local averages. The *lines* themselves, which recorded local, physical quantities, rather than mathematical expressions of them, were the principal manifestation of nature's order; Humboldt even offered the editors of the *Dictionnaire des sciences naturelles* an entire article on 'Lines', in an effort to promote terrestrial physics.[23]

Thus lines, and the notion of equilibrium they embodied, prescribed a particular organization and dynamic of science, a particular notion of how terrestrial physics progressed. Order emerged from the landscape, so to speak, in a process of territorial expansion; under the regime of terrestrial physics, natural philosophy progressed with the gradual extension of precise, disciplined measurement, the spread of mean values over an accurate map. The type of enterprise which could establish and track natural equilibria was an organized network of observers, or better, corresponding physical observatories, dispersed over large expanses of the earth's surface, using comparable instruments and standard protocols. In this sense, terrestrial physics was natural history conducted in an observatory.

Mean values and the lines which expressed them guaranteed another kind of order: they gave an exact, physical account of the sensitive layman's intuitions of Nature's constancy, in terms of the 'co-operation of physical forces'. Each mean value is irreducibly historical, local, and characteristic, and yet attached to an identifiable 'type' to which it is bound within fixed limits. Sober, accurate, and global measurement of the proper quantities led to recognition of unity and order, 'analogies' between distant climates, vegetations, topographies, even cultures. Steppes and deserts are analogous phenomena which occur in all climates, on every continent; the fluctuations of isovegetational lines determine a limited number of vegetational types, which vary within strict limits. 'The characteristic traits of nations are like the internal structures of plants, spread across the earth's surface. Everywhere, one sees the stamp of a primitive type, despite the differences produced by the nature of the climate, the soil, and the combination of several accidental causes.'[24] Nature effected this conservation of type, in anthropology and philology as in geognosy, plant geography, and climatology, through mutual compensation. Humboldt demonstrated 'algebraically' that in the Peruvian Andes a stratum of sandstone replaces the stratum of porphyry-schist found in the Italian Alps (according to the principle of the identity of strata or independence of formation); that the relative abundance of genera represented in Arctic flora is replaced by an abundance of species within comparatively few genera at the equator; that along a single isothermal line 'extreme' and 'moderate' climates can be distinguished; 'that the distribution of annual heat between winter and summer follows a definite type [*Typus*] along each isothermal line; that the deviations from the type are confined to certain limits and that they are subject to one and the same law in the zones which cross the concave or convex summits of the isothermal lines'.[25] This may seem to share in the general mania for comparative morphology, a Romantic 'organicism', typified usually by Goethe and Cuvier. In Humboldt's 'physics', however, the conservation of type was not a governing heuristic *principle* but an empirical *result*, a consequence of Nature's regularity that connected the numerical equilibration of physical forces with the aesthetic sensibilities of the 'cultivated individual'.

The progress of civilization and the appeal of Humboldtian science

In the crucial period which saw the emergence of natural science out of natural philosophy, in the first decades of the nineteenth century, Humboldt developed and promoted a comprehensive science (a *physics*), which associated the separate branches of natural philosophy under a dynamic equilibrium model of natural order and developed the concepts and techniques for realizing that order. 'Average' and 'mean' supplied physical–mathematical, organizational, and aesthetic concepts of how order and equilibrium progressively emerged from apparent conflict. Under the aegis of 'general equilibrium', 'terrestrial physics' gave a unified identity to natural philosophy, demonstrated how its parts might be placed in service of global historical progress, and made natural philosophy the privileged custodian and anchor of order.

Humboldt's project attracted a broad spectrum of participants and informants stationed around the world. The lengthy notes to successive editions of his *Ansichten der Natur* ('Views of Nature') (1808, 1826, 1849) and to *Kosmos* (1845–62) record the personal communication of observations from French naval officers, East India Company physicians, Russian provincial administrators, Spanish military commanders, and German diplomats, among others; the Humboldt Nachlaß in Berlin preserves many fragments of those communications, cannibalized by Humboldt and often left without name or date, many of them apparently unsolicited. An admiring Charles Darwin sent observations of atmospheric refraction, excerpted from Captain Fitzroy's logbook, from aboard HMS *Beagle* in 1833.[26] Francis Beaufort, soon to become chief hydrographer to the British Admiralty, expressed himself nobly and typically on the occasion of conveying measurements of horary barometric variations in 1826: 'I dare not flatter myself that these ambulatory observations can be of much service – but the endeavour to contribute to the stock of facts, which you already possess on this interesting subject, should be considered as a public duty by every body – and your acceptance of them will be felt as a personal honor.'[27]

Beaufort's sentiments and the sheer extent of participation in Humboldt's grand equilibrium, especially by colonial officials and bureaucrats, suggest that as a programme for the organization and progress of science, terrestrial physics served Europe's 'civilizing mission'. Humboldt's informants joined Helen Maria Williams, Humboldt's first English translator, in the 'sympathy' aroused by the traveller, 'while he imprints the first step that leads to civilization and all its boundless blessings, along the trackless desert, and, struggling with the savageness of the untamed wilderness, obtains a victory that belongs to mankind.'[28] Contributing to Humboldt's lines was a philanthropic, even heroic act, as frequent retellings and representations of Humboldt's ascent of Chimborazo, barometer in hand, made clear. To extend the reach of mean values was to extend European civilization in its best and most progressive aspect.

In this sense, terrestrial physics had its most direct and visible impact on the former Spanish colonies of South and Central America, where violent struggles for independence between 1808 and 1825 vented previously subterranean social forces, only too apparent to Humboldt during his travels. The forces of intellectual and material commerce that were gradually eroding Madrid's influence in its own colonies were inseparable from the geological forces that had riven the continent with its vast river systems: 'Whoever is master of the Orinoco easily penetrates the provinces of Cumana, Caracas, Barinas, indeed, via the Rio Meta, all the way to Santa Fé de Bogotá. Via this river, ideas can quickly be put into circulation in South America, ideas and goods, and here, where everyone is so hungry for cheap English goods, a muslin blouse makes a revolution as easily as a book in France.'[29] For three decades during the upheavals in Spanish America, from his study in Paris, Humboldt was churning out volume after volume of his account of nature and society in equatorial America. Humboldt's natural philosophical projects both rode and fed the wave of speculative projects that bore European capital and travellers to Spanish America in increasing numbers in the first decades of the century. As Humboldt pointed out to the Spanish colonial administration, and to advocates of Western expansion in the new North American republic, his terrestrial physics provided relatively cheap methods for surveying extensive territories with sufficient accuracy for 'politic' administration, using barometers, chronometers, and thermometers rather than strict and painfully slow geodetic methods.[30] Humboldt's political essays on Mexico and Cuba, meticulous and voluminous compendia of statistics culled from colonial archives and informants, his improved maps, his survey of possible sites for a trans-isthmus canal, attracted the interest of British and French investors, eager to exploit abandoned mines and new markets (markets, Humboldt pointed out to his readers in 1825, which 'have increased to 70 million *piastres* annually'). The Geological Society of London published extracts from Humboldt's *Political Essay on the Kingdom of New Spain* (London, 1811) to encourage mining ventures in Mexico. In turn, Humboldt regarded large, government-sponsored commercial ventures as potential vehicles for his own projects, new expeditions in India and Siberia or a research institute in Mexico.[31]

Figure 17.3 Frontispiece to the *Atlas géographique et physique du Nouveau Continent* (Paris, 1814), engraved by Barthèlemy Roger after a drawing by François Gérard. Mercury, god of commerce, helps the fallen Aztec prince to his feet, while Minerva, goddess of letters, extends an olive branch. The ruins of Mexican culture, the monuments of political upheaval, occupy fore- and background; they are dominated by the most sublime monument of natural upheaval, Chimborazo, the volcanic Andean peak represented on Humboldt's 'Physical profile of the Andes', with its luminous and constant cap of snow. The motto 'Humanitas. Literae. Fruges.' is taken from the *Letters* of Pliny the Younger (book VIII, letter 24), where the gifts of Greece to the civilized nations of Europe are enumerated: 'Liberal Arts, Science, Agriculture'. The plate graces an atlas dedicated to geographic, hydrographic and topographic maps, and hypsometric profiles of large landmasses. Humboldt's terrestrial physics offered America a restoration to law and order, via the forces of commerce and civilization.

But the implication of terrestrial physics in the resurgence of imperial projects in the early nineteenth century went deeper than supplying eager projectors and governments with useful maps and statistics. Mary Louise Pratt, supplying a perspective from the history of travel narrative, with particular attention to the function of natural history in those narratives, has shown recently how the 'co-operation of physical forces' produced by terrestrial physics provided both European and Creole societies with the tools for 'reimagining' America.[32] Humboldt's replacement of the *botaniste nomenclateur* with the *botaniste physicien*, the Linnaean collector with the physical observatory, represents both the disappearance and apotheosis of the naturalist in a Nature constituted by

Figure 17.4 Humboldt in his study in Oranienburger Straße, Berlin. Watercolour by Eduard Hildebrandt, 1856; from Douglas Botting, *Humboldt and the Cosmos* (London, 1973), p. 271.

dramatically conflicting forces, which 'deeply impress[es] us with a sense with a sense of her greatness' and 'speak[s] to us forcibly'.[33] Humboldt's own self-presentation in the *Personal Narrative*, and the ideal traveller from whose perspective we enjoy the *Views of Nature*, is the mere 'vestige of a narrative persona', overwhelmed and replaced by forces of nature: '[The traveller] feels at every step that he is . . . in the centre of the torrid zone . . . on a vast continent, where every thing is gigantic, the mountains, the rivers, and the mass of vegetation . . . He can scarcely define the various emotions, which crowd upon his mind; he can scarcely distinguish what most excites his admiration.'[34] The forces themselves supply the narrative and the drama, and the story they tell is one of dynamic balance, equilibrium, and progress, told by the Humboldtian terrestrial physicist and his lines.

Pratt looks at Humboldt's science as a lens through which European and Creole audiences could view America; but it is clear that terrestrial physics was a means not only for reconstructing 'the New World', but for reforming and renovating the Old as well. Humboldt's America provided a mirror in which Europe might see itself, changed. This account of Humboldtian science, focusing on its all-embracing concept of Nature's lawfulness and progressiveness and its practical roots and realization in various mechanical, chemical and graphic 'balancing' techniques, has the advantage of preserving Cannon's emphasis on practice while getting at what others found useful and attractive in it. Thereby, 'Humboldtian science' illuminates the reorganization of knowledge and disciplines in the early nineteenth century that defined the emergence of natural science out of natural philosophy.

Further reading

Botting, Douglas, *Humboldt and the Cosmos* (New York, 1973).

Bruhns, Karl (ed.), *Life of Alexander von Humboldt*, trans. Jane and Caroline Lassell, 2 vols. (London, 1873).

Cannon, Susan Faye, *Science in Culture: The Early Victorian Period* (New York, 1978), ch. 4.

Cawood, John, 'Terrestrial magnetism and the development of international scientific cooperation in the early 19th century', *Annals of Science*, 34 (1977), pp. 551–87.

Goetzmann, William, *Army Exploration in the American West 1803–1863* (New Haven, 1959).

Hagen-Hein, Wolfgang (ed.), *Alexander von Humboldt: Life and Work*, trans. John Cumming (Ingelheim am Rhein, 1987).

Home, R. W. 'Humboldtian science revisited: an Australian case study', *History of Science*, 33 (1995), pp. 1–22.

Nicolson, Malcolm, 'Alexander von Humboldt, Humboldtian science, and the origins of the study of vegetation', *History of Science*, 25 (1987), pp. 167–94.

Pratt, Mary Louise, *Imperial Eyes: Travel Writing and Transculturation* (London and New York, 1992).

18 Biogeography and empire

With its emphasis on the regions or 'nations' of plants and animals, its concern with the discovery and utilization of species useful to the homeland, its written surveys of the living beings of different countries, and the administrative network of colonial botanic gardens, timber forests, tea-gardens, museums and menageries that served both centre and periphery, the study of animal and plant geography in nineteenth-century Britain was one of the most obviously imperial sciences in an age of increasing imperialism.[1] At that time, when concepts of 'empire' and 'imperialism' were being dramatically forged on the anvil of colonization, the conceptual framework, methodologies and practical techniques developed to deal with foreign animals and plants took their tone directly from those used in national expansion. The insistence of Charles Darwin (1809–82) and Alfred Russel Wallace (1823–1913) on competitive struggle as an engine of change, for example, was only the most prominent and long-lasting expression of a deep-seated cultural ideology that had already found its premises in the study of animal and plant distribution over the globe.

This chapter describes some prominent features of British biogeographical thought in the half-century before Darwin,[2] focusing in particular on the topic as a colonial activity. The fragmentary, individualized nature of the field needs to be emphasized, since it played a major role in neutralizing any movement towards disciplinary status.

Searching for foreign specimens

Taking the study of animal and plant geography at its most basic level, it is readily apparent that little attempt at understanding the regionality of the natural world could have been undertaken in Britain without some knowledge – either accurate or misleading – about the organisms in different areas. Ever since the great age of European geographical exploration that opened with the epic figure of Henry the Navigator and encompassed Vasco da Gama, Columbus, Magellan, and Drake, and led on to the voyages, among others, of La Condamine, Pallas, Cook, Vancouver and Humboldt, great emphasis was laid in Europe on the significance of collecting the native fauna and flora of new regions and on bringing specimens back for cataloguing and future assessment, particularly after the time of Linnaeus.[3] Without these raw materials naturalists had little to go on.

Much more than any other branch of the natural history sciences, early biogeography in Britain was therefore intimately tied up with access to exotic specimens. Ideas of what might constitute a regional fauna or flora, and complementary ideas about the patterns they might make over the globe, were at heart rooted in a generalized concern with acquiring and understanding foreign species. It seems important, then, for historians to ask how collections were made and what they represented: what in other sciences would be called the means of production.

The great majority of British naturalists of the eighteenth and nineteenth centuries in fact considered foreign organisms much more exciting and interesting than those found at home.[4] The inexhaustible lure of travel and the anticipated pleasures of foreign lands, both mental, moral, and physical, were important components in the history of this subject – indeed they became crucial factors as the relaxed aura of eighteenth-century social life metamorphosed into the straitlaced Victorian era.

Nevertheless, a love for natural history and a desire to travel are not sufficient reasons in themselves to account for the expansion of overseas activity among naturalists. Far more significant was the hierarchical structure of British society and the expansionist national ethos. Collectors during the years 1770–1840, the period bounded by the voyage of Joseph Banks (1743–1820) and that of Alfred Russel Wallace, can best be understood in social terms; for it was these social terms that determined the kind of expedition and in what capacity the naturalist travelled as well as regulating the institutions to which specimens were ultimately sent.

Three main types of travelling naturalists and collectors can be discerned. Firstly, there was a certain number of freelance entrepreneurs, some wealthy, like Joseph Banks, independently pursuing their own interests, and others entirely self-supported by the income anticipated on the sale of specimens after their return. In both cases, the resulting specimens belonged entirely to the collector, his or hers to give away to a museum, friend, or specialist, or to keep or to sell.

Secondly, and more centrally for the historian of British biogeography, there were naval and military personnel, particularly surgeons, and other official employees, such as colonial administrators and diplomats, each appointed by a relevant government department and commissioned onto a succession of land or sea expeditions according to their place on the promotion ladder or

sent to officiate at colonial institutions overseas. Surgeons had often trained at the noted medical school at Edinburgh University, with its excellent natural history facilities run by Robert Jameson (1774–1854), who was constantly involved with supplying the Admiralty with suitable surgeon-naturalists,[5] some also having spent time learning the best of contemporary medical and biological thought in Paris.[6] Naval and military officers had undergone a rigorous technical education, and diplomats and civil servants invariably had a university background. Many such functionaries of the state went on to become established – even celebrated – naturalists of the period.

Thirdly, there were people employed to collect: the gardeners, seedsmen, bankclerks, taxidermists, weavers, booksellers and sporting entrepreneurs who managed to get themselves a roving commission for a national institution, such as the Royal Botanic Gardens at Kew, or for an ambitious new society such as the Horticultural Society of London. Others found wealthy patrons. The eighteenth-century London merchants James Petiver (1663–1718) and Peter Collinson (1694–1768) were two remarkable natural history patrons who sponsored several collecting expeditions in North America and elsewhere: Collinson was more responsible than any other man for introducing American plants to Europe's botanists and gardeners.[7] Aristocrats such as Sir Hans Sloane (1660–1753) or Sir Joseph Banks, and great landowners like Lord Derby or the

Figure 18.1 Collecting as a form of sport. Acquiring foreign specimens frequently merged into big-game hunting. Paul du Chaillu's sensational account of his travels first introduced the idea of gorillas as ferocious animals. From Paul du Chaillu, *Explorations and Adventures in Equatorial Africa* (London 1861).

Duke of Northumberland, were equally instrumental in later years. Dispatched on merchant ships to Australia, Africa and the Americas, paid by results, their services dispensed with once the job was completed, many of these collecting employees remain obscure and little known to scholars.[8]

These three groups necessarily overlapped and shaded into each other at the points where freelance naturalists acquired a commission to collect and where government professionals found it useful to act in an entrepreneurial fashion. In a social world where few paid positions in science existed, flexibility was all.

Colonial officials

Professional naval and military men, colonial officials, government surveyors, railway and dockyard engineers, merchants and administrators may, however, justly be considered the people who, by furthering the nationalistic aims of their parent community, brought many of the wonders of the rest of the world to the attention of developed societies. Britain, France, Portugal and the United States were predominantly expansionist nations at that time, located at the centre of ever-increasing commercial and administrative networks held together by the trading routes of the oceans. War only intensified activities overseas. After the Berlin decrees of 1805, Napoleon and the British, for example, resorted to a world-wide programme of blockade and counter-blockade which brought profound changes in international relations: Britain struck out at the French Caribbean colonies, demolished the Dutch East Indian empire, and permanently annexed the Cape of Good Hope as a linking point between the two hemispheres. Canada was split; the Far East divided out; the Mediterranean and Pacific apportioned. South America and the scramble for Africa were soon to come.

Consequently, an extraordinary range of careers could be established around the incessant activity of travel, not merely in the obvious areas of expertise, such as the naval and military arms of government and the War Office, railways, canals, port and harbour authorities, shipping agencies, insurance, customs and excise, and import/export companies dealing with a whole range of commodities from luxuries like tea, fur and porcelain, to the mundane, like South American guano and cement, but also in less obvious places where travel was seen as an entrée into society, as a mark of patriotism, learning or experience, as an important venture for identifying the physical resources of far-flung territories, or as an economic buttress to fiscal programmes at home. In this latter sense, colonial officals who travelled or lived overseas almost always subscribed to the social order in their home country, demonstrating

Figure 18.2 Conservatories and greenhouses displayed the plant treasures of the world. The great conservatory at Chatsworth, designed by Joseph Paxton and Decimus Burton, took four years to build and covered just over three-quarters of an acre. People came from all over Europe to marvel at the 'tropical scene with a glass sky'. From *Illustrated London News* (1844).

how much they approved of and wished to endorse the society that sent them forth. It is here at this intersection between travel, commerce and public service that official bodies like the British Admiralty, the War Office, the Hydrographer's Office, the Board of Longitude, the Ordnance Survey and its offshoot the Geological Survey, the East India Company and its training college at Haileybury, the Royal Naval College at Portsmouth, the Observatory at Greenwich, the Royal Dockyards, the Royal Military Academy at Woolwich and the Royal Engineers' Institution at Chatham take on fresh significance. The occupations that these institutions offered often permitted, sometimes even encouraged and expected, an interest in science and natural history, and they were for many people an acceptable way of combining inclination with salary and prospects.

The job of naval surgeon, for instance, famously represents a long-standing tradition reaching back at least as far as the seventeenth century, of taking advantage of a voyage into some distant part of the world to study natural history or collect specimens – be they plant, animal, mineral, medicinal or ethnographic. On many occasions the ship's doctor was the only member of the company, besides the captain, who had received a scientific training of sorts (although the Admiralty was never too fussy about actual qualifications, issuing its own medical diploma if necessary) and was expected to know about man and the natural world.

A naval or military appointment also permitted individuals to take leave of absence or to go on to half-pay after their return to Britain, creating valuable time in which to work on their collections and write up results. Specimens brought home by these men were invariably Crown property and ultimately were deposited in the British Museum or some other significant institution like the herbarium at Kew Gardens, the East India Company Museum or the Royal College of Surgeons, which included a growing collection of palaeontological remains; and government appointees usually gave their collections directly to the experts associated with these bodies for identification and display.[9] Richard Owen (1804–92) and John Gould (1804–81) founded their high-flying careers on precisely such a museum-based service in comparative anatomy and taxonomy. Neither could have hoped to become famous without such institutional opportunities to classify for empire.[10] But many of these travelling servants of the Crown identified and described the material themselves, petitioning the Army or Navy for funds, and obtaining government grants for illustrated publications. A great number of seminal texts in the history of biology were produced in this manner, ranging from James Kingston Tuckey's *Narrative of an Expedition to Explore the River Zaire* (1818) to Darwin's lavishly illustrated *Zoology of the Voyage of HMS* Beagle (1838–43). The British Treasury granted Darwin the sum of £1,000 in order to produce coloured pictures of the new species described.[11] Joseph Hooker's *Botany of the Antarctic Voyage of HM Discovery Ships* Erebus *and* Terror *in the Years 1839–1843* (1844–60) was similarly subsidized by £1,000 from Treasury coffers. As might be expected, careers could founder without this kind of support. Johann Reinhold Forster (1729–98), for instance, always at odds with Captain Cook and the Admiralty, was banned from publishing the Australasian materials he collected during the *Resolution* voyage and was left grantless; his son's, and then his own, self-financed ventures in 1777 and 1778 were notorious publishing disasters.[12]

Many such government-sponsored travellers, particularly those associated with the British army or navy, were consequently recognized as 'experts' upon their return – if not in modern terms, at least in the eyes of their contemporaries. To accompany a voyage of exploration provided an opportunity for becoming an authority on a particular group of organisms – as did Robert Brown (1773–1858) from the *Investigator* expedition to Australia (1801–5), who specialized in the plants of the southern hemisphere – or in a specific area, as did Andrew Smith (1797–1872) with the Cape Colony.

Britons further appointed to botanic gardens or tea-plantations, forestry managers, or agents in the Hudson's Bay Company, for instance, were similarly highly regarded. Countless expatriates and colonial officials who would otherwise be lost to history became – for a short while – authorities on some point relating to the natural history of their area, and contributed information to the natural history bodies and journals proliferating in London and elsewhere.[13] Hugh Falconer (1808–65), the versatile naturalist stationed first in the Bengal establishment of the East India Company as an assistant-surgeon and then in Saharanpur, where he took over the running of the botanic garden from John Forbes Royle, followed just such a trajectory. As superintendent of the gardens he advised the British government – and his friends at Kew – on the best locations for the new colonial enterprise of tea-plantations, explored the Sivalik Hills for fossils, traced the Indus to its source, and played a central role in the scientific life of Calcutta. On returning briefly to England in 1843, he was appointed to superintend the mounting of his dramatic Sivalik fossils for exhibition in the British Museum – a position funded by the East India Company as if he were still in post. Back in India again after 1847 he ran the Calcutta Botanic Garden, the prime institution in the Kew-run network of colonial gardens, and was instrumental in all kinds of government action in exploiting economic botanical resources, especially relocating plantation crops like tea, cinchona, and rubber. Here, as elsewhere, the plantation as a form of agricultural organization could hardly be exceeded as a mark of European colonization.[14]

Civil servants and diplomats were equally involved. Robert Swinhoe was persuaded to collect fish for Albert Gunther (1830–1914) at the British Museum during a diplomatic mission to China; and John Pretherick, the consul of Khartoum sent Gunther further specimens from Africa. In 1867 another British consul co-authored a book with Gunther on the fishes of Zanzibar.

The advantages offered by an intimate acquaintance with nature outside Europe, perhaps even a practical monopoly on certain key topics, supported by a large collection of specimens, therefore went a long way towards helping professional men who were also enthusiasts for natural history join the community of savants on their own terms. A large proportion of the efflorescence of natural history work during this period is thus directly attributable to the diverse threads and networks of the British colonial system.

Science of empire

More than this, naval, military, and colonial collecting endeavours fully reflected the developing infrastructure of empire. Expeditions were drawn up to fulfil complex administrative and national purposes in which geographical exploration and the glamour of discovery were often only secondary elements. Cook's voyages, though replete with scientific achievements, including the astronomical observation of the transit of Venus, were made to claim the southern continent for the Crown. Banks's botanical activities were instrumental in establishing the potential for new British settlements – and a penal colony – in New South Wales.[15] Subsequent expeditions to the area claimed further land for Britain, including Tasmania, New Zealand and as many of the Pacific islands as possible.[16]

But it is often forgotten that all such British expeditions had a colonial purpose. Robert Brown, for example, sailing with Matthew Flinders on the *Investigator* in 1801, was accompanying a ship expressly sent out to establish a British presence in Western and Southern Australia before the French. Flinders unexpectedly met Captain Baudin halfway round the coast (commemorated in the name Encounter Bay) and found, to his relief, that scientific courtesy transcended national concerns at that point. Later, on returning to England to commission a new ship to replace the disintegrating *Investigator*, Flinders was less lucky. The Napoleonic Wars had intensified and he was imprisoned on Mauritius for seven years. The crew left behind in Port Jackson (Sydney) had more opportunity than most to explore the Australian hinterland – to major botanical effect in Brown's case.[17]

In similar vein, the *Beagle* voyage, under the command of Robert FitzRoy (1805–65), was an exercise to ascertain the commercial and strategic potential of the east coast of South America in the light of the Argentine states having been released in 1825 from their commitment to trade only with Spain and Portugal.[18] One of FitzRoy's further duties was to reclaim the Falkland Islands from Argentina, literally showing the British flag. Another was to extract a government fine from the queen of Tahiti. His route around the globe was in essence determined by geopolitical and national factors rather than by any strictly scientific questions. Arctic exploration – so much a part of the early Victorian experience – and the long-lived search for the North-west Passage ignited by Cook's third voyage in 1778 further mirrored these combined purposes.

At the other end of the scale, Joseph Hooker (1817–1911) opened his two major taxonomic works on the New Zealand and

Figure 18.3 Joseph Dalton Hooker in the Himalayas. Hooker believed that imperial territories should look to the mother country for scientific guidance. In this symbolic sketch he is accepting rhododendrons unknown to European botanists as colonial tributes. By William Tayler, 1849.

Indian floras with severely worded instructions to colonial botanists. Overseas workers, he thundered, were not equipped to understand their own plants: they frequently believed species to be new, when research in the extensive collections at Kew revealed them as geographical variants of a single, widespread form; they had inadequate reference works; they inconvenienced other naturalists (Hooker was thinking of himself) with a proliferation of local geographic or personal names.[19] His message was simple. Plants were to be sent to Kew – the hub of the colonial scientific network. Hooker instinctively saw himself as the linchpin of a centre-and-periphery situation where freedom on the boundaries was strongly discouraged.

So the collections of natural objects gathered during various expeditions, and those sent back to Britain from the nation's far-flung outposts, represented, by their very presence on British soil, the whole culture of imperial enterprise. They represented the fact that expeditions were made to claim sovereignty; that naturalists and professional men with natural history interests were appointed to assess the resources found during such expeditions; that interested parties were stationed in colonial outposts; that the diplomatic corps and other personnel of empire contributed in a material way to the expansion of knowledge about foreign animals, plants and minerals; and that the information emerging was an essential element in government decisions about expansion.

Assessing the world's patterns

Theories of distribution based on such collections were equally grounded in geopolitical concerns. These too reflected the ethos of empire. The sixty or seventy years before the *Origin of Species* (1859) was published were remarkable in Britain, at least, for the number and variety of theories relating to the geography of animals and plants. It seems that almost everyone who worked with natural material – including fossils – had something to say about patterns, dispersal or habitat in relation to the species, genera or families discussed. 'The extent of this parcelling out of the globe,' said Charles Lyell (1797–1875) in 1832, 'may be considered as one of the most interesting facts clearly established by the advance of modern science'.[20]

To some degree, this ought to be expected. The location where a specimen was first found was enshrined in the Linnaean taxonomic scheme as a fundamental diagnostic character, and the given name often reflected its geographic situation – in colloquial terms just as much as in scientific binomials. This was so much the case by the 1840s and 1850s that British naturalists began to rebel against it. As Hooker predicted, they found species were much more widely spread than formerly supposed and that names which included geographic elements ('Europaeus' or 'Africanus', for example) were frequently misleading. When the British Association for the Advancement of Science set up a regulatory committee for nomenclature in 1842, at the instigation of Hugh Strickland (1811–53), these geographic names were among the first to be censured.[21]

Yet the idea of geographical *regions* also came to the fore in the mid-eighteenth century, crystallized in part by Linnaeus's emphasis on the discrete, self-contained nature of faunas and floras, and given concrete force by the long-standing public awareness that different kinds of animals and plants were specific to different areas. So when British naturalists undertook a geographical assessment of foreign organisms their usual aim was to establish which forms were indigenous to a particular country or region. Numerous catalogues of regional floras and faunas were produced during this period.

And underneath the dry textual lists in these systematic works lay a strong commitment to the idea of 'nations' of animals and plants. Parallels with indigenous human populations were constantly drawn through the metaphors used to convey ideas of animal or plant units: terms like 'state', 'kingdom', 'province' and 'nation', and nationalistic expressions like 'motherland', 'outpost' and 'station' can all be found. Significantly, the word most used by early nineteenth-century students of animal and plant distribution patterns was 'colonist'. This was the muscular language of expansionist power.

Moreover, the actual way that animal and plant 'nations' were discussed reflected general principles of empire. Ever since Alexander von Humboldt and Augustin de Candolle had pioneered the technique of 'botanical arithmetic', it was commonplace to describe plant regions – floras – in numerical terms just like a human population survey.[22] Animal groupings, especially insect populations, were similarly described. In this procedure, the number of species and genera was counted, just as Humboldt directed, and simple correlations and proportions calculated, as in the number of species to genus or the overall statistical relationship of the most prominent families. These numerical techniques were firmly based on the traditional Germanic science of enumerating political states: of assessing the numbers of individuals in different social groups, the physical resources and industries of a region, the manners and customs of rural folk, the distribution and character of the ruling classes, and so forth. It is no coincidence that British state departments compiled extensive regional statistics at the same time: both of Britain, as represented by Sir John Sinclair's statistical survey of Scotland (1814), and abroad, such as Francis Buchanan's vast report on the Presidency of Bengal compiled for the East India Company.[23]

Such techniques were immensely appealing in Britain. They became second nature to thinkers like Thomas Vernon Wollaston (1822–78), whose *Insecta Maderensia* (1854) provided a comprehensive account of Madeiran insects. Wollaston was intrigued by the relationship between aberrant or otherwise endemic species and the geographic isolation of this mid-Atlantic island. Islands, he believed, experienced more 'creative' activity than continental landmasses: a point that his comparative statistics could bring out with dramatic force. In Hewett Cottrell Watson's hands, plants, too, were readily organized into statistical data. Watson's terminology reverberates with the bureaucratic enthusiasms of a British administrator overseas. His *Cybele Britannica; or, British Plants and Their Geographical Relations* (1847–59) divided British species into precisely demarcated 'provinces', 'counties' and 'vice-counties', and devised a test 'to show the number of *spaces* (not the number of precise *places*) in which each one occurs'.[24]

Not surprisingly, maps became an essential element in the representation and development of concepts like these. Charts, maps, and surveys were the very stuff of expansion in the days when the British empire was colouring half the globe pink.[25] Lines were drawn on existing geographic maps to indicate the topographic spread of faunas and floras. This doubling up told naturalists a great deal about organic distribution, particularly in the way that mountain ranges, deserts and waterways limited living units. Abstract concepts about the geographical range of living beings were thus fused – in a purely visual form – with existing information about landforms.[26]

Interest in what might be called the socio-political factors in regional units also ran high. Augustin de Candolle's (1778–1841) definition of the 'stations' and 'habitations' of plants in his *Essai élémentaire de géographie botanique* (1820) encouraged many European naturalists to investigate the physiological and environmental conditions required for life, both among plants and animals. From these studies emerged a general consensus about dependency distribution, which, at its most basic, was concerned with the nature of the soil, climate, frost, elevation and sunshine. Later on, Candolle's son Alphonse brought work in this area to a climax in his mighty *Géographie botanique raisonnée* (1855). Sophisticated assessments of vegetation zones and attempts to pinpoint characteristic species or life-forms soon followed.[27] Less obvious is the fact that these dependency studies paralleled similar concern with the medical problems of expatriate Englishmen abroad. Fevers, contagious diseases and non-specific disabilities related to tropical life were not known as 'the white man's burden' for nothing, and much medical and official government anxiety was expressed during this period over the Europeans' apparent inability to cope physically with foreign climes.

To this, some naturalists such as Darwin added extensive research into the means by which animals and plants were dispersed over the globe: the sea-transport of seeds, carcasses and larvae; the possession of hooks and sticky attachments; the effect of wind, rivers and ocean currents; and in Darwin's case, the unwitting transport of organisms by mankind from country to country.[28] Important new concepts of 'habitat' and 'station' emerged out of these and comparable researches.

Studies in acclimatization were a further significant aspect of this kind of practical distribution-based natural history. Commerce, colonialism and scientific enterprise intersected dramatically here, particularly in the activities of the early Zoological Society of

Figure 18.4 The variation of types of vegetation with altitude. From A. Dupis et al., *Le Règne végetal, divisé en traité de botanique générale, Flore médicale et usuelle, horticulture botanique et pratique* (Paris, 1864–9), vol. 1, plate 6.

London, the Horticultural Society, and the Society for the Encouragement of Arts, Manufactures, and Commerce.[29] William Walton, a leading enthusiast for the introduction of the South American guanaco into Britain, argued that woollen cloth made from their hair could rival the Indian paisley and calico trade. Sir Stamford Raffles (1781–1826) and Lord Derby had much the same idea in relation to breeding 'useful' animals from the menagerie and farm attached to the Zoological Society. In a world where rubber plants could be taken from Brazil and relocated in Malaya, where sugarcane, cotton and potatoes became staple crops in regions formerly

Figure 18.5 Visiting the zoo. The London Zoological Society opened its gardens to the public one day a week. One notice asked ladies not to touch any of the animals with their parasols, 'considerable injury having arisen from this practice'.

foreign to them, and where rabbits, goats and rats followed in man's wake, it must have seemed that, for some species, at least, the natural boundaries had no hard and fast meaning.

Most important, however, was the elder Candolle's apotheosis of a war in nature. Competitive struggle among individuals for space, water, and other physical needs, and between groups of species for the occupancy of geographical areas caught at the conceptual heart of British natural history. Each fauna and flora could be regarded as a colonizing force, with its entrepreneurs and forward lookouts, its army and navy, merchants and supply teams, its dominant classes and plebeian masses. Competition – 'war' as Candolle memorably stated – between and among these units was seen as the primary cause in determining the topographic extent and shape of faunistic and floristic boundaries. Living beings were thought to possess an inherent, almost inexhaustible, ability to disperse and expand their range. Only physical barriers and competition could stop them (Figure 18.6). Like colonizing human

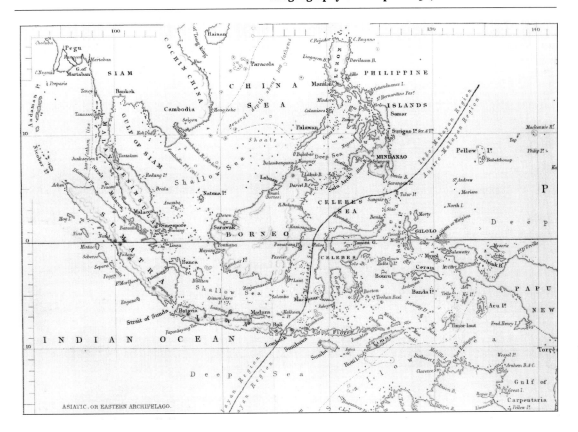

ASIATIC, OR EASTERN ARCHIPELAGO.

beings, it was felt, plant and animal species lived life as a battle: a battle with the elements and with themselves.

Political and social metaphors are not hard to discern here. Joseph Hooker even included the category of 'tramp' in a systematic work to account for the appearance of European weeds in New Zealand. Yet Charles Lyell brought the motif home forcefully to British readers with his talk of the 'economy of nature' in which 'foes' and 'allies' undergo 'continual strife'. Lyell saw the parallels between human and organic struggle just as clearly as Darwin was to do some ten years later, perhaps more clearly. 'If we wield the sword of extermination as we advance,' he went on to say, 'we have no reason to repine at the havoc committed':

we have only to reflect, that in thus obtaining possession of the earth by conquest, and defending our acquisitions by force, we exercise no exclusive prerogative. Every species which has spread itself from a small point over a wide area, must, in like manner, have marked its progress by the diminution, or the entire extirpation, of some other, and must maintain its ground by a successful struggle against the encroachments of other plants and animals.[30]

Figure 18.6 Wallace's Line. After many years collecting in the Malaysian Archipelago, Alfred Russel Wallace divided the islands into two groups on the basis of their animals and plants. The 'line' between them signified an absolute boundary between separate faunas. From *Proceedings of the Royal Geographical Society* (1863).

Conclusion

The ideology of empire as expressed by Lyell, Darwin, Wallace, and others was integral and essential to biogeography. Practically and intellectually, notions about distribution patterns were grounded in contemporary colonial activities and concepts. In the same way that the imperial ethos dictated a formula of centre and periphery, motherland and distant outstations, so animal and plant distribution studies in early nineteenth-century Britain were almost entirely carried out by British personnel travelling to the scene of interest, others remaining at home ready to synthesize the results for British use. That these individuals were a mixture of entrepreneurs and professional men associated with government and national institutions should surprise no one familiar with the general history of Britain in the century after the Industrial Revolution. This was how the empire was built.

So, too, the nature of the assumptions and explanations put forward to account for biogeographical regions can be attributed to the overriding ethos of colonization. The ethos gave purpose to naturalists' endeavours. It provided metaphors and a rationale; the raw materials and a way to understand.

Further reading

Arnold, David (ed.), *Imperial Medicine and Indigenous Societies* (Manchester, 1988).

Brockway, Lucile H., *Science and Colonial Expansion: The Role of the British Royal Botanic Gardens* (New York, 1979).

Browne, Janet, *The Secular Ark: Studies in the History of Biogeography* (New Haven, 1983).

Cameron, Ian, *To the Farthest Ends of the Earth: 150 Years of World Exploration by the Royal Geographical Society* (New York, 1980).

Crosby, Alfred W., *Ecological Imperialism: The Biological Expansion of Europe, 900–1900* (Cambridge, 1986).

Egerton, F. N., 'Studies of animal populations from Lamarck to Darwin', *Journal of the History of Biology*, 1 (1968), pp. 225–59.

Levere, Trevor, *Science and the Canadian Arctic: A Century of Exploration, 1818–1920* (Cambridge, 1992).

Livingstone, David, *The Geographical Tradition: Episodes in the History of a Contested Enterprise* (Oxford, 1992).

MacKenzie, John M. (ed.), *Imperialism and the Natural World* (Manchester, 1990).

MacLeod, Roy (ed.), *Government and Expertise: Specialists, Administrators and Professionals, 1860–1919* (Cambridge, 1988).

MacLeod, Roy and Lewis, M. (eds.), *Disease, Medicine, and Empire: Perspectives on Western Medicine and the Experience of European Expansion* (London, 1988).

Reingold, Nathan and Rothenberg, Marc (eds.), *Scientific Colonialism: A Cross-Cultural Comparison* (Washington, DC, 1987).

Sheets-Pyenson, Susan, *Cathedrals of Science: The Development of Colonial Natural History Museums during the Late Nineteenth Century* (Kingston and Montreal, 1988).

Stafford, Barbara M., *Voyage into Substance: Art, Science, Nature, and the Illustrated Travel Account, 1760–1840* (Cambridge, MA, 1984).

19 Travelling the other way

Travel narratives and truth claims

All narratives take the reader or listener on a journey, and many of them tell the story of a journey too. Narratives are organized to move through time, to transport the reader, and to bring us home again, augmented by the experience and by the knowledge we have acquired. This narrative motion is enacted as much in non-fictional accounts, like the many records of nineteenth-century surveying voyages, as it is in the *Odyssey*, or *Gulliver's Travels*. The differences begin when we examine the freight the reader gains by the expedition.

Almost all accounts of travels offer wonders, as well as a record of hardship endured by the narrator and his (usually his) companions. When Othello woos Desdemona he does so with his 'traveller's history' which is also a 'travaillous history': the ear does not discriminate between hardship and travel; their identity is part of the pleasure of the story told and part of its verification. The travel narrative, published or recounted, is a record of survival: the narrator is *here* to tell it in retrospect even as the reader sets out on the journey. That double motion offers reassurance: the experiences undergone, the knowledge gained, the treasure or the specimens preserved, are all trophies for the returning traveller and are proffered also to the reader. Material specimens and treasures are to be had only by proxy, but knowledge is more portable and may become part of the reader's own experience by reading the book. The publication of the 'history' affirms the traveller's re-entry into his initial culture, one presented as shared with the reader. After a spell as alien, the narrator is again homely, caught into current society's processes of exchange and affirmation.

These are some of the implicit assurances offered by travel narrative. But the question of its claim to truth is always ambiguous: travellers' tales are notorious for their self-serving exaggeration. The reader cannot check the authenticity of the monsters described, the adventures lived through. Confusions between actual and imagined voyages are frequent. Some writers, like Daniel Defoe (1661–1731), ransacked the accounts of others and adopted the plain style of authentic record, the careful downplaying of crisis, to such effect that his account of *A New Voyage Round the World By a Course Never Sailed Before* (London, 1725) was accepted as a true record for years after his death, to say nothing of the convincing (but fictional) record of his Captain Singleton's travels in Africa or Robinson Crusoe's make do and mend on his fortunately fertile island.

The tradition of imaginary voyages is ancient and continuous and natural historians on their travels therefore found themselves writing within rhetorical modes that were both enabling and dangerous to their project: enabling because detailed sensory description was valued in the genre, dangerous because such description was easily melded into fantasy and received as playful exaggeration not controlled observation. Diaries, field notes, samples and specimens, all the local and immediate evidence of encounter and categorization, therefore became particularly important in vouching for the objectivity of record. Yet at the same time the phenomenological, the personal moment, the record of what is smelt, touched, tasted, seen and heard by the subject, provides other convincing written evidence of the authenticity of what is told. It also translates as pleasure, one of the most compelling persuasive registers that writing can reach.

The question of the personal becomes a key issue: who sees? what is seen? what are the conditions of observation? The personal both vouches for and limits the scope of observation, and its authority. So the persona adopted by the narrator, or the taken-for-granted social contract between narrator and readers, must be exploited to give the widest possible scope and the greatest possible appearance of objectivity and disinterested access. Freiherr Alexander von Humboldt's (1769–1859) *Personal Narrative of Travels to the Equinoctial Regions of the New Continent, during the Years 1799–1804* (tr. H. Williams, 7 vols., London, 1814–29) provided a formidable example for nineteenth-century naturalists. His eminent eye, his range of sensuous description, his human contacts, impelled the reader into an identification that was physical and intellectual, specific and generalizing. For Charles Darwin (1809–82), the *Personal Narrative* fuelled his enthusiasm for travel and observation alike and provides a constant point of reference in his description of the *Beagle* voyage: Alcide d'Orbigny is 'second only to Humboldt' (vol. III, p. 110), and the phrase 'Humboldt has observed' recurs frequently (e.g. vol. III, p. 477).[1] Humboldt's presence in citation also indicated the international gentlemanly community of enquirers.

The title-pages of many a travel narrative reveal the strategies of endorsement adopted. Henry Walter Bates (1825–92), Alfred Russel Wallace's friend and collaborator, in his title suggests an all-inclusive cornucopia without too much self-consciousness about categories of knowledge; the whole is sustained, and contained, by his self-description as 'the Naturalist': *The Naturalist on the River Amazon: A Record of Adventures, Habits of Animals, Sketches of Brazilian and Indian Life, and Aspects of Nature under the Equator during eleven years of travel* (2 vols., London, 1869). Bates emphasizes the length of his sojourn ('eleven years') and the range of his interests. He includes events ('adventures'); systemizing ('habits'); and amateur and contingent observations of other peoples, not claimed as ethnographic studies ('sketches'). He acknowledges different levels and styles of knowledge-acquisition within one person's experience. Though social class is often important in the traveller's self-presentation, it is not so in the Bates example: here, in the late 1860s, his profession ('naturalist') is emphasized.

Earlier, the gentlemanly traveller, such as Louis Antoine, comte de Bougainville (1729–1811), is assumed in his *Voyage autour du monde* (1771) to have a synthesizing gaze that can encompass the discrete findings of fieldworkers and accommodate them to his own large vision. The trust of the authorities that send the traveller out is also an imprimatur: the King, the Admiralty, the Royal Society, and, in the period that most concerns me here, the support of the British Association for the Advancement of Science (so we find, for example, in 1831 an account of 'His Majesty's sloop *Chanticleer's* voyage under Commodore Henry Foster, FRS').

Underpinning these social and intellectual claims to authority, and seemingly largely unobserved by initial writer and reader, is a formation we can afford at present to be sharp-eyed about: that of the imperial and colonizing enterprise. That process may have a religious function: missionaries going forth to convert individuals and cultures. It may be part of land acquisition and conquest. It may also be part of a concern to map the seas, the sounds, the minerals, the rocks, the rivers, the interiors, the peoples of the place. When that exploration is condensed with a pattern of expectation that assumes a range of development in human societies moving from primitive to civilized a particularly ample authority can be claimed, since the civilized is assumed to be the place from which the writer starts and to which he returns. The writing is itself civilization at work on the unruly.

Yet current readers would be self-aggrandizing if they believed themselves to be the first to offer sceptical critique of the motivation, outcome, and justification of voyages of exploration. Near the end of *Travels into Several Remote Nations of the World, In Four Parts. By Lemuel Gulliver, first a Surgeon, and then a Captain of Several Ships* (London, 1726), Jonathan Swift (1667–1745) has Gulliver muse on the 'distributive justice of princes':

For instance, a crew of pirates are driven by a storm they know not whither, at length a boy discovers land from the topmast, they go on shore to rob and plunder, they see a harmless people, are entertained with kindness, they give the country a new name, they take formal possession of it for the king, they set up a rotten plank or a stone for a memorial, they murder two or three dozen of the natives, bring away a couple more by force as a sample, return home, and get their pardon. Here commences a new dominion acquired with a title *by divine right*.

And Denis Diderot (1713–84) inverts Bougainville's *Voyage autour du monde par la frégate du roi* La Boudeuse, *et la flûte* L'Etoile *en 1766, 1767, 1768, et 1769* (1771) in *Supplément au voyage de Bougainville* (1773) with an imagined Tahitian's telling account of *his* journey to Europe: a voyage in the reverse direction that was by no means fictional only, as Gulliver's reference to 'samples' suggests, and as the account of the two voyages of the *Beagle* later in this essay will make clear.

The voyages with which this essay is chiefly concerned were those whose prize was represented as knowledge rather than treasure. The categories are, however, not altogether separate. Although the nineteenth-century journeys that set out from Britain to survey the seas and coasts around the world were not piratical, not part of that unconcerned predation that earlier centuries justified as exploration or discovery, they were nevertheless an expression of the will to control, categorize, occupy and bring home the prize of samples and of strategic information. Natural history and national future were closely interlocked. And natural history was usually a sub-genre in the programme of the enterprise, subordinate to the search for sea-passages or the mapping of feasible routes and harbours.

The prosaic quality of such voyages' enterprise could also become part of the claim to truth-telling. Here, there is to be no surplus adventuring, no high-flown description, but an accurate record of observation and encounter. This level aim assumes that the ship's company has one coherent project and plan in view. But the disturbances manifest in the narrative language describing such voyages undermines that assumption. Fascination with the unfamiliar, fear and loathing, the longing for stable systems of communication, sickness, religious fervour and the physical pleasures of exploration all pressed across and became part of enquiry and left their traces in the writing. And so, particularly, do the peoples encountered.

Natural history and indigenous peoples

He expressed his regret that so little attention was given to Ethnography, or the natural history of the human race, while the opportunities for

Figure 19.1 Conrad Marten's drawing of a Fuegian from the Tekeenica tribe. Frontispiece to R. Fitzroy, *Narrative* (London, 1839), vol. 2.

observation are every day passing away; and concluded by an appeal in favour of the Aborigines Protection Society.

Among the records of the ninth meeting of the British Association for the Advancement of Science, held at Birmingham in August 1839 (London, 1840, 'Transactions of the Sections', p. 89) occurs this summary of a paper by the ethnographer James Cowles

Prichard (1786–1848). It suggests one of the most pressing issues raised by travels and their narratives in the nineteenth century and pursued particularly by Prichard. What are the boundaries of natural history? Are human beings within its scope? Are they one species or several? Are they separate from all other species because created as souls by God? And do all, all savages, have souls? Or are they – here danger lies – a kind of animal? (If they, then we?)

Elsewhere in this same volume, in the 'Synopsis of grants of money at Birmingham' and under the heading 'Zoology and botany' we discover the response to Prichard's appeal:

For Printing and Circulating a Series of Questions and Suggestions for the use of travellers and others, with a view to procure Information respecting the different races of Men, and more especially of those which are in an uncivilized state: the questions to be drawn up by Dr. Pritchard [*sic*], Dr. Hodgkin, Mr. J. Yates, Mr. Gray, Mr. Darwin [etc.] . . . £5.00 (p. 27).

The grant is among the smallest given, but it is there. The Aborigines Protection Society seems to be less a humanitarian than a natural historical enterprise, though one of an unusual kind since the subjects observed are also to be the questionnaire's informants (not something that other botanic or zoological subjects such as kelp or Ascidiae, sloths, ant-eaters, or armadillos, can be expected to perform). That double role of being the scrutinized subject and the independent respondent is peculiar to human beings. Over and over again the narratives of voyages demonstrate how the borders of natural history were blurred by human encounter and how evolutionary theory profited from that growing uncertainty about the status of the human in knowledge and in nature. The zoological and the linguistic appear side by side as parallel kinds of evidence: so, the surgeon Wilson on the *Beagle*, writes: 'The Fuegian, like a Cetaceous animal which circulates red blood in a cold medium, has in his covering an admirable non-conductor of heat' (vol. II, Appendix 16, p. 143), while Appendix 15 (pp. 135–42) is a vocabulary of Fuegian languages.

A review in the *Gentleman's Magazine* (March 1831), the year that Darwin joined the *Beagle*'s expedition, shows a typical squirm of argument about these matters. The book reviewed is a compendious *Narrative of Discovery and Adventure in Africa, from the earliest ages to the present time: with illustrations of the Geology, Mineralogy, and Zoology. By Professor Jameson, James Wilson, Esq. FRSE and Hugh Murray, Esq. FRSE*. It is not the geology, zoology or mineralogy that fascinates and irritates the reviewer but the peoples encountered. The reviewer is writing within the journal's pro-slavery stance:

All savages present to us, in certain respects, tricks, habits, and oddities like monkeys; and it is certain that in artificial acquirements they do not reach the elevation of dancing dogs . . . We are no advocates for the abduction of Africans, because it is robbery, and sometimes consequentially murder . . . but [like impressment or conscription it is] . . . assuredly a means of rendering idle and worthless people useful members of the community. That the African cannot become such useful members at home, is evident from the following tokens of their degrading characteristics as human beings (p. 237).

Whose 'community' is that to be served? The question does not bear answering. The argument presents 'savages' simultaneously as at the very least *like* animals and yet as degraded human beings. The indigenous inhabitant is here, squeamishly though it is put, to be 'improved' only by removal from home and subjection to 'civilized masters'.

Travellers who would have believed themselves heartily against slavery nevertheless commandeered individuals to function as pilots or translators. For example, Robert Fitzroy (1805–65) records a 'boy, whose name, among the sealers, was Bob'. 'Mr. Low had a Fuegian boy on board the *Adeona*, who learned to speak English very tolerably, during eighteen months that he staid on board as a pilot and interpreter' (vol. II, p. 188). Taking local people aboard ship for a time was not altogether unusual: motives varied. One of the most benign was that of Frederick Beechey (1796–1856), as he recounts it in his *Narrative of a Voyage to the Pacific and Bering's Strait, to Co-operate with The Polar Expeditions: performed in His Majesty's Ship* Blossom, *under the command of Captain F. W. Beechey, RN, FRS, etc. in the years 1825, 26, 27, 28* (London, 2 vols., 1831). Beechey came upon a shipwrecked group of Otaheitean islanders who had drifted 600 miles by canoe from their home. He took on board a family, Tuwarri and his wife and children, and the chief of the group and delivered them back to Chain Island. Tuwarri behaves according to the expectations of natural virtue: he is grateful, regrets parting, and is sorry not to be able to 'send some little token of his gratitude'.

These feelings, so highly creditable to Tuwarri, were not participated by his wife, who, on the contrary, showed no concern at her departure, expressed neither thanks nor regrets, nor turned to any person to bid him farewell; and while Tuwarri was suppressing his tears, she was laughing at the exposure which she thought she would make going into the boat without an accommodation-ladder (p. 236).

Beechey is mortified by the wife's insouciance and eagerness to be gone. Gratitude had, conveniently, long been assumed to be a natural virtue. He also criticizes Tuwarri's lack of curiosity while on board, though he praises his 'strong sense of right and wrong', and he is mildly shocked when Tuwarri 'was not received by his countrymen with the surprise and pleasure which might have been expected; but this may, perhaps, be explained by there being no one on the beach to whom he was particularly attached' (p. 237).

The absence of wonder or surprise was one of the phenomena that most disconcerted Western travellers in their encounters with indigenous people and which they described as most animal-like. Curiosity was so strong a driving force in Western expeditions, and so valued as a disinterested or 'scientific' incentive as opposed to the search for material gain, that the absence of an answering curiosity was felt as rebuff or even insult. Moreover, the reader of the narrative is likely, functionally, to agree with this view unless alerted, since the reading of natural historical travel narratives is an intensified form of that zealous curiosity that drives all reading.

Captain P. Parker King (1793–1856), the writer of the first volume of the three-volume set that includes Fitzroy and Darwin's accounts of the *Beagle* voyages, is in the main an astute and sympathetic observer. He describes his first encounter with the Fuegians in January 1827, after visiting the Patagonian Indians, linking the terms 'brutes' and 'want of curiosity':

They appeared to be a most miserable, squalid race, very inferior, in every respect, to the Patagonians. They did not evince the least uneasiness at Mr. Sholl's presence, or at our ships being close to them; neither did they interfere with him, but remained squatting round their fire while he staid near. This seeming indifference, and total want of curiosity, gave us no favourable opinion of their character as intellectual beings; indeed, they appeared to be very little removed from brutes; but our subsequent knowledge of them has convinced us that they are not usually deficient in intellect. This party was perhaps stupified by the unusual size of our ships, for the vessels which frequent this Strait are seldom one hundred tons in burden (p. 24).

King does not quite settle between cowed passivity or, more romantically, dignified remoteness: indeed he makes the point that interpreting the behaviour of other groups is always risky and unreliable. He repeatedly records the degree to which his own group had to correct their initial impressions and emphasizes that analogies with either animal or Western behaviour patterns are liable to mislead. As Anthony Pagden puts it in *European Encounters with the New World* (1993), one discovery made by travellers is that 'it is incommensurability itself which is, ultimately, the only certainty' (p. 41).

The need for native interpreters of local languages was therefore pressing. Marshall Sahlins in *Islands of History* (1985) has opened up the ways in which entire cultural systems of reference got disastrously caught across each other in Cook's final encounter with the Hawaiians. Western travellers, whether natural historians or not, soon discovered that the apparently universal repertoire of the body and its gestural systems is dangerously unreliable as a measure of meaning and intent. Language, though limited, is less volatile. With our current emphasis on the indeterminacies of language it is striking to realize the degree to which in travel narrative gesture is treacherous, language (even a few words) a blessedly stable resource and coin.

Beechey, for example, gives an account that is amusing if you are not on the spot, of what happened when he tried to persuade the Gambier islanders to dance, having admired their musical instruments. Beechey, hoping to get the islanders to offer their dances in exchange, gets the marines to 'go through some of their manoeuvres . . . this, however, had a very different effect from what was intended; for the motions of the marines were misinterpreted, and so alarmed some of the bystanders, that several made off, while others put themselves into an attitude of defence, so that I speedily dismissed the party' (p. 176). Darwin's much later *The Expression of the Emotions in Man and Animals* (1872) may owe much to his puzzling experiences on the *Beagle* and be in part a final attempt to regulate the irregularities he had there encountered.

Darwin and the Fuegians

On board the *Beagle* when Darwin joined the ship's surveying expedition in 1831 as companion to Captain Fitzroy and additional natural historian were seventy-four persons, seventy-three of them men. A girl of twelve or thirteen was the only female aboard. Her own name was 'yok'cushlu', but to discover that one must go to a fragment of a vocabulary on page 135 of the appendix to volume II of the three-volume *Narrative of the Surveying Voyages of his Majesty's Ships* Adventure *and* Beagle. Elsewhere in the text, by both Fitzroy (vol. II) and Darwin (vol. III), she is always called 'Fuegia Basket' as she had been named on the 1830 passage of the *Beagle* to England. She was one of three Fuegians whose return to their homeland was part of the second expedition's purpose and, so far as Captain Fitzroy was concerned, a compelling incentive. The other two were men, 'el'leparu' or 'York Minster', and 'o'run-del'lico' or 'Jemmy Button'. A fourth Fuegian, whose own name is lost but who was hauntingly called 'Boat Memory', had died of smallpox shortly after his arrival in England. It is worth dwelling on the story of the Fuegians' journey to England and return to Fuegia del Tierra, and the subsequent return upon return of Darwin and Fitzroy to the same locality to check their whereabouts, welfare and behaviour. Later travellers, such as W. Parker Snow, came on Jemmy Button again twenty years later. The story, pieced together, raises a number of important issues that recur in travel narratives, including issues that go into the formation of Darwin's thought. What was the impact of his meeting and acquaintance with these three young people long before he had encountered what he thought of as 'a savage'? And what did Jemmy Button's subsequent history suggest concerning adaptability, survival and cultural diversity and inheritance?

Captain Fitzroy later glosses his reasons for taking the Fuegians to England in different ways. In his letter to the Admiralty (23 May 1831), seeking support for his expedition to return the Fuegians to their own country, he emphasizes that 'I hoped to have seen these people become useful as interpreters, and be the means of establishing a friendly disposition towards Englishmen on the part of their countrymen, if not a regular intercourse with them' (Appendix to vol. II, p. 91). That secular and diplomatic explanation is overlaid by a religious one as support comes in from a missionary society who send a young man Matthews to accompany the Fuegians, learn language from them, and then use them as a group from which to promote the conversion of their countrymen. (That goes almost disastrously wrong and Matthews has to be rescued by the returning *Beagle*). Fitzroy back-projects a lofty humanitarian motive, which he universalizes defensively as a 'natural emotion' not capricious behaviour on his part.

Initially, however, it is clear that the various young people were annexed as hostages. Boat Memory and York Minster certainly were so. Fuegia Basket was left behind on board with two other children when women who had been detained escaped by swimming. The other children were returned but Fuegia, showing, Fitzroy asserts, no particular desire to go, was retained. Jemmy Button was conceived as something between hostage and 'interpreter and guide' (vol. II, p. 5). While seeking a boat stolen from the *Beagle* 'accidentally meeting two canoes . . . I prevailed on their occupants to put one of the party, a stout boy, into my boat, and in return I gave them beads, buttons, and other trifles' (vol. II, p. 6). Hence the name 'Jemmy Button'.

This shockingly cavalier appropriation of human beings rings ironically alongside Fitzroy's constant complaints about the Fuegians' pilfering. Yet, that fundamental error acknowledged, Fitzroy cared for the young people. He did not display them. He introduced them to his family and friends. He had them educated and looked after and taught certain mechanical skills. He disliked them being called 'savages'. Fitzroy claimed, oddly, that even their features improved with education. Describing Tekeenica and Alikhoolip people (both Fuegian groups, Jemmy being Tekeenica, the others Alikhoolip) he writes:

The nose is always narrow between the eyes, and, except in a few curious instances, is hollow, in profile outline, or almost flat. The mouth is coarsely formed (I speak of them in their savage state, and not of those who were in England, whose features were much improved by altered habits, and by education) (vol. II, p. 175).

Fitzroy's account is riven with contradictions provoked by the gap between his general assertions and his individual reactions: for example, he considers the Tekeenica particularly degraded, yet praises Jemmy Button most highly of his charges. During their stay in England Jemmy Button and Fuegia Basket appeared to adapt to their new conditions, though York Minster remained recalcitrant. Jemmy Button relished shoes and gloves; Queen Adelaide gave Fuegia one of her own bonnets. A phrenological examination performed in London unsurprisingly confirmed 'objectively' the judgements already made on their personalities (Appendix 17, pp. 148–9). These are the assurances the reader is offered, apparently bolstered by the 'factual', sample-like, materials of the various appendices freed from narrative process.

Darwin was not present on the first voyage nor implicated in the decision to take the Fuegians away. He joined the ship on which they were to be returned to their native land. He therefore met the Fuegians in their Western demeanour and outfits. Perhaps the degree of shock that Darwin felt when confronted with 'the inhabitants of this savage land', Tierra del Fuegia, was because he was habituated to the Westernized young people on board ship; certainly his remarks suggest so: 'I could not have believed how wide was the difference between savage and civilized man. It is greater than between a wild and domesticated animal, in as much as in man there is a greater power of improvement' (vol. III, p. 228). 'Jemmy understood very little of their language, and was, moreover, thoroughly ashamed of his countrymen' (vol. III, p. 230).

Darwin observes that the appalling physical conditions in which the Fuegians live – the cold, the lack of food and heating, the smoke half-blinding them in their wigwams – means that all their energies are engrossed in animal survival:

It is a common subject of conjecture what pleasure in life some of the less gifted animals can enjoy: how much more reasonably the same question may be asked with respect to these barbarians. At night, five or six human beings, naked and scarcely protected from the wind and rain of this tempestuous climate, sleep on the wet ground coiled up like animals (vol. III, p. 236).

The terms in the surrounding passage swerve, registering a disturbance that will not settle: he calls them 'poor wretches', 'barbarians', 'human beings': 'Viewing such men, one can hardly make oneself believe they are fellow-creatures, and inhabitants of the same world' (p. 235). But in a footnote Darwin argues that their lack of attainment does not imply inferior capabilities: 'Indeed, from what we saw of the Fuegians, who were taken to

England, I should think the case was the reverse' (vol. III, p. 235).

Darwin's encounters with Fuegians in their native place gave him a way of closing the gap between the human and other primates, a move necessary to the theories he was in the process of reaching. But it came after his experience of Fuegians abroad, since shipboard was a stylized form of England. The individuals, Jemmy, York and Fuegia, seemed to provide evidence of the human capacity for physical adaptation within the individual lifecycle. What was left moot was the question of descent. What would happen to the next generation? Would Jemmy's children be 'improved' by their father's experience? The answer was unexpected and disconcerting, and is much developed in the second, 1845, edition of the *Journal of Researches*.

When Darwin and Fitzroy returned in the February of the succeeding year (1834) they were appalled. A canoe appeared 'with one of the men in it washing the paint off his face'.

This man was poor Jemmy, – now a thin, haggard savage, with long disordered hair, and naked, except a bit of blanket round his waist. We did not recognise him till he was close to us; for he was ashamed of himself, and turned his back to the ship. We had left him plump, fat, clean, and well dressed; – I never saw so complete and grievous a change (2nd edn., p. 227).

Figure 19.2 'Fuegians going to trade in Zapallos with the Patagonians'. From R. Fitzroy, *Narrative* (London, 1839), vol. 2.

The native has gone native

Instead of admiring Jemmy's survival skills Darwin, like Fitzroy, reads this as degradation. Jemmy tells them 'that he did not wish to go back to England'. Darwin is more satisfied when 'we found out the cause of this great change in Jemmy's feelings, in the arrival of his young and nice-looking wife' (p. 227). York Minster and Fuegia Basket have disappeared, York Minster having robbed Jemmy Button. Only distant and indirect tidings are ever heard of Fuegia Basket: a sealer in 1842 'was astonished by a native woman coming on board, who could talk some English. Without doubt this was Fuegia Basket. She lived (I fear the term probably bears a double interpretation) some days on board' (footnote to p. 227).

So the hope of finding interpreters and guides, of sending back missionaries to convert their own people, came to very little. Instead the Fuegians acquired some macaronic skills and learnt to trade with language. Jemmy introduced English terms into the vocabulary of his group. Darwin's sentimental (but wary) hope concerning Jemmy's descendants is mocked by actual events:

Every one must sincerely hope that Captain Fitz Roy's noble hope may be fulfilled, of being rewarded for the many generous sacrifices which he made for these Fuegians, by some ship-wrecked sailor being protected by the descendants of Jemmy Button and his tribe! (pp. 227–8)

It is easy now to see the absurdity of this high-flown hope, for the Fuegians had no reason to see Fitzroy as a benefactor for stealing them from their homeland.

One further travel narrative gives a wry later vignette of transculturation: W. Parker Snow's *A Two Years' Cruise off Tierra del Fuego, the Falkland Islands, Patagonia and the River Plate: A Narrative of Life in the Southern Seas* (London, 1857). Snow met Jemmy Button 'quite naked, having his hair long and matted at the sides, cropped in front, and his eyes affected by smoke'. He still spoke English and on board the ship wanted to put on clothes and requested a 'knife to cut meat'. Jemmy proved himself adept at immediately reframing himself in either context. Snow later comments that Jemmy's tribe was the least to be relied on in any dealings, having learnt a double language and behaviour.

I had never before fallen in with one who had been transplanted to the highest fields of intellectual knowledge, and then restored to his original and barren state (vol. II, pp. 36–7).

Snow shows Jemmy his picture in Fitzroy's *Narrative* (Figure 19.3):

The portraits of himself and the other Fuegians made him laugh and look sad alternately, as the two characters he was represented in, savage

Figure 19.3 The page of comparative portraits that Parker Snow and o'rundel'lico/Jemmy Button looked at together. From R. Fitzroy, *Narrative* (London, 1839), vol. 2.

and civilized, came before his eye. Perhaps he was calling to mind his combed hair, washed face, and dandy dress, with the polished boots it is said he so much delighted in: perhaps he was asking himself which, after all, was the best? – the prim and starchy, or the rough and shaggy? Which he thought, he did not choose to say; but which I inferred he thought was gathered from his refusal to go anywhere again with us (vol. II, pp. 37–8).

The Fuegian encounter continued to raise questions for Western travel narrative about cultural choice and helped to undermine assumptions about improvement and authority.

While the rest of the *Beagle*'s company left familiar England, sailed around the world, explored its people, geology, flora and fauna, and returned home, the Fuegians were performing a counter-motion from Tierra del Fuegia to England and back. They had not chosen to explore the world or discover its properties. They were taken away from familiar Tierra del Fuego to exotic England by guile or force and then, after many adventures, returned more than two years later to a place not quite that from which they had set out. The two journeys, that of Darwin and that of the Fuegians, share one long lap: for him setting forth, for them returning. The Fuegians could compare England and Fuegia: Darwin could not yet do so. They could be not only objects of encounter but his informants and authority; and Jemmy Button, at least, we know was so.

It is difficult – impossible – to enter the Fuegians' experience of this story, though certainly as sentimental to imagine that they enjoyed nothing as that they relished everything and were grateful for their kidnap. Telling and knowing are ungrounded by this attempted exchange: what are the natural historical categories for these subjects? The reader can know el'leparu, o'rundel'lico and yok'cushlu, if at all, only under the sign of their Western soubriquets as York Minster, Jemmy Button and Fuegia Basket. Their own narratives were oral, in another tongue, another place. That process of familiarization and estrangement is repeated in relation to the language of the *Beagle* voyagers themselves. Trying to understand the sensibility expressed by the British sailors in that act of re-naming the Fuegians, renaming them moreover with those particular nicknames, is likely to make us register our baffled distance from the shipboard community of the 1830s more intensely than does anything in the rest of Fitzroy's urbane or Darwin's ardent prose.

Further reading

Adams, Percy G., *Travellers and Travel Liars, 1660–1800* (Cambridge, 1962).

Beer, Gillian, 'Four bodies on the "Beagle"': touch, sight and writing in a Darwin letter', in J. Still and M. Worton (eds.), *Textuality and Sexuality* (Manchester, 1993).

Carter, Paul, *The Road to Botany Bay: An Essay in Spatial History* (London, 1987).

Keynes, R. D. (ed.), *The* Beagle *Record: Selections from the Original Pictorial Records and Written Accounts of the Voyage of HMS* Beagle (Cambridge, 1979).

Pagden, Antony, *European Encounters with the New World: From Renaissance to Romanticism* (New Haven, 1993).

Pratt, Mary Louise, *Imperial Eyes: Travel Writing and Transculturation* (London, 1992).

Prichard, James, *The Natural History of Man; Comprising Inquiries into the Modifying Influence of Physical and Moral Agencies on the Different Tribes of the Human Family* (London, 1843).

Sahlins, Marshall, *Islands of History* (London, 1987).

Smith, Bernard, *European Vision and the South Pacific*, 2nd edn. (New Haven, 1988).

Stocking, George W., *Victorian Anthropology* (London, 1987).

Todorov, Tzvetan, *The Conquest of America: The Question of the Other*, trans. Richard Howard (New York, 1984).

20 Ethnological encounters

Victorians visiting museums and exhibitions regularly encountered foreign peoples displayed in their native costumes, adorned with ornaments and surrounded by their artefacts in their reconstructed environments. Some of these 'natives' came to the natural history stage of their own volition, others through coercion or deception, and still others as corpses, valued for the quality of their skeletal remains. Such displays made very clear that these people were there to be observed as an integral part of the spectacle of natural history. What remained very unclear for most of the century was whether animals and humans were physically, morally or intellectually related – and whether the different human races belonged to the same species. Victorian museum-goers pondered and reflected on exactly these sorts of questions.

Writing about the history of ethnology or anthropology raises important ethical questions peculiar to this branch of natural history. The subject calls for more than a history of the ideas of great European thinkers. Curiously, many present-day historians are rather more prone to treating 'non-Western' peoples as specimens, like plants or minerals, than their Enlightenment counterparts who were deeply inquisitive about the humanity of newly 'discovered' peoples. A better alternative, on ethical grounds, is to explain the production of ethnological knowledge in terms of the beliefs, actions and intentions of all the human groups involved. We can do this by adopting a more global or decentralized framework to consider encounters between members of diverse cultures in different parts of the world – kings, merchants, naval officers, travellers, nomads and pastoralists, to name but a few.

In the course of this essay, I will clarify the meaning of ethnology as distinct from anthropology, examine some of the strategies used by travellers to acquire ethnological information and artefacts, and discuss the importance of the geographical gift as a method that underpinned much of nineteenth-century ethnographic investigation. Then I shall address the question of how historians can present peoples on the other side of the encounter – those being observed and described. More generally I shall examine the status of this knowledge in Britain, though restricting the scope of my enquiry to the work of the philanthropically minded ethnological societies.

The history of nations

It is undoubtedly true that the tradition of Europeans recording systematic descriptions of other peoples has a very long history, reaching back to the medieval travels of Marco Polo and beyond. However, the terms 'ethnology', 'ethnographic' and 'ethnological' are of recent origin, introduced into the English language only in the 1830s and 1840s. James Cowles Prichard (1786–1848) and his colleagues in the Ethnological Society (f. 1842) used these words quite specifically to refer to the 'history of nations'.[1] The self-conscious use of the word 'ethnology' stressed the importance of studying both the physical and civil history of foreign and particularly non-Christian peoples.[2] These self-styled ethnologicals viewed the distribution of races across the globe as the product of civil (i.e. human, legal and social) and climatic processes. The foremost challenge facing the science of ethnology was to explain how the offspring of a single human pair at the time of the Biblical Deluge had diverged through migration into separate 'tribes' or 'nations' with diverse customs, physical features and, particularly, beliefs.

In charting the histories of nations, the study of language was the discipline's foundation. For humanist scholars of the seventeenth century and Enlightenment philosophers, the acquisition of language was the criterion of sociability, thought to raise barbarous peoples out of the state of nature towards a state of civility. The discovery of esoteric scriptural texts in lost ancient languages such as Sanskrit, lying outside of Christendom, helped to overturn this view of 'other' peoples as predominantly alien. For most Victorians the idea of a completely savage state of nature was simply an anachronism of bygone days. For the ethnologicals, successes such as the discovery of Sanskrit and the decoding of the hieroglyphics on the Rosetta stone reaffirmed their belief that language held the key to unravelling the story of humanity. Although comparative studies of evidence such as skulls and other remains could illuminate the effects of climate on the physical history of the races, these could only serve as a subsidiary criterion to that of language.

Comparative studies of mythology and language helped to breed a measure of cultural relativism. Differences in language and vocabularies were used to account for diversity, while similarities were taken as evidence for the unity of humanity. The driving force behind ethnology was a group of Quaker philanthropists (predominantly middle-class manufacturers and industrialists), whose commitment to religious tolerance (in contrast with the orthodox Anglicans of the day) inclined them to be relativistic and sympathetic to peoples holding religious beliefs other than their own. This attitude towards helping classes of people less fortunate than themselves motivated them to support 'good causes' with great zeal. The philanthropic mood of the country was driven by the widespread movement to abolish slavery throughout the world, led by the eloquent evangelical, the Reverend William Wilberforce (1759–1833). However, as the focus of the anti-slavery movement shifted away from Britain where it was legally abolished in 1807, towards the United States, another evil seemed to fill its place. The abuses, massacres and extinction in British colonies of indigenous peoples such as the Tasmans, Aborigines and Maoris became a *cause célèbre* for the Quaker philanthropists, who founded the Aborigines Protection Society (APS) in 1837. Not coincidentally, its founding members induded the most important ethnologists such as James Prichard and Thomas Hodgkin (1798–1866). Quaker philanthropy and ethnology shared one crucial theoretical feature, the doctrine of monogenesis, which stated that all human races were varieties of the same one (i.e. mono) species. In principle, monogenesis provided philanthropists with a scientific basis for declaring a common racial heritage for all peoples, and hence for legitimating their claim that indigenous peoples should be entitled to legal and civil rights, and for insisting that the colonial state was morally obliged to safeguard them.

The relativism derived from comparative studies could be used for other, and even opposing, causes. The comparative anatomist, Robert Knox (1793–1862), argued that the physical and moral differences between races were so pronounced that they could never be bridged. In his view colonialism was imposed and was therefore doomed to failure. From this it followed that the philanthropic movements could never be more than a conscience-clearing exercise and a mistaken imposition of one race's ideas on another. This also had its scientific justification in the doctrine of polygenesis, which stated that the human races were in fact different species and hence more distantly and irreconcilably related than monogenesists were willing to concede.[3] These anatomists of the human body, some of whom founded the Anthropological Society in 1863, asserted the corresponding superiority of the material evidence of human remains over linguistic evidence, which they associated with an older generation committed to ideals of the past.

The ambiguous implications of scientific knowledge for the political status of slaves and indigenous peoples were not restricted to the level of theory. Debates over the abolition of slavery and the governing of India attracted the attention of leading political philosophers for well over a century, starting most conspicuously in the 1780s with the evangelical revival and the rise of the abolitionist movements. One of the reasons for this longevity is that

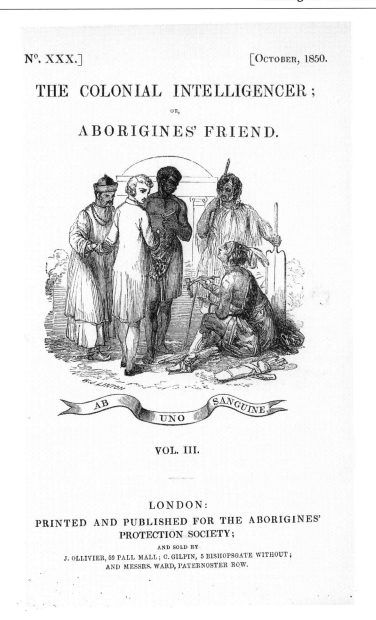

Figure 20.1 Emblem of the Aborigines Protection Society. The Society's *desideratum* was to extend the European hand of friendship to the other races of the world in the form of peace and commerce. The significance of ethnological science for the society is shown in their motto 'Ab uno sanguine', meaning 'from one blood', the doctrine of monogenesis.

the slave trade had its own powerful supporters, not least among some of the elite in the scientific institutions of natural history. Take, for example, Sir Joseph Banks (1743–1820), the pivotal figure of English natural history during his long-time presidency of the Royal Society (1778–1820). One of Banks's many exploits was to found the African Association (f. 1788), the antecedent to the Royal Geographical Society (f. 1830), whose basic purpose was to explore and to map the interior of Africa with the general aim of promoting future commerce. In the same year (1788), Captain

William Bligh (1754–1817) with Banks's support, attempted to transport breadfruit, a rugged, nutritious staple food, from Tahiti to the West Indies, with the aim of providing cheaper food to make the slave plantations more profitable. Knowledge in the sciences of geography and natural history provided crucial resources for the alliance of savants, plant breeders and plantation owners. Many politicians and civil servants in positions of power, who publicly damned the slave trade, refused to take action against it, where they might well have been able to. For example, John Barrow (1764–1848), the Admiralty Secretary and staunch Tory, who did so much to promote African exploration, attacked the abolitionist movement for campaigning against the interests of the plantation owners.[4]

Collecting language

The success of ethnological research was dependent on gaining reliable access to information about the people of other nations: the civil history of their form of government, their religious beliefs, language, manners, customs and habits. In order to gain access to this kind of knowledge, it was vital to find reliable and productive sources throughout the world, and particularly in areas where there were British political and economic interests at stake.

The Aborigines Protection Society established a network of agents throughout the British colonies. Individual ethnologists took advantage of natural history correspondence networks to exchange information about vocabularies and to acquire artefacts. However, many ethnologists were physicians with practices, and therefore rarely ventured further afield than the English Channel. Others did most of the travelling, recording, translating, describing and collecting. Whether employed as merchants, engaged as colonial officers or commissioned on Admiralty voyages of discovery, these travellers often cast their nets wide searching for anything whose value carried a premium at home – new medicinal herbs or valuable mineral deposits as well as colourful and exotic costumes and artefacts. Many travellers collected written descriptions and made sketches of the peoples and places they visited, hoping to build a reputation and make a tidy profit out of their eventual publication. The enormous travel literature industry, which had blossomed in the second half of the eighteenth century, provided ethnologists with a rich and varied set of resources. Besides catering to the better-off classes, publishers with the help of editors and gentlemen scientists performed the task of helping their authors to reorganize their travel descriptions systematically into the different branches and sub-divisions of natural and civil history.

By the late eighteenth century the expansion of travel literature, the proliferation of missionaries' reports, the evangelical sentiments that motivated the abolitionist movement and the publication of the tracts of freed slaves had greatly enhanced the general awareness and level of knowledge among the reading public. Concurrently, ethnologists like Johann Friedrich Blumenbach (1752–1840), were complaining about the variability and frequent inaccuracy of travellers' observations. The elite standards set by government-sponsored expeditions to the South Seas under Captain Cook (1728–79) were difficult for others to match. The new level of co-operation between naturalists, such as Banks or Johann Reinhold Forster (1729–98), and their artists, such as Sydney Parkinson (1745?–71) or William Hodges (1744–97), helped usher in a new era of descriptive natural history. These men were riding the crest of a huge wave of imperial expansion and they capitalized on it, in a grab-and-run fashion as they 'discovered' and foraged new floras and faunas. In turn these collections brought them fame and influence. Relatively few members of the extensive colonial botanical networks could dream of emulating the Pacific collections of Banks, with his powerful backing from scientific and political elites. Many did, however, make relatively modest contributions by bringing new plants to the attention of the superintendent of a colonial botanical garden or by sending seeds to Banks or his librarian Robert Brown (1773–1858).

Collecting was no trivial activity and posed difficult problems for all curious travellers. Preparing zoological and botanical specimens for long voyages back to Europe was a difficult business. Many of the dangers of climatic and ecological change that inflicted death on ships' crews were meted out to the flora and fauna. The constant spray of sea water across a ship's deck and the crushing of its timbers during a full gale were enough to finish off many a collection. Individual natives of foreign nations coming to Europe either to be exhibited, educated or in some cases for adventure, had to share the crew's physical degradation and malnutrition before the real ordeal of encountering life in a port like Bristol, Liverpool or London. Of course many specimens were dead before the voyages had begun. Skeletons and crania, the most coveted artefacts of all, could get damaged if the jars broke, or decay if the preservative leaked away, and one marvels that rebellious sailors did not always jettison offending collections overboard.

Frustrated ethnologists also criticized collectors for asking the wrong questions, omitting crucial information necessary for comparative studies and describing exotic and fanciful phenomena, where more thorough and banal observations might have been more informative. As a remedy, ethnologists (like other naturalists) made attempts throughout the nineteenth century to organize and discipline 'field' observers in their observations by printing instructions or questionnaires outlining precisely what to observe and record. For example, Jean-Marie Degérando's instructions produced for the Société des Observateurs de l'Homme, published in 1800, provided an itemized and detailed list of questions.[5] The Ethnological Society produced and distributed a *Manual of Ethnological Inquiry* (2nd edn., 1852) for missionaries, military officers, men of science in the colonies, and other travellers. Evaluating the effectiveness over the short term was difficult. The mixed evidence was enough to convince the British Association for the Advancement of Science, who was footing the bill, to cancel the project, but not before Prichard had revised it as a contribution to the Admiralty's *Manual of Scientific Enquiry* (1849), a handbook covering all of the field sciences, with contributions from the leading gentlemen of science of the day.[6] The prestige enjoyed by this manual owed as much to the reputation of the surveyors and explorers in the Royal Navy as reliable suppliers of 'field' observations as to its list of illustrious authors.

Prichard's philological instructions were intended to help travellers cope with what was recognized as a uniquely difficult area of research. Unlike other branches of natural history such as botany, visual observation was by no means the primary mode of research. Oral and audial skills were more important. The investigator's capacities to listen, remember, mimic and articulate foreign sounds and new words were essential skills for building vocabulary lists and recording the phonetic elements of other languages. But a good ear and a sound memory were not very useful without a systematic means of transcribing the sounds into a written form. Philological research required developing orthographies, sets of phonetic codes from which the original sounds could be reconstructed. The better educated, sponsored or wealthier travellers could purchase and carry with them the orthographies and vocabulary lists of previous travellers who had visited neighbouring regions or other peoples within the same family of languages. Comparing orthographies and common words was a prerequisite for travellers hoping to contribute new sounds and words into the existing records. Hence they would typically use the orthography and vocabularies as a primer for learning the language and gradually reach the stage of adding new words, clarifying the existing orthographies and very occasionally adding new sounds to the lists.

From the end of the eighteenth century onwards, European scholars embraced the study of religious, legal and mythological texts written in ancient Oriental languages such as Sanskrit. An emerging science of language provided a crucial set of intellectual tools for this form of enquiry. William Jones (1746–94) and Friedrich Schlegel (1772–1829) laid the foundations for 'the new philology' with their focus on grammatical structures and genealogies.

Philological debate in England raged well into the nineteenth century, as the traditionalist school of antiquarians were attacked for
their superficial use of affinities between the transcriptions of
foreign words. The criticism was thoroughly polemical and not
always deserved. However, the new philologists like John Kemble
(1807–57) advocated 'a comparative anatomy of language' whose
principal methods were to collect 'specimens' of language, to study
the rules of grammar and parts of speech and to pay closer attention to the phonetics behind the transcribed words.[7]

There was a recognition amongst travellers that certain kinds
of vocabulary such as numbers, geographical labels, names for
animals and personal pronouns were particularly desirable for
understanding a people's way of life and communicating with
them. There was, however, no universal system of taxonomy for
language – like those developed by Linnaeus (1707–78) for describing the external appearance of botanical or mineralogical species –
until the 1840s. Ethnological instructions began to provide phonetic symbols for encoding foreign sounds based on well-known
English words – similar to the phonetic guides that one sees in
the front of dictionaries today. Using these effectively required an
extended study both of a foreign language and the phonetic codes.
Furthermore one must remember that travellers of all classes and
countries were sporting an incredibly wide range of accents which
further complicated the task of the comparative philologists. Since
this was long before the invention of 'BBC English', there was little
that could pass for standard speech. Differences in the results of
philological research sometimes raised more questions about the
backgrounds of travellers than they answered about questions of
human origins and migrations.

Edward Parry (1790–1855) and John Ross (1777–1856), both
Arctic explorers having extensive contact with Inuit (Eskimoan
peoples), provide a useful contrast. Parry worked hard to acquire
a basic knowledge of Inuktitut, to record an extensive vocabulary
and to grasp the subtleties of the orthography borrowed from
descriptions of Greenlandic Inuit. As a consequence he produced
a comprehensive and relatively sympathetic description of their
manners, customs and habits. In contrast, John Ross made a poor
showing of learning a little Inuktitut, and by the time of his second
voyage (1829–1833) had delegated the ethnographic duties to his
nephew, James Clark Ross (1800–62), who had greatly improved
his command of the language with Parry's encouragement. John
Ross thought that the Inuit would become extinct. Believing that
their condition was much like that of his working-class able
seamen, he administered liberal quantities of grog on Christmas
Day, supposing that this would lessen their misery. Philanthropy,
whether in the guise of missionary activity or travel diaries

Table 20.1. *Vocabulary list taken from the Khond people in the Gomsur mountains of Orissa in India*

mother	ma	aya
grandfather	bodda boppa	akenja
grandmother	bodda ma	atha
grandfather, mother's side	wodja	akey
elder brother	nunna	dada
younger brother	sana bhai	apo
elder sister	nanni	bhai
younger sister	bhouni	budi
what is your name?	tuora naum kiso	anni padatatti inu
where is your village?	toura gau koture utchi	mi naju embam unni
come early in the morning	boddu soccale aso	daasay vam *or* bam
how far is your village?	tuora gau kete door utchee	mi naju esse duro
my village is ten coss	humara gau dos kos utchee	mai naju dos kossu ane
how far is Patlingeah?	Patlinga kete dur utchi	Patlingaki essee dur ane
have you cooked food?	bhatta randuloki	veyha vasiti ginnaa
have you eat food?	bhatta khailoki	veyha tisi ginnaa
how many seers of rice did you eat?	kete ser chaulo bhatto khailo	pranga esse adiah tinge
I eat three seers of rice	tin ser chaulo bhatto khaila	tin adiah veyha tiee
do you eat meat?	Maunsa khailoki	inu onga tinji ginnaa
what meat will you eat?	tu ki maunsa khaibo	ami onga tinji
I will eat mutton	cheli maunsa khaibu	odu onga anu tiee
I will eat fowl	kukuda maunsa khaibu	kodju ongha tiee
I will eat fish	matcho khaibu	minu acu tiee
I will eat dry fish	sukwa matcho khaibu	nore minu tiee
I will eat pig's flesh	gusri maunsa khaibu	anu paji onga tice
who is your God?	tumara kau Deybota	mindi amni peynu
what is his name?	tumara Deybota namo kiso	mi peynu anni peynu
how many gods have you?	tumara kete Deybota utchi	mi peynu esse peynu munne
Takurani is the greatest of all	samos teko bodda takurani	gulletiki dumalurri durani
how do you pray to the God?	ki mante Deybota puja korocho	peynuki singhi lakideru
what is the name of this month?	ye masuro namo kiso	iree anni danju
February	Mhago	Keydu
how many people do you sacrifice in one month?	gute masure kete manuso katuso	oro danju esso anni mirimmi pakidirru
amongst the khonds how many castes?	khondare kete jati utchi	khoi jati esse jati munney

Vocabulary lists served multiple purposes. They assisted East India Company administrators in revenue collecting, provided evidence justifying stronger colonial intervention in this region and added grist for the philologist's mill. From W. Taylor, 'On the manners, customs and rites of the Khoondis, or Khoi Jati, of the Goomsoor Mountains', *The Madras Journal of Literature and Science*, 6 (1837). p. 30.

necessitated prolonged periods of contact and linguistic interchange.[8] Developing orthographies and recording vocabularies demanded an adeptness at cross-cultural communication and a measure of intimacy far more elaborate than was necessary for the crude task of carrying off indigenous corpses. Consequently, the ability of Victorian travellers to study and master foreign languages became a measure of their expertise. The image of the educated Victorian traveller found its apotheosis in such figures as the explorer, scholar and poet, Richard Burton (1821–90). He exemplified the art of 'going native' which, besides recapturing the romantic mystique of the Orientalists, could express the existence of the 'other' in the self.

The gift of mapping

A surprisingly small number of explorers charted vast areas of land and sea throughout the globe in the eighteenth and nineteenth centuries. The help of native informants to acquire 'the geographical gift' was crucial to this project. Deception or coercion were frequent tactics that complemented this exchange of gifts. Simon Lucas, travelling on behalf of the African Association, provides an archetypal example. He set out for the North African kingdom of the Fezzan on the pretence of collecting medicinal plants and Roman antiquities. Through the British Consul and the Turkish Pasha in Tripoli, he arranged to be guided by two Shereefs of the King of the Fezzan. When it became apparent that he would not reach Fezzan, he convinced one of the emissaries that they should jointly prepare a map of Northern Africa as a gift for the Bey – only that Lucas should keep a copy for his own use. For the young informant, it may have provided him with a gift for the King. For the African Association, it filled in large areas of the African puzzle, another step in planning and maintaining trading regimes along the African coast. The significance of the explorer's deception was to hide the nature of the asymmetry of the gift of geographical knowledge. The extent to which his informant was actually deceived is difficult to determine.[9]

These explorers' maps were powerful objects because they could be combined systematically with other geographical knowledge, including descriptions of manners, customs and habits. As the African Association's imperial geographer James Rennell (1742–1830) emphasizes, geography was in his view a *topographical* science. Reading signs on the *surface* of the landscape (the sources of rivers, oases, cloud patterns) provided the key for piecing together the landscape's inner propensities for imperial commerce – the direction and flow of its waters, the moral qualities of its populations and the caravan routes for the traffic in humans.

Rennell likened African topography, lacking coastal rivers and ubiquitous in its expanses of desert, to 'the skin of a leopard'.[10] In so doing he helped invent the racial stereotype of Africa as a wild and dangerous animal. This kind of systematic comparative geographical knowledge was not vanquished by the formation of scientific disciplines in the nineteenth century. On the contrary, it provided a template to underpin ethnological enquiries for the next century.

One reason for the longevity of the paradigm of the topographical map is that it could be flexibly incorporated into different political ideologies. One could, for instance, compare the acquisition of native maps by Francis Hamilton Buchanan (1762–1829), the Edinburgh-trained surgeon who travelled widely through India and Burma with Edward Parry, the philanthropically minded Royal Navy captain, Arctic explorer and son of a prominent physician from Bath. Buchanan acquired the map of Ava (Figure 20.2), an area extending out from both shores of the Irrawaddy river, while he was based with the British mission in Burma in 1795, carrying out a botanical study of the region. This, and his other surveys of Mysore and Bengal, had been commissioned by Richard Wellesley (Lord Mornington), the incumbent Governor-General of India, as a means of justifying an aggressive policy of expansion to his critics in the Colonial Office and the East India Company. Buchanan never received the recognition that such extensive surveys might have earned him, owing chiefly to the dramatic political downfall of his patron. Buchanan's publication of his map, some twenty years later after he had retired to Scotland, can be seen as an attempt to improve his reputation as a natural historian, an aim to which he had dedicated his entire life. Parry's politics, by contrast, were paternalistic and philanthropic, influenced by his father. When Parry published maps acquired from Inuit during his north-west passage voyages, he credited his sources, commented on their impressive range of knowledge and espoused a cultural relativism concomitant with his philanthropic, evangelical ideas. Notwithstanding this ideological contrast, Buchanan and Parry were both very concerned with eliciting the indigenous geographical gift.[11]

Topographical mapping entered the realm of the human mind with the development of phrenology in the late eighteenth century by a Viennese physician, Franz Joseph Gall (1757–1828). This science was predicated on the assumption that particular areas of the surface of the skull corresponded to particular mental characteristics. Hence an examination of an individual's head could supposedly provide an outline of the abilities or essential features of that person's mind (Figure 20.3). Concurrently, the physical researches on the history of the globe by Alexander von

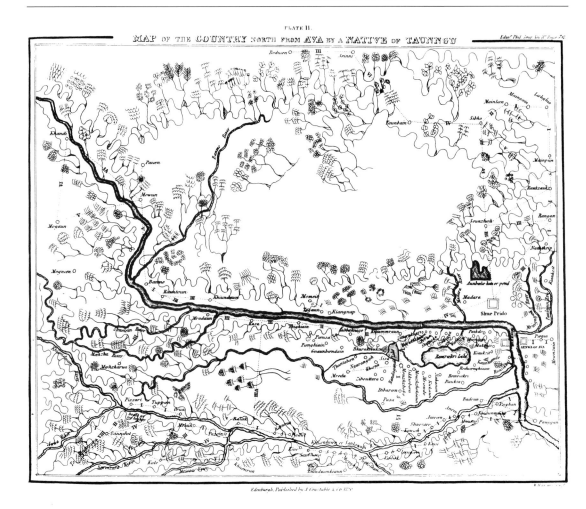

Figure 20.2 Map of Ava, *Edinburgh Philosophical Journal* 4 (1821), p. 76. The Edinburgh-trained surgeon, Francis Hamilton Buchanan, spent much of his career based in India. While accompanying the Governor-General's mission to the Kingdom of Ava (Burma) in 1795, Buchanan solicited this extraordinary map from a relative of a local diplomat. The double lines refer to rivers (the largest being the Irrawaddy), the wavy lines to mountains, and the smaller cluster of waves to trees or forests. Maps like these often accompanied larger botanical or statistical surveys. For example, Buchanan's subsequent survey of Mysore was commissioned by the Governor-General, Lord Mornington, to justify his defeat of Tipu Sultan and conquest of Mysore to his critics in the East India Company and Britain.

Humboldt (1769–1859) and his desire to revolutionize and replace natural history as a science of collecting with a new topographical physiognomy of the earth based on precision measurement and analysis, had a tremendous effect in terms of the proliferation of the use of maps as analytical tools in new disciplines. His creative use of graphical methods and, in particular, his extension of the

Figure 20.3 'Dr. Spurzheim in his consulting room measuring the head of a peculiar looking patient'. Coloured aquatint by J. Kennerly, 1816, after R. Cocking. Phrenology was practised both in the British Isles and in the colonies. The ease with which a wide range of people could practise phrenology contributed to the controversy surrounding its use and interpretation.

use of iso-lines for mapping a wide range of natural variables such as mean temperatures and barometric pressures, played a crucial role in promoting the importance of geographical distributions. In the hands of Humboldtian admirers such as Augustin-Pyramus de Candolle (1778–1841), these were crucial in the formation of the new field of biogeography. By the same token, these graphical methods helped to legitimate the scientific status of ethnologists who defined their subject in terms of the historical and geographical distribution of peoples. Taken out of its philological context, the Humboldtian project found its reception in figures like the controversial anatomist Robert Knox and his student James Hunt (1833–69) (a speech therapist), as well as the evolutionist Alfred Russel Wallace (1823–1913) and the founder of the science of eugenics and the promoter of physical anthropology, Francis Galton (1822–1911).[12]

The second voyage of Captain John Ross, in which his nephew, Lieutenant James Clark Ross, first located one of the two North magnetic Poles, presents an interesting example of mapping beliefs. John Ross, a high Tory and a promoter of phrenology, believed that the Copper Inuit he encountered were akin to the working class, and that they would in all likelihood become extinct. His interest in mapping people's minds rather than studying their languages and trying to understand their beliefs is indicative of a growing trend of a materialistic natural history of man and a social determinism. For Ross, the phrenological mapping of exemplary heads, which were chosen from those of the powerful elite, offered a pedagogical means for inspiring respect for authority among the lower deck of seamen. In other words, it justified a very conservative model of society which approximated the extreme hierarchy of one of His Majesty's ships.

The historiography of 'the other'

One challenge facing historians is to examine the varied perspectives of ethnology's human subjects. This suggests a shift in emphasis towards cross-cultural contact, the worlds of indigenous peoples, the social basis of knowledge in oral cultures and the history of legal representation for colonized peoples. The priority usually given to the ethnologist's perspective over that of the human subject should not necessarily be assumed or taken for granted. To return for a moment to Captain Parry and his north-west passage voyages, one of the best examples of an interactive co-existence was his period of contact with Inuit hunter–gatherers (Aivilingmiut and Iglulingmiut, who live in what is today part of the Arctic archipelago of Canada) from 1821 to 1823. Because the expedition is well known on account of the quality of its ethnological descriptions, one might suppose that the navy officers were the instigators of the contact. In fact, the critical meeting with the ships' crews was actually initiated by Inuit bringing whale-bone to the ships, with the intention of trading. The philanthropically motivated officers cultivated the relationship, in part for company and diversion during the long Arctic winter, and in part for the production of ethnological knowledge about the Inuit.[13] The linguistic competence of the officers not only enabled them to speculate on the linguistic basis of Inuktitut and hence the migrations and origins of the people, but also to excel in acquiring the gift of Inuit geographical knowledge. This was accompanied by the production of Inuit oral accounts about the British naval expedition. In terminating this episode of contact at Winter Island, the evidence suggests that the Inuit, being semi-nomadic, made a planned seasonal move towards Igloolik. Therefore, as well as being indebted to his instructions, training, family tradition in medicine, and library on board his ship, Captain Parry's ethnology was thoroughly contingent on the willingness of the Inuit to co-operate and help him – which they did for reasons culturally beyond Parry's control.

Another important question for historians is whether it is possible to generalize about the role of the subject in producing ethnographic knowledge. If one acknowledges that ethnological knowledge is the product of the interaction of cultures, one is forced to admit that historians describing contact should be attentive to the concepts and categories of all constituent cultures. Given the extraordinary diversity of cultures and economic modes of production throughout the world, making such generalizations is much more easily said than done! The answers are as yet still inconclusive. More attempts to reconstruct detailed descriptions of contact drawing on historical evidence from both sides of the contact are needed to provide a sounder basis for making an assessment. Such studies are currently forthcoming from revisionist accounts of contact in the South Pacific and the Americas. They hold forth the possibility of explaining cross-cultural contact in terms of kinds of social organization (e.g. hunter–gatherers, pastoralists) or ethnicity (e.g. constructed identities), in addition to the more traditional categories of social historians (e.g. interests, game theory).

Historians of contact regularly face paradoxical situations of different peoples from radically different cultural worlds contributing to each other's history and identity through trade, colonization, warfare or even explicit avoidance. The importance of issues of land tenure for understanding these conflicts can scarcely be exaggerated. The cultural diversity of concepts of land tenure helps to illustrate why important colonial treaties dealing with the sovereignty and settlement of aboriginal or indigenous lands have been so ambiguous and contested right up to the present day.

Many cultures have never entertained European notions of land ownership as an individual's private property. The semi-nomadic Inuit for example, traditionally held that people belong to the 'land' and have responsibilities in caring for it, but in no sense, claimed to 'own' it to the exclusion of others. On the other hand, the Maoris (of what is now New Zealand) had very definite ideas of ownership. Victorian treaties annexing vast areas of land for the British Crown can be seen as a further extension of the asymmetry of the geographical gift. For example, the Treaty of Waitanga (1840), negotiated between the British Crown and the Maoris, enshrined radically different understandings of sovereignty – to the advantage of the colonials. According to the Maoris' version, they were merely ceding *kawanatanga*, the root 'kawana' being invented by the missionaries as a direct transliteration of the English word 'govern', in return for the Crown's protection. For the Maoris in no way implied surrendering their sovereignty, their system of chieftainships (*rangatiratanga*), or their rights to the tenure of their land, which they considered to be unalienable.[14] The oppositions between the English and Maori languages and concepts, between the spoken and the written word and between oral transmission and the technologies of printing provided the means of amplifying the inherent differences in the Treaty. The contested role of language as a means of defining sovereignty has been central to the ongoing resolution of colonial treaties in New Zealand, Australia, Canada and the United States.

Ethnology and colonialism

The Quaker-led anti-slavery societies emerged at just the time in the late eighteenth century when slavery was becoming less profitable owing to the over-production of sugar and falling prices. By the same token, it is easy to overplay the importance of the ethnological and anthropological societies as a cause of race prejudices in Victorian England. The debates and publications of these learned societies are better thought of as a running commentary on issues of race and colonization. Language provided a critical focus for ethnological enquiry. At the theoretical level, studies of language provided a framework for a coherent history of humanity and a benevolent justification for colonization. At a practical level, language was a principal battlefield in class struggles for political representation. The right of the working classes to send petitions to Parliament had traditionally been denied on the grounds of their ungentlemanly style of English. Similarly, Tom Paine's struggle to find a vocabulary for expressing radical political ideas in a vernacular idiom reminds us that free political expression was limited to a relatively small privileged elite. Most of the Quakers were very sensitive to this issue because they had themselves been classed as religious dissenters, and hence denied the right to vote, or to study at Oxford or Cambridge. However, far from supporting working-class or aboriginal emancipation, their general aim was to promote a peaceful and prosperous process of colonization in place of the waves of colonial violence inflicted on indigenous peoples. Their role was to criticize the government's colonial policy and to demand greater intervention in, and colonial responsibility for, protecting aboriginal rights. This implied that the government should be more open to the political voice of dissenters and that colonial bureaucrats should use their positions more responsibly.

One of the roles of the Aborigines Protection Society was to use its ethnological knowledge to shed light on colonial politics. Once again, consider New Zealand, where the Maoris had resisted selling their land and had engaged in bloody battles to defend it. One major source of friction were the woodlands which Maoris claimed to own, but did not actually cultivate. Colonial settlers attempting to seize the land argued that it was merely unoccupied 'wasteland'. The Aborigines Protection Society defended the position of these Maoris by discussing the importance of the forests as a source of timber as well as food for their animals. This implied that the forests were the legal property of the Maoris under the Treaty of Waitanga, which guaranteed them the full possession of their 'land and

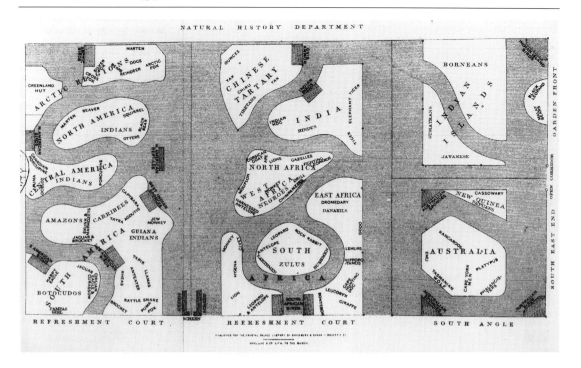

NATURAL HISTORY DEPARTMENT

Figure 20.4 Plan of the ethnological exhibit at Sydenham in S. Phillips, *Guide to the Crystal Palace and Park* (London, 2nd edn., 1854). The ethnographic section of the exhibition stands in contrast to the manufactures and arts of the industrialized nations. These 'other peoples' are mapped out in the exhibition roughly according to their geographical location.

estates, forests, fisheries and other properties which they may collectively and individually possess'.[15] The advice given to the Colonial Office was that the restriction of Maori rights was only impeding colonial development. Instead they advocated that Maori property rights should be respected, and that if the restriction of the right to vote to '*those who can read and write the English language*' were lifted, Maoris would be more willing to sell some of their private property.

Ethnological testimonies and reports from the colonies also provided important evidence for ethnological debates in England. Were indigenous peoples like the Maoris and Aborigines doomed to extinction, as the followers of Thomas Arnold (1795–1842) were arguing? According to the Aborigines Protection Society, the Arnoldians were mistaking colonial mismanagement for evidence of necessary extinction. Their acknowledgement of the uncultivated woodlands as 'wilderness and waste' was deemed to be proof of their ethnological ignorance.[16] Whereas the APS claimed that a just colonial administration would not only lead to colonial progress but generate a more truthful ethnological record. By setting 'the minds of the Natives . . . at rest, and native agency employed, a survey like that of the *Doomesday Book* might soon be produced . . . The needs of European surveyors to make the record would soon be felt, and their operations, now opposed, would be craved and assisted.'[17]

244   DR. A. W. BELL—*On the*

bottom of the deep cañons, which debouch upon the Colorado
Cañon.   Both tribes were encountered by our parties about the
35th parallel; they are comparatively harmless, and much resemble the Pai-utes of the Great Basin.   The valleys of the
Colorado, from the end of the Black Cañon almost to the head
of the gulf, are inhabited by Indian tribes who occupy an intermediate position between the semicivilized Pueblo Indians and
the wild Apache races.   They have for some time kept peace
with the whites; but contact with them appears to be rapidly
hastening their extinction.

Mojave Indians.

As there is no special interest attaching to these savages I will
leave the accompanying woodcut, copied from a photograph
taken at Fort Mojave, to speak for itself.   The Mojaves are the
largest tribe, and once numbered ten thousand.

I must now say a few words about the ruins which are to be
found scattered throughout New Mexico, Arizona, and Northern
Mexico.   There is scarcely a valley in the Rio Grande basin in
which the stone or adobe foundations of villages are not to be
found; there is scarcely a spring, a laguna, or a marsh upon the

Figure 20.5 A page taken from the *Journal of the Ethnological Society of London*, new ser. 1 (1869), p. 244, showing a woodcut of the Mojave Indians. According to the text, this inexpensive woodcut, copied from a photograph, 'speaks for itself'. This air of resignation implicitly relieves the settler-woman of responsibility for the situation she finds herself in. The gesture of her left hand is meant to show colonial benevolence towards the Mojaves. Their dejected facial expressions testify to the apparent inevitability that 'contact with the whites . . . appears to be rapidly hastening their extinction.' Visual images like these are indicative both of the plight of indigenous peoples and of the loss of hope in the philanthropic project in the second half of the nineteenth century.

The eclipse of philanthropic ethnology

Philanthropic ethnology lost much of its impetus by the 1860s. A
whole series of political developments in Britain and throughout
the empire overtook them and crowded them out of their niche

of benevolent commerce. A new era of imperialism, much less sympathetic to the *mœurs* and interests of colonized peoples was setting in. Middle-class Englishmen, many of them professionals, were adopting a staunchly nationalist self-image, to the detriment of the working classes and other races. Philanthropy was looking much less attractive to the majority of the middle classes who had adopted ambition, progress and self-help as their moral creed. The rapid demographic decline, and in some cases extinction, of colonized peoples was no longer inspiring credible alternatives. The destruction wreaked upon them in the 1840s and 1850s, if anything, had shown that the machinery of European expansion was unstoppable. The accumulation of wealth from the economic exploitation of the colonies was accepted at face value.

When the Anthropological Society broke away from the Ethnological Society in 1863, the concept of 'race' had shed its historical and moral connotations in favour of a decidedly more physical and anatomical view. The new society's membership was dominated by professionals – lawyers, clerics and physicians – many of whom (having protested against the admittance of women to the Ethnological Society) wore their racial prejudices with greater openness and arrogance. The new evolutionary theory to some extent replaced the debate over monogenesis and polygenesis as the scientific justification of racial inequality. Ethnological questions addressing the distribution of races and the customs, manners and habits of foreign nations were not so much rendered obsolete as subsumed by the new anthropology, with its growing emphasis on techniques based on measurement and classification. Likewise, antiquarians having championed the literature of the Anglo-Saxons as evidence of a national past for well over a century, turned to the archaeological evidence of Iron Age pits to fathom the past. As they did so, new taxonomic systems for classifying peoples were introduced based on the evolution of their tools. The balance of the weight of evidence shifted towards anatomical specimens and other material artefacts.

If any one event signalled the ethnologists' loss of faith in a unified world, it was perhaps the ethnological exhibit at Sydenham, in the wake of the Great Exhibition of 1851 (see Figure 20.4). When critics argued that the juxtaposition of 'the savage man' with the world's most advanced instruments of industrial power was abrupt and useless, the reply of John Conolly (1794–1866), the President of the Ethnological Society, was feeble. While making the usual deference to the unity of God's designs, he conceded that the 'well-executed specimens' might help to draw popular attention to ethnology, a science 'far from being widely known'.[18]

Further reading

Belich, J., *The New Zealand Wars and the Victorian Interpretation of Racial Conflict* (Auckland, 1988).

Bravo, M. T., 'Science and discovery in the search for a North-west Passage, 1815–1825', Ph.D. dissertation (Cambridge University, 1992).

Brockway, L. H., *Science and Colonial Expansion: The Role of the British Royal Botanic Gardens* (London, 1979).

Greenblatt, S., *Marvellous Possessions: The Wonder of the New World* (Chicago, 1991).

Greenhalgh, P., *Ephemeral Vistas: The Expositions Universelles, Great Exhibitions and World's Fairs, 1851–1939* (Manchester, 1988).

Lorimer, D., *Colour, Class and the Victorians* (Leicester, 1978).

Mackay, D., *In the Wake of Cook: Exploration, Science and Empire, 1780–1801* (London, 1985).

McKenzie, D. F., 'The sociology of a text: oral culture, literacy and print in early New Zealand', in P. Burke and R. Porter (eds.), *The Social History of Language* (Cambridge, 1987), pp. 161–97.

Morrell, J. and Thackray, A., *Gentlemen of Science: Early Years of the British Association for the Advancement of Science* (Oxford, 1981).

Pratt, M. L., *Imperial Eyes: Travel Writing and Transculturation* (London and New York, 1992).

Sahlins, M., *Islands of History* (Chicago, 1985).

Stocking, G., *Victorian Anthropology* (New York, 1987).

Thomas, N., *Entangled Objects: Exchange, Material Culture, and Colonialism in the Pacific* (Cambridge, MA, 1991).

2I Equipment for the field

Anyone who has ever admired a scarab's iridescence, an orchid's convolutions, a fluorite crystal's geometry or a crinoid's intricate stem can appreciate the aesthetic power of natural forms. For many eighteenth- and nineteenth-century naturalists, the intrinsic beauty of these forms was enhanced by the belief that they held clues that would unlock the secret patterns linking all of creation. By gathering together and comparing representatives of as many species of animals and plants as they could, some naturalists hoped to discover the natural classification system that related all living things to one another. Other naturalists hoped that samples of geological and fossil formations would help them discern the processes that shaped the Earth's surface over time. The search for these fundamental patterns meant that natural history in this period was a science based on *specimens*: objects of natural origin that had been prepared in ways that allowed them to be examined, compared to similar objects and described in a concise, informative manner. They were manageable pieces of the natural world that could be bought, sold, exchanged, transported, catalogued, displayed and consulted by many people. Specimens were not, however, natural objects: they were artificial things designed and constructed by naturalists to answer various scientific needs.

An orderly cabinet of accurately identified geological specimens contributed much to the geologist's comprehension of a subject as vast and intractable as a landscape. A well-labelled piece of shale small enough to fit in one's palm could stand in for a gigantic formation of the same kind of stone. Similarly, zoologists and botanists coped with the complexity of living organisms by reducing them to simpler, static elements, namely, a selected set of physical remains, field notes and drawings. Dried plants, bird skins and pinned beetles in a naturalist's study collection were convenient stand-ins for their dynamic, living cousins.

Animal specimens often bore little resemblance to the organisms from which they had been made, however. Certain animal parts such as shells, beaks, bones and skins could be conveniently stored in a dry state with a relatively low risk that they would be destroyed by insect scavengers or mildew. As a result, hard tissues like these formed a disproportionate part of most specimens. The soft internal organs of vertebrates and the fragile bodies of molluscs, caterpillars, sea anemones and the like were impossible to preserve unless they were stored wet in jars or tins of alcohol. Unfortunately, alcohol tended to bleach and toughen these structures; the necessary containers distorted one's view of the object and required a great deal of storage space and were often difficult to transport; and spirits evaporated quickly unless the seal of a jar was perfectly airtight, a tricky thing to achieve using the materials available, namely, sealing wax, bladders, pitch and turpentine. Furthermore, a collection of pale, isolated organs and pickled sea creatures suspended in jars looked ghastly and was not a welcome addition to the average naturalist's drawing room. As a result, wet specimens were more commonly kept in the collections of institutions such as universities and the Muséum d'Histoire Naturelle in Paris. The logistics of preservation meant that those animals that had sturdy, attractive body parts were far more popular as subjects than those creatures that did not; even among the former, the durable bits got considerably more attention than highly perishable items like digestive systems and musculature.

The kind of specimen that one created from a living animal or plant had important scientific ramifications. Zoologists and botanists use a *method* when sorting a group of similar specimens such as aquatic insects or mosses: that is, they choose one or more anatomical features (like wings or spores) that all the specimens have in common and arrange the group according to comparisons of that part. The choice of the part is critical to the resulting classification system. For instance, one could end up with a single large group of allied species by comparing wings, or several small groups of allied species by comparing mouth parts. Of course one could not choose a part that was not present in all the specimens in question, and the parts that *were* present were often those that were easiest to preserve, not those that would be the most useful for classification. Since there was no definite way of determining which method would lead to the natural order that God had had in mind at the moment of creation, there was a great deal of acrimonious debate over the choice of anatomical features, the methods based on them and the resulting classification systems.

Naturalists in different countries had different resources available to them for the creation and preservation of specimens. The French government, for instance, devoted a great deal of money and space to storing complete animal bodies in spirit, and this gave French naturalists a wider range of anatomical characters to choose from than their colleagues in Britain, who did not have good institutional support. This meant that French classification methods were often difficult for English naturalists to apply in the eighteenth and through much of the nineteenth centuries, because the English did not have the same kind of specimen resources to examine. It also meant that French and English naturalists tended to specialize in the study of different groups of animals; the former could explore the forms of perishable, soft-bodied creatures as well as those which were more durable, while the English generally concentrated on those items that could be preserved as dry specimens.

While the kind of specimens used by a naturalist often directly influenced his or her classification work, the classifying naturalist's scientific agenda could also affect what parts of an animal would be preserved because they were considered important. This two-way interaction between the specimen material and the practice of classification means that it is important for historians to consider the specimens themselves and the way they were created when studying the development of taxonomy in this period. For instance, an ornithologist who classified birds according to how and what they ate would make an effort to preserve the digestive organs as well as the skin and bones; an ornithologist who felt that external characters were sufficient for classification would discard all of the internal organs and focus on the animal's surface features. Sometimes factors unrelated to science influenced the shape of specimens. English naturalists had to pay a stiff tax on glass jars until 1845; even those zoologists who wanted to study soft tissues and animals often could not afford the jars necessary to store them in spirit.[1] This meant that English zoologists studying molluscs lavished their attention on the features of the animals' shells and disregarded the actual animal. Soft-bodied invertebrates generally received very little attention in England until the second half of the century, when it was cheaper to maintain a cabinet of wet specimens. In contrast, French naturalists had taken pains since the 1780s to collect and preserve starfish, coral polyps, nudibranchs and other fascinating creatures. They also preserved the internal organs of vertebrates. As a result, the French had considerably more material to incorporate into their classification systems than did their English counterparts, and this led to two very different styles of zoological enquiry and classification.

In addition to their fundamental role in the scientific agenda of natural history, specimens held other attractions for naturalists. We have already mentioned the aesthetic qualities of many natural forms. Specimens that were especially rare or even unique often had considerable pecuniary value. Owning such a specimen gave a naturalist and his cabinet a satisfying cachet that was increased if the object was the *type* for its species and set the standard to which all similar specimens had to be compared for identification.

For many naturalists, the process of collecting specimen material in the field was an integral part of their commitment to natural history. Collecting combined healthy outdoor exercise with intellectual rigour; there was often a dose of evangelical fervour as well. The Romantic literary movement of the early nineteenth century gave additional sanction to the contemplation and exploration of rural and wilderness landscapes. Collecting fulfilled Victorian society's criteria for legitimate leisure activity, and the steadily increasing number of naturalists prowling the English hedgerows and the British Empire's distant perimeters as the century progressed confirmed that, in spite of the need for special equipment, many hours in the hot sun, long hikes with wet feet, gory procedures, bookkeeping chores and a certain amount of ridicule from those less scientifically inclined, collecting specimens was eminently enjoyable.

Natural history's acquisitive aspects owed much to an older tradition of gathering together objects for contemplation and study. Britons in a financial position to collect had been enthusiastically gathering art objects, coins, archaeological artefacts, monstrosities, novel mechanisms and other interesting things into cabinets of curiosities since the seventeenth century. It was generally expected that well-to-do travellers would return from trips abroad with their luggage swollen by new treasures.[2] Natural objects made up a significant percentage of the contents of many cabinets, and by the middle of the eighteenth century there was a demand for instruction manuals that explained how one could collect impressive tropical animals, plants and geological material and get it all home more or less intact. These manuals described the necessary equipment and how to use it to create specimens for every branch of natural history. The books were conveniently sized and were carried in the pockets of persons whose collecting grounds were restricted to their home parish as well as in the travel kits of missionaries, civil servants, military personnel and tourists. By the early 1800s there was a steady supply of instruction materials available to beginners in every branch of natural history.

From the start, authors of comprehensive guidebooks such as Dr John Lettsom in his *Naturalist's and Traveller's Companion* (1774), Edward Donovan in his *Instructions for Collecting and Preserving Various Subjects of Natural History* (1799) and George Graves in his *Naturalist's Pocket-Book and Tourist's Companion* (1818) stressed the importance of keeping a field notebook in which to record details about the objects one collected. They wanted budding naturalists to collect much more than rock fragments and stacks of pressed plants: they wanted data about these things that were not intrinsic to the physical remains of an organism or to an isolated rock sample. The uses of a particular plant, the elevation of a fossil bed, the feeding habits of an insect, the eye colour of a tropical bird – information of this kind could only be preserved in the memory and, more importantly, in the field notes, of the person who found the item in its native setting. At the very least, every item needed to be labelled with its point of origin. It was the combination of the physical object with data about its life, uses, location – in short, its *nature* – that gave a specimen its scientific value. In *The Naturalist's and Traveller's Companion*, Lettsom (1744–1815) emphasized this important difference between the naturalist's agenda and the random acquisition of souvenirs:

[It] is of more consequence to discover the natural history of one destructive or useful insect, than merely to collect and bring over [to England] twenty in their perfect state; the former, at the same time that it makes the science more entertaining, bids fair to benefit mankind, while the latter serves only to fill the cabinets of the curious.[3]

Natural history's scientific goals gave it greater focus than earlier collecting traditions, and intellectually it required far more of its participants.

One category of information that these authors were especially interested in was the usefulness of any natural object. A plant that could cure syphilis or a new natural dye would provide the British Empire with a very marketable commodity, and there was a certain amount of pressure on patriotic naturalists to discover such resources. George Graves (1784–1839), writing in the tradition of Francis Bacon, believed that the exploration of nature's bounty for the benefit of humankind was the central tenet of natural history. He wrote in his *Naturalist's Pocket-Book or Tourist's Companion*:

[I]t is not by the mere accumulating [of] a large variety of curious species that the science is advanced, but it is by acquiring a knowledge of the habits and propensities, the contrasts, the similarities, the uses or injuries they offer to mankind, that gives life and spirit to the science; and in fact is the true and only real use of the study.[4]

In order to encourage his readers to collect these data along with their physical booty, Graves included several sample pages from his own well-organized field notebook that clearly illustrated his point (Figure 21.1). Carrying a notebook with neat columns labelled for each kind of information encouraged the neophyte to put down his observations about each item while the details were still fresh and before proceeding to the next object.[5] In this way, the information necessary for accurate labelling, fruitful examination and any formal description of the specimens would be readily available when the naturalist returned home from Borneo after months of voyaging or from the next parish that same afternoon. When naturalists published descriptions of new species or reported sightings of rare ones, they often included excerpts from their field

1818	Memorandum.	Length.	Breadth	Wt.	Food.	Places of resort	To what purpose applicable	Nest and Eggs, and times of migration.	Note.	Colour of eyes	Colour	Proper name.
9	We were exceedingly amused this day, in observing a number of small birds running up the trunks and branches of some large Elms, seeming to pry with anxious curiosity into every crevice of the bark; we shot several, but could not discover any sexual difference. The species belonged to Order 1. Genus 6. Titmouse.	Five inches and a half, of which the tail alone measured two inches and a half.	6 inches and a half from tip to tip of the wings when expanded.	3½ drams.	insects, their eggs and larvæ; flesh, young birds, carrion and grain.	Woody places, principally large old trees; whose trunks are mossy.	The Natives use the wing and tail feathers, as ornaments to their head dress; as in their wars they only use the black feathers, but at other times both the red and black. The skins and feathers are of considerable importance, as they are among the principal objects of barter with several tribes of the Natives; the flesh of the young birds is eaten, as also their eggs.	Nest composed of moss, wool, hair and feathers, mostly pendant, or else placed in the forked part of a branch of some small tree, the entrance to one side, Eggs transparent white, fine-ly sprinkled with red; in number from 13 to 20. The species is a constant resident.	A continual twitter.	Dark hazel.	A confused mixture of Ash colour and brown, with some dark patches.	Bottle Tit. Ragged Robbin.
10	Shot a large bird on the border of a lake, on dissection we found it was a male. This was the first of the kind we had seen. It belonged to Order 9. Genus 3. Flamingo.	From the tip of the bill to the tail, 4 ft. 5 in. from the tip of the bill to the toes, 6 feet.	From tip to tip of wings 10 feet 2 inches.	About six pounds.	Fish, aquatic insects, reptiles, & vegetables	The shores of the sea, rivers, on the banks of large lakes, and salt water inlets.		Their nest is composed of sticks and decaying vegetable, placed on a hillock in some shallow pool, the birds sit on the nest with their legs hanging over its sides. Eggs two, perfectly white. Migrate to the West in autumn, and returns East, early in the spring.	A harsh scream.	Dark hazel, soon after death, becoming cover-ed with a white film.	Large quill feathers black, all other parts a bright scarlet.	Tococo, and by the European inhabitants Flamingo.

22	Saw an immense flock of the last named bird.
Nov 4	We shot two Swallows, which exactly resembled the common Chimney Swallow, we were informed by the inhabitants that they arrive about the end of October, and quit towards the end of February.
18	We observed large numbers of wild fowl on a salt water lake, but were unable to procure any, as the instant we approached they took wing; the next morning we repaired before day-light to the same spot, and were so fortunate as to kill several, they were Mergansers, and from the confused mixture of their colours, were evidently either young birds, or deeply in moult, they are only seen in these parts after unusually wet seasons; we could learn nothing of their history.
1819 Feb	Observed the Swallows congregating, as they do in England, (previous to their leaving it,) this continued for several days, but on the wind shifting to the South-east, they all departed.
22 Mar 9	We found a nest with several greenish white eggs, minutely spotted with brown, but though we did not touch it we never observed any bird frequenting the spot for several weeks; we therefore possessed ourselves of it, and found each of the eggs had a small hole in one end, and on removing them on discovered the bones of some bird beneath, which was probably the parent, but how destroyed or the eggs, punctured, we could never ascertain.

Figure 21.1 A sample page from George Graves's *Naturalist's Pocket-Book or Tourist's Companion* (London, 1819) showing beginners how to organize their field notes.

notes, particularly the location of the find. These details gave the descriptions a sense of validity: they also helped naturalists with similar interests to collect the organism or explore the geological feature themselves. One of the most important books published in early nineteenth-century natural history was an edited field record: Charles Darwin's fascinating narrative of his journey around the world, *Journal of Researches into the Geology and Natural History of the Various Countries Visited by HMS* Beagle (1839). Darwin carried a pocket-sized notebook with him at all times to jot down ideas and observations. He filled eighteen of these in the five years of the voyage, and drew on the scientific data they contained for his later work *On the Origin of Species* (1859).

The field notebook was the one piece of equipment used for work in all the branches of natural history. There certainly were people collecting and studying natural objects who did not use this tool as often as they might have, and many cabinets in Britain lacked the proper labelling so essential to sound scientific work.[6] However, anyone who hoped to make a lasting contribution to the description of new species, the theories about the formation of the Earth's surface or the search for the natural order had to keep meticulous records of their collecting activities.

Each of the three main branches of natural history – geology, botany and zoology – had its own particular arsenal of collecting equipment. In the sections that follow we will briefly survey the basic tools and the skills needed to apply them. Each branch could be pursued to some degree with a minimal financial investment; and this encouraged newcomers to explore the possibilities before focusing their attention on one or two areas. It also allowed those of modest means to share in the delights of the science, a fact that contributed to natural history's heterogeneous demographic profile. The wide range of physical demands and of necessary skills meant that beginners could easily suit their own tastes. For instance, those persons who objected to bloodshed in the pursuit of leisure activities could study plants and rocks. Women were well represented in botany, conchology and entomology, although there were fewer of them in geology and vertebrate zoology.

Geology

Writing about the tools necessary for a geologist in the field, the science journalist J. E. Taylor (1837–95) asserted: 'he cannot take *too few*! It is a great mistake to imagine that a full set of scientific instruments makes a scientific man.'[7] For Taylor, the most important instruments were hammers and stout footwear. One hammer was relatively light for breaking off small chunks of rock, prying into crevices, trimming fossils, and so on.[8] The second hammer needed a larger, heavy cylindrical head appropriate for real pounding. He also recommended a third shaped like a small pickaxe with one of the points flattened horizontally, so that it earned the name 'platypus pick'. The shape of the head was especially useful for working in clay, and layered or flaky deposits. Hammers like these were available from dealers in scientific instruments at reasonable prices (Figure 21.2). Next on Taylor's list came thick-soled shoes or boots; he added that geologists always stood out at meetings of the British Association for the Advancement of Science because of their distinctive footwear.

Having covered the basics, Taylor included a few optional items, starting with a pipe and tobacco, which he enjoyed most while 'solitarily disinterring organic remains which had slumbered in the heart of the rock for myriads of ages'. He also noted that a flask of beer tasted better on a geological expedition than at any other time. Finally, after the creature comforts, he bowed to logistics and acknowledged that a leather bag for schoolbooks was useful for carrying one's finds. Persons especially interested in gathering fossils should carry a few narrow chisels, a putty knife, and some paper, cotton wadding or sawdust to prevent damage to delicate items during the hike home. Taylor believed that a bit of preparation in advance would save any geologist a great deal of trouble and disappointment. In 1819 the mineral surveyor William Smith (1769–1839) set about publishing geological maps of England by county; he covered twenty-one of them by 1824 when he ran out of money for the project. These maps gave such a good description of the terrain that they remained in print until 1911.[9] By the

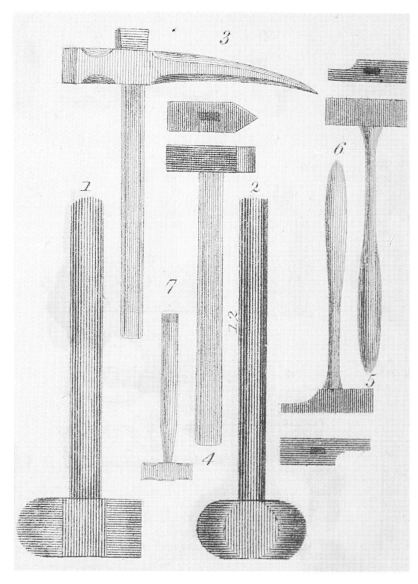

Figure 21.2 A selection of geological hammers from an advertisement in W. D. Conybeare and W. Phillips, *Outlines of the Geology of England and Wales*, Part I (London 1822).

second half of the century the Geological Survey maps were also readily available and widely consulted.

The geologist's goal was to collect rock samples roughly four inches by six inches, and about an inch thick. Mineral samples, particularly crystals, were often smaller. Fossils varied enormously in size, with a logical upward limit being what one could carry; if the find was especially important one could return with help. Sand and other fine material fit neatly into glass vials or small wooden boxes. All of these items had to be accompanied by the appropriate notes, including the exact location of the find in relation to the formation as a whole, and in regard to the strata surrounding it. One also recorded the formation's thickness; whether it was horizontal, vertical or inclined; in which direction it ran; whether it was fractured, disturbed or overturned; any characteristic fossils or minerals, and so on. One could draw few meaningful conclusions from an isolated fragment, but an accumulation of well-documented samples could reveal the history of a landscape.

Botany

Like geology, botany had a spare, inexpensive set of tools. The most basic was the vasculum, a tin box with a tightly fitting lid and a strap that allowed it to be carried slung across the back. The earliest vascula were candle boxes pressed into service in 1704.[10] Those for sale in the late eighteenth and early nineteenth century were square and called 'botany boxes' or 'sandwich boxes'; many of them did double duty as guardians of a packed lunch.[11] The box's dual uses could lead uninformed bystanders to speculate on the mental condition of botanists, as shown by this anecdote related by one devoted collector of fungi:

Country labourers are often sorely puzzled by the acts of cryptogamic botanists; they stand agape in utter amazement to witness poisonous 'frog-stools' bagged by the score. Oft-times one gets warned that the plants are 'deadly pisin'; but collectors are usually looked upon as harmless lunatics, a climax in this direction generally being reached if a gentleman in search of *Ascoboli* and the dung-borne *Pezizae*, sits down, and after making a promising collection of horse or cow-dung, carefully wraps these treasures in tissue paper, and puts them in his 'sandwich-box'.[12]

Over time these collecting boxes increased in size, making it possible to bring home plants with fewer broken parts. Once gathered, plants were often packed between sheets of paper; half-sheets of newspaper worked quite well. Botanists also carried one or two smaller boxes for delicate items; a knife or small trowel for getting at roots and rhizomes; and perhaps a hooked stick to pick up water plants and pull down branches. There were a number of fine identification books available by the 1830s, many of which were small enough to be taken along for consultation in the field.

Botanical specimens consisted of the entire plant, including the roots, bulbs or rhizomes, whenever possible. Young plants with their cotyledons (seed leaves) still visible were important for identification because the embryo often revealed relationships that were hidden in the mature plant. The various stages of the flower were also useful. The colours usually succumbed to the drying process, but a careful watercolour sketch made in the field or from a fresh cutting could stand in for the original. Fruit, seeds and spores (for

1.—The "Church in Danger" at Dinmore.   2.—Gathering the "Vegetable Beef Steak" and other Tree Fungi.   3.—Bog Fungi at Lyonshall Wood.   4.—The Drive to Downton.
5.—The Drive to Moor Court.   6.—Raking for Truffles.   7.—M. J. de Seynes, " Professeur agrégé à la Faculté de Médicine de Paris."   8.—M. Maxime Cornu, of the " Jardin des Plantes," Paris.

THE GREAT FUNGUS MEETING AT HEREFORD.

ferns, mosses and fungi) accompanied the mature plant whenever possible. Because naturalists had not yet identified the larger underground organism, mushroom-producing fungi were only represented by the relatively tiny fruiting bodies that appeared above the surface. Of all the organic specimens, plants were unique in their ability to replicate themselves from seeds, bulbs and cuttings. Seeds collected abroad could be shipped packed in a moisture-retaining medium such as sand or raisins. It was often possible to raise an entire plant from these parts if one had accurate notes about its growing needs. As a result, botanical specimens were often cultivated in the botanist's garden (or later, in the greenhouse), as well as being arranged and dried on to sheets of paper and stored in a herbarium.

Figure 21.3 A humorous but realistic illustration of fungus collectors drawn for a popular periodical by Worthington G. Smith. The figures on the left and right are carrying typical botanical field gear, including a vasculum worn on the belt and one carried over the shoulder. Note the woman in the group of truffle hunters.

Zoology: conchology

Conchology, the study and classification of mollusc shells, survived in Britain long after it had been transformed on the Continent into malacology, the study of molluscs in their entirety. This odd

Figure 21.4 Worthington G. Smith, who specialized in fungi, recorded these impressions of a collecting expedition for readers of the *Graphic*.

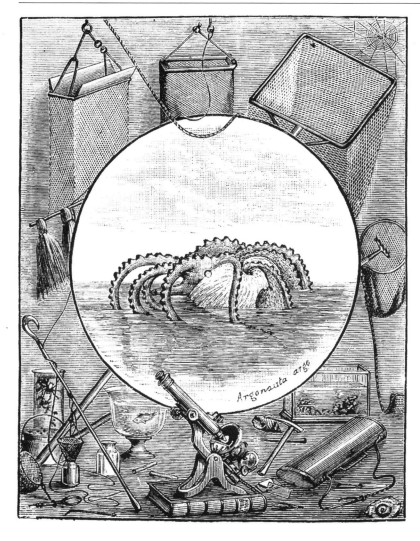

Argonauta argo

Figure 21.5 A selection of equipment from several branches of natural history, including a number of dredges. The dredge on the left includes lengths of unravelled rope called 'tangles' for snagging spiny animals like crabs and sea urchins. There is a ring-net for butterflies on the right, several aquaria, botanical tools, a microscope, and a rock hammer along the bottom. Note that the book on which the entire collection is centred is not a field guide or instruction manual, but the Bible. Frontispiece by F. T. Law from Rev. N. Curnock, *Nature-Musings on Holy-Days and Holidays* (London, 1886).

division of the animal from its artefact was due partly to the difficulties of preserving the soft tissues of the animal and partly to the striking aesthetic qualities of shells; indeed, conchology was one of the most popular branches of natural history. Holidays at the seashore meant that many people had opportunities to collect; even land snails and freshwater molluscs received attention.

The well-equiped conchologist carried a stout knife for scraping uncooperative limpets and chitons from rocks at low tide, a trowel for digging in the sand, a large metal spoon with holes for straining the bottom of tide pools and a box or bucket to hold her captures until they could be dispatched in a pot of hot water. It was important to gather living molluscs because their shells were in the finest condition. Since the great majority of these animals live on the sea floor, conchologists developed a number of dredges for scraping them up (Figure 21.5). Dredging required the help of at least one person to manage the boat, and another to help with the lines and lifting. It was best done from a large vessel with a soft breeze behind it, and was often a co-operative venture among several conchologists. Naturalists on board naval and merchant vessels had excellent opportunities for dredging and often brought home rare and unique specimens.

Zoology: birds and other vertebrates

Birds made up the vast majority of vertebrate specimens in British collections. Small mammals lacked the visual appeal necessary to win a following, and large mammals took up too much storage space and required extensive and costly taxidermic work. Fish lost their brilliant colours immediately after death, and could only be preserved in expensive jars of spirit. Some zoologists tried making plaster casts of the fresh fish and painting them with realistic colors, but this was not satisfactory for either scientific or aesthetic purposes. Reptiles, too, were poorly represented.

In contrast, birds offered naturalists the benefits of small size, brilliant plumage and the delights of collecting with a rifle. The number and quality of large, exquisitely illustrated books on birds published before 1840 reflected their popularity as subjects of study. Starting in the late eighteenth century, French naturalists working at the Muséum in Paris developed new techniques for protecting the skins of birds and other animals from the destructive activities of insects. They saturated the skins with various pesticides, mounted the skins on false bodies and sealed them inside glass boxes, an approach that could preserve bird and animal skins for several decades (Figures 21.6 and 21.7). Birds looked especially striking prepared in this way, and soon cases of stuffed specimens became popular items for decorating middle-class drawing rooms. These specimens were of limited scientific value, however, because they could not be handled for measurement and close examination. Around 1820 the zoologist William Swainson (1789–1855) introduced an important innovation when he began to store bird skins as stuffed cylinders in flat drawers, enabling him to store hundreds of accessible specimens in one piece of furniture.[13] In all of his instruction manuals, he was adamant about careful labelling to ensure that each skin could be connected with the appropriate field notes.

When collecting birds with a gun, it was helpful to keep one barrel loaded with large shot and the other with dust shot so that one could take down any bird that came into the sights without damaging it too badly. Once the bird was dead, all the orifices and the wounds had to be carefully stopped up with cotton to prevent the feathers from being soiled; it was then placed head downward

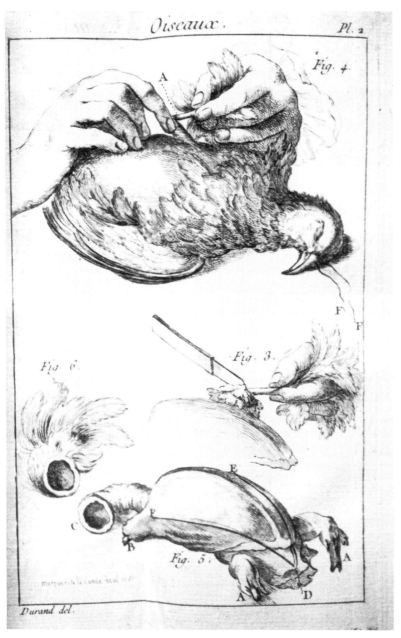

Figure 21.6 A fold-out plate from Etienne-François Turgot's *Mémoire instructif sur la manière de rassembler, de préparer, de conserver et d'envoyer les diverses curiosités d'histoire naturelle* (Paris, 1758), showing how to skin a bird that will be mounted as a specimen.

in a paper cone that kept it clean and immobile in the game basket. Ornithologists collecting in the tropics had to act within a few hours to preserve birds against scavenging insects and putrefaction by skinning them and preparing the clean skin with a heavy dose of pesticide, usually a combination of arsenic and soap, although tobacco, ashes, pepper and corrosive sublimate were also used. The rifle's appeal to the middle class and the aristocracy meant that most avian specimens had bullet holes in them somewhere, although Graves and others also recommended the use of old

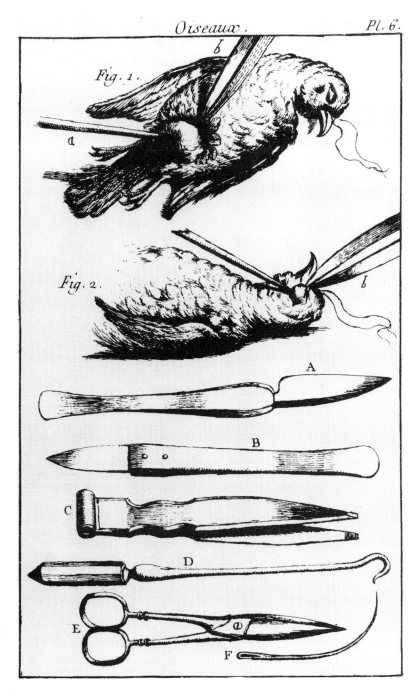

Figure 21.7 Another plate from Turgot's *Mémoire* demonstrating the final stitches that close the mounted skin, the placement of a glass eye, and several tools essential to the taxidermist.

poaching tricks like poisoned seeds and fruit, foot snares, nets to capture diving waterfowl and glue smeared on branches.[14]

Bird specimens were usually just the skin with the skull, feet, and beak attached; often the last two were painted to approximate their living colours. Species in which the male, female and juveniles had different plumage sometimes found themselves split accordingly. Nests and eggs were popular items, and it was usual to kill at least one of the parents to secure accurate identification of the eggs, which were then stored as empty shells. Toward the end of the century a general shift in popular sentiment brought field glasses and telescopes into the hands of naturalists and the complexities of migration, song, mating patterns and other behaviours began to trickle into ornithological field notebooks and bring a new depth to the physical specimens.[15]

Zoology: entomology

Of all the branches of natural history, entomology developed the largest arsenal of specialized field equipment and the most detailed library of instructive literature for beginners. Pursuing and capturing insects for study was a novel practice indeed. Hunting birds, mammals, fish and molluscs had a legitimacy rooted in gastronomic necessity; as did the search for useful plants and fungi. But insects were seldom even noticed unless they stung, bit or threatened food plants, and their diminutive size made them seem like a frivolous subject for the serious mind. As a result, naturalists interested in this diverse, ubiquitous tribe suffered a certain degree of ridicule.

Collecting insects was not a straightforward process. There were literally thousands of nuances of habitat, season and, for the majority of insects, the complication of three complete changes of form within a lifetime. The physical part of entomological specimens usually consisted of the adult form only; larvae and pupae were rarely gathered. Killing insects in a way that kept their fragile extremities intact required considerable finesse and the techniques varied widely from one class of insects to another: what worked well for beetles would destroy the fragile wings of a butterfly. In addition, insects like sphinx moths, mantids and dragonflies were extremely tenacious of life, and only the advent of potassium cyanide killing jars in the 1840s ended decades of doing them in with steam, red-hot needles and other draconian methods.

One of the most thorough introductory manuals was George Samouelle's *Entomologist's Useful Compendium*, published in 1819. Samouelle (d. 1846) was a London bookseller who spent his free daylight hours collecting in open areas around the city. In addition to step-by-step explanations of each stage in the collecting process, Samouelle provided detailed descriptions of productive hunting grounds, including directions to specific heaths, sand-pits and woodlands within easy reach of London naturalists. A large portion of the book consisted of an entomological calendar that indicated when and where thousands of different species could be found throughout the year. Helpfully, Samouelle arranged to sell collecting gear through bookshops, so that a reader could purchase her nets and pins conveniently. The *Compendium* sold briskly for years, and if plagiarism can be considered an indication of positive peer review, it was widely respected by the entomological community for the next several decades.

The basic field outfit of an entomologist included sturdy clothes, a jacket with ample pockets, several pill boxes, a vial or two of alcohol for dispatching beetles, a stock of good German pins in several sizes, a knife for prying up bark, a trowel for digging and a cork-lined box (Figure 21.8). After 1830 it was also common to carry a bottle with some deadly chemical: ammonia, ether, chloroform, oxalic acid, and prussic acid were all tried and happily discarded in favour of the relative safety and predictability of a layer of plaster saturated with potassium cyanide in the bottom of an airtight jar.

The premier piece of entomological equipment was the net, of which there were two basic kinds. The first, called a 'fly-net' or a 'clap-net', was a tool unique to England. Using it, the entomologist could enclose a very large zone of insect-inhabited air and claim everything inside. One could lay it on the ground under shrubs and trees to catch anything that fell as the foliage above was beaten with a stick; and one could walk through tall meadow grass sweeping the net from side to side to capture all the insects for several feet on either side of one's path. It was very difficult to deploy in a discriminating manner, but on the other hand, it was easy for the beginner to use and it had a higher rate of captures per sweep than its cousin, the ring-net. The ring-net, a circle of wire trailing a cone of muslin and attached to sticks of various lengths was not widely used in England until the mid-1830s. It had long been the tool of choice in France, and the English allegiance to the imprecise but voluminous clap-net may have been due in part to a feeling that the relatively tiny ring-net was a Continental affectation. The main reason had more to do with the fact that using the ring-net well required considerable practice – during which one would inevitably lose at least one precious rarity – to master the right combination of aim, lunge and twist necessary to pluck a quick, erratic target out of the air and keep it in the small muslin pocket long enough for it to be brought to the hand. Since both the failure rate and the technique itself were conspicuous, the neophyte English entomologist afflicted with spectators probably felt uncomfortable using this net.

Nets worked well on airborne insects, but they were clumsy and unreliable used on insects perched on surfaces. In these situations, entomologists gently pinched their prey between the gauze panels of a pair of forceps. Generally, entomologists made good use of

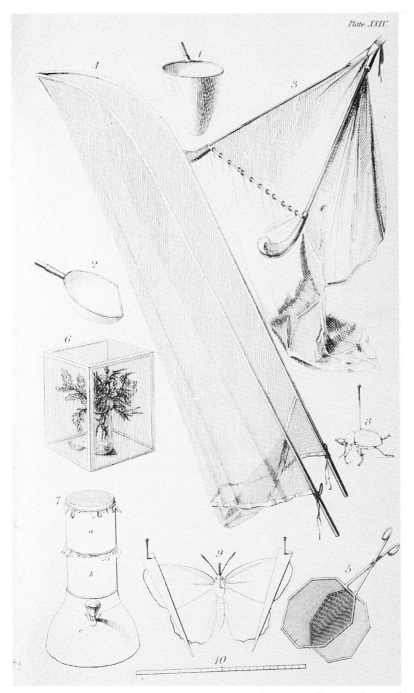

Figure 21.8 A selection of entomological equipment from the *Introduction to Entomology* by the Rev. W. Kirby and W. Spence, vol. IV (London, 1826). Item 1 is a bag- or ring-net; 2 is a landing-net for aquatic insects; 3 is a net for sweeping long grass; 4 is a clap-net or fly-net; 5 is a pair of forceps; 6 is a cage for raising caterpillars to maturity; 7 is a device for killing moths with steam; 8 indicates the correct way to pin a beetle, and 9 the proper way to prepare a butterfly.

whatever was handy. The thumb and forefinger – often moistened with spit – worked as well as forceps if one was quick enough. Umbrellas worked much like clap-nets and had the added benefit of being less conspicuous to carry around. Top hats lined with cork made a convenient place to pin an unexpected capture until one could get it home.

There was a limit to the number of insects one might encounter simply by walking about and looking for them, since many species live out their lives hidden by darkness, water and soil. In addition, most entomologists' collecting time was a precious thing, and any device that increased the number of species captured in a given hour had a definite attraction. These two factors combined to encourage the use of several ingenious kinds of baits that lured large numbers of insects into the hands of vigilant collectors. The simplest approach took advantage of existing resources: the rotting body of a mole thrown on the rubbish heap, or the parasites attracted to cattle and horses. It was a small step from these things to deliberately setting baits, like half-buried jars baited with a bit of rotting meat. One could capture quite a few moths by sitting in a room at night with the window open, a lamp lit, and a ready net. There were even lanterns that could be worn in front of the body so one could walk about at night attracting prey. Two of the most successful lures exploited the fact that many nocturnal creatures will flock to the promise of free beer, abundant sweets and the possibility of sexual liaisons. 'Sugaring', discovered in 1833, involved smearing posts and tree trunks with sticky mixtures of beer or rum and molasses. Moths and other insects would come in droves to partake, and could be gently plucked from the buffet by an entomologist lurking nearby.[16] Several authors recommended setting a sex trap, in which a captive female moth is used as bait for males. In 1803 Adrian Haworth described taking a caged female who was 'yet a virgin' into the woods where her presence attracted hordes of males who became 'so enamoured of their fair and chaste relation as absolutely to lose all kind of fear for their own personal safety.'[17] While this technique may have led to an excess of male butterflies and moths in study collections, it did bring many rare species under scientific scrutiny.

In conclusion, the amount of information about the natural order of the created world that a naturalist could glean from any specimen was a function of its physical completeness and its documentation. Both of these factors were dependent upon the skills, resources and agenda of the person preparing the specimen, so it is important to consider collecting techniques and the specimens that they produced when considering any naturalist's theories about classification or the forces that shaped a landscape. Geological specimens were especially dependent upon meticulous notebook entries for their scientific value, because unless an item could be related to a larger formation it was merely a sample of a particular mineral, fossil or rock and did not contribute much to an understanding of geological processes over time. Organic specimens had more information intrinsic to them, especially if their structure was relatively intact, such as that of a living plant, the body of an adult insect or an entire animal preserved in spirits. When naturalists worked with specimens this complete, the resulting classification systems tended to be more comprehensive and enduring than those based on more fragmentary materials, although complete physical remains did not eliminate controversies over method because one could still argue over the priority assigned to any given anatomical feature. It is especially important to consider a naturalist's specimen data set when the material examined included only a small part of the original organism, such as a fossil or the skull of a mammal. The naturalist's inferences about the whole creature and its relationship to other creatures were closely tied to the object at hand, and that object might be misleading because it lacked some essential features such as the teeth or the larval form. In addition to their central role in the science of natural history, specimens played a crucial part in what might be called the spirit of natural history: they were the interface between the naturalist and the natural world in all its variety. The aesthetic power of specimens and the adventure of collecting them in the field were motivations as basic to most naturalists as their desire to discover the order behind nature's wonderful complexity.

Further reading

Allen, David E., *The Naturalist in Britain: A Social History* (London, 1976; 2nd edn., Princeton, 1995).

Darwin, Charles, *The Voyage of the* Beagle, ed. Leonard Engle (New York, 1962).

Farber, Paul, 'The development of taxidermy and the history of ornithology', *Isis*, 68 (1977), pp. 550–66.

Frost, Christopher, *A History of British Taxidermy* (Lavenham, Suffolk, 1987).

Graves, George, *Naturalist's Pocket-Book and Tourist's Companion* (London, 1818).

Kirby, William and Spence, William, *Introduction to Entomology* (London, 1815–26).

Larsen, A. L., 'Not since Noah: English scientific zoologists and the craft of collecting, 1800–1840', Ph.D. dissertation (Princeton University, 1993, order no. 9328049).

Samouelle, George, *Entomologist's Useful Compendium* (London, 1819).

Swainson, William, *Taxidermy* (London, 1840).

Winsor, Mary P., *Starfish, Jellyfish, and the Order of Life* (New Haven, 1976).

22 Artisan botany

Historians have often been frustrated in recovering the perspective of working-class participants in the pursuit of knowledge. With a few notable exceptions, the analysis of working-class science has revealed more about the dominant middle-class ideology than its supposed subject matter. Such studies reflect a model of popular science that Whig social reformers believed in: a diffusionist model in which knowledge was handed down via Mechanics' Institutes or the Society for the Diffusion of Useful Knowledge. Heroic accounts of autodidacts in 'improving' literature purported to show the moral benefits of scientific pursuits and to promote the philosophy of individual self-help.

Largely informed by this literature, historians have noted the involvement of artisans and operatives as natural history became increasingly popular in Britain during the late eighteenth and nineteenth centuries. That manual workers could participate in this area has been attributed to the accessibility of nature and natural productions; the gathering of specimens, it seems to be assumed, required little expertise. Amusing accounts of working men using Latin names while speaking in broad dialect are quoted, but little attention has been paid to how or why these men acquired this terminology in the first place.

In contrast, this essay investigates the practice of science from a working-class point of view. The artisan botanists of Lancashire provide an unusually rich source for such a study.[1] However, this has not been generally perceived because of the failure to place their activities in context. It is only once we establish that artisans practised their science in the pub that it becomes possible to attempt to recover the experience of participants – from the illiterate gatherer of plants to the most expert with a command of the Linnaean system – and to show how the entire range of this activity was a communal concern.

My focus on the artisan botanists as a group is therefore deliberate. Inevitably, most evidence derives from the botanical 'experts' among the artisans. To discuss these men in individual terms, however, obscures the way in which they embodied and managed the tense relationship between oral and literary cultures.

It also reinforces the assumption that working-class women were not interested in such pursuits, rather than challenging us to investigate why artisan botany was associated with working *men* in the early nineteenth century. Aspects of working-class botanical practice become clear only in the light of artisanal notions of skill.

Pub botany provides a particularly good example of the intersection of 'popular' and 'elite' culture. But in order to analyse this relationship, we need to dispel the notion that 'popular' and 'learned', or 'high' and 'low', culture are fixed categories defined by their content, and see them instead as emergent social constructs of the dominant class in the early nineteenth century.[2] It is not music, science or poetry that defines polite or working-class culture, but rather the function of these activities, in terms of their practice and meaning, in different social spheres. In the period when science came to be imbued with cultural authority, the practice of artisan botanists enables us to investigate popular science from a fresh perspective.

Placing artisan botany

Although the pub was by its very definition *public*, it became an exclusively working-class location. By 1850 'no respectable urban Englishman entered an ordinary public-house, and by the late 1830s the village inn, where all classes drank together, had become a nostalgic memory'.[3] This change came about as much from the withdrawal of upper-class patronage of sports such as cock-fighting and the middle-class preference for rational recreation, as from the development of working-class organization derived from the collective values of the workplace. The pub operated as an extension of the workplace: not only were working men paid there, but many houses were trade-specific. They operated as houses of call for tramps, and were the location of friendly societies, benefit clubs and illegal trade unions. In addition, publicans sponsored popular recreations ranging from bull-baits to gooseberry shows.

Place assumed immense cultural and political importance as space itself became class-specific. Enclosure, the game laws and the geographical demarcation of towns like Manchester effectively excluded the poor from space that was formerly public. The middle class spent its leisure in exclusive cultural locations or in the privacy of their own homes. During the Chartist period, the pub was an important part of the 'free zone' available for working-class activities.[4] As far as the middle class was concerned the pub had become a private and dangerous place and needed to be rendered more public.

Intrusive policing of pubs continued in force in 1850, backed up by a strong Sabbatarian movement. In a spate of convictions, the *Manchester Guardian* reported that Joshua Barge, publican in Prestwich, had been fined £5 for serving liquor to working-men botanists during the hours of divine service. Barge's conviction served to make public a practice that was extremely private, for the incident provoked Thomas Heywood, stationer from Cheetham Hill and a member of the Prestwich Botanical Society, and John Horsefield (1792–1854), handloom weaver and president of the society, to defend Sunday botanical meetings in the *Manchester Guardian*.

Artisans met in pubs because this enabled them to develop an organizational structure whereby botanical knowledge could be acquired and extended. Minimal education and lack of time posed severe problems for artisans wanting to learn more about plants. In most cases, an artisan's interest in botany was first stimulated by a herbal. Richard Buxton (1786–1865), apprenticed to a shoe-maker in Ancoats in 1798, taught himself to read at the age of sixteen. When a journeyman, Buxton accompanied his master on country walks to search for herbs to prepare 'diet drinks'. Puzzled by plants whose names they did not know, Buxton purchased a copy of Culpeper's *Herbal*. Not satisfied with this, in 1808 he bought William Meyrick's *New Family Herbal* (1789) from which he learned the Linnaean system.[5] When a youth, Horsefield became fascinated by Culpeper's descriptions of plants and first encountered the twenty-four classes of the Linnaean system in James Lee's *Introduction to Botany* (1760). Over forty years later, Horsefield could still 'distinctly recollect the determination that actuated me, to overcome the difficulties that lay in the way of learning them'. 'I wrote these 24 names down on a sheet of paper', he recalled, 'and fixed it to my loom-post, so that when seated at my work, I could always have opportunities of looking it over'.[6]

Individual autodidacts usually sustained their initial efforts by finding fellow enthusiasts. Gradually, they formed associations based on the Methodists' organizational structure. They held large meetings on Sundays and founded local societies. Horsefield started to attend Sunday botanical meetings from 1808 – a time when he had no access to books – and found that the difficulties he encountered in trying to apply the Linnaean principles were 'by my attending these general meetings, removed by some practical botanist or other, better acquainted with the subject than myself; and many such persons were often in attendance'.

The lack of books was one of the main reasons that artisans established local botanical societies in public houses. Members paid sixpence a month, of which two or three pence went into a book fund. In this way, the Oldham Botanical Society purchased twenty volumes between 1775 and 1795, and between 1820 and 1850 the Prestwich society bought 131 volumes. The remaining amount of each member's fee went towards liquor. Publicans allowed the use of a room for meetings and were usually responsible for the safe keeping of the society's library and the 'box' or funds. In return, members were expected to spend enough on drink to satisfy the publican. This method of payment was commonly called the 'wet rent'.[7]

The purpose of local society meetings was to inspect plants and to borrow and return books. At the end of each meeting, specimens were selected for the society's herbarium which was also kept in the pub. Rules were imposed and fines were incurred for turning up without plants, swearing, pinching specimens and *arriving* at a meeting in an intoxicated state. Most importantly, of course, there was the obligation to pay the monthly membership fee.

As with friendly societies, drink was conducive to a sense of conviviality. The social function of the pub sustained the botanists' communal effort and helped to recruit new members. At the Prestwich Society it became common practice that 'after the more serious business of the meeting had been disposed of' the assembled company would remain into the night drinking and singing. Horsefield composed 'The Botanists' Song' commemorating how 'science circles with the glass'. James Crowther (1768–1847), a porter in Manchester, claimed that his specimens always looked 'best through a glass'.[8]

The knowledge of one becomes the knowledge of all

Books, plants and drink were not sufficient to hold a society together. Above all, a president was required in each society and at the Sunday meetings who could give plants their Linnaean names and instruct less experienced members. While the office of president may have been the preserve of an individual, who usually served for life, his function was to promote a communal activity in which his role was just one part. As Horsefield stressed, 'we instruct one another by continually meeting together; so that the knowledge of one becomes the knowledge of all, and we make up for the deficiency of education by constant application to the subject'.

The means by which the 'knowledge of one' became 'the knowledge of all' revolved around the Sunday pub meetings, the location of which was varied so that 'persons having to travel a great distance one meeting, have it nearer home another'. The artisans' aim was 'specific discrimination and accuracy in botanical

Figure 22.1 Artisans' pub meetings not only fulfilled a didactic purpose but also allowed the more expert botanists to accumulate information rapidly. Several artisans were interested in particular groups of plants and used the general meetings to acquire specimens and learn of new habitats. In December 1811, for example, Edward Hobson, who specialized in bryology, asked for mosses to be brought to a Sunday meeting. John Horsefield gathered more than twenty species for this occasion, but admitted there was 'scarcely one of which I knew by name'. A few years later he discovered 'that rare little species Funaria Templetoni, as it was then called; a moss that has not yet . . . been found in any other part of England' (*Manchester Guardian*, 21 Dec. 1850, p. 5). The illustration shows actual specimens (1/2″ to 3/4″ high) in Edward Hobson's *Musci Britannici* (exsiccatae) 2 vols. (Manchester, 1818–22).

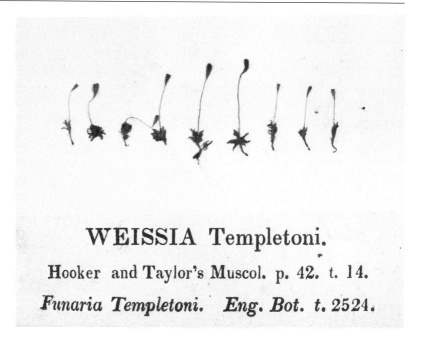

WEISSIA Templetoni.

Hooker and Taylor's Muscol. p. 42. t. 14.

Funaria Templetoni.   Eng. Bot. t. 2524.

nomenclature'.[9] Those attending were required to bring plant specimens, which were randomly piled on a table before being named aloud by the president (Figure 22.1). This oral transmission of names was essential when some members were illiterate. In his recollections as an errand boy on the Oldham Road, James Middleton recorded that:

A botanical society existed in Failsworth about 1818 . . . The members of it were all working men, mostly weavers, some of whom could not read . . . They went into the fields and gathered what they called 'speciments', which they brought to the meeting of the society to have them named.[10]

The early method of learning Linnaean names relied on repetition as a memory device, but some time after 1830 there was a shift away from this communal activity. Originally, as Horsefield recalled, the president.

taking a specimen off the table . . . gave it to the man on his left hand, telling him at the same time its generic and specific name; he passed it on to another, and so on round the room; and all the other specimens followed in a similar manner. But, from the noise and confusion caused by each person telling his neighbour the name of the specimen, some being unable to pronounce it, some garbling it, and all talking at once, we have been constrained of late years to adopt another method.

By 1850, the president named the specimens while the company remained silent.

Nevertheless, this change did not alter the fact that the president's skill was judged according to an old, informal and internally governed structure of authority within the *oral* tradition, with those who remembered the most claiming precedence over those with most to learn. What was being judged, however, was a complex mixture of the literary and the oral embodied in a particular individual.[11] When Horsefield took over this role, he discovered that it 'requires no small degree of skill, together with a good memory, to name a heap of specimens amounting sometimes to several hundreds'.

Intersecting social worlds

The artisans' interest in 'scientific botany' had developed alongside the widespead practices of floriculture, horticulture and herbalism. Herbalists and gardeners attended artisans' botanical meetings and were members of societies. Herbalists may not have been interested in the Linnaean principles of classification, but they did want to know how to recognize plants that were the basis of cures and to distinguish them from similar, often poisonous, ones when gathering specimens. The method of bringing plants to a meeting to be named and handing them round for inspection served this purpose well. Gardeners and nurserymen, working for the gentry, had good reason to be acquainted with the Linnaean names of plants, for in the eighteenth century horticulture was second only to local floras in establishing the Linnaean system in British botany.

The resulting mix of interests in botanical societies created tensions when participants placed different values on the information and specimens exchanged.[12] Plant names and habitats were useful for varying reasons. Botanists required plants with roots for herbarium specimens and, sometimes, for growing in their own gardens, but wished primarily to preserve plants in their native habitats. Herbalists also wished to preserve a growing supply but since their power resided in possessing access to particular plants, they were often secretive about locations. They too dug up plants in order to grow their own private supplies. Nurserymen and dealers were the least scrupulous in hunting for plants. In a competitive market, the value of a rare plant could best be maintained by having a monopoly over its sale, and this was achieved only by digging up the entire stock.

The accumulation of knowledge for individual gain became more problematic as the century progressed. Safeguards against excessive secrecy were taken by a combined benefit and botanical society founded in 1849: the Hyde Faithful Botanical Society. The 'Botanical Rules' warned that: 'Any member having a competent knowledge of botany or any particular plant or plants then and there produced, shall freely instruct his fellow members, or be excluded [from] this society' and that 'Any member who meets with a scarce, useful, and valuable plant, is requested to take it to the person appointed to cure them, or to cure it himself, that nothing may be lost which may be useful to the society.'[13]

Within working-class culture the issue of plants as property created tensions between botanists, herbalists and horticulturists; these tensions were exacerbated when artisan botanists interacted with practitioners from the middle and upper classes. Collaboration and exchange of information between artisans with varying degrees of knowledge of a uniform labelling system were made possible by frequent meetings and the naming of huge piles of plants. Most gentlemen botanists, however, had little patience with this method and relied on classificatory systems to provide a standard language. When William Wilson (1799–1871), a gentleman bryologist in Warrington, was visited by a collector of 'medicinal herbs', he noted impatiently that the man 'knows very little of Mosses or indeed about any thing except English names – his latin is excessively bad – quite an illiterate man –'.[14] The herbalist was useless because he had no way of *knowing* which plants Wilson wanted.

In contrast, when John Martin (*c.* 1783–1855), weaver from Tyldesley, visited Wilson to request confirmation for his identification of a rare moss, Wilson was delighted with him. Immediately reporting the find to William Jackson Hooker (1785–1865), Professor of Botany at Glasgow, Wilson later remarked that Martin was 'void of conceit and offensive familiarity: intelligent without arrogance: studious yet unassuming'. He 'confesse[d] his poverty without shame' and had an 'air of decency' about him.[15]

Martin, however, initially approached Wilson with a typical artisanal distrust of a 'middleman' who might appropriate the botanical discovery. Handloom weavers like Martin had suffered a great loss of status by 1830 resulting from their diminished control over the products of their labour. Increasingly dependent on merchants for the sale of their cloth, weavers suffered destitution as lack of work allowed middlemen continuously to lower wages. To the degraded artisan, this system was resented not so much for its exploitation, but rather because it was an unequal exchange.[16]

In Martin's opinion, fair exchange for his specimens and information was acknowledgement by gentlemen botanists. He was not defending his interests alone: the moss, though identified by Martin, had been discovered by the late William Evans (d.1828), weaver and herbalist, who had served as president of the Tyldesley Botanical Society. 'Should you communicate the discovery to any of your botanic correspondents, especially to Professor *Hooker*', Martin stipulated to Wilson, 'I hope you will give my deceased

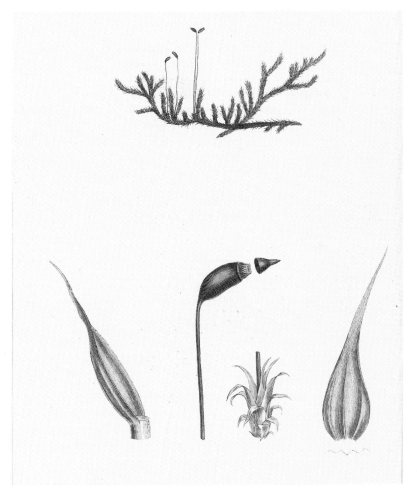

Figure 22.2 The illustration shows *Hypnum salebrosum*, the 'Pale shining Feather-moss', discovered by Edward Hobson, and depicted in vol. 3 of W. J. Hooker's *Supplement to the English Botany of the late Sir J. E. Smith and Mr. Sowerby* (London, 1843).

friend the honour of the discovery.'[17] Gentlemen like Hooker and Wilson knew that their research depended on collectors and consequently gave artisans the credit they requested. Hooker duly published the history of the moss giving full acknowledgement to both Evans and Martin, 'a zealous and accurate Botanist'.[18]

Judgements based on such acknowledgements alone make it difficult for historians to get beyond the role of artisans merely as providers of specimens and local information for elite botanists, to an understanding of the meaning of such exchanges to artisan participants in natural history (Figure 22.2). Only in times of dire need did working men sell their collections or collect for gentlemen in return for payment; usually they requested specimens, identifications or other information in exchange. This was particularly true of bryological specimens as few working men had access to a microscope, which was necessary for observing the minute characters used in classifying mosses. The use of Linnaean nomenclature allowed artisans to set up networks beyond the range of their meetings and allowed communication with gentlemen. Although both groups spoke a common Linnaean language, their attitude towards botanical property gave the specimens strikingly different meanings.

As far as the working class was concerned, the upper classes wielded their power in science largely through the printed word. Intellectual property as an individual possession secured through print was of great importance at a time when the discovery of a nondescript, or even a rare plant, could secure a gentleman fellowship of the Linnean Society. Thus, Heywood claimed, not only were individual discoveries by artisans sometimes appropriated by gentlemen, but this lack of printed recognition resulted in the loss of the collective history of the Lancashire botanists. For artisans, the issue of acknowledgement was a group concern, not just an individual one.

The collective nature of artisan botany was most clearly expressed in their pub meetings. On the few occasions that gentlemen attended, their social superiority did not entitle them to special privileges. Within the pub, the middle- and upper-class expectation of deference from their social inferiors was pointedly ignored by working men. Here, the artisans were in control of the social situation – not because their meetings were private and exclusive, but because the pub was a specific cultural location that belonged to the people. If gentlemen went away without the information they sought, Heywood warned, it was because

they are too proud to ask for any information, they think it should be given to them from such poor persons as they consider the botanists to be, without the trouble of asking for it, or even of stating what kind of information they require . . .

In the pub, gentlemen had to play by the artisans' rules.

The property of botanical skill

'Independence' and 'property of skill' were defining characteristics of the artisan mentality and were so deeply embedded in artisan culture that they continued to be defended by deskilled workers, such as handloom weavers and shoemakers.[19] These were the values that artisan botanists displayed in their interactions with gentlemen. Independence was asserted through concern with status, refusal to adopt subservient attitudes and an aversion to charity. But the most fundamental concept underlying the consciousness of artisans was their sense of possessing a 'property of skill'. This was a source of pride to an artisan and entitled him to the respect of others. For at least some of the deskilled artisans of Lancashire, botanical skill served to restore a sense of status and respectability.

While individuals possessed botanical skill, the artisanal practice of botany was regarded as part of a collective craft. For an artisan, the 'Mystery', or property, of any craft 'belonged to no individual but to the body of craftsmen past, present and future'.[20] The collective values of the artisan botanists were maintained through their association in pubs, for here the 'Art and Mystery' of botany was conveyed by precept and example just as crafts were learned in the workplace.

The emergence of the pub as a male world in the early nineteenth century was associated with these newly articulated artisanal attitudes towards their trades. As the exclusivity of skilled crafts was threatened in the period of manufacture, artisans, including domestic handloom weavers, combined in order to defend their rights against cheap unskilled labour. Gender distinctions overrode considerations of technical aptitude: skill belonged to men. Although there is evidence of female participation in botanical meetings around 1812, after this period women are conspicuous by their absence. It seems likely that women were excluded from societies on the grounds of who was entitled to possess 'botanical skill', even though they were no less qualified than men to participate in botanical meetings and the methods used to disseminate knowledge were not exclusive in themselves.

The possession of skill imposed on its holder 'the obligation of the proper performance of his craft'.[21] For the artisan botanists, this 'proper performance' was the achievement of the aim of their meetings: specific discrimination and accuracy in botanical nomenclature. The possession of botanical skill was most clearly displayed by the use of Linnaean taxonomic names. The artisans defended their use of such terms, for without this knowledge they could not properly be admitted to the practice of botany. Sensitive to criticisms that Latin terms 'cannot be learned by the common or uneducated people', Horsefield did not explicitly defend the artisan botanists in terms of class, but simply stated that the Linnaean nomenclature should be used by 'all genuine botanists'. Few artisans understood the literal meaning of the Latin words, but this did not prevent them from learning the nomenclature. As Horsefield admitted: 'Of Latin, as a language, we know very little; but this we know, that a uniform nomenclature is far preferable to a mixed one . . . A Latin name is . . . as easy to learn as a mere English one'. For the artisan the *meaning* of the Linnaean names lay in the display of botanical skill.

Styling themselves as botanists, the artisans maintained a strong group identity. They expected fair exchanges of specimens and information (as much from one another as from gentlemen) and to receive recognition for their discoveries. Moreover, they felt that the skill they possessed entitled them to respect from the scientific community at large. In 1843, William Bentley (1815–81), blacksmith in Royton, obviously saw no incongruity in asking whether the Linnean Society of London had 'a fund at its disposal to assist decayed persons, who, by their industry have rendered themselves eminent in any branch of Science, which comes under its notice'. By this means, he hoped, the elderly and destitute John Mellor (1767–1848), handloom weaver and gardener, could receive financial help. Although Bentley's tone was deferential, the collectivist value of mutual aid, particularly that extended to others in the same craft, is reflected in his view that Mellor's botanical skill entitled him to help from fellow botanists at a time of need. Bentley was, in effect, treating the Linnean Society (one of the most elitist of the London scientific societies) as a botanists' benefit club.[22]

For the artisan, then, skill represented 'an "honour", the possession of which entitled its holder to dignity and respect'.[23] The transference of these values to the property of botanical skill is most clearly demonstrated by Horsefield. On reading a comment on the 'ignorance' and 'degradation' of Lancashire operatives in the *Gardener's Magazine* of 1829, he found himself *'vexed'* enough to take up his pen and challenge the author:

What your ideas of our 'dreadful state of degradation' may be, I cannot positively say: had you used the word destitution instead of degradation, you would have been more correct . . . If to be half-employed, half-paid, and half-fed, constitute 'dreadful degradation', I can sincerely assure you we are *now* dreadfully degraded indeed.

As for our 'ignorance', I don't think we are more ignorant than any other class of His Majesty's subjects. The intricate paths of science are seldom sought for by any man, whatever his station in life may be . . . and even amongst *us* you might find some instances of devotedness to literature and science . . . botany being a favourite pursuit . . . and botanical meetings frequent and well attended.[24]

The politics of participation

The pub, of course, was not the only place that artisan botanists practised their science. However, an understanding of the pub as one of the main sources of sustenance of the collective values of artisanal culture allows us to interpret their practice of botany. For although the middle-class portrayal of the artisans, and, indeed, the material products of their botanical practice, make them look as if they had absorbed the bourgeois credo of individual self-improvement, their scientific practices can be understood only by drawing on work that explicates artisanal attitudes.

This also provides a way of understanding the experience of ordinary people engaged in scientific pursuits that goes beyond generalizations derived from the public statements of radical or Chartist leaders. The latter have led to characterizations of artisan naturalists as 'solitary and superior' autodidacts who 'kept a proud distance from the pub'.[25]

But these radical leaders made literacy an integral part of social and political change by dissociating it from the culture of drink. This marked a sharp contrast with the eighteenth century and those like the 'widely read Ashton apprentice who entertained primarily in alehouses'.[26] However, the emphasis on the discontinuity between the radical use of literacy in the early nineteenth century and popular traditions is too pronounced. Iain McCalman has shown that 'humour, escapism, sex, profit, conviviality, entertainment and saturnalia should be admitted to the popular radical tradition' not least because these elements probably did more for its survival than the 'sober, strenuous and heroic aspects which are more customarily described'.[27] So too did drinking, singing and the communal values of artisanal culture sustain the scientific pursuits of the working class. Science can be associated with the pub as much as with the radical unstamped press, Owenite Halls of Science, and Mechanics' Institutes.

In cases where there is little evidence of overtly political or ideological agendas, an emphasis on skills and place is as revealing (and more representative) of the practice of working-class science. This does not mean that the activity of the artisan botanists – and their claim that they *were* botanists – had no political meaning. By looking at the way in which artisans actually did science, and not just the ways in which they talked about its potential power, we can start to produce a political spectrum of working-class science that prises apart the extreme models of radical confrontation and incorporation into the bourgeois sphere. We can thus break away from the view that 'genuine' working-class science is 'alternative' or 'fringe' science. The political challenge from the artisan botanists consisted in participation by working men in what was rapidly being appropriated as the cultural property of the educated and leisured classes (Figure 22.3).

Defining popular science

Science was contested territory in the first half of the nineteenth century. For many, even within the elite, the contest revolved around the issue of whether science should be popular, in the sense

Figure 22.3 The artisan belief that possession of skill entitled them to be cultural producers was perhaps best demonstrated by their struggle to participate in those areas of knowledge that were considered the province of the upper and middle classes. Brian Maidment, *The Poorhouse Fugitives* (Manchester, 1987), pp. 14–17, argues that ignoring 'Parnassian poetry' – poetry that deliberately uses the most complex and elaborate traditional forms within the British literary tradition – in favour of Chartist, radical and dialect poetry is to overlook an important aspect of working-class aspiration: the ambition to participate on equal terms at a high cultural level. It is entirely appropriate that the self-taught Parnassian poet Charles Swain composed the epitaph on Horsefield's gravestone, to be seen in the graveyard of St Mary the Virgin, Prestwich.

of being open to a wide range of participants who could contribute to and benefit from the production of knowledge. Increasingly, however, the term 'popular science' was used by the dominant culture to signify scientific literature and activities that had little or no interaction with elite science. Thus, the meaning of 'popular' in 'popular science' by the mid-nineteenth century was 'that which is excluded from institutions of legitimation, either because of the material conditions of its production, or because its general accessibility lays it open to charges of debasement or simplicity'.[28]

This was true for other cultural activities, but in science the argument is particularly clear. More than any other area of knowledge production, scientific practice became increasingly associated with specific sites from which 'the people' were excluded. By defining the laboratory and the experimental station as the sites of legitimation of botany and zoology from the mid-nineteenth century (and thereby increasing their status), the place of science became strictly defined and popular science was marginalized. This marginalization did not, of course, exclude artisans alone. On the contrary, those to be excluded were clergymen, women and 'non-professionals' in general, thus producing the category of the 'amateur' whose work was rarely of use to the professional scientist.

The impotence embedded in the definition of popular science by the 1850s, and the emphasis of most twentieth-century histories of botany on tracing a progressive path to an experimental science, has obscured the effects of artisan participation. Just as oppositional uses of science affected the ways in which those with authority constructed the boundaries around knowledge, so participation by artisans in natural history also shaped the response of the elite.

The accessibility of the Linnaean classification in particular made it a powerful tool for the production of reliable information. This led some botanists to continue arranging their textbooks and monographs according to the Linnaean system well after they had accepted the superiority of the natural classification proposed by Antoine-Laurent de Jussieu at the end of the eighteenth century. Hooker continued to use the Linnaean system in new editions of his *British Flora* (1830) until the late 1840s, although he thought the natural system superior for experts. In 1843, the naturalist Edward Forbes praised Linnaeus for endowing botany with a 'universal language'. More literally than he realized, Forbes celebrated the Linnaean system because an 'easy means of acquiring and arranging information is a great help to the workmen of science'.[29] For the same reason, and possibly because he had heard that the Royton blacksmith Bentley and other 'young would be Botanists' were embarking on the study of bryology, in 1846 Hooker bemoaned the fact that Wilson would have to adopt the latest classificatory system in his *Bryologia Britannica* (1855), as 'with the mosses especially, in proportion as you depart from the artificial system, you increase the difficulty . . . & you cannot render the study *popular*'.[30]

Conclusion

The historical recovery of the practice of artisan botany is not possible from the literary traces of elite botanists, nor even from the few texts produced by the artisans themselves. For one of the functions of scientific texts is to obscure the work involved in the production of knowledge in order to render such knowledge objective.

The careful construction of literary accounts aims to produce a scientific community bound by particular forms and conventions.[31] Participation necessarily involves conforming to these conventions, and the more successful the participation, the less obvious the local concerns. It is therefore essential to look beyond the literary conventions of science, which draw us towards characterizations of the artisans as mere providers of specimens to elite botanists. Instead, we need to regard science, like other aspects of culture, as a practice in order to understand its meaning in an artisanal context.[32] When science is viewed as an activity in which literary production is just a part, it becomes no less exact to call John Horsefield a botanist as to apply the term to the inept manager of a Suffolk brewery, a Liverpool banker or a Dante scholar – W. J. Hooker, William Roscoe and Charles Lyell Sr, respectively.

The contest over science in the early nineteenth century was a contest about who could participate and on what terms. The *result* of this contest was the redefinition of popular science. The middle class increasingly rendered working-class scientific activity politically neutral through control over printed texts, a preoccupation with producing accounts of the lives of autodidacts to put forward moral lessons, and by giving natural history a central role in rational recreation. Unless we look at how the dominant culture reconstructed the popular by these means, we can never escape the mid-nineteenth-century middle-class view of the artisan botanists as 'harmless and industrious'.[33] In recovering popular science, then, we need to restore 'the relations between social power, political democracy and cultural production' which are part of the history of definitions of 'the popular'.[34] Only in this way can we see that artisans were part of the making and shaping of science in the early nineteenth century.

Further reading

Allen, D. E., 'Life sciences: natural history', in P. Corsi and P. Weindling (eds.), *Information Sources in the History of Science and Medicine* (London, 1983), pp. 349–60.

Bailey, P., *Leisure and Class in Victorian England: Rational Recreation and the Contest for Control, 1830–1885* (London, 1978).

Cooter, R. and Pumfrey, S., 'Separate spheres and public places: reflexions on the history of science popularisation and science in popular culture', *History of Science*, 32 (1994), pp. 237–67.

Desmond, A., 'Artisan resistance and evolution in Britain, 1819–1848', *Osiris*, 2nd ser., 3 (1987), pp. 77–110.

Johnson, R., ' "Really useful knowledge": radical education and working-class culture, 1790–1848', in J. Clarke, C. Critcher and R. Johnson (eds.), *Working-Class Culture: Studies in History and Theory* (London, 1979), pp. 75–102.

Ophir, A. and Shapin, S., 'The place of knowledge: a methodological survey', *Science in Context*, 4 (1991), pp. 3–21.

Secord, A., 'Science in the pub: artisan botanists in early nineteenth-century Lancashire', *History of Science*, 32 (1994), pp. 269–315.

'Corresponding interests: artisans and gentlemen in natural history exchange networks', *British Journal for the History of Science*, 27 (1994), pp. 383–408.

Shiach, M., *Discourse on Popular Culture: Class, Gender and History in Cultural Analysis, 1730 to the Present* (Cambridge, 1989).

Thompson, E. P., *The Making of the English Working Class* (London, 1980).

Vincent, D., *Bread, Knowledge and Freedom: A Study of Nineteenth-Century Working Class Autobiography* (London, 1981).

23 Tastes and crazes

Fashions, in the strict sense, do not occur in intellectual matters, nor can they. For they are light-hearted products of imitation and show, necessarily transient and shallow in order to fulfil their function of expressing a merely temporary inclination or mood. They rise, they fall and in their turn are then replaced by something similar.[1] It is common to speak of intellectual fashions, but by that is meant no more than that some set of facts or a theory has caught on and become popular (and not necessarily only temporarily): being non-visual, it cannot serve as a vehicle for eye-catching display; being serious, it is not bound to be discarded once adopted too widely and persisted in too long.

Natural history, however, is not and never has been a purely intellectual pursuit. It has a considerable aesthetic component as well, of varying strength at different periods and in different individuals. Many people are attracted to it primarily for visual reasons, rather than to study behaviour, work out distributions or formulate concepts. Even in its most primitive manifestation, collecting, there can be a delight in shapes and colours and patterns which co-exists with the mere pleasure of acquisition or the sheer satisfaction of having the evidence for some additional item of knowledge.

Once this extra-intellectual interest goes beyond a certain point, natural history is liable to take on an additional dimension: to be drawn on for reasons that are purely aesthetic-cum-social, to become the prey of genuine fashion. At the extreme, the very subject itself may become the plaything of fashion, as happened in the eighteenth century and more especially in the Paris salons. More usually, though, one particular facet is fastened upon and inflated out of all proportion to its intrinsic importance as an area of study. Aspects of natural history which have some obvious potential as vehicles for symbolism are particularly vulnerable to being raided like this, and it is no accident that plants and shells have formed the subjects of the most salient instances that have occurred. In the words of one Victorian magazine, shells 'are so brightly clean, so ornamental to a boudoir', while the special attractiveness of plants, by virtue of their foliage no less than of their flowers, is in need of no emphasizing. Both possess the additional advantage of easily recognized features which readily lend themselves to adoption as design motifs, by which means they can become completely incorporated into décor, totally abstracted from nature. Indeed, it may well be that natural objects normally become the focus of powerful fashions only if they are subtly in accord with the wider artistic expression of a particular outlook associated with a particular span of years – only if, in their own small way, they serve to reflect that elusive entity, the 'spirit of the age'.

Conchyliomanie

Shells first became noticeably the subject of an elegant pursuit in the seventeenth-century Netherlands. As early as 1607 artists there were being commissioned to paint leading collectors of these with choice specimens in front of them. Just like the postage stamps of later generations, shells came in a gratifying assortment of shapes and colours; many were pleasing to the eye and some even beautiful; they were obtainable from all corners of the globe, but from many of them only rarely and only with great difficulty; a beginner could start with a presentable array, put together by his own unaided efforts and without any expenditure of money, while for the connoisseur there was a challenging hinterland of scarcity, conferring monetary value accordingly. For one of these Dutchmen, Pierre Lyonet, shell collecting was indeed no more and no less than a branch of art collecting, the finest specimens being purchased with all the care and discrimination that went into his buying of paintings.

By the early eighteenth century there were cabinets full of shells to be found in the houses of the wealthy over much of Europe. Two of the finest collections drawn from all parts of the then-known world, were those of the London physician, Sir Hans Sloane (1660–1753), and of an Amsterdam apothecary, Albert Seba (1665–1736). Some of these cabinet owners went to great lengths to arrange their specimens artistically (Figure 23.1) while others went further still by trying to improve on nature's efforts by paying to have individual specimens 'beautified' artificially (a fate similarly visited in the next century on the tougher kinds of birds' eggs). 'Shell-doctoring', as this was called, developed into quite a trade in the Netherlands, providing a living for numerous practitioners. Yet a very much fatter living was to be had by the specialist auctioneers: for collectors were continually dying, or simply tiring of their hobby, or disposing of one collection preparatory to starting all over afresh. At these sales impressively high prices, sometimes even absurdly high ones, were increasingly reached as the pursuit began to assume the proportions of a craze – the *conchyliomanie*, as it was dubbed by the French.

Figure 23.1 A shell 'portrait' in the cabinet of a Dutch collector. Albertus Seba, *Thesaurus* (Leiden, 1758), vol. 3, plate 37.

Like the fern craze of the following century (of which more below), with its suggestive echoes of the Gothic Revival, the shell craze is under strong suspicion of having owed its super-normal vitality to acting as a side expression of a dominant trend in contemporary art. Indeed, one authority on the art of the period has gone so far as to claim that the craze precisely paralleled in its rises and falls the concurrent fashion for *rocaille*, the shell-like motif that became so ubiquitous from 1719 onwards.[2] However much truth there may be in that, it is certainly striking how suddenly the '*conchyliomanie*' came to an end: there was one final, immensely lucrative auction in 1757, that of the vast collection of the French ambassador at The Hague, the Marquis de Bonnac, and then prices abruptly dropped – and never afterwards recovered. 'All fashions end in excess', according to a dictum of the great couturier

Poiret, and maybe it was simply greed that did the shell collecting craze to death. Or maybe it was that the eyes of everyone in the salons had at last finally wearied of *rocaille* and of anything that resembled it.

Seaweeds

Closely akin to the shell craze, at least to the extent that it was similarly a product of the searching of beaches, was the somewhat later, but decidedly lower-key, fashionable concern with seaweeds. In this case, untypically, what began largely as an artistic vogue went on to open up a lasting field of serious study: more usually the reverse occurs, or the vogue and the study develop in tandem.

In 1751–2 John Ellis (1710–76), a London merchant active in botanical circles, began receiving collections of 'sea plants' and corallines from correspondents of his who lived on the coast. The reason for this is unclear, but it led him into making a pioneer classificatory study of those hitherto neglected groups, in the course of which he was to rediscover that corallines were animals, not plants (as generally supposed up till then) and, in a series of publications, to remove once and for all the misconception that they were an intermediate link between the two. This scientific work, however, had an unlikely by-product, for, apparently on a whim, he one day made a miniature seascape out of his specimens, giving this the modish name of 'grotto-work'[3] – in evident allusion to the then fashionable taste for caves, crags and ruins and their 'picturesque' accompaniment of greenery. The existence of this came to the knowledge of one of his scientist friends, the Rev. Stephen Hales, who thereupon asked him to make something similar for the Princess Dowager of Wales, to whom Hales was Clerk of the Closet.[4] After that a salon fashion seems to have arisen for this delicate type of fancy-work, a fashion which was to continue in some degree all through the century following (and to enjoy a revival in recent years as part of the taste for 'pressed-flower' pictures, in the guise of 'underwater scenes' in which seaweeds do duty as foliage).

It may have been this fashion and the demand for the necessary raw material that it generated that caused the Clerk to the Royal Society, Emanuel Mendes da Costa (1717–91), to write in those same months to a leading antiquary in Essex in the following (surprisingly peremptory) terms: 'Send me a small box of the seaweeds or corallines found on your coast. You have only need to lay them in a heap, damp as they are.'[5] About the same time a Mrs Le Coq, down at Weymouth, was likewise being pressed into service by that inveterate collector of almost everything, the Duchess of Portland; and presently quite a few others were joining in.

In the surviving letter-book[6] of Dru Drury, a wealthy London silversmith, are to be found, dated April 1764, some 'Directions given to Mr Warr, Cap. Mayle, etc. to be sent into Devonshire and Cornwall for Collecting Sea Weeds'. While still crudely minimal, these at least improved on Da Costa's by insisting that the specimens be washed in fresh water on being gathered, before being put moist into a box or barrel and packed very tight.

It was this very simplicity that seaweeds shared with shells (and, later, ferns) as items to take up and preserve that doubtless accounted for a good deal of their popularity with collectors. In this respect they were notably different from so many other items with which naturalists found it necessary to concern themselves. Like shells, too, seaweeds could be sought after without embarrassment, for provided they were picked up off the foreshore – and did not have to be waded for or hammered off the side of a rock-pool – they did not involve unseemly postures or any stare-inducing equipment. Once collectors left the safety of the foreshore, by contrast, they were liable to meet with problems. 'When we first began Sea Weeds,' the Berwick naturalist Dr George Johnston (1797–1855) once confided to a correspondent, 'my wife carried a larger muff than the present fashion would commend, and many a heavy stone and well-filled bottle has therein been smuggled.'[7] Not everyone, though, was willing to operate with such furtiveness. In the view of the no-nonsense Mrs Margaret Gatty (1807–73), 'any one really intending to *work* in the matter must lay aside for a time all thought of conventional appearances'. For her own forays along the edge of the tide she favoured a pair of boy's shooting boots, rendered waterproof with a thin coat of neat's-foot oil; above those merino rather than cotton stockings, and petticoats, also preferably of wool, that never reached below the ankle; over those a ladies' yachting costume (in her opinion, 'as near perfection for shore-work as anything that could be devised'); and, to complete the ensemble, a hat instead of a bonnet. Cloaks, shawls and all millinery she warned against as hopelessly impractical. Even she, though, otherwise dauntless though she was, was forced to admit that 'a low-water-mark expedition is more comfortably taken under the protection of a gentleman'.[8]

It was presumably because seaweeds were firmly associated in the public mind with a highly respectable kind of handicraft that so many women were emboldened to take up forming collections of them. Very much a minority fashion though this one remained all along, it was to become increasingly noteworthy for the prominence attained in it by that otherwise then so generally diffident sex. The fact that in its early stages the pursuit had gained a strong foothold among the landed gentry, and the male landed gentry at that, must also have conferred on it a certain social cachet which

Figure 23.2 Shore collecting: 'a low-water-mark expedition is more comfortably taken under the protection of a gentleman'. G. H. Lewes, *Sea-side Studies* (London, 1868).

no doubt provided them with additional encouragement. That aristocratic influx, indeed, temporary though it proved, appears at first sight an even more striking and unexpected feature of the fashion than the later prominence of women. Those were the years, though, in which the uppermost layer of society was deeply in thrall to the earliest manifestations of Romanticism, in which a taste for a feathery green covering of stone or rocks (as in the grottoes which provided John Ellis with a kind of code-word to flourish as a cover) was one conspicuous ingredient. The connection between the modish mediaevalism and the focus on seaweeds is well exemplified by the action of one of those landowners, John Stackhouse (1742–1819), whose Cornish estates included some marine frontage, in having a castle-like folly put up to serve as a base while he worked on this group of plants specifically. Some twenty years later, in 1795, his *Nereis Britannica* appeared as the outcome, one of several superb folios to be devoted to the subject by affluent enthusiasts at that period.

The women collectors were all much too self-effacing to contemplate such feats; indeed, with only one or two exceptions none of them were ever to venture into any form of print. How then did it come about that so many of them achieved a considerable measure of genuine botanical renown, ending up with new species and even genera named after them? The answer to that question is to be found in the fact that most were condemned to lives of boring uneventfulness in small, relatively isolated seaside towns, in which a regular walk along the beach was one of the few kinds of outdoor recreation permissible. Short of dredging, an elaborate procedure which called for specialized equipment, the only way any collector had of acquiring specimens of the little-known kinds restricted to the deeper waters was to keep a watch on the beaches for any stray examples of them that happened to have been dislodged and cast up by storms. This elementary task was one for which such women were peculiarly well-situated. One by one, over the years, they were brought into touch as a result with specialists with monographs under way on these plants; and delighted to learn that their patrolling could be put to such wider and loftier ends, they scanned their local shores all the more diligently, periodically packing off by post consignments of their gleanings. In this way the seaweed fashion was to become the classic instance of the harnessing to the shafts of scholarship of what originally started out as no more than a tasteful diversion.

The fern craze

To a more limited extent such a claim could be put forward too in some mitigation of the greatest and ultimately most destructive natural history fashion of all: the Victorian fern craze.[9] A British Isles phenomenon more or less exclusively, this was remarkable both for the hurricane-like force of its impact when it eventually took off and for the length of the preceding gestation. In some form or other, though at very different levels of popularity, ferns were the subject of fashionable interest throughout the whole of the nineteenth century. This was primarily horticultural in its inspiration and expression and can truly be ascribed to natural history only very secondarily. It is the classic case in its turn of how natural history has always been potentially subject to the powerful gravitational pull of neighbouring cultural realms, the boundary between gardening and field botany being one that is particularly ill-defined and porous.

Like seaweeds, ferns originally caught the fancy of cultivated circles when Romantic tastes were beginning to stir, and almost certainly because of the similar appeal of their fronds when clothing old walls and weathered rocks. Unlike seaweeds, however, at the outset they were frustratingly lacking in diversity. The species found wild in Britain were few in number and the larger of those were unspectacular and in the main insufficiently unalike. People consequently looked to the wealth of ferns in the tropics for greater aesthetic enrichment. Unfortunately, though, until the closing

Figure 23.3 N. B. Ward's own personal fern-case, designed as a window of Tintern Abbey. Frontispiece to N. B. Ward, *On the Growth of Plants in Closely Glazed Cases* (2nd edn, London, 1852).

years of the eighteenth century hardly any of the tropical species had found their way into the hothouses of Britain, for the simple reason that till then no one had any idea how to raise ferns from spores and the vicissitudes of the sea voyages were generally fatal to the chances of successfully importing the living plants whether young or mature. Once the secret of spore reproduction was discovered and made known, in the 1790s, professional gardeners on the staff of some of the more horticulturally adventurous establishments started competing with one another in a race to grow the widest selection. Hothouses, however, were rich men's indulgences and that fashion seemed destined to be limited just to this tiny elite.

Then, early in the 1830s, the remarkable properties of more or less airtight glass cases were accidentally discovered by a London general practitioner, Nathaniel Bagshaw Ward (1791–1868). Wardian Cases (as they came to be called), the equivalent of today's 'bottle gardens', at last allowed living plants to be transferred without problems between regions with quite different climates; but, of more immediate importance, they provided micro-environments in which vegetation could flourish indoors (or on balconies) seemingly indefinitely, impervious to the fumes of the gas-lighting by then in increasingly general use. As the chief aim in having these cases was to brighten one's house with greenery all the year round, ferns were at once the favourites for this purpose, and would have been even had they not possessed a faintly Romantic resonance. At the same time glass at that period was formidably expensive, as a result of heavy duties originally imposed to help pay for defending the country against Napoleon, and that made ownership of such a case something of an extravagance. For a long time, therefore, this décor fashion stayed necessarily confined to the comparatively well-to-do.

Towards the end of the 1830s an enthusiasm for the native wild ferns arose in the ranks of field botany, at that time a suddenly fast-growing pursuit. For this, two books were mainly responsible, both by gifted freelance writers: *An Analysis of the British Ferns and their Allies* (1837), by George William Francis, which made up for a humdrum text with an attractive set of copperplate drawings, then a novelty in a work brought out at only a modest price; and *A History of British Ferns* (1840), by Edward Newman, which had first been published serially in a magazine and, despite making do with woodcuts, was written in a bubbling style which still reads irresistibly. Both books emphasized in their very titles that there was more scope than generally realized in learning to distinguish the relatively few native species and – as the justification for seeking these out was still primarily horticultural – in forming a collection of them on the garden rockery or, if Wardian Cases were beyond one's means, in the living-room under inexpensive bell-glasses.

As the botanists poked around in the lanes and coombes of the more westerly parts of Britain (where the damper climate gives rise to ferns in greater profusion), they began to light upon districts where some of the species had 'sported' with unusual frequency, putting out fronds with gross irregularities, sometimes of considerable beauty. The fashion for collecting the wild species was thereupon rescued from dying of banality by a fresh surge of enthusiasm for hunting and growing these 'varieties' (as they were called, though in strict scientific terms they were merely monstrosities). Handbooks duly appeared in response, in which the variants were given Latin names and described and the places in which they had been found were reverently listed.

Then, just after 1850, this obscure and decidedly recherché little fashion exploded all of a sudden into a craze of nation-wide proportions, of quite extraordinary vehemence. Every other person in the country, it soon began to seem, wanted ferns to grow in their gardens or in the rooms of their houses; and as the supply of exotic species was as yet strictly limited, it was the ones that grew wild in Britain that were inevitably the principal victims. As new handbooks were rushed out and existing ones hurriedly reprinted, nurserymen signed up agents on commission to scour the countryside for every fern of sale value that they could find. Fern touts meanwhile sprang into existence, hawking the roots of choice rarities in the streets of cities and even on the summit of Snowdon. Half-starved country folk, seeing to their astonishment a source of ready money growing all around them, joined in with no less abandon and even less discretion. Whole hillsides were stripped bare; woods were cleared of every frond; even private estates were invaded and plundered. It has taken the best part of a century for the native fern flora to recover.

There were two main causes for this huge, dramatic outburst, one obvious, the other less so. The obvious one was the arrival on the market of mass-produced sheet glass, consequent upon the repeal of the glass duties (after persistent, high-level lobbying) in 1845. Those duties had not only made that material unduly expensive: they had also discouraged technical innovation throughout the industry. The Crystal Palace, housing the Great Exhibition of 1851, was the immediate, stunning outcome. All at once everyone wanted their own miniature versions of that – and greenhouse manufacturers and glaziers proceeded to make fortunes as a result.

The other cause was a major switch in taste. As the leading trade magazine, the *Gardeners' Chronicle*, observed in a perceptive editorial in 1856, a liking for 'exquisitely beautiful foliage' was rapidly replacing 'merely gaudy flowers' in the public favour. The latter taste, by implication, was less sophisticated because less subtle. 'Lovers of plants', the editorial went on, 'begin to prefer graceful form to mere spots of colour . . . Dress, furniture, architecture, are all now moving upon the same road side by side' – in the direction of ornamental intricacy and the loving elaboration of detail. By no coincidence, 1851, the year of the Great Exhibition, had also been the year in which John Ruskin first championed the Pre-Raphaelites. In 1855, in the second edition of his *Seven Lamps of Architecture*, he had gone on to sound the opening trumpet-blast of what was to be the Gothic Revival.

It was not long before ferns were breaking out like a rash in almost every conceivable decorative medium. Meanwhile people pressed the actual fronds, fixed them on white paper and hung them up on walls in frames. They also arranged them in pleasing patterns, again on white paper, and then sprayed them with indian ink to obtain silhouettes of a pleasing delicacy – an accomplishment known as 'spatter-work'. This last was thus, ironically, a late by-product of the fashion, instead of its initiator, as the equivalent 'seaweed pictures' had been a century earlier.

So quintessentially were ferns sensed as embodying later-Victorian taste that they were still being taken from the wilds in horrifying quantities right up until the time of the First World War. By then, though, the fashion had become vulgarized, and was largely being kept alive by a commerce determined to squeeze from it every last penny. As if in obedience to the rules identified by fashion theorists, nurserymen ended up by offering the horticultural equivalent of 'hypertrophy', 'over-extension' and 'sartorial hysteria': fern varieties exhibiting such an extreme of deformity as to be tantamount to caricatures. When such a point is reached, it is a signal that a fashion has exhausted its stylistic possibilities before it has fully exhausted its social energy, with no alternative then left to it but to waste itself in flailing against a barrier as invisible as it must be impassable.[10]

Aquaria

There was one further major fashion which descended on British natural history in the mid-nineteenth century – with a similarly jarring abruptness and leaving behind it a similar trail of damage. In some ways a side-branch of the fern craze, it sprang from one of the same roots as that, was equally dependent on relatively cheap glass and essentially consisted of introducing into the Wardian Case movement and animation. Unlike the other fashions already described, however, it had no discernible stylistic associations or symbolical import: it seems to have been purely the product of a technical development. And although it was in part a décor fashion, that might well not have been the case had it not fitted into a niche prepared for it already by the fern craze. Much of the enthusiasm it engendered may have been owed, rather, to its sheer novelty and its impact as a curiosity.

The technical development in question was the extension to the animal world of the oxygenating principle which made possible the Wardian Case: the chemical effect produced by growing plants tightly enclosed in glass not only enables them to stay alive indefinitely, but also sustains any creatures that are placed in with them. Although Ward had taken this step himself as early as 1841, converting one of his plant-cases into what he called an 'aqua-vivarium',[11] he was not concerned to publicize it and it was not for another ten years that its existence was made generally known (in the official catalogue of the Great Exhibition). By that time a professional chemist, Robert Warington, had begun undertaking a series of experiments in which he succeeded in demonstrating, and reported in specialist journals with exemplary thoroughness, the scientific basis of the principle.[12] Additionally, he was able to prove that it held for salt water as well. After that it merely remained for someone with the necessary journalistic skill to alert the wider world to the exciting possibilities that this opened up.

Such a one quickly materialized in the person of Philip Henry Gosse (1810–88). Gosse wrote on natural history for a living and had already produced several of what was eventually to be a long line of beautifully illustrated books on the subject, some of which became bestsellers; he is better known today, though, as the oppressively religiose parent in that classic of autobiography, *Father and Son* (1907). Apart from his skill as a writer, his special contribution was to identify the 'aqua-vivarium' – or the 'aquarium', to which that had speedily become contracted – with the study of marine life more or less exclusively: for he chanced just at that time to have lighted upon the fauna of rock-pools and promptly succumbed to a passion for this. In his very next book, *A Naturalist's Rambles on the Devonshire Coast* (1853), he not only trumpeted his delight in that newly discovered miniature world, but also provided instructions on how to create a marine aquarium.[13]

Another nation-wide craze thereupon followed. All around the coasts rock-pools were pounced on and stripped of their inhabitants. Shops specially catering for the aquarist sprang up. The wealthy had palatial tanks erected in their drawing-rooms. Marine menageries pulled in the crowds in city after city and town after town.

But like all exaggerated fashions, this one had too much energy invested in it to be capable of lasting very long. Unlike the ferneries, though, aquaria were not expressive of Victorianism:

Figure 23.4 The marine aquarium. Engraving by F. W. Keyl in George Kearley's *Links in the Chain; or Popular Chapters on the Curiosities of Animal Life* (London, 1862), facing p. 111.

they bore no burden of symbolism that dictated that they should vanish once that symbolism lost its force. Aquaria have consequently continued in use down to the present day more or less uninterruptedly, though a good deal less ubiquitously and with that one-time halo of novelty long since forgotten.

For the scientific world there was one notable outcome: the renowned Stazione Zoologica at Naples, a bold speculative venture embarked upon in 1870 and the scene of much important work at the cutting-edge of biology subsequently, was conceived on the assumption that it could be substantially financed from the fees charged for admission to a high-quality exhibition aquarium. Appropriately, the person engaged to build this was an Englishman, William Alford Lloyd, who owed his reputation and indeed his very career to the strong demand for the highly specialized form of construction work that the craze had engendered in Britain.

The fern craze, too, had its windfalls for science, albeit of a more modest character. They included the discovery of the value of spore characters for distinguishing between species, the proving of fern hybridity and the harnessing of nature-printing (a technique which the craze largely fostered) for the study of venation in fossil plants. Bud-propagation and apospory were similarly legacies to horticulture. All of these would doubtless have come to pass in time, but the intensity of focus that the craze induced caused such discoveries to be made much earlier than would otherwise have been the case. Whatever view one may hold of fashion as a process – and it has its admirers just as it has its detractors – there can be no denying that it does at least provide one good service to humanity in speeding up the adoption of useful practices and knowledge.

Further reading

Allen, D. E., *The Victorian Fern Craze* (London, 1969).

The Naturalist in Britain: A Social History (orig. London, 1976; Princeton, 1995).

'Natural history and visual taste: some parallel tendencies', in A. Ellenius (ed.), *The Natural Sciences and the Arts: Aspects of their Interaction from the Renaissance to the 20th Century* (Uppsala, 1985), pp. 32–45.

'The Victorian fern craze: Pteridomania revisited', in J. M. Ide, A. C. Jermy and A. M. Paul (eds.), *Fern Horticulture: Past, Present, and Future Perspectives* (Andover, 1992), pp. 9–19.

Brock, W. H., '*Glaucus*: Kingsley and the seaside naturalists', *Cahiers victoriens et edouardiens*, 3 (1976), pp. 25–36.

Dance, S. P., *A History of Shell Collecting* (Leiden, 1986).

Rehbock, P. F., 'The Victorian aquarium in ecological and social perspective', in M. Sears and D. Merriman (eds.), *Oceanography: The Past* (New York, 1980), pp. 522–39.

Taylor, J. E., *The Aquarium: Its Inhabitants, Structure, and Management* (London, 1876).

24 Nature for the people

Natural history seems to enjoy a unique status among the natural sciences. While growing in reputation as an academic discipline throughout the eighteenth century, it was still viewed as a science close to the public domain, open and egalitarian. It has even been argued that during the French Revolution natural history embodied the ideal of a democratic science. In contrast with Newtonian mechanics, held to be a form of an aristocratic knowledge, the natural science promoted under the influence of Denis Diderot (1713–84) and the Encyclopaedists, and by the writings of Jean-Jacques Rousseau (1712–78) as retailed by Jacques-Henri Bernardin de Saint-Pierre (1737–1814), enabled people to discover that everyone was on an equal footing in the 'kingdom of natural beings'. The egalitarian 'ideology' associated with natural history could account for the contrast between the successful foundation of the Muséum d'Histoire Naturelle in 1793 and the rebuffs suffered by the Académie des Sciences in the same period.[1] For instance, between 1794 and 1799 there were seventy-five leading articles on natural history in the two hundred issues of *La Décade philosophique*, a journal devoted to science, politics, arts, the theatre, and literature.[2]

After reaching a peak in the first half of the nineteenth century, the social status of natural history underwent a relative decline with the rise of laboratory biology in the second half of the century. During the same period the popularization of science developed as a large-scale enterprise, disseminating scientific information of all kinds to the public at large. Natural history was, and still is, one of the favourite topics of popular books and magazines, and was also widely diffused through lectures, exhibitions, and museums. The popularity of natural history can be gauged from the success of Gilbert White's bestseller *The Natural History of Selborne* (1789), which went through approximately 200 new editions and translations between 1789 and 1970,[3] and that of Jean-Henri Fabre's *Souvenirs entomologiques* (1879–1907), which are still in print, a century after the first edition. Studies of a sampling of popular scientific magazines indicate that, with the exception of medicine, more articles were about natural history than any other scientific discipline.

This longstanding commercial success seems to distinguish natural history as the most popular of all the sciences. Does this mean that it is more accessible to lay people than mechanics, physics, or chemistry? It has, in fact, been generally argued that the popularization of astronomy, physics, or chemistry requires great efforts to adapt their contents for a lay audience. Because it is assumed that any real understanding requires mathematical knowledge, the popularizers' task is usually described as a kind of translation process, which inevitably alters the truth. Since natural history did not rest on mathematical prerequisites, it would seem to have enjoyed the unique privilege of being directly understood, of being academic and popular at the same time.

In calling into question the privileged status which this idealistic view appears to confer on natural history, this chapter will try to give a more complex picture and a more realistic view of popularizing practices.

What exactly is meant by a 'popular science'? It would be naive to infer that natural history was spontaneously interesting and attractive to everybody from the fact that it could produce bestsellers or enjoy a wide circulation. This chapter will try to present the alleged 'popularity' of natural history as the result of a number of strategies for raising interest in the subject among various social groups. Rather than simply assuming that natural history was uniquely able to bridge the growing gap between expert knowledge and the non-expert population, we will show that these two categories gradually emerged through tensions and debates amongst naturalists. The question of the boundaries between these two categories will first be examined by confronting the differences between the technical and popular literature, and then by comparing the various practices of natural history. The final section will characterize popular natural history within a broader cultural context as an intellectual, moral, and aesthetic pursuit: in a word, as a way of life.

A bucolic science?

When speaking of natural history, most people have in mind lively, picturesque descriptions of the behaviour of birds and insects, or of beautiful flowers. Indeed, for most audiences these are more appealing subjects than arguments about electromagnetism or chemical reactions.

However, one must not forget that this bucolic view of natural history has largely been shaped by its popularizers. By concentrating on particular subjects – such as animal behaviour or floral display, in exotic or familiar surroundings – and by portraying the naturalist as an adventurer or an explorer of the treasures of nature, they have perpetuated belief in a traditional ideal of natural history as both pleasant and useful. With improved printing and woodcut technologies, illustrations became crucial to this strategy, as did documentary films in the twentieth century. A book like *Le Monde des fleurs* by Henri Lecoq (1802–71) combines scientific material with sentimental prose, illustrated by bucolic or exotic scenes (Figures 24.1 and 24.2).[4]

These mythical images flourished in the abundant literature of natural history for children. These works magnified the general tendencies of the literary genre of popularization, and provide a good 'observatory' for our purpose. Books for children made use of narrative devices such as family conversations on the hearthrug or instructive strolls in a pleasant countryside which offered many opportunities for teaching and moralizing. In order to make their works appear more attractive to younger readers, a number of authors used the traditional structure of fairy tales for their educative purposes: in such cases they did not hesitate to humanize and sentimentalize the animal and vegetable kingdoms.[5] Some of them even claimed that stories about the natural world could better satisfy the appetite of children for wonders than traditional fairy tales, without the risk of stimulating superstitions or magical thinking.[6] This view, however, rested on a use of the term 'history' in the phrase 'natural history' that was no longer generally accepted. In its post-Linnaean meaning, 'history' referred to arid descriptions of morphological features, usually adorned with long columns of Latin terms, rather than to pleasing narratives.

One striking point which deserves attention is that the notion of a 'general public' (however vague it could be) seems to have emerged before the status of professional naturalist had been firmly established all over Europe. In the early nineteenth century, a number of naturalists – especially in France where a cluster of institutions accelerated the process of professionalization[7] – daily confronted with the difficulties of nomenclature and taxonomy, became aware of a divorce or even a conflict of interests related to natural history. They shaped a clear distinction between science and what they demarcated as the interest of the general public. They thus forged the fiction of an undifferentiated 'general public', referring to all those who did not share their own approach to natural history.

In fact, an overview of late-eighteenth-century learned publications such as the *Philosophical Transactions* or the *Comptes-rendus de l'Académie des Sciences* shows that natural history articles were as much governed by strict literary conventions as articles in the experimental sciences.[8] Both attempted to maintain a sharp

Figure 24.1 'The verge of a beautiful forest is like the outskirts of a new world'. From Henri Lecoq, *Le Monde des fleurs* (Paris, 1870), p. 268.

distinction between reporting and interpreting, both claimed to give exhaustive accounts of their subjects. For instance, description of species was considered as *reporting*, while critical revision of a genus could be regarded as *interpreting*. As for exhaustiveness, the aim of any survey of flora and fauna was to include all the species in a given area.

Naturalists such as Georges Cuvier (1769–1832) were so involved in their specialized research field that they viewed public interest as a possible danger to the advancement of science. Cuvier

Figure 24.2 'Brazilian landscape of the Sierra dos Orgoas with *Araucaria Brasiliensis*'. From Henri Lecoq, *Le Monde des fleurs* (Paris, 1870), p. 412.

complained that René-Antoine Ferchault de Réaumur and other eighteenth-century naturalists were insufficiently precise in their descriptions because they cared about the public interest. He praised his colleagues for having the courage to give minute descriptions of species in their writings. In other words, Cuvier claimed the right for naturalists to bore their readers: naturalists had to be allowed the tedium of minute descriptions of species in the interests of utility.[9]

Thus Cuvier legitimated the academic status of his own research

Figure 24.3 The botany of
the old gardener. From
Emile Desbeaux, *Le Jardin
de Mademoiselle Jeanne:
Botanique du vieux jardinier*
(Paris, 1887), frontispiece.

by contrasting the public interest and the interest of science. Ironically, in doing so he encouraged the emergence of popular scientific writings. Popularizers were prompt to use such rationales to
legitimate their own enterprise, as distinct from more learned writings. The editor of *Dictionnaire pittoresque d'histoire naturelle* (1834–
1840), who pretended to introduce 'the masses to the beautiful discoveries that savants have made in the realm of natural sciences',
equally emphasized the distance between 'history', which could be
found in technical literature such as dictionaries intended for

advanced readers, and 'stories' which were intended for the pleasure of the public.

[These dictionaries] are full of the most rarified scientific words and definitions, and for this reason are out of reach of gentlemen and students who do not wish to know, for example, that such and such an author has divided the genus fly, which everybody knows, into 2,000 groups, according to one hair more or less in the jaws, but who would learn with pleasure how flies breed, what their habits are, what tricks they use to escape their foes. A dictionary specially intended to present natural history from this point of view would be very popular.[10]

Two principal characteristics of popularization emerge from this programme: picturesque narratives and a concern for utility. It might be expected that features of this kind would typify writings for the general public, as ways of making the classifications and technical descriptions of natural history appear more attractive and accessible. In fact, a reliance on stories and utility – and on literary devices such as artificial dialogues, conversations and use of common language – is present in relatively few popular writings. Moreover, some of them can also be found in 'scientific' works. Charles Darwin's *Journal of Researches* (1839), based on his voyage on the *Beagle*, is the best-known example, but is only one of many books relevant to science which could also be read as narratives by a literate public.

Popular writings often pretended to be more concise and more synthetic than scientific treatises on a given subject. For instance, Louis Figuier's *Tableaux de la nature* (1862–71) were intended to provide a glance over natural history by recording the most important and most picturesque 'scenes from the organic world'.[11] However, the project resulted in ten heavy volumes, ranging from mineralogy (two volumes) to the vegetable and animal kingdoms, including one volume on micro-organisms. Figuier (1819–94), like many other popular writers, described at length various different kinds of fossils and plants. In this case, the 'glance' was mainly a commercial device to entice the public.

Turning to more theoretical commitments, the difference between popular and technical literature is more subtle than usually believed. If in 1863 Figuier refrained from mentioning the controversy about the age of the earth raised by Charles Lyell and other geologists, by contrast, many other popularizers, such as Victor Meusnier (1817–1903) were quick to commit themselves in scientific controversies. Far from simply reflecting specialist opinions, popular writings sometimes challenged the authorities. The hostile reactions to Robert Chambers's *Vestiges of the Natural History of Creation* (1844) arose from a perception of the effectiveness of this strategy.[12] Some academic naturalists involved in popularization, like Paolo Mantegazza (1831–1910) – who was at the same time a champion of Darwin in Italy and a prolific popular writer – used popularization as a sounding-board for their theories.[13]

Nor were minute technical details a distinctive feature of the technical literature. They were equally common in popular writings, while professional scientists were also conscious that they had to 'wrap up technical details in a pleasant cover'. For instance, the French zoologist Armand de Quatrefages (1810–92) wrote:

> In general, I have discarded overly technical details and dealt almost exclusively with general issues. I often tried to imitate the physician who uses honey to disguise the medicine whose taste would otherwise disgust the patient, hence the descriptive or historical details which accompany nearly all the chapters of this work.[14]

Indeed, technical details did not play the same function in the textual structure of scientific and popular works: whereas in scientific writings they answered a requirement for exhaustiveness, in popular writings they were more often intended to enhance the picturesque effect. Nevertheless, except by relying upon declared and undeclared intentions, it is not easy to make a clear-cut distinction between these two genres in natural history. Both scientific and popular works used specific rhetorical devices according to the particular audiences they were aimed at.

Are we going to find a better criterion by considering the nomenclature employed? Each animal or vegetable species has a scientific Latin name: for example, *Taraxacum officinale* Weber (Friedrich Weber, 1781–1823) or *Grus grus* L. (L. for Carl Linnaeus), and some species – especially flowering plants, mushrooms, and vertebrates – also have several names in the languages of the countries where they live.[15] It might be expected that Latin names would be used in the technical literature and common names in the popular literature. Indeed, some popular books used only the common names.[16] For instance, among the renowned series of floras directed by Gaston Bonnier (1853–1922), one little volume issued in 1889 gave only the French names.[17] Some academic writers used only scientific names. Alphonse de Candolle (1806–93), in the foreword of his *Géographie botanique raisonnée* (1855), commended the modern tendency to substitute one international name coined according to definite rules for a crowd of common names.[18] He argued that scientific names look strange to the people because they were unfamiliar, not because they were Latin. Thus Latin terms like *Geranium, Hortensia, Rhododendron,* or *Fuchsia* were in common use, while if you spoke to an English worker about the plant named Nipplewort in English, he was as likely to prefer the botanical name *Lampsana*, or 'he will laugh at

Figure 24.4 Crane, *Grus cinerea*. From E. Le Maout, *Histoire naturelle des oiseaux* (Paris, 1855), p. 349.

both of them because he has never heard either of them before'.[19] De Candolle concluded by urging botanists to promote, and the lay public to accept, the use of Latin names.[20]

Most works, however, gave both the common and scientific names. This was the case in White's *Natural History of Selborne* (1789), as well as in *Nature Notes for 1906*, the country diary of Edith Holden (1871–1920), and many other books from the nineteenth century to the present.[21] Even books devoted to birds – a group where the number of species is relatively small and the common names unambiguous – give the Latin names.

The difficulty in tracing a definite borderline in this field can be partly related to the fact that despite the clear-cut distinction made by practitioners between the 'active' naturalists and the 'passive' general public, there were at least three distinct audiences for natural history. Between the full-time naturalists – a few paid professionals with positions in universities and museums or gentlemanly specialists devoting their lives and fortune to science – and the more or less literate lay public, there were also occasional practioners who did not content themselves with reading but *practised* natural history by collecting specimens of plants, insects, or sea-shells. And presumably, if archival materials made it possible to establish a detailed picture of those who bought and read the volumes mentioned, historians would find a number of other intermediate categories.

One category in particular can be identified. Women were active cultivators of natural history, especially botany. They attended public lectures and their presence was sometimes mentioned as an indicator of the popular success of the lecture.[22] Women became a favourite target of a whole range of books which were not basically different from other popular textbooks, except in claiming to avoid overly explicit statements about sex.[23] Women like Jane Marcet (1769–1858), who had family links through husband or father with scientific circles, wrote a number of popular books, though some chose to remain anonymous.[24] However, the impact of women in this field would be underestimated if one only considers publications. Beyond their contribution on the public stage women played a significant – though invisible – role inside the domestic sphere. Using more or less formal means, mothers daily taught a basic knowledge of botany to their children. These familial practices which counterbalanced the relative absence of women from learned societies and public institutions have been well documented for the British case.[25] For the French case, there are indications of a similar involvement by women. For instance, Jean-Jacques Rousseau wrote his *Lettres sur la botanique* (1771–73) to Madeleine Catherine Delessert in order to help her teaching her daughter botany.[26]

A science open to lay practices?

Unlike the laboratories of physics or chemistry, the natural world appears to belong to everybody. At least in principle, anyone with normal capacities is supposed to be able to contribute to the advancement of natural history. Moreover, since the field of investigation extends all over the world, professional naturalists have long been aware that they need the help of well-disposed volunteers to collect rocks and fossils, report on birds' migrations, and similar phenomena. Despite – or perhaps because of – this need, co-operation between these groups generated increasing tensions. These are epitomized by the semantic switch – at least in the English language – in the term 'amateur' toward the end of the

Figure 24.5 'Look at this nest in the ivy, Mamma says'. From Tante Jane, *Les Vacances de Suzanne* (Paris, 1896), p. 12.

nineteenth century. The old positive meaning of 'connoisseur' has been gradually overthrown by the pejorative sense of 'dilettante' emphasizing a lack of seriousness and reliability.

As early as 1828, part-time cultivators of botany were sometimes perceived as obstacles to the advancement of natural history by professionals. Probably impressed by the rapid triumph of the new chemistry shaped by Antoine Lavoisier all over Europe, Augustin-Pyramus de Candolle complained that a revolution in the methods of botany could not take place along the lines of the chemical revolution, because over a long period, botany had been in the hands of those who saw only its practical applications, or who practised it only for their own pleasure.[27]

Nevertheless, harmonious co-operation between occasional practitioners and professional naturalists was possible through a clear division of labour, with the volunteers wandering over hills and mountains in order to provide the 'professionals' with raw materials for taxonomic skills and interpretative minds to work upon, as they sat in their academic chairs and arranged natural history collections in museums. Toward the end of the nineteenth century, with the development of laboratory biology, collecting lost some of its prestige, and the contrast between 'amateur' and 'professional' practices was reinforced. As long as specialized naturalists were engaged in descriptive studies, they continued to need the collaboration of field naturalists. The complementarity of the two groups has often been proudly emphasized by the non-professionals: thus in his presidential address to the Société Linnénne de Provence in 1909, the barrister and occasional entomologist Elzear Abeille de Perrin (1843–1910), claimed that 'in descriptive botany and descriptive zoology, it is the amateurs who have done most to advance the science, leaving to academics better equipped for the task of the laboratory studies which they carry out so well'.[28]

This division of labour was embodied in the methods for determining species. When Linnaeus's system, which offered an easy and accessible way for beginners to find the names of plants, began to be given up in the early nineteenth century, botanists showed considerable ingenuity in devising ways of making floras useful for different levels of expertise.[29] For instance, the third edition of Jean Baptiste Lamarck's *Flore française* revised by A.-P. de Candolle (1805), consisted of a first volume intended for amateurs and beginners, which provided a convenient key for identifying plants, invented by Lamarck (1744–1829) for the first edition (1778); the following three volumes offered a lengthy description of species in their natural order, intended for learned botanists. This device proved to be a success. The third edition of the *Flore française* not only enjoyed scientific fame but also had a widespread circulation; according to de Candolle 5,000 copies were printed and sold in twenty years.[30] Since then, all floras and some other field guides for birds or insects have been designed on the same pattern.

The difficulty was not to get the cultivators of natural history to work, since they volunteered and worked eagerly. Nor was it to gather reports, information, and collections from them, since many local natural history societies included both volunteers of various degrees of training and a few paid naturalists, and were able to provide such materials. The main problem was that the cultivators of natural history formed an undisciplined crowd which the professionals would have liked to keep under their control. In this respect, the botanic garden as viewed by de Candolle in his *Dictionnaire des sciences naturelles* (1822), was an ideal space within which the general public, non-professional, and professional naturalists could associate freely.[31] The botanic garden was designed as a setting for experiments in plant cultivation, as a 'living book' which made the elements of botanical knowledge available to anyone, and as a public space for promenading.

Natural history as a way of life

One can presume that cultivators of natural history were motivated not only for the benefit of professional science but for many other reasons. It has often been noticed that the development of public lectures, books, and magazines of popular science in the nineteenth century served purposes of social and moral control. But were there any specific social norms or religious beliefs attached to popular natural history? Were they explicit or implicit messages?

In his introductory lecture to a course of geology delivered at the University of Oxford in 1852, Hugh Edwin Strickland (1811–53) – best known for being killed by a train while geologizing in a railway cutting – emphasized the social and moral advantages of the practice of geology for clergymen:

The clergy who form the larger portion of the students annually reared and sent forth from this place, will derive much benefit, and no detriment, by an acquaintance with the principles of geological science. Natural theology is an appropriate adjunct to Revealed Religion, and a study of the works of the deity in antecedent ages of the earth, cannot be inconsistent with the devout contemplation of the dealings of God with man. It is also no small advantage to a studious or laborious ecclesiastic to have an inducement for taking exersize in the open air, and refreshing the mental faculties by the contact with external nature. Many of our clergymen too, are destined to pass their lives in remote parishes, far from educated neighbours, and where their parochial duties are too light to occupy their whole time. The monotony of such a position will often, in spite of the best principles, react upon the nerves, and render a person listless, if not discontented. To a mind thus diseased, I can prescribe no better remedy than this: Make a geological map of your parish. Form a collection of all its animal, vegetable, and mineral productions. When this is done, extend your researches to the neighbouring parishes, or to the whole county.[32]

In the early nineteenth century popular natural history fostered the Romantic sense of a secret harmony between human states of mind and natural landscapes. Humboldt clearly intended to develop a Romantic aesthetics through his famous *Ansichten der Natur* ('Views of Nature') (1808), which was simultaneously published in French and German, and later translated into English.[33]

A few years later it became apparent that this emphasis on the aesthetic dimension of natural history was part of a longstanding struggle against the influence of *Naturphilosophie*. In a letter to François Arago in 1827, Humboldt mentioned, as one of several reasons for staying in Prussia, the influence he expected to exert on young German readers who had not been shown 'how without deviating from physical truths, one can still address the imagination and the mind'. Aesthetics, in short, was not solely the property of *Naturphilosophie*, but could be used in the service of empirical studies.[34]

As well as being perceived as a source of aesthetic pleasure, through bringing the soul into a closer rapport with nature, natural history was said to stimulate the growth of reason. While mathematics and physics helped to develop an analytical spirit among young people, natural history was indispensable for developing more qualitative forms of reasoning. Advocating the popularization of zoology, in 1854 Armand de Quatrefages emphasized the benefit to the mind of the practice of observing, classifying, and 'combining great amounts of facts and the ideas based on them, in order to grasp their true relations and more distant consequences'.[35]

Another concern among popularizers of natural history was religion. Natural history, much more easily than sciences like chemistry, gave rise to a host of apologetic arguments. Palaeontology, which dealt with the origin of species and bordered on the realm of myth and scriptural creation stories, obviously touched a religious chord. Extinct species of reptiles, as well as prehistoric tribes, were favourite topics for nineteenth-century popularizers. Whatever the religion of the public, the order of nature suggested the idea of a deity. Even after the emergence of stratigraphy, which called the biblical chronology for the age of the earth into question, many popular writers attempted to reconcile religious with scientific findings. Figuier, for instance, claimed that there was complete harmony between geology and religion: 'Nothing is better suited than the study of geology to display the eternity and unity of the deity. It shows us, so to speak, the creative power of God at work.'[36]

In many cases, assertions of this kind were no more than rhetorical devices in the foreword of a volume, designed to legitimize popular science as 'good reading', a guarantee of social and moral order. Sometimes, however, the apologetic purpose went further. Between 1844 and 1873, a French cleric, the Abbé Migne (1800–76) edited a monumental *Encyclopédie théologique*, intended for the education of parish priests. Among the 171 volumes, the *Dictionnaire de botanique*, by Jéhan de Saint-Clavien (1803–71), happened to be among the best suited for apologetics. By means of extensive use of citations and journalistic plagiarism, Jéhan managed to give a relatively good survey of the botanical knowledge of his time, but without making any effort to stimulate lay practice. From the thousand entries in this volume the central idea that emerges is one of a common ground between science and religion. Jéhan did not content himself with grasping every opportunity to celebrate the wisdom of the Creator as seen in the admirable economy of nature. When mentioning popular beliefs about water

Figure 24.6 'Ideal view of the earth in the Lias period'. From Louis Figuier, *La Terre avant le déluge* (Paris, 1864), p. 176.

diviners, in the entry 'Hazel-tree', he argued that science needed the help of religion to save itself from seductive superstitions.[37]

If the popularization of natural history was so efficiently conscripted in the service of Christianity, was it equally capable of serving other religions? Camille Flammarion's prolific works show that natural history could also nourish a more idiosyncratic religion. Whereas popular astronomy only helped Flammarion to re-evaluate the place of man within the immensity of the universe, natural history explicitly allowed him to advocate a religious sense of the immanent presence of God within the order of nature.[38] The religious messages derived from nature were so flexible that Ernst Haeckel (1834–1919) could even enrol natural history in his fight against established religion. Popular writings sometimes supported extreme anti-religious attitudes. In France, for instance, Clémence Royer (1830–1902), the translator of the third edition of *Origin of Species*, reproached Darwin for his reticence on religious issues. The popularizer Victor Meusnier (1817–1903), a hardline socialist, denounced the religious prejudices underlying the hostility of the French scientific establishment to Darwinism and spontaneous generation.

An ironical response to the quest for religious or anti-religious meanings in nature was later given by J. B. S. Haldane (1892–1964). When theologians asked what conclusions could be drawn about the Creator from a study of His creation, Haldane is said to have answered, 'an inordinate fondness for beetles'.[39]

Even if popular natural history did not really threaten or strengthen religious creeds, it was supposed to provide the basis for an ethics or, more broadly, a way of life. It encouraged outdoor activities as well as reasoning. It was also a source of aesthetic pleasures, as expressed by Thomas Henry Huxley (1825–95): 'To a person uninstructed in natural history, his country or sea-side stroll is a walk through a gallery filled with fine art works, nine-tenths of which have their faces turned to the wall'.[40] Despite their different aims, nineteenth-century popularizers would have agreed.

How are we to assess the legacy of the popular tradition in natural history? From a commercial standpoint, natural history undoubtedly made an important contribution to the rise of the popular press in the nineteenth century and to the emergence of publishing empires such as Flammarion, Hetzel, Hachette, Cassell, Knight, and Chambers.[41] From the social point of view, it can be assumed that the popular clichés of rural life and exotic landscapes helped to create a consensus on potentially controversial issues, such as rural depopulation and colonization.

From a scientific standpoint, it is clear that ecology inherited many of the characteristics of the nineteenth-century tradition of natural history. However, in the early twentieth century ecologists developed ambivalent feelings towards naturalists. While acknowledging the common origin of their fields, they stressed the contrast between natural history and ecology. In this disciplinary strategy, the 'popular' image of natural history played a key part. For instance, Charles Elton (1900–91) wrote that ecology 'simply means a scientific natural history'.[42] Later, in an essay on 'Natural history', the American ecologist Aldo Leopold (1887–1948) described the emergence of laboratory biology as the result of competition with a declining naturalistic programme:

Laboratory biology came into existence at about the time when amateur natural history was of the dickey-bird variety and when professional natural history consisted of labelling species and amassing facts about food-habits without interpreting them. In short a growing and vital laboratory technique was at that time placed in competition with a stagnated outdoor technique. It was quite natural that laboratory biology soon came to be regarded as the superior form of science.[43]

However, Leopold went on to say that this gloomy picture was no longer true. 'In the interim, field studies have developed techniques and ideas quite as scientific as those of the laboratory.' The amateur has new opportunities to contribute to science 'if he has imagination and persistence' and Leopold stressed the civic importance for the ordinary men and women of an understanding of the living world. In spite of the conspicuous changes which natural history has undergone, there is a remarkable continuity in the rationales given for its popularization, among which the simple desire to open the eyes of the public to the wonders of the living world remains foremost.

In conclusion, the popular natural history tradition can be seen as a major aspect which contributed to shaping the discipline. Though natural history was never 'popular in itself', it remained open to various degrees of non-specialized languages and to various forms of practice. A clear-cut distinction between science and the public could not be strictly defined around a territory, like the space of the laboratory in sciences such as physics or chemistry. However, through tensions and debates between various writers and practitioners, the growth of popular natural history both created an image of the public and, at the same time, stabilized the identity of the 'natural scientist'.

Further reading

Ainley, Mariane Gostonyi, 'The contribution of the amateur to North-American ornithology: a historical perspective', *The Living Bird*, 18 (1980), pp. 161–77.

Allen, David Elliston, *The Naturalist in Britain: A Social History* (London, 1976; Princeton, 1995).

The Botanists. A History of the Botanical Society of the British Isles through a Hundred and Fifty Years (Winchester, 1986).

Drouin, Jean-Marc, 'Comprendre et dominer le monde végétal. La vulgarisation de la géographie botanique au XIXe siècle', in Pascal Acot (ed.), *La Maîtrise du milieu* (Paris, 1994), pp. 39–57.

Duris, Pascal, *Linné et la France 1780–1850* (Genève, 1993).

Govoni, Paola, 'Divulgazione scientifica: un genere marginale?', *Intersezioni*, 11(3) (1991), pp. 553–64.

Kohlstedt, Sally Gregory, 'The nineteenth-century amateur tradition: the case of the Boston Society of Natural History', *Boston Studies in the Philosophy of Science*, 33 (1976), pp. 173–90.

Lowe, P. D., 'Amateurs and professionals: the institutional emergence of British plant ecology', *Journal of the Society of Bibliography of Natural History*, 7(4) (1976), pp. 517–34.

Lucas, A. M., 'Scientific literacy and informal learning', *Studies in Science Education*, 19 (1983), pp. 1–36.

Mornet, Daniel, *Les Sciences de la nature en France au XVIIIe siècle* (Paris, 1911).

Ritvo, Harriet, 'Learning from animals: natural history for children in the eighteenth and nineteenth centuries', *Children's Literature*, vol. XIII (1985), pp. 72–93.

Secord, Anne, 'Science in the pub: artisan botanists in early nineteenth-century Lancashire', *History of Science*, 32 (1979), pp. 269–315.

Sheets-Pyenson, Susan, 'Popular science periodicals in Paris and London: the emergence of a low scientific culture, 1820–1875', *Annals of Science*, 42 (1985), pp. 549–72.

Shteir, Ann B., 'Linnaeus's daughters: women and British botany', in Barbara J. Harris and JoAnn K. MacNamara (eds.), *Women and the Structure of Society* (Durham, NC, 1984), pp. 67–73.

   'Botany in the breakfast room: women and early nineteenth-century British plant study', in Pnina G. Abir-Am and Dorinda Outram (eds.), *Uneasy Careers and Intimate Lives: Women in Science, 1789–1979* (London, 1987), pp. 31–44.

25 Natural history and the 'new' biology

Historians of biology have made much of the great transformation that took place in biology in the decades around the turn of the twentieth century. In that period a new sort of biology appeared, in which emerging sciences such as genetics and experimental embryology opened up new vistas for understanding living nature. The contrast between these sciences and 'traditional' natural history appears stark. Natural historians sought to uncover the large-scale pattern of living nature, through collecting in the field and classifying in the museum; 'modern' biologists in their laboratories sought to penetrate the internal workings of the living organism to discover their fundamental causes. Natural history was accessible to amateurs and amenable to popularization; the new biology depended on highly technical microscopic analysis and experimentation and was open only to professional scientists.

These contrasts between nineteenth-century natural history and the new biology of the turn of the twentieth century have often been conceptualized in terms of a *progression* from the former to the latter, involving an intellectual ascent from mere collecting and description to a higher sort of science involving experiments and explanations. The standard story also implies that, by the early twentieth century, natural history had been replaced by modern experimental biology, or at least that natural history survived only as a vestige of its former self, conducted by ageing amateurs and lesser scientists who were incapable of keeping up with the times. Indeed, the very contrast between natural history and 'modern' biology suggests that natural history was inherently old-fashioned and destined to be superseded.

There are reasons to be suspicious of this story, for it has clearly been written from the perspective of the 'new' biologists and their chroniclers. For those interested primarily in the rise of the modern approaches to biology, the story of natural history after the mid-nineteenth century might seem at best backward-looking, and more probably irrelevant to understanding the new biology. As a result, the fate of natural history in the late nineteenth century remains poorly understood. Was natural history somehow transformed into the newer science? Or was the former simply supplanted by the latter? Or did natural history find itself displaced from its former centres of activity and remade in new ones?

To answer these questions it helps to start from the vantage point of natural history research itself. This chapter focuses on the history of one broad area of research in the late nineteenth century that had traditionally been considered central to natural history: what we might call 'life-history studies' of animals, which undertook to understand all aspects of individual species, including their life-cycles, distribution, habits and behaviour, and connections to the past (see Figure 25.1). The essay first traces the shifting alignments of this problem cluster with two other parts of zoology, namely systematics, which studied relationships among species and their organization into the larger systematic order of nature (and which, of course, was also considered a main pillar of natural history), and morphology, the study of anatomical form and its development. We shall see that as zoologists developed new intellectual research programmes, they recast these relationships more than once. I shall then consider the sharp distinction often made between experimental zoology and natural history; here I will suggest that there was more continuity between life-history studies and this new approach than meets the eye. Finally, a look at the changing institutional locations available for all of these orientations will help us understand more literally 'where natural history went' by the early twentieth century. Although my story focuses primarily on Germany, we may take it as a starting point for a broader rethinking of the fate of natural history in the late nineteenth century.[1]

Life-history studies in and out of natural history

In 1835, when the naturalist Arend Friedrich August Wiegmann (1802–41) began publishing the new journal *Archiv für Naturgeschichte* ('Archives of Natural History'), natural history was generally confined to plants, animals and fossil remains, the rest of mineralogy having split off largely into its own realm. At this point, life histories were clearly part of natural history, along with classification, the other main branch. In addition to studies of plants and fossil organisms, Wiegmann noted in the unpaginated 'Prospectus' to the first issue:

Essays in descriptive zoology, descriptions of new genera and species, reports on the mental abilities [*Seelenfähigkeiten*], modes of life and geographical distribution of known animal species, even anatomical [*zootomische*] communications, insofar as these justify or secure the systematic position of an animal or an entire group, will find a suitable place here.[2]

Figure 25.1 'Life history of a simplest organism', *Protomyxa aurantiaca*. Plate I from Ernst Haeckel, *History of Creation* (New York, 1887), showing Haeckel's understanding of the complex life-cycle of this single-celled organism. The top three figures show internal differentiation; spores then emerge from the original cell (4–5, lower left) to swim freely in the ocean. Landing on hard surfaces, they pull in their tails and gradually change their forms while staying in contact with these surfaces (6–8), absorbing nutrients from around them (9–10). They then develop into their mature forms either through simple growth (13) or amalgamation (14). The mature form extends tiny strands when hungry (11), which act as a net for catching other tiny organisms that serve as food (12). When sated, the Protomyxa reabsorbs its tiny tentacles in preparation for the renewal of its cycle (15, 16, 1). Although Haeckel conducted life-history research early in his career, his later work generally neglected the organism's interaction with its environment.

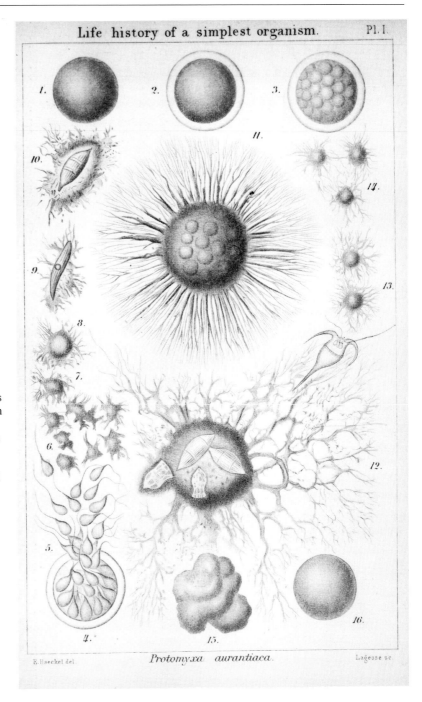

Life history of a simplest organism.   Pl. I.

Protomyxa aurantiaca.

Not much over a decade later, however, the premisses of Wiegmann's journal were challenged by a younger group of researchers who claimed a more scientific approach to their subject. The men who founded the *Zeitschrift für wissenschaftliche Zoologie* ('Journal for Scientific Zoology') in 1848 directly challenged the scientific character of natural history, as is clear from their prospectus:

We desire to give our journal the most scientific character possible . . . To this purpose we exclude all announcements of new genera and species that do not relate to this task, unless these offer us a more thorough-going insight into plant and animal structure [*Bau*], into the life-history of animals and plants, or in the lawful organization of the organic realms. For the same reason we will exclude any kind of simple notes and natural history news . . . On the other hand, from the truly scientific side of botany and zoology nothing will be excluded.

The editors also noted that they would exclude all applied topics and concentrate on 'those parts of the science that are currently in need of cultivation' – comparative anatomy, histology, and embryology (collectively known as morphology), and physiology.[3]

Clearly the main difference between the *Zeitschrift* and the *Archiv* lay in the newer journal's emphasis on morphology and physiology. But in distinguishing their own approach from that of natural history, the editors also *appropriated* certain aspects of natural history as part of 'scientific' zoology, while excluding other aspects, which by implication were *not* scientific. Thus life histories were to be included as scientific, while applied topics and purely taxonomic contributions were not. This is significant, because life histories comprised a fairly broad portion of the overall territory of natural history. When researchers choosing to separate themselves from natural history identified the part (taxonomy) with the whole (natural history), they made invisible much of the intellectual work that gave the taxonomic work its life.

In the German-speaking states in the 1850s and 1860s, 'scientific' zoologists and systematists viewed each other across a great divide. Systematists feared the younger generation would neglect the collections in favor of morphological studies, and the 'scientific' zoologists were busy trying to prove to themselves, their colleagues, university administrators, and state officials that their new combination of life history studies and morphology was indeed worthy of support. In the Prussian universities the systematists won out, while in a number of other German states, most notably Bavaria and Baden, the scientific zoologists tended to be appointed to university chairs.[4]

If life-history studies were severed from natural history and appropriated by 'scientific zoologists' in the 1850s and 1860s, however, this arrangement – and the attendant eclipse of systematics – was only temporary. Charles Darwin's *Origin of Species* appeared in 1859 and was soon translated into German. In the wake of Darwin's theory, many zoologists during the 1860s and 1870s once again revised the relationships and boundaries between life-history studies, morphology and systematics.

The leading architect of this rearrangement was the zoologist Ernst Haeckel (1834–1919). Educated in the tradition of scientific zoology, Haeckel seized upon Darwin's theory as a way of reuniting morphology with systematics in a new, deeper way. With his famous phrase 'ontogeny recapitulates phylogeny', also called the 'biogenetic law', Haeckel claimed that one could trace the ancestral history (phylogeny) of a species by looking at the embryological development (ontogeny) of present-day individuals. Early stages of development, according to Haeckel, corresponded to the most ancient ancestral forms, and the developing embryo passed through the stages of its own evolutionary history. Drawing in part on this method, 'evolutionary morphology', which sought to establish systematic relationships among present organisms by tracing them back to common ancestors through a combination of comparative anatomy, embryology and palaeontology, rapidly came into vogue in Germany and England in the 1870s. In the work of Haeckel and many of his followers, classification and morphology became joined into a common enterprise, independent of life-history studies. As younger scientists took up Haeckel's programme in the 1870s and 1880s, they turned increasingly to the close study of embryonic stages, using new microscopic techniques of staining, fixing, and slicing up embryological material to trace and compare the development of embryological structures in the service of phylogenetic speculation (Figure 25.2).

Haeckel's approach has often been viewed as dominating late nineteenth-century biological research, especially in Germany. That he attracted numerous followers there can be no doubt, but others were troubled by the speculative character of his work and by the dogmatic fervour with which he preached his evolutionary doctrines to the general public. Moreover, Haeckel's biogenetic law appeared to neglect an aspect of biology that held great importance in the work of both the scientific zoologists and Darwin himself, namely, adaptation. As a number of his colleagues pointed out, Haeckel's programme of evolutionary morphology did not treat very seriously the organism as a living being in contact with its environment – an area of considerable significance to those interested in understanding animals' life histories.

Thus despite Haeckel's vogue, many German zoologists continued to pursue life-history topics such as the effects of the conditions of existence on animals' habits, forms and geographical distribution, and the nature of generation. However, with the appearance of 'evolutionary morphology' on the intellectual landscape, the programme of 'scientific' zoology and its relationship to systematics changed. Whereas in the 1850s 'scientific' zoologists united life-history studies and morphology in contrast to systematics, in the 1870s and 1880s life-history studies and morphology

Figure 25.2 Frontispiece to Ernst Haeckel's *History of Creation* (New York, 1887). Haeckel considered this sponge the primitive form of all sponges, and believed that the developmental stages enumerated here corresponded to the transitional organisms that evolved from a single-celled creature into the sponge. Note the use of cross-sections to show the internal development of cells, a technique increasingly used from the mid-nineteenth century on that was much aided by new stains and fixatives.

themselves came to be perceived as somewhat distinct strands of scientific zoology. Both branches claimed a relationship to systematics, but they drew their systematic conclusions from quite different methods: evolutionary morphologists undertook anatomical and developmental studies based primarily on preserved and fixed material, developing species distinctions purely from comparing adult and embryonic forms, while life-history researchers were more apt to make an effort to observe the living creature in its natural surroundings, and to base explanations of species relationships on their adaptation to different conditions of existence. The difference was expressed with some exasperation by Carl Theodor von Siebold (1804–85), the oldest of the scientific zoologists and a long-time life history researcher, in a letter to a colleague in 1879. Although the new evolutionary morphologists were right to turn up their noses at the older systematists' endless and boring construction of bad species, Siebold wrote, in focusing so closely on their prepared sections they neglected the living animal itself. 'But where is the observation of the way of life of these animals, why does one learn so little of the activities of those very animals whose [anatomical] organization is known with the utmost precision?'[5]

In 1887 the emerging fault-line was recognized more formally in the division of a new periodical, the *Zoologische Jahrbücher* ('Zoological Yearbooks'), into two sections, one devoted to 'animal anatomy and ontogeny,' the other to 'animal systematics, geography, and biology'. ('Biology' here is used in the typical late nineteenth-century German meaning of 'the study of the organism in relation to its physical environment', rather than the science of life as a whole). The former category represents pure morphology, while the latter represents the continuation of natural history, now re-emerging from its submersion into 'scientific zoology'. In the 1880s and 1890s both halves of the journal flourished, suggesting that, far from dying out, life-history studies actually increased in visibility. As we will see later on, this impression is borne out in the expansion of institutional homes for such studies in the same period.

Experiments in natural history and biology

But what did zoologists conducting 'life-history studies' actually do? Once again, standard histories of biology do not help us out much – they do not usually recognize this research at all. Instead, they focus on the challenges posed to *both* morphology and natural history by a new group of researchers, who relegated the two to the same category of 'merely descriptive', non-explanatory science. Although they often politely acknowledged the important empirical base provided by the descriptive and comparative methods, advocates of experimental studies emphasized that solid causal explanations of living nature would only come from the higher, more scientific method of experiment. As Wilhelm Roux, the founder and persistent champion of *Entwicklungsmechanik* (developmental mechanics), said in a speech introducing his programme in 1889, even if descriptive and comparative approaches to understanding animal form were completely perfected, the human drive for 'causal understanding' would remain unsatisfied; the only route to satisfying that thirst was through analytical experiments.[6]

The implication, so widely assumed that it was rarely stated directly, was that natural history was by its very nature descriptive. But just as the scientific zoologists in the late 1840s wrote life histories out of natural history and into 'scientific zoology', so too did polemical experimentalists in Germany and America around the turn of the century represent natural history as something less than it was. In claiming for themselves the unique ownership of experiment, these polemicists airbrushed experiments out of the history of natural history.

Experiments in natural history? The very phrase sounds like a contradiction in terms. After all, natural history, we have learned for over a century now, is a matter of *description* and classification. As Thomas Henry Huxley (1825–95) put it in his 1876 lecture 'On the study of biology', already by the eighteenth century the term 'natural history' had come to be applied to 'those phenomena which were not, at that time, susceptible of mathematical or experimental treatment'.[7] But my point is just this: we must be sceptical of histories written by men such as Huxley, who himself was involved in reforming natural history into a new kind of biology.[8] What happens if, instead, we look at the actual work of people engaged in life-history studies? If we do this, it appears that natural historians saw nothing contradictory in conducting experiments to answer questions that interested them.

To depart from Germany for a moment, we may take as a case in point the nineteenth century's most famous naturalist, Charles Darwin (1809–82). In the *Origin of Species* Darwin described immersing the seeds of various plants in diverse stages of desiccation in sea-water for different periods of time to see whether they would float and later germinate. At another point he described setting up an aquarium with just-hatching molluscs, and then immersing a duck's feet in the water for a period of time. After ascertaining that many of the molluscs attached themselves to the duck's feet, he removed the duck from the water and measured how long they survived out of water.[9] In both cases the point was to determine the ways that organisms could become dispersed well beyond their immediate neighbourhood, a point of crucial importance for understanding biogeographical distribution. Now, these experiments do not require elaborate equipment, and they are not described in a formal protocol, but none the less they are experiments, involving manipulation, hypothesis-testing and controlled circumstances – activities we would recognize as experimental, and that naturalists at the time recognized as such too.

Nor were Darwin's questions about dispersion the only problems in life-history studies amenable to experimental investigation. Parasitic worms, which typically went through different developmental stages while living inside the bodies of their hosts, offered a particular challenge to establishing their life-cycles and classification, for it was often difficult for the naturalist to know whether he was faced with two different parasites or the same parasite at different stages. To solve this problem, naturalists seeking to follow the life history of parasitic worms developed a method of 'feeding experiments', in which they fed different stages of the worms to host animals (typically ducks, dogs, pigs or calves) and then slaughtered the hosts after different amounts of time to see where they now lived in the host and what they looked like.[10]

Naturalists conducted experiments to study various aspects of the life histories of other animals as well. In 1870 Carl Theodor von Siebold published experimental research establishing the existence in paper wasps of true parthenogenesis – the production of offspring by a female with no contribution from the male; his experiments involved controlling the development of eggs and protecting them from fertilization by any outside males. In the early 1870s the prominent evolutionary theorist August Weismann (1834–1914) conducted a series of experiments altering the temperature surrounding butterfly eggs to understand the effects of seasonal temperature differences on butterfly form. And in the same period, the Kiel zoologist Karl Möbius (1825–1908) conducted aquarium experiments to answer the question of where the food for deep-sea animals comes from. He set up an elaborate protocol combining different sorts of sand, mud, and ooze containing bottom-dwelling creatures and periodically stirred in water of different temperatures to try to model the internal currents that might carry foodstuffs down from the surface.[11]

Should these experiments 'count' as natural history or not? Certainly they were undertaken in the service of basic questions about the life history of the organisms in question, and some of them also yielded information important for classification. In this sense, they formed part of a continuous tradition of research on species histories. In Germany in the 1870s, however, they would be more likely to have been considered 'biological' research, most often by this time distinguished from 'morphological' studies. Most of the researchers under discussion appear not to have cared what it was called, confident that it was good and interesting research. By the early twentieth century, however, 'experiment' had been so successfully appropriated by promoters of laboratory biology that the idea of 'experimental natural history' was much more contentious.[12]

Natural history and the institutional expansion of zoology

The last important question to ask in regard to life-history studies is where they were carried out: what sorts of institutions supported such studies, and how did this support change over the late nineteenth and early twentieth centuries? Again, the standard histories, while largely ignoring life-history studies, give weight to the idea of a transformation from museums to laboratories. In Germany in the 1880s and early 1890s, a number of leading universities raised new buildings or substantially refurbished older ones to house laboratories for zoological teaching and research. New laboratories were opened at Leipzig (1880), Freiburg (1886), Berlin (1888), Würzburg (1889) and Heidelberg (1893/4), all leading zoological research centres. What was entailed physically was not minor: whereas museums housing natural history collections required drawers and cabinets protecting specimens from the damaging influence of light, laboratories, especially in an age before electric lighting, required large windows so that microscopists could take advantage of natural light (Figure 25.3). The new buildings represent nicely the intellectual transformation from natural history to scientific zoology, as scientists inside them sought to delve into the structures, functions, and development of animal bodies, slicing them up and then throwing the remains away. In the United States as well, a similar change occurred (Figure 25.4), as university after university introduced laboratory research in the late nineteenth century.[13]

To represent this as *the* institutional transformation in biology, however, is to miss the bigger picture. To begin with, at most of the very same universities where new laboratories were built, natural history research continued. Some of this research took place within the new laboratories themselves: classification continued to be an important part of research more often categorized as 'morphological', and life-cycle studies were certainly flourishing in university laboratories as well. More importantly, however, within a number of larger universities the natural history museum did not disappear but rather gained autonomy, becoming an entity independent of the (laboratory-based) zoology institute, and no longer having to compete directly for funding with its sister science. At Berlin, for example, the professorship of zoology was divided in 1883 into two, with one professor designated head of the zoological institute (laboratory) and the other designated conservator of the museum. At American universities as well, the museums persisted: to take a prominent example, at Harvard the Museum of Comparative Zoology maintained nearly complete independence from the

Figure 25.3 Assistants and advanced researchers working at the University of Heidelberg's Zoology Institute, 1908. As was typical, microscopical research took place at tables set up in front of large windows to take advantage of natural light. The professor's office lay through the door in a parallel room behind the assistants' room.

university's zoology department, where laboratory research and instruction became more prominent.[14]

If we focus exclusively on university zoology, a good case can be made for a dramatic change, even if it is better represented as the expansion and differentiation of university zoological research into museum and laboratory approaches, rather than the replacement of the former by the latter. But the story of what was going on in universities forms just one small part of the institutional expansion of zoological research in the late nineteenth and early twentieth centuries. The growth of non-university sites of research, including independent municipal and state natural history museums, zoological gardens, and research institutes funded by particular economic interests in the late nineteenth century is a phenomenon at least as important as the expansion of university-based research, but curiously it has been neglected by most historians of biology. As a result we know little about these institutions and the research that went on in them. And yet, for natural history questions in zoology, these institutions were crucial.

Consider natural history museums, which had long been the primary institutions for research not only in systematics but also in related life-history areas such as geographical distribution.

Although cabinets and museums are typically associated with 'old-fashioned' nineteenth-century natural history, in fact the majority of museums (in Germany, anyway) extant by the mid-twentieth century were founded in the decades around 1900.[15] Indeed, in the late nineteenth century, just when 'biology' was said to be moving *out* of the museum, away from systematics and natural history and into laboratory-based biology, museums were undergoing a boom.

These public museums were not just sites for education, but also places for conducting scholarly research. Part of the justification for new museum construction was to fulfil what came to be called the 'New Museum Idea', the principle that collections for research should be absolutely separated from those given over to public display. According to Sir William Henry Flower (1831–99), who became superintendent of the British Museum (Natural History) soon after it completed its move to its spacious new quarters in South Kensington in 1883, the great natural history museums in Vienna and Berlin, as well as the American National Museum in Washington, DC and the Museum of Comparative Zoology at Harvard (not to mention the British Museum itself), all adhered to this

Figure 25.4 Students in a biology laboratory class in Science Hall at the co-educational University of Wisconsin, *c*.1900. In 1882 the zoology professor E. A. Birge brought the latest microtome back from Leipzig and introduced the construction and microscopical study of thin sections to his undergraduate students. By the turn of the century laboratory exercises using microscopes were well established as a typical part of undergraduate biology education. Over the blackboard in the front of the room are large wall posters showing the life-cycles of various invertebrates.

plan.[16] Although invisible to the public, the 'working collections' made the large public museums the central loci of research in systematics, just as they had been a century earlier at places like the French Muséum d'Histoire Naturelle. The important difference was that in the intervening century university-based laboratory biology had come to the fore, casting museum biology into its shadow.

If systematists now found suitable research homes in museums outside of universities, zoologists interested in life-history questions had other options. In Germany a surprising number of zoologists ended up working for the fisheries industry in the decades around the turn of the century. As the founder of the 'Section for coastal and high seas fisheries' of the German Fisheries Union told subscribers to his journal in 1891, the future of the salt-water fishing industry depended on learning as much as possible about economically significant food fish such as the herring. Important tasks such as learning migration patterns and the lifeways of herring depended in turn on an even more basic research problem, identifying the varieties of herring.[17] With this sort of problem in mind, the Fisheries Union, the Imperial government and the Prussian state government jointly sponsored considerable research on the flora and fauna of the North Sea and, with the strong support of the newly founded German Zoological Society, they co-operated to found the marine station at Helgoland in 1892 (Figures 25.5 and 25.6). Work at the Helgoland station was aimed at both 'pure' research and instruction on the biology of the sea as well as research more directly in the interests of the fisheries industry, studying the life cycles of important food-fish, the North Sea as a physical environment for these fish, the role of plankton in the ecology of the North Sea, and artificial culture of certain marine food fishes and lobster. The station and its various sponsors thus acted as major patrons of life-history research in marine biology, as smaller-scale inland stations did for the freshwater fisheries.[18]

Public natural history museums and marine stations were probably not the only locations for important research into natural history questions in late nineteenth-century Germany: we still do not know much about what research went on in zoos or in schools of agriculture, forestry, and veterinary medicine. But the case of the fisheries industry suggests that we might well look to institutions such as these, which focused on applied problems, to trace more closely the direction that life-history studies took in the later nineteenth century. Certainly a comparison with the United States suggests that these sorts of institutions would be the place to look; there, researchers at agricultural experiment stations and state and federal entomological offices conducted considerable natural historical research in the interests of agricultural improvement.[19] In any case, it should be evident that the universities, although they

may have been the primary sites of morphological research and perhaps some sorts of experimental zoology, were by no means the only locations for natural historical research in the late nineteenth century.

Figure 25.5 Photo of Helgoland Biological Research Station, facing the sea. The main building with public aquaria is toward the left; the four adjacent buildings also belonged to the station. From Charles Kofoid, *Biological Stations of Europe. US Bureau of Education Bulletin*, 4 (1910), facing p. 224.

Conclusion: representing the history of natural history

If we stand back from the details of the story here and ask again, 'What happened to professional natural history in the latter part of the nineteenth century?' it seems clear that is inappropriate to talk of natural history's decline and replacement by a newer, better sort of science. It is equally inappropriate, however, to view natural history as holding the same place in the biological sciences in 1900 as it did in 1835, when Wiegmann began the *Archiv für Naturgeschichte*. The most consequential change in the overall landscape lay in the gradual expansion and diversification of the biological sciences, which made natural history just one of several orientations that a biologist could pursue. By 1900, in his presidential address to the American Society of Naturalists, the cytologist

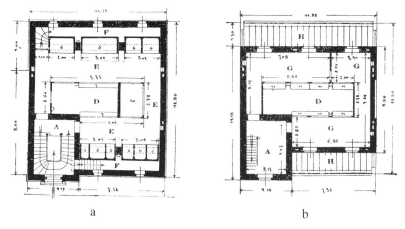

a   b

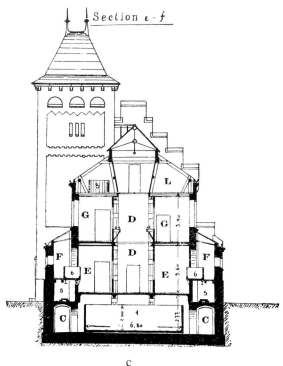

Figure 25.6 Plans showing layout of Helgoland Biological Research Station. As in the 'New Museum Idea', the biological station combined public exhibitions (on ground floor, plan a: exhibition aquaria numbered 6, 7, and 10) with ample but separate research space (on upper floor, plan b: small research aquaria numbered 11). Plan c shows cross-section of building, with low-level reservoir for sea-water labelled 1 and filter beds labelled 5. The central atrium D allows light into the research and exhibition areas. A, stairwell. C, corridor. E, exhibition room. F, service corridor to aquaria. G, laboratories. H, overhead windows. L, attic room. All measures in metres. From Charles Kofoid, *Biological Stations of Europe. US Bureau of Education Bulletin*, 4 (1910), plate XL.

c

Edmund B. Wilson (1856–1939) could jocularly divide biologists into three general classes – 'bug-hunters' (field naturalists), 'worm-slicers' (morphologists) and 'egg-shakers' (experimentalists) – even though he contended that the boundaries between these groups were artificial and that the goal of biology was a unified view of life that joined all perspectives.[20] Even in 1900 this list would have been incomplete, ignoring breeders interested in genetics, ecologists who combined laboratory and field studies, and museum taxonomists. The list could undoubtedly be expanded still further.

As the story of life-history studies suggests, however, the diversity of biological sub-specialities evident in 1900 was neither the straightforward result of a branching of a more general, inchoate 'natural history' into separate biological disciplines, nor a consequence of the simple addition of new specialities such as experimental embryology or genetics. Proponents of new orientations often sought to capture part of the older problem areas and recast them into a new framework. Thus the German 'scientific' zoologists of the 1850s successfully (if temporarily) lifted life-history studies out of natural history and recast them in a closer relationship to morphology. As morphology took on new hues in the 1870s and 1880s in response to Darwinian evolutionary theory and Ernst Haeckel's recapitulation doctrine, life histories re-emerged as a distinct biological problem area. Then in the 1880s and 1890s, laboratory experimentalists sought to claim exclusive jurisdiction over experiment and remove it from the context of life-history studies. Like the 'scientific' zoologists before them, they did this in large part without acknowledging the place in the previous configuration held by the territory appropriated.

This example suggests that the story of natural history's role in the history of biology is best represented metaphorically not by a tree, as it often is, with 'natural history' at the bottom and various biological specialties emerging naturally out of it. Rather biology might more appropriately be thought of as a landscape in which territory is contested, divided, reunited, and its boundaries redrawn over time by competing (and sometimes co-operating) groups. And yet this metaphor, too, is incomplete, for it does not take into account the institutional expansion that made it not a question of *either* natural history *or* the 'new' biology surviving in the early twentieth century. Both flourished in a period of rapid and dramatic expansion, and while natural historians no longer exclusively controlled the intellectual and institutional terrain within the universities, they were able to open up other locations for research, such as independent museums and zoos, government bureaus, and privately sponsored institutes. The different agendas of these institutions' sponsors undoubtedly affected the research conducted under their auspices, but as yet historians do not know just what the effects were.

In representing the history of natural history to ourselves, then, to accommodate intellectual reconfigurations and diversification as well as institutional growth, we might modify our geopolitical metaphor in two ways. First, the 'territory' of biology is not stable, but (in this period) growing. Second, we might best consider the landscape of biology as not a 'natural' one but one undergoing cultivation. In addition to the exploration and tilling of the unknown edges of the territory, new theoretical and methodological events can lead to the cultivation of different intellectual 'crops' even within an apparently stable region.

If in 1835 natural history was the crop cultivated by most students of living things, by 1900 this was no longer the case. Simultaneously, however, the overall landscape under cultivation had grown. This combination of factors left natural historians with a territory that at least in institutional terms was larger, but that also constituted a relatively smaller portion of the total area under cultivation. In this sense, natural history was *both* growing *and* declining in the early twentieth century.

Which view of this dual process a naturalist perceived depended in part upon where he or she was standing in the field. He or she might be in a spot at the growing edge of natural history (for example, in fisheries-based life-history studies that would become a growing point of ecology), or in one that was being encroached upon by another biological specialty. Someone in the latter position (for example, in a university zoology department where laboratory approaches were expanding rapidly) might respond in several different ways. He might watch in dismay as his local area under cultivation was gradually surrounded; he might fight back the 'noxious weeds' taking over; he might transplant his crop into a more hospitable location; he might come to an agreement with the other crops' cultivators as to an allowable growing area for his own crop; or he might engage with them in a co-operative act of hybridization.

This way of representing natural history – and the history of science more broadly – gives us some sense of the complexity of change, as well as the importance of paying attention to multiple perspectives in the history of science. Because historians interested in biology at the turn of the century have largely focused on the university-based cultivators of the 'new' biology, they have limited themselves to scrutinizing one part of the biological terrain then under cultivation. From this perspective, natural history has undergone a steady decline since the late nineteenth century. And yet, as this essay suggests, it seems likely that if we explore other areas of the biological terrain, especially beyond the universities, we may well yet discover important pockets of it cultivated by natural historians.

Further reading

Allen, David Elliston, *The Naturalist in Britain. A Social History* (London, 1976; Princeton, 1995).

Allen, Garland, *Life Science in the Twentieth Century* (Cambridge, 1978).

Caron, Joseph A., ' "Biology" in the life sciences: a historiographical contribution', *History of Science*, 26 (1988), pp. 223–68.

Cittadino, Eugene, *Nature as the Laboratory. Darwinian Plant Ecology in the German Empire*, 1880–1900 (Cambridge, 1990).

Flower, William Henry, *Essays on Museums and Other Subjects Connected with Natural History* (London, 1898).

Kofoid, Charles A., *The Biological Stations of Europe*. US Bureau of Education Bulletin, 1910, no. 4, whole number 440 (Washington, DC, 1910).

Magnus, David, 'Down the primrose path: competing epistemologies in early twentieth century biology', *Isis* (forthcoming).

Maienschein, Jane, Rainger Ronald, and Benson, Keith R. (eds.), 'Special section on American morphology at the turn of the century', *Journal of the History of Biology*, 14 (1981), pp. 83–191.

Nyhart, Lynn K., *Biology Takes Form: Animal Morphology and the German Universities, 1800–1900* (Chicago, 1995).

Rainger, Ronald, Benson, Keith R. and Maienschein, Jane (eds.), *The American Development of Biology* (New Brunswick, NJ, 1991).

Epilogue

26 The crisis of nature

This collection of essays has explored the principal cultures of natural history from the Renaissance to the opening of the twentieth century. These display the extraordinary variety of ways in which Europeans have ordered both nature and themselves. Moreover, these cultures have laid the foundations for enquiries into nature during our own time. We have inherited a patchwork of institutions, ranging from herb gardens of the sixteenth century to the great museums created in the age of high imperialism just before the First World War. Specimens in our collections once graced the cabinets of Renaissance princes and learned virtuosi. We can still see Conrad Gesner's shells, Johann Beringer's formed stones, Linnaeus's dried plants and Anton Dohrn's preserved marine organisms. When we visit the botanical garden at Padua, the Smithsonian Museum in Washington, the Australian Museum in Sydney or the American Museum of Natural History in New York, we walk on paths laid by our predecessors.

A century or more ago, naturalists imagined that the growth in natural history which they had witnessed during their lifetimes would never end. Louis Agassiz made sure that his Museum of Comparative Zoology at Harvard was built far from the rest of the College, 'on such ground as will never be an impediment to its indefinite increase'. The English comparative anatomist Richard Owen dreamed in 1859 that the collections of the British nation would eventually exhibit all the species on the globe – every elephant, every whale, every insect. These were imperial fantasies, utopian projects for a culture of natural history that would expand forever.[1] Such optimistic predictions, however, have proved illusory.

In the eighteenth century, Europeans saw themselves occupying a position in the economy of nature that was both beneficial and benign; at the end of the twentieth, nature embodies purity and wholeness, but capitalist economic relations produce habitat destruction, global warming and mass extinction. Issues surrounding 'nature' are typically portrayed in terms of decay, decline and crisis. Nature is strong only in the final reckoning, when the human species succumbs to the iron law of extinction. Against this

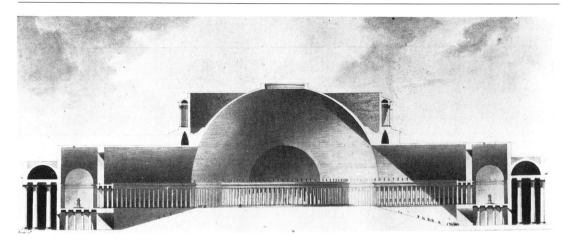

Figure 26.1 'Projet de Muséum', designed in 1783 by the French architect Etienne-Louis Boullée, in response to a competition set by the Académie des Sciences. This vast structure was planned to hold all the objects of art and science, including comprehensive collections of animals, plants, shells, medals, coins, paintings, maps, sculptures and books. With a gallery of statues of great men, the Muséum would serve as a temple to nature and genius. The figures ascending the stairway in the lower right-hand corner give an idea of the scale. See J.-M. Pérouse de Montclos, *Etienne-Louis Boulée: De l'architecture classique à l'architecture révolutionnaire* (Paris, 1968), pp. 163–4, plate 54.

background, one might expect a massive increase in support for the successor sciences to natural history – those involving organic diversity, classification and distribution. Yet these subjects are themselves in crisis, with funding slashed and recruitment in decline.

This epilogue uses the perspective provided by the earlier chapters to analyse current debates about the future of natural history. It asks why the sciences of classification occupy so low a place in the hierarchy of disciplines; examines the difficulties involved in adapting established scientific institutions and practices to the political priorities of the late twentieth century; and discusses recent changes in the presentation of natural history in the public sphere, especially in relation to environmental concerns. Natural history, viewed as a focus of media attention and as a form of scientific practice, is at a pivotal point in its career.

The ends of natural history

What is natural history? As the chapters of this book have demonstrated, the term never has had a single meaning, and has covered a range of different subjects. The same remains true today. As defined by contemporary television and the publishing industry, natural history is an inclusive subject that captures a large market share, more than any other area of science. Natural history is at a high point of public interest as the twentieth century draws to a close. Wildlife programmes are some of the most popular on television, such as David Attenborough's *Life on Earth* series and those sponsored by *National Geographic* magazine. These have audiences that would have astounded Louis Figuier and J. G. Wood, two of the most successful popularizers in the nineteenth century. The essays of Stephen Jay Gould, which are modelled on a late Victorian genre cultivated by Frank Buckland and E. Ray Lankester, are among the bestselling non-fiction works of the post-war era.

This upsurge in popular interest, however, has been accompanied by a precipitous loss of status within the sciences, among students and among science policy-makers. As is shown in the final section of this book, the late nineteenth century witnessed fragmentation on the part of practices which had been labelled 'natural history'. This was accompanied by a challenge from many of those who wished to make science a paid professional career controlled by disciplined experts. The new biology was housed not only in museums and gardens, the traditional sites of study, but also in laboratories. Even the field was colonized by the laboratory approach, through the sciences of ecology, geophysics and oceanography.[2] Studies of classification were broken up, with systematic zoology, botany, palaeontology and mineralogy as their disciplinary inheritors. Even these are now disappearing. Botany is encompassed by 'plant sciences', geology by 'earth sciences'.

Many taxonomists self-deprecatingly agree that their activity is somehow 'not a science'. Within the realm of scientific practice, the term 'natural history' is now itself something of a museum specimen, persisting largely through the names of institutions like those created by Agassiz and Owen: to call someone a 'natural historian' sounds quaintly old-fashioned or even abusive. Systematics, as the invertebrate palaeontologist Simon Conway Morris has said, is 'deeply unsexy'.[3]

The process of pushing once-dominant fields to the margins of science has a long history. At the time of Georges Cuvier, comparative anatomists, geologists and physiologists carved out new roles for themselves, as followers of a specialist vocation, and often claimed to put natural history behind them. From this perspective, the eighteenth century is the age of natural history, a subject involving static descriptions, while the nineteenth pursues process and history. In many ways, when authors such as Michel Foucault locate an epistemic break around 1800, they simply underwrite the definitional strategies of Cuvier and his colleagues. Defining natural history always involves acts of exclusion and inclusion, what is real knowledge and what is merely popular.[4] That is why the status of natural history has been constantly questioned during the past four centuries, and why, in retrospect, it seems to have come to an end so often.

Taxonomists have, like the members of any speciality, created barriers around their subject which emphasize its difficulty and distinctive identity. As James Cullen of the Royal Botanic Garden in Edinburgh has said:

The introversion built up leads inevitably to ritualization, and a mystery cult develops, with its arcana, its sacred texts and rules of procedure, its priesthood and prophets, its orthodoxy, heretics and schisms. Like any cult, its sacred texts are not for the understanding of the uninitiated, so they are written in specialist language under the restriction of evil-eye-averting formalities. Again, like any cult, there are controversies about how many angels can sit on the head of a pin, sectarianism and ancestor-worship.[5]

Founding heroes are useful in establishing credibility, for they can underwrite claims that the subject 'became a science' through an organizing theory or law. Cuvier, Darwin and E. O. Wilson have been hailed as Isaac Newtons of the living world – despite Immanuel Kant's claim that there could never be a Newton of the grassblade.[6]

Frequently, such efforts to establish intellectual authority involve embracing the style of sciences further up the disciplinary pecking order. Debates among taxonomists about the utility of computer hardware, genetic technologies or sociobiological laws can be seen in this light. Notably, Wilson – an ant systematist and student of animal behaviour – began to develop sociobiology as a counter to the advancing hegemony of molecular biology at Harvard. He envied the reductionist triumphs of the double helix, seeing 'in its style and outrageous success the sword that might cut one Gordian knot in biology after another'.[7] Distancing techniques of this kind maintain disciplinary status when a subject verges too close to popular knowledge.

For thirty years, reports on the state of systematics in Europe and the United States have emphasized problems with funding, recruitment and perceptions by other scientists. These problems are particularly ironic, in that there is no sign of any loss of intellectual vitality or sense of purpose within taxonomy. The field has been transformed by debates over cladistics, numerical taxonomy, molecular genetics and other new techniques.[8] Studies of invertebrate fossils such as trilobites, for example, have been revolutionized by functional analysis involving attention to details best revealed in specimens from collections decades old. Such research, however, is continuously in danger of being cut off for want of funds.[9]

Immediately after the Second World War, Gavin de Beer spoke of 'the descending spiral of the training of taxonomers'.[10] According to an American survey, the number of Ph.D.s awarded in entomology, which averaged about 170 per year in the 1970s, fell to 133 in 1988. In the same period, Ph.D.s awarded in molecular biology rose from 136 to 362.[11] Policies based on short-term financial considerations have only brought a serious situation to a crux.

It is paradoxical, but no real surprise, that natural history is today both hugely popular and at the bottom of the scientific hierarchy. Those areas which might once have been part of it, but which have established high status – such as mathematical population ecology – are precisely those where the division between 'scientific' and 'popular' is best defined. On the other hand, there are many areas of natural history where a wide variety of practitioners can become involved, such as the ornithological or botanical field surveys sponsored by groups like the Audubon Society and the Botanical Society of the British Isles.[12] Such groups have played vital roles in research and in wildlife conservation. However, within the world of high science, the status of such subjects remains low. This is the age of the Human Genome Project, not of a co-ordinated survey of the world's fauna and flora.

Adapting the past

Part of the difficulty lies in adapting the existing structures of natural history to new priorities. The post-war landscape of science funding was dominated by mission-oriented laboratories and centred on machines: CERN, the Space Telescope, the Lawrence–Livermore Laboratory. Collections, associated with a focus on studies of difference, diversity and behaviour, seem to have little place in the world of 'big science'.

Up to the final decades of the nineteenth century, however, the situation was very different. Laboratories in natural philosophy were little more than cabinets of equipment belonging to ill-paid professors; the national centres for science were museums and the fleets of sailing vessels equipped to collect for them. These missions of natural history conquest were the largest scientific enterprises of their time. Our herbaria, for example, are in many respects relics of the era of colonialism. Over half of the herbarium sheets in the world are in Western Europe, and the role of institutions like Kew Gardens, the Natural History Museum and the Muséum National d'Histoire Naturelle is clearly an inheritance from an age when much of the world was ruled from London and Paris.

Three Australian botanists have argued that only a few of these collections should be retained, and the rest thrown away so that money can be spent on other projects. 'What would be lost if label data on all sheets was recorded,' they ask, 'a careful selection of sheets kept and the rest pulped?'[13] Some molecular biologists go much further, and believe that chromosomal studies have rendered pointless the whole apparatus of Linnaean procedure and Latin names. Understandably, the latter position has found little support from those who create and use botanical knowledge. But debate is bound to continue about the need for specimen-based collections in taxonomic research, especially in an era of declining funds and changing public priorities.

Zoos, like herbaria, have suffered a crisis of confidence, and have met the challenge in various ways. Just as botany students no longer necessarily want to study 'dried corpses' on herbarium sheets, so too zoologists do not always need (or want) to see specimens skinned, stuffed or in cages. From the point of view of scientific conservation, animals in zoos are increasingly valued not as individuals but as samples of endangered gene pools held in trust for future generations. The public, for its part, prefers to see the bearers of these rare genes as though they were wandering wild in a native habitat, whether this be a nature park or on a video screen.[14]

The problems for natural history institutions have been most acute in post-imperial Britain, where both the London Zoo and the Natural History Museum have had to radically re-think their functions. The Natural History Museum, like all British cultural institutions, has been forced by cuts in government funding to redefine its priorities in narrowly commercial terms. Jettisoning an agenda organized around the collections as a long-term strategic resource, a 'corporate plan' announced in 1990 replaced these with work on 'biodiversity, environmental quality, mineral resources, agricultural resources, human evolution and human health'.[15]

Such changes created a storm of controversy, not least because the actual cuts often seemed random and opportunistic. Many critics within the systematics community felt that the rationale behind the institution's existence did not appear to have been thought through. If these were the priorities, why have a research museum with comprehensive collections at all? Perhaps, as one correspondent suggested, the specimens could be sold to the Germans, Americans or Japanese.[16] There is, of course, no inherent reason why collections of natural objects should remain sacrosanct. On grounds of historical precedent it is highly unlikely that they should do so. But it appeared that, in this case, decisions with major international consequences were being left to the day-to-day workings of the market.

This is particularly unfortunate, especially in that almost all commentators (including Margaret Thatcher, the Prime Minister at the time) call for more strategic planning in taxonomic research. With some 90 per cent of species still undescribed, there is an obvious need to establish priorities, standardize recording techniques and develop channels for international co-operation and more effective use of public interest. Countries such as Costa Rica, where a corps of locally trained 'parataxonomists' has been created as part of an initiative to inventory the country's flora and fauna, show promising routes forward. To be effective parts of the taxonomic network, however, such programmes depend upon the established collections in Europe and America, and the specialized experts who staff them.[17] Because natural history developed

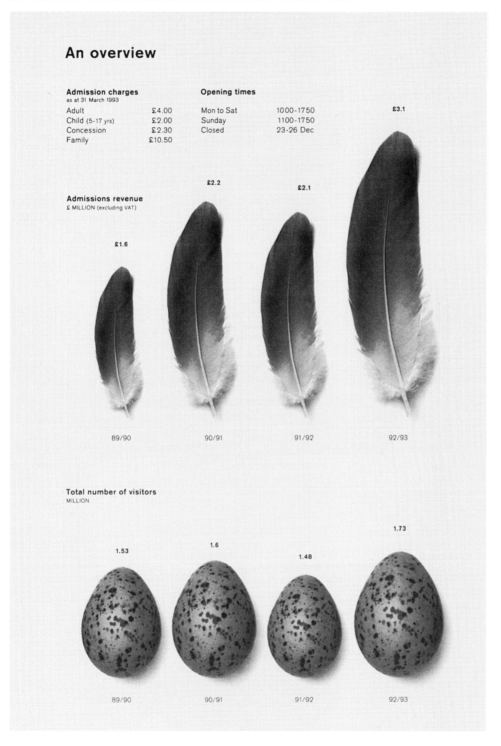

Figure 26.2 New uses for natural history collections in the 1990s. From *The Natural History Museum Triennial Report* (London, 1993), p. 2.

through empire, with historic centres of research located in Europe, cuts in London inevitably affect the rest of the world.

The dilemmas faced by museums, zoos, herbaria and related collection-based institutions go well beyond the peculiarities of the British case, and are consequent on economic changes associated with the end of colonialism and the creation of a global corporate economy. Power over the natural world no longer seems to rest – as it did before the twentieth century – in centralized inventories of nature. Instead, the modern bureaucratic state works through the laboratory. Disease is thought more likely to be cured by genetic manipulation and computer modelling than by discovery of new substances in species on the verge of extinction. At least since Louis Pasteur, the slogan has been (in Bruno Latour's terms) 'give me a laboratory and I will raise the world'.[18] No one now raises the world with museums or herbaria, which have come to be associated with conservation, tradition, preservation – not with innovation, economic growth or survival.

Consuming nature

Perhaps the most striking irony of the decline in status of sciences dealing with diversity and classification is that it has coincided with intensive coverage of the environmental crisis by the mass media. Thus despite public fascination with the diversity of life, research into systematics and taxonomy is badly funded and has a poor image.

In part, the problem is common to many areas of the sciences. Journalists work to a schedule of spectacles and news stories, so that focus tends to be on high risk, catastrophe, competition and the race to discovery. Post-war science journalism, especially in the United States, emerged in conjunction with the space programme, and most of those assigned to such stories preferred to cover the physical sciences, especially cosmology and particle physics, which are easier to package in newsworthy terms.[19] Environmental and conservation stories are typically triggered by disaster (seals dying in the North Sea), or through skilled interventions by concerned parties such as Greenpeace.

Scientists working on long-term schedules – in the case of taxonomy, measured in decades – remain out of the limelight. They are consulted, often frequently, about events as they arise; but they respond to a media agenda set largely by others. When sulphur dioxide emissions reach the news, bryologists suddenly appear on television; lichenologists had a brief moment of fame in the wake of the Chernobyl disaster. The personalities consistently in the public eye, such as Gould and Wilson, are notable both for their rarity and for the controversy surrounding what they do. The

Flora Europaea (Cambridge, 1964–80) or *Mammals of the Neotropics* (Chicago, 1989–) are major accomplishments of science in the post-war era, but they do not sell newspapers.

The issue is particularly acute in museums, botanical gardens and zoos, which have since the nineteenth century been engaged in balancing behind-the-scenes research with public display. In recent years, this delicate arrangement shows signs of coming apart. The commercial imperatives imposed by financial stringency have forced such institutions to increase shop space and admission charges, and to follow an exhibition schedule driven by turnstile receipts. Exhibitions are contracted out to commercial firms, which draw on the staff expertise only fitfully.[20]

Many museums increasingly keep their collections in storage, and feature video screens, hands-on experience and live-action robots in their public space. Only a handful of tropical butterflies are on display, but a travelling exhibition features gigantically enlarged mechanical insects. Here, political pressures for commercialization and the often naive campaign for the 'public understanding of science' have the same effect: the public itself is distanced from the actual processes of research. Museums look more like department stores, and knowledge is presented as a finished commodity. In the controversy over staff cuts at the Natural History Museum in London, Walter Bodmer celebrated the commercial imperative: we should 'perhaps learn to some extent from the wonders of Disneyland'.[21]

The point is not whether the objects on display are 'real' or not. All museum specimens are inevitably artificial – skinned, trimmed, pressed, bottled, stained.[22] Even the original bones of dinosaurs have to be carefully chiselled out of their matrix and soaked in artificial preservative before they can be studied. The substantive issue is the distance between the displays and the social world of the work of research. When we see dinosaur bones – even as plaster casts – we see the pieces of evidence used by specialist palaeontologists. Difficulties of interpretation, problems in reconstruction, are at least potentially open to the public gaze. When we look at rubber models powered by motors, the world of practice becomes invisible, for the shared object – the specimen – has been covered up. We may feel able to imagine dinosaurs more vividly, but our understanding of how scientists work is impoverished.[23]

Viewing the public for knowledge as consumers, and knowledge itself as a commodity, has serious consequences for the image of science. Consumers expect finished products, backed by something akin to a guarantee. When knowledge fails to deliver on its promises, by proving (as it eventually does) to be provisional, uncertain and open to controversy, the public all too easily becomes disillusioned. The problem is especially acute with

Figure 26.3 *Nature*, 341 (12 October 1989), p. 477.

sciences dealing with the complex interactions of environmental systems. Here, an open-ended, historically informed view of the processes of scientific work is essential to public understanding.[24]

At its best, television can be highly effective in conveying this sense of science in action and in controversy. However, the medium is overwhelmingly conservative; indeed, it is hard to imagine how such a pervasive element of culture could be otherwise. Science documentaries depend on familiar narratives, notably detective fiction and the western, with scientists unlocking the mysteries of the universe. In programmes on the environment, scientists are typically outside experts rather than engaged participants in a process of debate. The best efforts of individual producers, however, forward a view of science embedded in politics and public controversy. Indeed, flagship series like *Nova* in the United States, *The Nature of Things* in Canada and *Horizon* in Britain provide opportunities for such programmes as part of the attempt to maintain journalistic balance.[25]

Natural history documentaries are the most popular forms of science programming, often brilliantly filmed by committed teams of camera personnel and producers. Even the earliest and most traditional of these programmes – Walt Disney's *True-Life Adventures*, Mutual of Omaha's *Wild Kingdom* and Jacques-Yves Cousteau's *Undersea World* – stress preservationist themes. There is considerable evidence that attitudes towards wolves, bears, sharks and other large carnivores have been transformed through the cumulative effect of filmed documentaries. At crucial moments television has played a powerful role in putting conservation issues on the immediate political agenda.

But as practitioners of television journalism have recognized for many years, the medium itself tends to induce passivity, by rendering everything displayed on the screen uniformly accessible. Viewers witness pastoral idylls or 'world in danger', often with no clear route to action – except perhaps for those suggested in the commercial breaks. Indeed, extinction is precisely the kind of spectacle that can sell programmes to potential advertisers. Once the cameras are out of sight, habitat is lost, poaching and corruption continue.[26]

Although popular natural history retains the high profile it has had for decades, closely related environmental issues (notably mass extinction, overpopulation and climatic change) have moved down the media agenda after a few years of headlines, special programmes and news stories.[27] In this respect, the period after the 1992 Earth Summit in Rio has been rather like the years after the first Earth Day in 1970, when public interest – which grew explosively after Rachel Carson's *Silent Spring* (New York, 1962) – almost entirely evaporated. Environmental degradation, habitat destruction and the loss of biodiversity will not become regular news items until they generate major catastrophes on a daily basis – by which time it will be too late.

A new culture for natural history?

The problem, all too obviously, is the lack of any long-term perspective. As environmental issues slip once again from the public agenda, as new priorities at museums and zoos seem to be taken for granted, and as the era of big-science physics begins to recede, it is appropriate to look at how we got to where we are. The essays in this book have shown that the cultures of natural history have always been embedded in settings particular to time and place; thus 'curiosity' was not originally a psychological attribute, but part of an early modern practice of collecting and display. The roots of our contemporary situation are to be found in a material history of practices, and any understanding will have to emerge from a recognition of the historical depth of current dilemmas.

How can the kind of work represented in these essays contribute towards this end? Above all, they suggest that the development of the sciences belongs in a history that locates human actions in nature, and not apart from it. Showing how dramatically the boundaries surrounding 'natural' and 'history' have themselves shifted is only one of the many ways that the study of the past can help to break down the debilitating divides which still separate nature and culture. The practices of natural history can be situated within the structures of material life, set in a world of historical circumstance which includes breadfruit, pigs, plantains, coal, water, ships, bricks, beer, books and glass. The history of natural history needs to become part of environmental history.[28]

The current public image of the sciences, which is based almost entirely on physics and astronomy, requires drastic revision. We need to learn how the hierarchy of sciences was established, and how natural history lost status. Histories of the life and earth sciences continue to be dominated by fields (notably evolution, plate tectonics, genetics and molecular biology) that claim a theoretical generality akin to that of physics. Similarly, a generation of historians who came to maturity in the late 1960s and early 1970s has brilliantly charted the intellectual origins of ecology, but little has yet been written on the practical day-to-day work of creating an inventory of the living world. As long as theoretical physics is seen (even implicitly) as the ideal form of knowledge, our understanding of the sciences will remain impoverished and reductionist.

Historical analysis can also show how these issues of status, and the practice of the natural sciences more generally, have always been embedded within wider political discussions about gender, class and expertise. Natural history and its successor disciplines are situated knowledges, part of rather than outside public controversy. Accordingly, there needs to be active, historically informed debate about the sciences and how they are depicted in museums, zoos, journalism and television. These issues are already being faced by curators, journalists and producers every day. Only through this kind of political process can the public sphere of the sciences avoid becoming a dusty irrelevance or a high-tech commercial theme park.

And finally, the history of the natural sciences needs to take an international perspective, centred on the creation of a global economy of market relations. Much of the current environmental crisis originates in structural inequalities created in the wake of de-colonization. Understanding the role of natural history in imperialism, racism and the world economy is a pressing need, which several of the chapters in this book have begun to explore. Here historical analysis can sharpen awareness of the place of science in the politics of development and the Third World.

History often seems irrelevant to contemporary issues. The history of natural history in particular can easily become a repository of anecdote and entertaining escape, a 'cabinet of curiosities'. In taxonomic studies, research into the older literature has traditionally had a more serious purpose, as part of the formal procedures for establishing nomenclature since the time of Linnaeus. But history is not just part of the process of naming species. Nor is it simply a search for appropriate ancestors, a backdrop to the emergence of modern ideas, or a nostalgic attempt to salvage a disappearing age.

Rather, a bold enquiry into the past can uncover the basic structures and large-scale patterns of change which lie behind our current dilemmas. We have inherited not just our institutions and practices, but our problems: and these can only be understood as products of history. A new culture of natural history will flourish only if it is effectively rooted in – and draws upon – a critical understanding of the past.

Further reading

Brennan, A., *Thinking about Nature* (London, 1988).

Grote, A. (ed.), *Macrocosmos in Microcosmo. Die Welt in der Stube. Zur Geschichte des Sammelns 1450 bis 1800* (Opladen, 1994).

Grove, R., *Green Imperialism: Colonial Expansion, Tropical Island Edens and the Origins of Environmentalism, 1600–1860* (Cambridge, 1995).

Hansen, A. (ed.), *The Mass Media and Environmental Issues* (Leicester, 1993).

Haraway, D. J., *Simians, Cyborgs, and Women: The Reinvention of Nature* (London, 1991).

Hawksworth, D. L. (ed.), *Prospects in Systematics* (Oxford, 1988).

Heywood, V. H. and Moore, D. M. (eds.), *Current Concepts in Plant Taxonomy* (London, 1984).

Latour, B., *We Have Never Been Modern* (New York, 1993).

Nelkin, D., *Selling Science: How the Press Covers Science and Technology* (New York, 1987, 2nd ed., 1995).

Ross, A., *Strange Weather: Culture, Science, and Technology in the Age of Limits* (London, 1991).

Secord, J. A. (ed.), 'The big picture'. Special issue of *The British Journal for the History of Science*, 26(4) (December 1993).

Silverstone, R., *Framing Science: The Making of a BBC Documentary* (London, 1985).

Wilson, A., *The Culture of Nature: North American Landscape from Disney to the Exxon Valdez* (Cambridge, MA, 1992).

Worster, D. (ed.), *The Ends of the Earth* (Cambridge, 1988).

Notes

1 The natures of cultural history

1 Quoted in Roger Chartier, *Cultural History: Between Practices and Representations*, trans. Lydia G. Cochrane (Cambridge, 1988), p. 7.

2 In *LES RUINES ou Méditation sur les Révolutions des Empires par M. Volney Député à l'Assemblée Nationale de 1789* (August 1791); in Volney, *Oeuvres*, vol. I (Paris, 1991), p. 195.

3 Novalis, *Werke*, H.-J. Mähl, R. Samuel (eds.) (Munich, 1978–87), vol. II, p. 234.

4 On the multiplicity of attitudes to nature in the eighteenth century see: D. G. Charlton, *New Images of the Natural in France: A Study in European Cultural History 1750–1800* (Cambridge, 1984); G. Gusdorf, *Naissance de la conscience romantique au siècle des Lumières* (Paris, 1976); J. Ehrard, *L'Idée de nature en France dans la première moitié du XVIIIᵉ siècle* (Paris, 1963); Keith Thomas, *Man and the Natural World: Changing Attitudes in England 1500–1800* (London, 1983); and J. C. Barrell, *The Idea of Landscape and the Sense of Place, 1739–1840* (Cambridge, 1972).

5 J. E. Lovelock, *Gaia: A New Look at Life on Earth* (Oxford, 1979).

6 On legitimatory uses of histories of the sciences see, for example, L. Graham, W. Lepenies and P. Weingart (eds.), *Functions and Uses of Disciplinary Histories* (Dordrecht, 1983).

7 A. Cain, 'The natural classification', *Proceedings of the Linnean Society of London*, 174 (1963), pp. 115–21; P. Sneath, 'Mathematics and classification from Adanson to the present', in *Adanson: The Bicentennial of Michel Adanson's* Familles des plantes (Pittsburgh, 1964), vol. II, pp. 471–98. M. Guédès in 'La Méthode taxinomique d'Adanson', *Revue d'histoire des sciences et de leurs applications*, 22 (1967), pp. 361–86, argued, however, that this was a misreading of Adanson, which originated in attacks on his methods by Georges Cuvier and Augustin Pyramus de Candolle. For a historically balanced appreciation of eighteenth-century quests for a natural system, see P. F. Stevens, *The Development of Biological Systematics: Antoine-Laurent de Jussieu, Nature, and the Natural System* (New York, 1994).

8 The replacement of a multiplicity of local conventions by international codes is outlined in 'Symposium on Linnaeus and nomenclatorial codes', *Systematic Zoology*, 8 (1959), pp. 1–47.

9 For a magisterial bibliography, see G. Bridson, *The History of Natural History: An Annotated Bibliography* (New York, 1994).

10 *The Structure of Scientific Revolutions* (Chicago, 1962), p. 3. On the effects of this new emphasis on social history see J. A. Secord, 'Natural history in depth', *Social Studies of Science*, 15 (1985), pp. 181–200.

11 On Foucault's claims about natural history see, for example, D. Stemerding, *Plants, Animals and Formulae. Natural History in the Light of Latour's* Science in Action *and Foucault's* The Order of Things (Twente, 1991).

12 M. Serres, *Préface*, in J.-M. Drouin, *L'Écologie et son histoire* (Paris, 1993).

13 See also Susan Leigh Star and James R. Griesemer, 'Institutional ecology, "translations", and boundary objects: amateurs and professionals in Berkeley's Museum of Vertebrate Zoology, 1907–39', *Social Studies of Science*, 19 (1989), pp. 387–420.

14 Laurie Nussdorfer, review essay in *History and Theory*, 32 (1993), pp. 74–83.

15 On the identity of cultural history see, for example, Lynn Hunt (ed.), *The New Cultural History* (Berkeley, 1989).

16 M. Arnold, *Culture and Anarchy* (London, 1869); W. Morris, *Hopes and Fears for Art* (London, 1882).

17 In this classification of practices we have drawn upon P. Bourdieu, *Outline of a Theory of Practice* (orig. 1972; Cambridge, 1977); S. Shapin, 'Pump and circumstance: Robert Boyle's literary technology', *Social Studies of Science*, 14 (1984), pp. 481–520; and B. Turner, *The Body and Society: Explorations in Social Theory* (Oxford, 1984).

18 On recruitment and delegation of allies in the conduct of disciplines, see B. Latour, *Science in Action* (Milton Keynes, 1987); on trust, trustworthiness and the division of labour see S. Shapin, *A Social Theory of Truth* (Chicago, 1994).

19 For a powerful critique see Roger Chartier, 'The chimera of the origin: archaeology, cultural history, and the French Revolution', in Jan Goldstein (ed.), *Foucault and the Writing of History* (Oxford, 1994), pp. 167–86.

20 N. Jardine, *Scenes of Inquiry: On the Reality of Questions in the Sciences* (Oxford, 1991).

21 See Bruno Latour, *We Have Never Been Modern* (New York, 1993).

22 Michael Lynch, 'Discipline and the material form of images: an analysis of scientific visibility', *Social Studies of Science*, 15 (1985), pp. 37–66; John Law and Michael Lynch, 'Lists, field guides and the descriptive organization of seeing: birdwatching as an exemplary observational activity', in M. Lynch and S. Woolgar (eds.), *Representation in Scientific Practice* (Cambridge, MA, 1990), pp. 267–300.

23 Novalis, *Werke*, H.-J. Mähl and R. Samuel (eds.), (Munich, 1978–87), vol. II, p. 234.

2   Emblematic natural history of the Renaissance

1 There are many surveys of early modern science that have chapters on Renaissance natural history, but a far better introduction to the topic is Michel Foucault, *The Order of Things: An Archaeology of the Human Sciences* (New York, 1970), pp. 17–45. Foucault is short on details but illuminating on Renaissance patterns of thought and organization.

2 Conrad Gesner, *Historia animalium lib. I. de quadrupedibus viviparis* (Zurich, 1551); *Historia animalium lib. II. de quadrupedibus oviparis* (Zurich, 1554); *Historia animalium lib. III. de avium* (Zurich, 1555); *Historia animalium lib. IIII. qui est de piscium et aquatilium animalium natura* (Zurich, 1558).

3 Gesner, *Historia animalium lib. I*, pp. 1081–96.

4 On Gesner, see Charles E. Raven, *Natural Religion and Christian Theology. The Gifford Lectures, 1951. First series: Science and Religion* (Cambridge, 1953), pp. 80–98; Hans Wellisch, 'Conrad Gesner: a bio-bibliography', *Journal of the Society for the Bibliography of Natural History*, 7 (1975), pp. 151–247; Henry Morley, 'Conrad Gesner', in Robert M. Palter (ed.), *Toward Modern Science* (New York, 1969), pp. 358–82; and a very good doctoral dissertation: Caroline Aleid Gmelig-Nijboer, *Conrad Gesner's Historia Animalium: An Inventory of Renaissance Zoology*, Communicationes biohistoricae ultrajectinae, 72 (Meppel, 1977).

5 Karen M. Reeds, 'Renaissance humanism and botany', *Annals of Science* 33 (1976), pp. 619–42. A similar study of humanism and zoology would be nice.

6 Readers who wish to encounter Gesner on their own but who do not read Latin might wish to search out Edward Topsell's *Historie of Four Footed Beastes* (London, 1607), which is largely a translation of Gesner. However, Topsell's chapters, including the one on the fox (pp. 220–8), are severely abridged, and they are usually completely rearranged, so that Gesner's organizational format is missing.

7 Desiderius Erasmus, *Adagiorum collectanea* (Paris, 1500); *Adagiorum chiliades* (Basel, 1536). On the adages, see Margaret Mann Phillips, *The 'Adages' of Erasmus: A Study with Translations* (Cambridge, 1964); and *Erasmus on his Times: A Shortened Version of the 'Adages' of Erasmus* (Cambridge, 1967). A modern translation of all

the adages is now being published: Desiderius Erasmus, *Adages*, trans. and annot. M. M. Phillips and R. A. B. Mynors (Toronto, 1982–); so far four of a projected seven volumes have appeared, containing the first 2,300 adages.

8 The best introduction to emblems and the emblem book genre is still Mario Praz, *Studies in Seventeenth-century Imagery*, 2 vols. (London, 1939–47). On Alciati's role in the founding of emblematics see Peter M. Daly (ed.), *Andrea Alciato and the Emblem Tradition: Essays in Honor of Virginia Woods Callahan* (New York, 1989), and several works by Daniel S. Russell: 'Alciati's emblems in Renaissance France', *Renaissance Quarterly*, 34 (1981) pp. 534–54; *The Emblem and Device in France* (Lexington, KY, 1985); 'Emblems and hieroglyphics: some observations on the beginnings and nature of emblematic forms', *Emblematica*, 1 (1986), pp. 227–43.

9 I have used the recent scholarly edition edited by Peter M. Daly, with Virginia W. Callahan, assisted by Simon Cuttler, *Andreas Alciatus*, vol. I, *The Latin Emblems, Indexes and Lists*; vol. II, *Emblems in Translation* (Toronto, 1985). The lynx is emblem 66, the hares with dead lion is emblem 154, the chameleon is emblem 53, and the fox with mask is emblem 189. The English translation of emblem 189 was used as the basis for the translation of Gesner's transcription of the Alciati emblem.

10 On the bestiary, see Wilma George and Brunsdon Yapp, *The Naming of the Beasts: Natural History in the Medieval Bestiary* (London, 1991). There is a fine selection of colour illustrations from various manuscripts in the British Library in Ann Payne, *Medieval Beasts* (New York, 1990).

11 *Der Gart der Gesundheit* (Mainz, 1485), compiled and printed by Peter Schoeffer. On the *Gart*, see Frank J. Anderson, *An Illustrated History of the Herbals* (New York, 1977), pp. 89–97.

12 On the Oppian manuscript used by Gesner, and on late classical zoological manuscripts in general, see Zoltán Kádár, *Survivals of Greek Zoological Illuminations in Byzantine Manuscripts* (Budapest, 1978).

13 See Anderson, *Herbals*, pp. 121–47.

14 For a look at just part of Gesner's vast correspondence network, see Vivian Nutton, 'Conrad Gesner and the English naturalists', *Medical History*, 29 (1985), pp. 93–7.

15 Pierre Belon, *De aquatilibus libri duo, cum eiconibus ad vivum ipsorum effigiem, quoad eius fieri potuit, expressis* (Paris, 1553); *L'histoire de la nature des oyseaux, avec leurs descriptions, & naifs portraicts retirez du naturel* (Paris, 1555); Guillaume Rondelet, *Libri de piscibus marinis in quibus verae piscium effigies expressae sunt* (Paris, 1554–5).

16 Nuremberg, 1595; the other three volumes were published in 1593 (but 1590 on title-page), 1596 and 1604.

17 The only scholar to argue that Camerarius's works are an integral part of Renaissance natural history is Wolfgang Harms, 'On natural history and emblematics in the sixteenth century', in Allan Ellenius (ed.), *The Natural Sciences and the Arts*, Acta universitatis Upsaliensis, Figura nova, 22 (Uppsala, 1985), pp. 67–83.

18 Most of the studies of Aldrovandi's natural history are in Italian; but see Paula Findlen, *Possessing Nature: Museums, Collecting, and Scientific Culture in Early Modern Italy* (Berkeley, 1994).

19 (Bologna, 1637), pp. 195–222.

20 Conjectures as to the cause of the demise are offered in William B. Ashworth, Jr., 'Natural history and the emblematic world view', in David C. Lindberg and Robert S. Westman (eds.), *Reappraisals of the Scientific Revolution* (Cambridge, 1990) pp. 303–32.

3   The culture of gardens

1 See the account by Robert Burton, *The Anatomy of Melancholy*, ed. T. Faulkner, N. Kiessling and R. Blair (orig. 1621; Oxford, 1989), p. 33.

2 Pliny, *Letters*, book V, letter VI, Loeb edn., vol. I, p. 343. For the fundamental Renaissance encapsulation of this and other ancient advice on the matter of gardens, see Leon Battista Alberti, *The Ten Books of Architecture* (1450), book IX, ch. 4;

see the reprint of the 1755 Leoni translation into English (New York, 1986), pp. 192–3.

3 Derek Clifford, *A History of Garden Design* (London, 1962), p. 75. The other two most influential gardens, in Clifford's view, are Versailles (designed by Le Nôtre) and Stowe (especially as designed by Capability Brown).

4 Arnaldo Bruschi, *Bramante* (orig. Italian edn., 1973; 1977), pp. 91–2. Nicholas had used much paradisiacal imagery to describe both the desirable structure of Rome and the Vatican, and also to portray his own role as governor of the Church: according to Westfall, he had been trying 'to give a proper setting for the pope, the Vicar of Christ, the gardener of the garden of wisdom'. Carroll William Westfall, *In This Most Perfect Paradise: Alberti, Nicholas V, and the Invention of Conscious Urban Planning in Rome, 1477–55* (University Park, PA, 1974), p. 161. 'In principalities, paradisiacal settings are used to show the acts of government that establish order and peace through the two principal means, letters and arms'. *Ibid.*, p. 156.

5 My account of the Belvedere Court and gardens is based on David R. Coffin, *Gardens and Gardening in Papal Rome* (Princeton, 1991); James S. Ackerman, *The Cortile del Belvedere* (Rome, 1954); Arnaldo Bruschi, *Bramante*; and Ellen Shultz (ed.), *The Vatican Collections: The Papacy and Art* (New York, 1982).

6 As translated by Ackerman in *The Cortile del Belvedere*, p. 32; he gives the original Italian on p. 145.

7 Clifford, *A History of Garden Design*, p. 66.

8 Coffin, *Gardens and Gardening*, p. 50.

9 Lucia Tongiorgi Tomasi, 'Projects for botanical and other gardens: a 16th century manual', *Journal of Garden History*, 3 (1983), pp. 1–34.

10 British Museum, Manuscript Sloane 801, fo. 1.

11 Dennis E. Rhodes, 'The botanical garden of Padua: the first hundred years', *Journal of Garden History*, 4 (1984), pp. 327–31, 330.

12 Georgina Masson, 'Italian flower collectors' gardens in seventeenth century Italy', in D. R. Coffin (ed.), *The Italian Garden* (Washington, DC, 1972), pp. 63–80.

13 *The Garden Book of Sir Thomas Hanmer, Bart*, transcribed by Ivy Elstob, introduction by Eleanour Sinclair Rohde (London, 1933; repr. Clwyd County Council, 1991), pp. xvii–xviii.

14 Wilfrid Blunt, *Tulipomania* (Harmondsworth, 1950).

15 See Coffin, *Gardens and Gardening*, ch. 13, 'Entertainment'.

16 Quoted by A. L. Rowse, *Simon Forman: Sex and Society in Shakespeare's Age* (London, 1974), p. 55, from Philip Stubbes, *Anatomy of Abuses* (1583), ed. F. J. Furnivall (New Shakespeare Society), Vol. I, p. 88, 'Horrible whordome in Ailgna [= England]'; spelling modernized by Rowse.

17 See Coffin, *Gardens and Gardening*, p. 215, discussing the Cardinal of Mantua, Francesco Gonzaga, in 1479.

18 For the frustration this social convention could engender, as late as the end of the nineteenth century, in a garden owner who wished to engage in the work of the garden, see Elizabeth von Arnim, *Elizabeth and her German Garden* (orig. 1898; London, 1985).

19 On this theme see, for example, John Harvey, *Early Nurserymen* (Chichester, 1974).

20 Thomas Hill, *The Gardener's Labyrinth* (orig. 1577), ed. with an introduction by Richard Mabey (Oxford, 1987), quoted on p. 8.

4   Courting nature

1 On Greek and Roman natural history, see G. E. R. Lloyd, *Science, Folklore and Ideology: Studies in the Life Sciences in Ancient Greece* (Cambridge, 1983), and Mary Beagon, *Roman Nature: The Thought of Pliny the Elder* (Oxford, 1992).

2 Archivio Isolani, Bologna, *Fondo Paleotti* 59 (F 30) 29/7, Aldrovandi to Camillo Paleotti, Bologna (11 December 1585).

3 On medieval natural history see Jerry Stannard, 'Natural history', in David C. Lindberg (ed.), *Science in the Middle Ages* (Chicago, 1976), pp. 429–66; Lindberg, 'Natural history', in *The Beginnings of Western Science* (Chicago, 1992), pp. 348–53; Willene B. Clark and Meradith T. McMunn (eds.), *Birds and Beasts in the Middle Ages: The Bestiary and Its Legacy* (Philadelphia, 1989); and Wilma George and Brunsdon Yapp, *The Naming of the Beasts: Natural History in the Medieval Bestiary* (London, 1991).

4 On the debates about Pliny see Arturo Castiglioni, 'The school of Ferrara and the controversy on Pliny', in E. Ashworth Underwood (ed.), *Science, Medicine and History*, vol. I (Oxford, 1953), pp. 269–79; and Charles G. Nauert, 'Humanists, scientists, and Pliny: changing approaches to a classical author', *American Historical Review*, 84 (1979), pp. 72–85. For a more general discussion of natural history as a humanistic discipline see Karen Reeds, 'Renaissance humanism and botany', *Annals of Science*, 33 (1976), pp. 519–42.

5 Gaspare Gabrieli, *Oratio habita Ferrariae in principio lectionum de simplicium medicamentarum facultatibus* (1543), in Felice Gioelli, 'Gaspare Gabrieli. Primo lettore dei semplici nello studio di Ferrara (1543)', *Atti e memorie della Deputazione Provinciale Ferrarese di Storia Patria*, ser. 3, 10 (1970), pp. 33–4, 37, 39.

6 See William B. Ashworth, Jr., 'Natural history and the emblematic world view', in David C. Lindberg and Robert S. Westman (eds.), *Reappraisals of the Scientific Revolution* (Cambridge, 1990), pp. 303–32; and Ashworth, 'Emblematic natural history of the Renaissance', in this volume.

7 Ambroise Paré, *On Monsters and Marvels* (orig. 1573; trans. Janis L. Pallister, Chicago, 1982), p. 128; Alice Stroup, *A Company of Scientists: Botany, Patronage, and Community at the Seventeenth-Century Parisian Royal Academy of Sciences* (Berkeley, 1990), p. 26. Further discussion of the fascination with nature's oddities can be found in Katharine Park and and Lorraine J. Daston, 'Unnatural conceptions: the study of monsters in sixteenth and seventeenth-century France and England', *Past and Present*, 92 (1981), pp. 20–54; and Paula Findlen, 'Jokes of nature and jokes of knowledge: the playfulness of scientific discourse in early modern Europe', *Renaissance Quarterly*, 43 (1990), pp. 292–331.

8 On botanizing as a gentlemanly pursuit, see Lucia Tongiorgi Tomasi, 'Gherardo Cibo: visions of landscape and the botanical sciences in a sixteenth-century artist', *Journal of Garden History*, 9 (1989), pp. 199–216; and Paula Findlen, *Possessing Nature: Museums, Collecting, and Scientific Culture in Early Modern Italy* (Berkeley, CA, 1994), ch. 4.

9 Paolo Boccone, *Museo di fisica e di esperienze* (Venice, 1697), p. 67. On the botanical garden in Pisa see Tongiorgi Tomasi, 'Projects for botanical and other gardens: a sixteenth-century Manual', *Journal of Garden History*, 3 (1983), pp. 1–34.

10 The material on Mattioli is drawn from Jerry Stannard, 'P. A. Mattioli: sixteenth century commentator on Dioscorides', *Bibliographic Contributions, University of Kansas Libraries*, vol. I (1969), pp. 59–81; and Richard Palmer, 'Medical botany in northern Italy in the Renaissance', *Journal of the Royal Society of Medicine*, 78 (1985), pp. 149–57 (quoted on p. 153). On natural history at the Habsburg court, see R. J. W. Evans, *Rudolf II and His World: A Study in Intellectual History 1576–1612* (Oxford, 1973), pp. 117–52, *passim*.

11 Evans, *Rudolf II and His World*, p. 119; Roger Schlesinger and Arthur P. Stabler (eds. and trans.), *André Thevet's North America* (Kingsland and Montreal, 1986), p. xxii.

12 See Paula Findlen, 'The economy of scientific exchange in early modern Italy', in Bruce Moran (ed.), *Patronage and Institutions: Science, Medicine and Technology at the European Courts, 1500–1750* (Woodbridge, Surrey, 1991), p. 19 (Ferdinando to Francesco I, Rome, 7 September 1576).

13 Of these naturalists, only Rauwolf has been treated extensively in English; see Karl H. Dannenfeldt, *Leonhardt Rauwolf: Sixteenth-Century Physician, Botanist and Traveller* (Cambridge, MA, 1968); Calzolari and Imperato are discussed in Findlen, *Possessing Nature*.

14 Thomas Platter's *Travels in England*, cited in Oliver Impey and Arthur MacGregor (eds.), *The Origins of Museums: The Cabinet of Curiosities in Sixteenth- and Seventeenth-Century Europe* (Oxford, 1985), p. 148.

15 As quoted in David S. Lux, *Patronage and Royal Science in Seventeenth-Century France: The Académie de Physique in Caen* (Ithaca, 1989), pp. 40, 68.

16 As quoted in 'John Gerard', *Dictionary of Scientific Biography*, vol. V, ed. Charles C. Gillispie (New York, 1972), p. 361.

17 Archivio Isolani, Bologna, *Fondo Paleotti*, 59 (F 30) 29/11, fo. 14; Walter E. Houghton, Jr., 'The English virtuoso in the seventeenth century', *Journal of the History of Ideas*, 3 (1942), p. 194.

18 Biblioteca Universitaria, Bologna, *Aldrovandi*, MS. 92, c. 11r; Ferrante Imperato, *Dell'historia naturale* (Naples, 1599), preface. On gift-giving, see Findlen, 'The economy of scientific exchange'.

19 On natural history as a form of 'civil conversation', see Jay Tribby, 'Cooking (with) Clio and Cleo: eloquence and experiment in seventeenth-century Florence', *Journal of the History of Ideas*, 52 (1991), pp. 417–39; and Tribby, 'Body/building: living the museum life in early modern Europe', *Rhetorica*, 10 (1992), pp. 139–63.

20 Richard J. Durling, 'Conrad Gesner's *Liber amicorum 1555–1565*', *Gesnerus*, 22 (1965), pp. 134–57. On Aldrovandi's catalogues of visitors, see Findlen, *Possessing Nature*, ch. 3.

21 The social uses of collecting are underscored in Giuseppe Olmi, 'Science–honour–metaphor: Italian cabinets of the sixteenth and seventeenth centuries', in Impey and MacGregor (eds.), *The Origins of Museums*, pp. 5–16; Olmi, *L'inventario del mondo. Catalogazione della natura e luoghi del sapere nella prima età moderna* (Bologna, 1992); and Krzysztof Pomian, *Collectors and Curiosities: Paris and Venice, 1500–1800*, trans. Elizabeth Wiles-Portier (London, 1990).

22 David C. Goodman, *Power and Penury: Government, Technology and Science in Philip II's Spain* (Cambridge, 1988), pp. 233–8 (quote on p. 235).

23 Paul Hultan, *America 1585: The Complete Drawings of John White* (Chapel Hill, NC, 1984), pp. 7–39, *passim*.

24 E. van den Boogaart, with H. R. Hoetink and P. J. P. Whitehead (eds.), *Johan Maurits van Nassau-Siegen 1604–1679: A Humanist Prince in Europe and Brazil* (The Hague, 1979); and P. J. P. Whitehead and M. Boeseman, *A Portrait of Dutch Seventeenth Century Brazil: Animals, Plants and People by the Artists of Johan Maurits of Nassau* (Amsterdam, 1989) (quote on p. 25).

25 Carolus Linnaeus, *Critica botanica*, trans. Arthur Hort (London, 1938), p. 58.

5 The culture of curiosity

1 For an account of curious travels on the Continent as part of the education of a gentleman, see Edward Leigh, *Guide for Travellers into Forein Parts* in his *Three Diatribes or Discourses* (London, 1671). Examples of such travels are described in John Raymond's *Itinerary Contayning a Voyage, Made through Italy* (London, 1648) and John Ray's *Observations Topographical, Moral, and Physiological* (London, 1673). The curious life-style can be seen clearly in the diaries of curiosi such as *The Diary of John Evelyn*, 6 vols., ed. E. S. De Beer (Oxford, 1955); *The Diary of Ralph Thoresby*, 2 vols., ed. Joseph Hunter (London, 1830); *The London Diaries of William Nicolson*, ed. Clyve Jones and Geoffrey Holmes (Oxford, 1985).

2 On the sublime, see S. H. Monk, *The Sublime: A Study of Critical Theories in XVIII-Century England* (Ann Arbor, MI, 1960); Marjorie Hope Nicolson, *Mountain Gloom and Mountain Glory: The Development of the Aesthetics of the Infinite* (Ithaca, NY, 1959). There is no corresponding study of the curious appreciation of nature and art apart from the present article.

3 *Diary of John Evelyn* (9 February 1665), vol. III, p. 399; (22 October 1684), vol. IV, pp. 389–90.

4 Ralph Thoresby, *Musæum Thoresbyanum* (London, 1713), pp. 440, 445; *London in*

1710 from the Travels of Zacharias Conrad von Uffenbach, ed. and trans. W. H. Quarrell and Margaret Mare (London, 1934), p. 101.

5 *Diary of Ralph Thoresby* (17 April 1679), vol. I, p. 29; (20 July 1681), vol. I, pp. 88–9.

6 *The Works of the Honourable Robert Boyle* (London, 1744), vol. III, pp. 202–12; *The Diary of Samuel Pepys*, 11 vols., ed. Robert Lathum and William Matthews (London, 1970–83), entry for 30 May 1667, vol. VIII, p. 243.

7 *Diary of Ralph Thoresby* (24 January 1682/3), vol. I, pp. 112–13; *London Diaries of William Nicolson* (4 January 1705), p. 268.

8 *Diary of John Evelyn* (14 July 1675), vol. IV, pp. 69–70; *Diary of Ralph Thoresby* (30 September 1680), vol. I, p. 65.

9 *Musæum Thoresbyanum*, p. 445.

10 *London Diaries of William Nicolson* (29 December 1704), p. 262.

11 Part 2, sections I and II.

12 *Diary of John Evelyn* (22 October 1684), vol. IV, pp. 389–90.

13 *London in 1710*, pp. 177, 186; *London Diaries of William Nicolson* (29 December 1702), p. 154; (9 November 1705), p. 300. 'Oakey-Hole' is Wookey Hole, a cave near Wells in Somersetshire.

14 *Diary of Ralph Thoresby* (11 June 1681), vol. I, pp. 85–6; letter from Lord Coleraine to William Courten (April 1694), Sloane MS 3962, fos. 233–4.

15 Henry Oldenburg, 'A preface to the third year of these tracts', *Philosophical Transactions*, 23 (11 March 1666), p. 413; Thomas Browne, *Pseudodoxia Epidemica* (London, 1646), pp. 43–5; *Diary of Ralph Thoresby* (20 July 1681), vol. I, pp. 88–9; Joshua Childrey, *Britannia Baconica* (London, 1661), dedication to Henry Somerset, Lord Herbert.

16 For circumstantial reporting in the new experimental philosophy, see Steven Shapin, 'Pump and circumstance: Robert Boyle's literary technology', *Social Studies of Science*, 14 (1984), pp. 481–520; Peter Dear, 'Totius in verba: rhetoric and authority in the early Royal Society', *Isis*, 76 (1985), pp. 145–61.

17 For criticisms of predecessors, see John Ray, *Catalogus Plantarum Angliae* (London, 1670), preface, and Nehemiah Grew, *Musæum Regalis Societatis* (London, 1681), preface.

18 Anthony Ashley Cooper, Third Earl of Shaftesbury, *Characteristicks of Men, Manners, Opinions, Times* (London, 1711), vol. III, pp. 1–4.

19 J. C. Harvey, 'The function of wonder and admiration in seventeenth century poetry and poetics', Ph.D. dissertation (Cambridge, 1975), especially pp. 117–18, 127; H. V. S. Ogden, 'The principles of variety and contrast in seventeenth century aesthetics, and Milton's poetry', *Journal of the History of Ideas*, 10 (1949), pp. 159–82.

20 *Diary of John Evelyn* (17 October 1671), vol. III, p. 594.

21 Krzysztof Pomian, *Collectors and Curiosities*, trans. Elizabeth Wiles–Portier (Cambridge, 1990), pp. 45–9; Antoine Schnapper, *Le géant, la licorne et la tulipe* (Paris, 1988), p. 11.

22 Entry for 17 January 1711/12 in Nicolson's 'Note Book' of antiquarian visits, *London Diaries of William Nicolson*, p. 699. Contraries were thought to be seen best when they were closely juxtaposed: H. V. S. Ogden, 'The principles of variety and contrast in seventeenth century aesthetics', pp. 167–71.

23 John Ray, *Historia Plantarum* (London, 1686–1704), vol. II, p. 1800, trans. in C. E. Raven, *John Ray, Naturalist: His Life and Works* (Cambridge, 1950), p. 229.

24 Letter from Lord Coleraine to William Courten (n.d., but with the note 'July 1687' in Courten's hand), Sloane MS 3962, fos. 225–6.

25 Sloane MS 1968, fo. 192.

26 A reference to Joshua telling the sun to stand still so that the Israelites could be avenged on their enemies (Joshua 10:12–13). I am grateful to Dr Andrew Cunningham for this reference.

27 Botanic gardens were also regarded as recreations of Paradise: see John Prest, *The Garden of Eden* (New Haven and London, 1981). For John Evelyn's 'Hortus Hyemalis', see *The Diary of Samuel Pepys* (5 November 1665), vol. VI, p. 289.

28 Presumably an animal from Borneo. I have been unable to find any other reference to it.
29 *Diary of Ralph Thoresby* (24 September 1694), vol. I, pp. 275–6.

6   Physicians and natural history

1 [Christopher Merrett?], *The Character of a Compleat Physician, or Naturalist* (London, 1680?), quotations from pp. 2–3, 6.
2 I am indebted for my knowledge of medieval medicine to Faye Getz. For a good general introduction to medieval herbals, see the introduction to Lynn Thorndike (ed.) and Francis S. Benjamin, Jr. (asst.), *The Herbal of Rufinus* (Chicago, 1946).
3 Quotation from Richard Palmer, 'Pharmacy in the Republic of Venice in the sixteenth century', in Andrew Wear, Roger K. French, and Ian M. Lonie (eds.), *The Medical Renaissance of the Sixteenth Century* (Cambridge, 1985), p. 110; also see Palmer, 'Medical botany in northern Italy in the Renaissance', *Journal of the Royal Society of Medicine*, 78 (1985), pp. 149–157; Gilbert Watson, *Theriac and Mithridatium: A Study in Therapeutics* (London, 1966); Jay Tribby, 'Cooking (with) Clio and Cleo: eloquence and experiment in seventeenth-century Florence', *Journal of the History of Ideas*, 52 (1991), pp. 417–39.
4 Clifford M. Foust, *Rhubarb: The Wondrous Drug* (Princeton, 1992).
5 C. H. H. Wake, 'The changing pattern of Europe's pepper and spice imports, c. 1400–1700', *Journal of European Economic History*, 8 (1979), pp. 361–403; Wake, 'The volume of European spice imports at the beginning and end of the 15th century', *Journal of European Economic History*, 15 (1986), pp. 621–35; R. S. Roberts, 'The early history of the import of drugs into Britain', in F. N. L. Poynter (ed.), *The Evolution of Pharmacy in Britain*, (London, 1965), pp. 165–85.
6 Wolfgang Schivelbusch, *Tastes of Paradise: A Social History of Spices, Stimulants, and Intoxicants*, trans. David Jacobson (New York, 1992).
7 This list is compiled from the pamphlets attributed to Peachey, and a few others, all of whose titles begin *Some Observations made upon the . . .*, and found in the British Library and Wellcome Institute Library.
8 A. J. J. van de Velde, 'Bijdrage tot de studie der werken van den Geneeskundige Cornelis Bontekoe', *Koninklijke Vlaamsche Academie voor Taal en Letterkunde, Verslagen en Mededeelingen* (1925), pp. 3–48.
9 Sigrid C. Jacobs, 'Guaiacum: history of a drug; a critico-analytical treatise', Ph.D. dissertation (University of Denver, 1974).
10 Saul Jarcho, *Quinine's Predecessor: Francesco Torti and the Early History of Cinchona* (Baltimore, 1993).
11 On Ghini (pp. 35–6), and more generally on Renaissance medical faculties and botany, see Karen Meier Reeds, *Botany in Medieval and Renaissance Universities* (New York, 1991).
12 Jean Fernel, *Methodus medendi*, quoted in Reeds, *Botany*, pp. 25–6.
13 The definitions are taken from the *Oxford English Dictionary*; also see Barbara J. Shapiro, *Probability and Certainty in Seventeenth-Century England* (Princeton, 1983); and Shapiro, *'Beyond Reasonable Doubt' and 'Probable Cause'* (Berkeley, 1991).
14 Hans Sloane, *A Voyage to the Islands Madera, Barbados, Nieves, S. Christophers and Jamaica, with the Natural History of the Herbs and Trees, Four-footed Beasts, Fishes, Birds, Insects, Reptiles, &c.* (London, 1707), sig. Bv.
15 Charlton T. Lewis and Charles Short, *A Latin Dictionary* (Oxford, 1975).
16 *An Intermediate Greek–English Lexicon, founded upon the Seventh Edition of Liddell and Scott's Greek–English Lexicon* (Oxford, 1968).
17 See Ian M. Lonie, 'The "Paris Hippocratics": teaching and research in Paris in the second half of the sixteenth century', in Andrew Wear, Roger K. French, and Ian M. Lonie (eds.), *The Medical Renaissance of the Sixteenth Century* (Cambridge, 1985), pp. 155–74, 318–26; Vivian Nutton, 'Hippocrates in the Renaissance', in *Die Hippokratischen Epidemien*, Sudhoffs Archiv, Beiheft 27, pp. 420–39; Wesley D. Smith, *The Hippocratic Tradition* (Ithaca, NY, 1979), pp. 15–18.

18 *Martin Lister's English Spiders, 1678*, trans. Malcolm Davies and Basil Harley, edited with introduction and notes by John Parker and Basil Harley (Colchester, 1992); A. Schierbeek, *Jan Swammerdam: Zijn Leven en Zijn Werken* (Lochem, 1946).

19 E. S. De Beer, *The Diary of John Evelyn*, 6 vols. (Oxford, 1955), vol. II, pp. 52, 53.

20 Jerome J. Bylebyl, 'The school of Padua: humanistic medicine in the sixteenth century', in Charles Webster (ed.), *Health, Medicine and Mortality in the Sixteenth Century* (Cambridge, 1979), pp. 335–70; Bylebyl, 'Medicine, philosophy, and humanism in Renaissance Italy', in John W. Shirley and F. David Hoeniger (eds.), *Science and the Arts in the Renaissance* (Washington, DC, 1985), pp. 27–49.

21 Reeds, *Botany*.

22 Rio Howard, 'Guy de La Brosse: botanique et chemie au début de la révolution scientifique', *Revue d'Histoire des Sciences*, 31 (1978), pp. 301–26; Howard, 'Guy de la Brosse and the Jardin des Plantes in Paris', in Harry Woolf (ed.), *The Analytic Spirit* (Ithaca, 1981), pp. 195–224; Howard Solomon, *Public Welfare, Science, and Propaganda in Seventeenth Century France* (Princeton, 1972); Alice Stroup, *A Company of Scientists* (Berkeley, 1990).

23 R. P. Stearns, 'The relations between science and society in the later seventeenth century', in *The Restoration of the Stuarts: Blessing or Disaster?* (Washington, DC, 1960), pp. 67–75; also see K. T. Hoppen, 'The nature of the early Royal Society', *British Journal for the History of Science*, 9 (1976), pp. 1–24, 243–73; Robert G. Frank, Jr., 'Institutional structure and scientific activity in the early Royal Society', *Proceedings of the XIVth International Congress of the History of Science* (Tokyo, 1975), vol. IV, pp. 82–101; Michael Hunter, *Establishing the New Science: The Experience of the Early Royal Society* (Woodbridge, 1989).

24 C. C. Gillispie, 'Physick and philosophy: a study of the influence of the College of Physicians of London upon the foundation of the Royal Society', *Journal of Modern History*, 19 (1947), p. 224.

25 M. Boas Hall, *The Scientific Renaissance* (orig. 1962; New York, 1966), p. 286; A. Rupert Hall, 'Medicine and the Royal Society', in A. G. Debus (ed.), *Medicine in Seventeenth Century England* (Berkeley, 1974), pp. 423, 421, 452. Also see A. R. Hall, 'English medicine in the Royal Society's correspondence: 1660–1677', *Medical History*, 15 (1971), pp. 111–25.

26 Michael Hunter, *The Royal Society and its Fellows 1660–1700: The Morphology of an Early Scientific Institution* (Chalfont St Giles, Bucks: The British Society for the History of Science, 1982), esp. pp. 22–32.

27 See Harold J. Cook, 'Physick and natural history in seventeenth-century England', in R. Ariew and P. Barker (eds.), *Revolution and Continuity: Essays in the History of Philosophy of Early Modern Science* (Washington, DC, 1991), pp. 63–80; and Cook, 'The cutting edge of a revolution? Medicine and natural history near the shores of the North Sea', in J. V. Field and Frank A. J. L. James (eds.), *Renaissance and Revolution: Humanists, Scholars, Craftsmen and Natural Philosophers in Early Modern Europe* (Cambridge, 1993), pp. 45–61.

7 Natural history as print culture

1 For example, L. Febvre and H.-J. Martin, *The Coming of the Book*, trans. D. Gerard (London, 1984); E. L. Eisenstein, *The Printing Press as an Agent of Change: Communications and Cultural Transformations in Early Modern Europe*, 2 vols. (Cambridge, 1979).

2 R. Chartier (ed.), *The Culture of Print: Power and the Uses of Print in Early Modern Europe*, trans. L. G. Cochrane (Cambridge, 1989), p. 3.

3 R. Atkyns, *The Vindication of Richard Atkyns, Esquire . . . with Certain Sighs or Ejaculations at the End of every Chapter* (London, 1669). For verisimilitude, the sighs in this chapter are derived virtually *verbatim* from this work: pp. 2 (1st and 2nd Sighs), 4 (3rd), 72 (4th), 9 (5th), 53 (6th), 3 (7th), 80 (8th).

4 J. Barnard, *Theologo–Historicus: Or, the True Life of the Most Reverend Divine, and*

Excellent Historian Peter Heylyn (London, 1683), sigs. [A5]ᵛ–[A6]ʳ, pp. 86–91, 215, 259–62; P. Heylyn, *Cosmographie* (2nd edn., London, 1657), frontispiece, title-page, sig. A3ᵛ, p. 28; Heylyn, Μικρόκοσμοσ (Oxford, 1625), pp. 17–19; J. Morrill, *The Nature of the English Revolution* (Harlow, 1993), p. 68.

5 Heylyn, *Cosmographie*, sig. A3ʳ; S. Shapin, '"A scholar and a gentleman": the problematic identity of the scientific practitioner in early modern England', *History of Science*, 29 (1991), pp. 279–327.

6 Barnard, *Theologo–Historicus*, pp. 86–9, 94; P. Heylyn, Κειμηλια 'Εκκλησιαστικα (London, 1681), p. iv; Heylyn, *Microcosmus* (Oxford, 1621), sigs. ¶2ʳ–¶3ʳ. The literature on early modern 'censorship' is extensive and embittered, and I do not mean to enter the lists for any particular side on this occasion. For opposing viewpoints, see C. Hill, 'Censorship and English literature', in his *Collected Essays, I: Writing and Revolution in Seventeenth-Century England* (Brighton, 1985), pp. 32–72, and S. Lambert, 'Richard Montagu, Arminianism and censorship', *Past and Present*, 124 (1989), pp. 36–68.

7 Barnard, *Theologo–Historicus*, pp. 94–5; M. Biagioli, *Galileo, Courtier: The Practice of Science in the Culture of Absolutism* (Chicago, 1993), pp. 149–54. See also R. S. Westman, 'Proof, poetics and patronage: Copernicus' Preface to *De Revolutionibus*', in R. S. Westman and D. C. Lindberg (eds.), *Reappraisals of the Scientific Revolution* (Cambridge, 1990), pp. 167–205.

8 G. Vernon, *The Life of the Learned and Reverend Dr. Peter Heylyn* (London, 1682), pp. 18–24.

9 Heylyn, Κειμηλια 'Εκκλησιαστικα, pp. ix–x; Barnard, *Theologo–Historicus*, p. 141; [Anon.], *Divine and Politike Observations . . . upon Some Lines in the Speech of the Arch. B. of Canterbury* (n.p., 1638), p. 10; W. W. Greg, *A Companion to Arber* (Oxford, 1967), p. 314; L. Jardine and A. Grafton, ' "Studied for action": How Gabriel Harvey read his Livy', *Past and Present*, 129 (1990), pp. 30–78.

10 Heylyn, Κειμηλια 'Εκκλησιαστικα, p. v; Barnard, *Theologo–Historicus*, pp. 94–101.

11 J. Moxon, *Mechanick Exercises on the Whole Art of Printing* (London, 1683–4), pp. 192–219, 382; W. Prynne, *The Unbishopping of Timothy and Titus* (London, 1636), pp. 176ff. D. F. McKenzie, *Cambridge University Press 1696–1712*, 2 vols. (Cambridge, 1966), vol. I, pp. 84–5, 141, 144; M. de Grazia, *Shakespeare Verbatim: The Reproduction of Authenticity and the 1790 Apparatus* (Oxford, 1991), ch. 1.

12 Barnard, *Theologo–Historicus*, pp. 102–5; [P. Heylyn], *France Painted to the Life* (London, 1656); Heylyn, *A Full Relation of Two Journeys* (London, 1656), sigs. aʳ–c2ᵛ.

13 Barnard, *Theologo–Historicus*, pp. 135–40; Heylyn, Κειμηλια 'Εκκλησιαστικα, p. xii; *Calendar of State Papers, Domestic, 1635–6*, p. 445.

14 Heylyn, *Cosmographie*, sig. A2ᵛ.

15 Barnard, *Theologo–Historicus*, pp. 258–64.

16 M. Macdonald, *Mystical Bedlam: Madness, Anxiety, and Healing in Seventeenth-Century England* (Cambridge, 1981), pp. 174, 181–3, 185–92, 288 n.59.

17 Heylyn, *Cosmographie*, sigs. A3ᵛ–A4ʳ.

18 See Table 7.1. The continuing use of the work by readers may be discerned from annotations in surviving copies: one in the British Library, for example (classmark 1505/19: the 1669 edition printed for Anne Seile), contains manuscript notes updating it from 1636, where the printed text ends, to at least 1712 (see p. 247).

19 H. Williams (ed.), *The Correspondence of Jonathan Swift*, 5 vols. (Oxford, 1963–5), vol. I, pp. 132–4; E. Bohun, *A Geographical Dictionary* (London, 1688), sigs. A3ʳ–[A5]ᵛ; Bohun/J. A. Bernard, *A Geographical Dictionary* (London, edns. of 1691, 1693, 1695, 1710), title-pages; A. à Wood, *Athenae Oxoniensis*, 4 vols. (London, 1813–20), vol. IV, p. 610; E. Bohun (ed. S. W. Rix), *Diary and Autobiography* (Read Crisp, Beccles, 1853), *passim*; L. Eachard, *A Most Compleat Compendium of Geography* (2nd edn. London, 1691), sigs. [A6]ʳ–[A7]ᵛ; Bohun, *Three Charges delivered at the General Quarter Sessions holden at Ipswich* (London, 1693), sigs. A2ʳ–[A4]ᵛ; [C. Blount], *Reasons Humbly offered for the Liberty of Unlicens'd*

Printing. To which is subjoin'd, The Just and True Character of Edmund Bohun, The Licenser of the Press (London, 1693), pp. 10–32; L. Moreri, *The Great Historical, Geographical and Poetical Dictionary* (London, 1694), preface; E. Bohun/J. A. Bernard, *Geographical Dictionary* (London, 1693 edn.), sigs. A2^r–[A4]^r; R. Astbury, 'The renewal of the Licensing Act in 1693 and its lapse in 1695', *The Library*, 5th ser., 33 (1978), pp. 296–322.

20 Heylyn, Κειμηλια ’Εκκλησιαστικα, p. xvii; Barnard, *Theologo–Historicus*, pp. 144–5, 202–3; [E. Bohun], *The Universal Historical Bibliotheque* (London, 1687). Bohun's letters to Petiver from Carolina are preserved as British Library MS Sloane 3321, fos. 33^r–74^r. The booksellers' call for subscriptions to the 1703 edition is in BL MS Harley 5946, pp. 118–19. For 'congers', see N. Hodgson and C. Blagden, *The Notebook of Thomas Bennet and Henry Clements (1686–1719), With Some Aspects of Book Trade Practice* (Oxford, 1956), p. 67ff. For transformations in reading practices, see R. Chartier, *The Order of Books: Readers, Authors, and Libraries in Europe between the Fourteenth and Eighteenth Centuries* (Cambridge, 1994), ch. 3, and Chartier, 'The practical impact of writing', in Chartier (ed.), *A History of Private Life. III: Passions of the Renaissance* (Cambridge, MA, 1989), pp. 111–59.

21 Atkyns, *Vindication*, p. 80; Job 19:22–4.

8   Natural history in the academies

1 J. Proust, *Diderot et l'Encyclopédie* (Paris, 1965).

2 R. Darnton, *The Business of Enlightenment. A Publishing History of the Encyclopédie, 1775–1800* (Cambridge, MA, 1979).

3 'Histoire naturelle', *Encyclopédie ou dictionnaire raisonné des sciences, des arts et des métiers*, vol. VIII, pp. 225–30.

4 D. Roche, *Le Siècle des Lumières en province: Académies et académiciens provinciaux, 1680–1789*, 2 vols. (Paris, 1978), vol. I, pp. 136–51.

5 C. Salomon-Bayet, *L'Institution de la science et l'expérience du vivant* (Paris, 1978), pp. 438–40.

6 *Patronage and Royal Science in Seventeenth-Century France: The Académie de Physique in Caen* (Ithaca, 1989), pp. 169–79.

7 Salomon-Bayet, *L'Institution de la science*, pp. 51–2.

8 Roche, *Le Siècle des Lumières*, vol. I, pp. 243–5.

9 Salomon-Bayet, *L'Institution de la science*, p. 53 and n. 4.

10 R. Hahn, *The Anatomy of a Scientific Institution: The Paris Academy of Sciences, 1666–1803* (Berkeley, 1971), pp. 2–34.

11 F. De Dainville, 'L'Enseignement scientifique chez les Jésuites', in R. Taton (ed.), *Enseignement et diffusion des sciences en France au XVIIIe siècle* (Paris, 1964), pp. 24–65; P. Costabel, 'L'Oratoire de France et ses collèges', in *idem*, pp. 67–100; R. Lemoine, 'L'Enseignement scientifique chez les Bénédictins', in *idem*, pp. 101–23.

12 Roche, *Le Siècle des Lumières*, pp. 94–6. 80 per cent of provincial academicians are known to have passed through the *collèges*, and 50 per cent undertook a university course.

13 P. Milsand, *Notes et documents pour servir à l'histoire de l'Académie des Sciences, Arts et Belles-Lettres de Dijon* (Dijon, 1970), pp. 180–5.

14 Salomon-Bayet, *L'Institution de la science*, pp. 104–5, 263–5.

15 Milsand, *Notes et documents*, p. 206.

16 M. Gosseaume, *Précis analytique des travaux de l'Académie des Sciences, Belles-Lettres et Arts de Rouen*, 5 vols. (Rouen, 1814), vol. I, pp. 167–8.

17 *Oeuvres philosophiques*, ed. P. Vernière (Paris, 1961), pp. 180–1.

18 G. Gusdorf, *Dieu, la nature, l'homme au siècle des Lumières* (Paris, 1977), pp. 243–59, 355–60.

19 E. Justin, *Les Sociétés royales d'agriculture au XVIIIe siècle, 1757–1793* (Saint-Lo, 1935).

20 F. Dagognet, *Des Révolutions vertes. Histoire et principes de l'agronomie* (Paris, 1973).

21 'Histoire Naturelle', p. 228.

22 D. Mornet, *Les Sciences de la nature en France au XVIIIe siècle* (Paris, 1911).

23 J. Torlais, 'La Physique expérimentale', in R. Taton (ed.), *Enseignement et diffusion des sciences*, pp. 620–45; C. Bedel, 'Les Cabinets de chimie', in *idem*, pp. 647–52; Y. Laissus, 'Les Cabinets d'histoire naturelle', in *idem*, pp. 659–712.

24 *Mémoire sur les cabinets d'histoire naturelle, et particulièrement celui du Jardin des Plantes* (n.p., n.d.), p. 2, cited in Laissus, 'Les Cabinets d'histoire naturelle', p. 669.

25 Salomon-Bayet, *L'Institution de la science*, pp. 373–95.

26 Salomon-Bayet, *L'Institution de la science*, pp. 392–5.

27 'Histoire naturelle', pp. 228–9.

28 B. M. Bordeaux, MS 828, Recueil de dissertation, MS 1699, pp. 3–4.

29 P. Barrière, *L'Academie de Bordeaux, centre de culture internationale au XVIIIe siècle, 1712–1792* (Bordeaux, 1951), pp. 83–96.

30 Roche, *Le Siècle des Lumières*, vol. I, pp. 355–75.

31 R. Tisserand, *Au temps de l'Encyclopédie, l'Académie de Dijon de 1740 à 1793* (Vesoul, 1936), pp. 265–6, 298–348.

32 J. Cousin, *L'Académie des Sciences, Belles-Lettres et Arts de Besançon, deux cents ans de vie Comtoise, 1752–1952* (Besançon, 1954), pp. 38–81.

33 Roche, *Le Siècle des Lumières*, vol. I, pp. 344–5, vol. II, pp. 138–48.

34 G. Lascault, *Le Monstre dans l'art occidental* (Paris, 1977).

35 Bachelard, *La Formation de l'esprit scientifique*, p. 27.

36 Gusdorf, *Dieu, la nature, l'homme*, pp. 258–9.

37 B. M. Bordeaux, MS 1299, Z, p. 24.

38 P. Duris, *Linné et la France, 1780–1850* (Geneva, 1993).

39 H. Daudin, *De Linné à Lamarck. Méthodes de la classification et idée de série en botanique et en zoologie, 1740–1790* (Paris, 1926–7), pp. 186–7.

9   Carl Linnaeus in his time and place

1 J. Beckmann (orig. 1765), quoted in R. Sernander, 'Hårleman och Linnaei Herbationes Upsalienses', *Svenska Linnésällskapets Årsskrift*, 9 (1926), pp. 78–86, 81.

2 *Monthly Review*, 3 (1750), p. 205.

3 Smith (1821), vol. II, p. 174. Letter from J. F. Gronovius to R. Richardson, (1738).

4 Carl Linnaeus, *Caroli Linnaei libellus amicorum* (1734–), *Valda avhandlingar av Carl von Linné*, trans. and ed. T. Fredbärj and A. H. Uggla, no. 30 (Ekenäs, 1958), p. 22, preface by Johan Browallius (15 February 1735).

5 Lars Roberg, quoted in R. Sernander, 'Linnaeus och Rudbeckarnes Hortus Botanicus', *Svenska Linnésällskapets Årsskrift*, 14 (1931), pp. 126–57, 147.

6 Letter from Linnaeus to Kilian Stobaeus (8 November 1728), quoted in M. Fries, *Linné. Lefnadsteckning*, 2 vols. (Stockholm, 1903), vol. I, p. 42.

7 Carl Linnaeus, *Deliciae Naturae* (Stockholm, 1773), reprinted as *Carl von Linné: Fyra Skrifter*, ed. A. H. Uggla (Stockholm, 1939), p. 108.

8 Carl Linnaeus, *Herbationes Upsalienses, I. Herbationerna 1747*, ed. [Å]. Berg, (Uppsala, 1952), p. 7, quoted from J. G. Acrel, 'Tal, vid praesidii nedläggande i Kungliga Vetenskaps Akademien' (1796).

9 W. T. Stearn, 'An introduction to the *Species Plantarum* and cognate botanical works of Carl Linnaeus', preface to vol. I of Carl Linnaeus, *Species plantarum* (fac. London, 1957); and John Lewis Heller, 'The early history of binomial nomenclature', *Huntia*, 1 (1964), pp. 33–70.

10 Fries, *Linné*, vol. I, p. 56.

11 Uppsala University Library, MS D.82a. Also cited in Fries, *Linné*, vol. I, p. 67.

12 *Bref och skrifvelser af och till Carl von Linné*, ed. T. M. Fries and J. M. Hulth (Stockholm and elsewhere, 1907–43), I:2, p. 59. Letter from Linnaeus to the Swedish Academy of Science, Uppsala (10 January 1746).

13 *Bref och skrifvelser*, I:2, p. 61. Letter from Linnaeus to Kalm via the Swedish Academy of Science, Uppsala (10 January 1746).

14 *Sveriges ekonomiska historia från Gustav Vasa* (Stockholm, 1949), II:1, p. 291.

15 *Bref och Skrifvelser*, I:7, pp. 137–8. Letter from Carl Hårleman to Linnaeus, Stockholm (21 July 1748).

16 Carl Linnaeus, 'Naturalicsamlingars ändamål och nytta', preface to *Museum S:ae R:ae M:tis Adolphi Friderici regis svecorum, gothorum, vandalorumque* . . . (Stockholm, 1754), repr., ed. T. M. Fries, in Carl Linnaeus, *Skrifter*, II (Stockholm, and Uppsala, 1906), p. 47.

17 P. F. Stevens, 'Metaphors and typology in the development of botanical systematics 1690–1960, or the art of putting new wine in old bottles', *Taxon*, 33(2) May 1984, pp. 169–211, esp. p. 172.

18 J. Lyon and P. R. Sloan, *From Natural History to the History of Nature* (Notre Dame, 1981), p. 106.

19 P. R. Sloan, 'The Buffon–Linnaeus controversy', *Isis*, 67 (1976), pp. 238, 356–75; J. L. Larson, 'Linné's French critics', *Svenska Linnésällskapets Årsskrift* (1978), pp. 67–79.

20 Quoted in R. Friedenthal, *Goethe. Sein Leben und seine Zeit* (Munich, 1963), p. 306.

21 Letter from Daniel Solander to Linnaeus. LC, Linnean Society, London. 'Paramaribo i Suriname' (11 July 1755).

22 Carl Linnaeus, *Skånska resa år 1749*, ed. C.-O. von Sydow (Stockholm, orig. 1751; 1975), p. 197.

23 Letter from Linnaeus to J. G. Gmelin (14 February 1747). Uppsala University Library, MS G 152al.

24 Linnaeus, 'Naturaliesamlingars ändamål och nytta' (1745), here quoted from Fries, *Linné*, vol. 2, Bilaga XXI, p. 20.

25 Linnaeus, from MS fragment, transcribed and quoted in G. Broberg, *Homo Sapiens L. Studier i Carl von Linnés naturuppfattning och människolära* (Motala, 1975), p. 144.

26 Carl Linnaeus, 'Tal, vid deras Kongl. Majesteters höga närvaro, hållit uti Upsala, på Stora Carolinska Lärosalen den 25 septemb. 1759', repr. in *Fyra skrifter*, p. 95.

27 S. Lindroth, 'Linné – legend och verklighet', *Lychnos* (1965–6), pp. 56–122, 66; Carl Linnaeus MS, 'Progressus botanices', Linnean Society, London.

28 Carl Linnaeus, *Nutrix noverca* (1752), in *Valda avhandlingar av Carl von Linné*, IV (Åbo, 1947), p. 15.

29 Carl Linnaeus, *Diaeta Naturalis* (1733), ed. A. H. Uggla (Uppsala, 1958), p. 11.

30 *Bref och skrifvelser*, I, pp. 5, 317. Letter from Linnaeus to Gustaf Cronhjelm (n.d., n.p., written in Stockholm, spring 1733).

31 Quoted from MS fragment, in C.-O. von Sydow, 'Linné och de lyckliga lapparna. Primitivistiska drag i Linnés Lapplandsuppfattning', in R. Granit (ed.), *Utur Stubbotan Rot. Essäer till 200-årsminnet av Carl von Linnés död* (Stockholm, 1978), pp. 72–78, 78.

32 Quoted from Adam Afzelius' eyewitness description, in R. E. Fries, '150-årsminnet av Linnés död', *Svenska Linnésällskapets Årsskrift*, II (1928), pp. 162–72, 168.

33 D. L. Hull, 'The effects of essentialism on taxonomy. Two thousand years of stasis', *The British Journal for the Philosophy of Science*, 15 (1965), pp. 314–66; 16 (1966), pp. 1–17.

34 *Göteborgs-Tidningen* (22 May 1957); from the collection at the Hunt Botanical Library, Pittsburgh.

35 *Uppsala Nya Tidning* (18 May 1957).

36 *Värnamo-Tidningen* (18 April 1969).

37 *Göteborgs Handels- och Sjöfarts-Tidning* (1969). I have seen the clipping, but it was not dated more precisely.

10 Gender and natural history

Portions of this essay have been adapted from L. Schiebinger, 'The private life of plants: sexual politics in Carl Linnaeus and Erasmus Darwin', in Marina Benjamin (ed.), *Science and Sensibility: Gender and Scientific Inquiry 1780–1945* (Oxford, 1991); and L. Schiebinger, *Nature's Body: Gender in the Making of Modern Science* (Boston, 1993).

1 Priapus is the Greco-Roman god of procreation and personification of the erect phallus.
2 *The Works of Aristotle*, trans. D'Arcy Thompson (Oxford, 1910), vol. IV, p. 572a. William Smellie, *The Philosophy of Natural History*, 2 vols. (Edinburgh, 1790), vol. I, p. 238. See also Nancy Cott, 'Passionlessness: an interpretation of Victorian sexual ideology, 1790–1850', *Signs: Journal of Women in Culture and Society*, 4 (1978), pp. 219–36; Ruth Yeazell, *Fictions of Modesty: Women and Courtship in the English Novel* (Chicago, 1991).
3 See Jennifer Bennett, *Lilies of the Hearth: The Historical Relation Between Women and Plants* (Camden, Ont., 1991).
4 Londa Schiebinger, *The Mind Has No Sex? Women in the Origins of Modern Science* (Cambridge, MA, 1989).
5 Ann Shteir, 'Botany in the breakfast room: women in early nineteenth-century British plant study', in Pnina G. Abir-Am and Dorinda Outram (eds.), *Uneasy Careers and Intimate Lives: Women in Science, 1789–1979* (New Brunswick, 1987), pp. 31–44.
6 William Smellie, 'Botany', *Encyclopaedia Britannica* (Edinburgh, 1771), vol. I, p. 646.
7 'Pistillum', *Dictionarium Britannicum*, ed. G. Gordon, P. Miller and N. Bailley (London, 1730). J. P. de Tournefort standardized the modern usage of the term 'pistil' (*pistile*) to designate the ovary, style and stigma of the flower in 1694; the term was introduced into botanical usage in England in the 1750s. The use of the term 'stamen' is much older; Pliny referred to the stamen of the lily. The technical use of the word in botany, however, dates to about 1625.
8 Smellie, 'Botany', *Encyclopaedia Britannica*, vol. I, p. 646.
9 Claude Geoffroy, 'Observations sur la structure et l'usage des principales parties des fleurs', *Memoires de l'Académie Royale des Sciences* (1711), p. 211.
10 Nehemiah Grew, *The Anatomy of Plants* (London, 1682), pp. 170–2.
11 Linnaeus, *Praeludia sponsaliorum plantarum*, section 16, cited in James Larson, 'Linnaeus and the natural method', *Isis*, 58 (1967), pp. 304–20, 306.
12 Lawrence Stone, *The Family, Sex and Marriage in England, 1500–1800* (New York, 1977) and *Road to Divorce: England 1530–1987* (Oxford, 1990).
13 Wilfrid Blunt, *The Compleat Naturalist: A Life of Linnaeus* (London, 1971), pp. 85, 167.
14 'Gamete', *Oxford English Dictionary*.
15 Margaret Jacob, 'The materialist world of pornography', paper delivered at the New York Area French History Seminar (September 1992).
16 Blunt, *The Compleat Naturalist*, pp. 176–7, 165–6.
17 Jeffrey Merrick, 'Royal Bees: the gender politics of the beehive in early modern Europe', *Studies in Eighteenth-Century Culture*, 18 (1988), pp. 7–37.
18 Thomas Laqueur, *Making Sex: Body and Gender from the Greeks to Freud* (Cambridge, MA, 1990).
19 See Christoph Trew, *Vermehrtes und verbessertes Blackwellisches Kräuter-Buch* (Nuremberg, 1750).
20 Carl Linnaeus, *Systema naturae* (1735) ed. M. S. J. Engel-Ledeboer and H. Engel (Nieuwkoop, 1964), 'Observationes in regnum vegetabile', nos. 7 and 12. See also Linnaeus to Haller (3 April 1737), in *The Correspondence of Linnaeus*, ed. James E. Smith, 2 vols. (London, 1821), vol. II, p. 229. For a discussion of Linnaeus's systematics, see James Larson, *Reason and Experience: The Representation of Natural Order in the Work of Carl von Linné* (Berkeley, 1971); and Frans A. Stafleu,

Linnaeus and the Linnaeans: The Spreading of Their Ideas in Systematic Botany, 1735–1789 (Utrecht, 1971).

21 Thomas Martyn, cited in David Allen, *The Naturalist in Britain: A Social History* (orig. 1976; Harmondsworth, 1978), p. 39.

22 Georges-Louis Leclerc de Buffon, *Histoire naturelle* (Paris, 1749), vol. I, pp. 13–20.

23 Allen, *The Naturalist in Britain*, chs. 1 and 2.

24 Smellie, 'Botany', *Encyclopaedia Britannica*, vol. I, pp. 627–53; and Smellie, *The Philosophy of Natural History*, vol. I, p. 246.

25 Smellie, 'Botany', *Encyclopaedia Britannica*, vol. I, p. 653.

26 Cited by Stearn in Blunt, *The Compleat Naturalist*, p. 245.

27 Erasmus Darwin, *The Loves of the Plants* (1789) in *The Botanic Garden* (fac. repr. Menston, 1973), canto I.

28 Erasmus Darwin, *A Plan for the Conduct of Female Education in Boarding Schools* (London, 1797).

29 Maureen McNeil, *Under the Banner of Science: Erasmus Darwin and His Age* (Manchester, 1987), pp. 133–5. See also Janet Browne, 'Botany for gentlemen: Erasmus Darwin and *The Loves of the Plants*', *Isis*, 80 (1989), pp. 593–621.

30 Roy Porter, 'Mixed feelings: the Enlightenment and sexuality in eighteenth-century Britain', in Paul-Gabriel Boucé (ed.), *Sexuality in Eighteenth-Century Britain* (Manchester, 1982), pp. 1–27.

31 Revd Richard Polwhele, *The Unsex'd Females; a Poem* (orig. 1798; New York, 1800), pp. 33–4.

32 Polwhele, *The Unsex'd Females*, pp. 10–20.

33 Priscilla Wakefield, *Introduction to Botany* (London, 1796).

34 Robert Proctor, *Value-Free Science? Purity and Power in Modern Knowledge* (Cambridge, MA, 1991).

35 See Schiebinger, *The Mind Has No Sex?*

11 Political, natural and bodily economies

The writing of this essay would not have been possible without the financial support of Girton College, Cambridge, and the personal support and advice of Paul White, Nick Jardine and Jim Secord.

1 D. G. Charlton, *New Images of the Natural in France* (Cambridge, 1984), ch. 4; Theodore Brown, *The Mechanical Philosophy and the Animal Oeconomy* (New York, 1981).

2 Charlton, *New Images of the Natural*, p. 15. The German term is *Bildung*; the French, *culture*.

3 The French *économistes*' views are outlined in Ronald L. Meek, *The Economics of Physiocracy: Essays and Translations* (London, 1962). Compare Myles Jackson, 'Goethe's economy of nature and the nature of his economy', *Accounting, Systems and Organisation*, 17 (1992), pp. 459–69.

4 Laurent-Benoît Desplaces de Montbron, *L'Agromanie, ou l'Agriculture réduite à ses vrais principes* (Paris, 1762).

5 Elizabeth L. Haigh, *Xavier Bichat and the Medical Theory of the Eighteenth Century* (London, 1984).

6 Leonora Cohen-Rosenfield, *From Beast-Machine to Man-Machine* (New York, 1968).

7 For a brief introduction to researches into 'Newtonian' forces see J. L. Heilbron, *Electricity in the Seventeenth and Eighteenth Centuries* (Berkeley, 1979).

8 Aram Vartanian, 'Trembley's polyp, La Mettrie, and eighteenth-century French materialism', *Journal for the History of Ideas*, 11 (1950), pp. 259–86.

9 Barbara Maria Stafford, *Body Criticism: Imaging the Unseen in Enlightenment Art and Medicine* (Cambridge, MA, 1991).

10 John Lyon and Phillip R. Sloan, *From Natural History to the History of Nature: Readings from Buffon and His Critics* (Notre Dame, 1981).

11 Phillip R. Sloan, 'The Buffon–Linnaeus controversy', *Isis*, 67 (1976), pp. 356–75.

12 Shirley A. Roe, *Matter, Life and Generation. Eighteenth-Century Embryology and the Haller–Wolff Debate* (Cambridge, 1981), pp. 27–9; Haller, preface to G.-L. L. de Buffon, *Allgemeine Historie der Natur*, vol. II (Hamburg, 1752).

13 H. Nadault de Buffon, *Buffon. Correspondance générale recueillie et annotée par H. Nadault de Buffon*, 2 vols. (Geneva, 1971), vol. I, pp. 327–31.

14 Lucile H. Brockway, *Science and Colonial Expansion. The Role of the British Royal Botanical Gardens* (New York, 1979); David Mackay, *In the Wake of Cook: Exploration, Science and Empire, 1780–1801* (London, 1985).

15 Gibbon, *The Decline and Fall of the Roman Empire* (orig. 1776–88; London, 1993); Ira O. Wade, *The Intellectual Origins of the French Enlightenment* (Princeton, 1971); Rousseau, *Discours sur l'origine et les fondements de l'inégalité parmi les hommes* (orig. 1755; Paris, 1965).

16 John H. Eddy, Jr., 'Buffon, organic alterations, and man', *Studies in the History of Biology*, 7 (1984), pp. 1–45.

17 J.-A.-C. de Condorcet, *Esquisse d'un tableau historique des progrès de l'esprit humain* (Paris, 1795); R. L. Meek, *Turgot on Progress, Sociology and Economics* (Cambridge, 1973); David Spadafora, *The Idea of Progress in Eighteenth-Century Britain* (New Haven, 1990).

18 La Mettrie, *Man a Machine* (London, 1750), p. 17.

19 R. K. French, *Robert Whytt, the Soul, and Medicine* (London, 1969), p. 11.

20 Compare Chris Lawrence, 'The nervous system and society in the Scottish Enlightenment', in B. Barnes and S. Shapin (eds.), *Natural Order. Historical Studies of Scientific Culture* (Beverly Hills, 1979), pp. 19–40.

21 P. Camper, *Oeuvres*, 3 vols. (Paris, 1803).

22 G. Barker-Benfield, *The Culture of Sensibility. Sex and Society in Eighteenth-Century Britain* (Chicago, 1992); A. Vincent-Buffault, *The History of Tears: Sensibility and Sentimentality in France* (Basingstoke, 1991); L. Jordanova (ed.) *Languages of Nature. Critical Essays on Science and Literature* (London, 1986).

23 Robert M. Maniquis, 'The puzzling mimosa: sensitivity and plant symbols in romanticism', *Studies in Romanticism*, 8 (1969), pp. 129–55; J. Browne, 'Botany for gentlemen. Erasmus Darwin and *The Loves of the Plants*', *Isis*, 80 (1989), pp. 593–621.

24 *Encyclopédie méthodique, Histoire naturelle des animaux* (Paris, 1782), article on 'Le hamster'.

25 Londa Schiebinger, 'Skeletons in the closet: the first illustrations of the female skeleton in nineteenth-century anatomy', in Thomas Laqueur and Catherine Gallagher (eds.), *The Making of the Modern Body: Sexuality and Society in the Nineteenth Century*, (Berkeley, 1987), pp. 1–38.

26 Joseph Priestley, *Experiments and Observations on Different Kinds of Air*, 3 vols. (London, 1774–7); Simon Schaffer, 'Measuring virtue: eudiometry, enlightenment and pneumatic medicine', in A. Cunningham and R. French (eds.), *The Medical Enlightenment of the Eighteenth Century* (Cambridge, 1990), pp. 281–318.

27 M. Norton Wise, 'Work and waste: political economy and natural philosophy in nineteenth-century Britain', part I, *History of Science*, 27 (1988), pp. 263–301.

12   The science of man

This essay is a considerably revised and condensed version of my article 'The natural history of man in the Scottish Enlightenment', *History of Science*, 27 (1989), pp. 89–123. Science History Publications have kindly allowed me to use portions of my text here.

1 While recognizing their gendered nature, throughout this paper I use the terms 'science of man' and 'man' in conformity with eighteenth-century usage. Sylvana Tomaselli explores eighteenth-century conceptions of the role of women in the history of humankind in 'The Enlightenment debate on women', *History Workshop*, 20 (1985), pp. 101–24.

2 John Gregory, 'A proposall for a Medicall Society. Written anno 1743', Aberdeen University Library (AUL) MS 2206–45, pp. 1, 4, 6–8.

3 John Gregory, *A Comparative View of the State and Faculties of Man with those of the Animal World*, 7th edn. (London, 1777), pp. 8–9.

4 However, as we have seen, Bacon did call for the compilation of natural histories of the mind in the *Parasceve*.

5 AUL MS 480, p. 3.

6 AUL MS 475, pp. 38–9, 47, 62–3, and MS 37, fo. 118^r.

7 On Skene as an inveterate classifier see B. P. Lenman and J. B. Kenworthy, 'Dr. David Skene, Linnaeus, and the applied geology of the Scottish Enlightenment', *Aberdeen University Review*, 47 (1977–8), p. 36.

8 [Alexander Gerard], 'Notes of lectures on moral philosophy. Taken by Robert Morgan [1758–59]', Edinburgh University Library (EUL) MS Dc.5.61, p. 2; *idem*, 'Lectures on moral philosophy', EUL MS doc.5.116, pp. 1–6.

9 On this issue see my '"Jolly Jack Phosphorous" in the Venice of the north; or, who was John Anderson', in Richard B. Sher and Andrew Hook (eds.), *The Glasgow Enlightenment* (Edinburgh, 1995).

10 For the revised curricula see Paul B. Wood, *The Aberdeen Enlightenment: The Arts Curriculum in the Eighteenth Century* (Aberdeen, 1993), ch. 3.

11 The museums established at the two Aberdeen universities are briefly discussed in Wood, *Aberdeen Enlightenment*, pp. 81, 86.

12 John Locke, *Essay*, ed. Peter H. Nidditch (Oxford, 1975), pp. 43–4, 161–2.

13 Locke, *Essay*, pp. 162, 402–4. On Locke's anthropology see William G. Batz, 'The historical anthropology of John Locke', *Journal of the History of Ideas*, 35 (1974), pp. 663–70.

14 Trans. Ernest Dilworth as *Philosophical Letters* (Indianapolis, 1961), pp. 53–4.

15 Jean Le Rond D'Alembert, *Preliminary Discourse to the Encyclopedia of Diderot*, trans. Richard N. Schwab and Walter E. Rex (Indianapolis, 1963), p. 84; D'Alembert's notion of an 'experimental physics' of the soul is very close to that of the natural history of the mind. Bentham's remarks are quoted in John Yolton, 'Schoolmen, logic and philosophy', in L. S. Sutherland and L. G. Mitchell (eds.), *The History of the University of Oxford*, vol. V, *The Eighteenth Century* (Oxford, 1986), p. 573.

16 George Turnbull, *The Principles of Moral Philosophy*, 2 vols. (London, 1740), vol. I, p. 9; Joseph Butler, *Fifteen Sermons Preached at the Rolls Chapel and a Dissertation upon the Nature of Virtue* (London, 1914), pp. 34–5.

17 On Hume see my 'Hume, Reid, and the science of the mind', in M. A. Stewart and John P. Wright (eds.), *Hume and Hume's Connexions* (Edinburgh, 1994), pp. 119–39.

18 Charles Louis de Secondat, baron de Montesquieu, *The Spirit of the Laws*, ed. and trans. Anne Cohler et al. (Cambridge, 1989), p. 316. On Buffon see Antonello Gerbi, *The Dispute of the New World: The History of a Polemic, 1750–1900*, trans. Jeremy Moyle, rev. edn. (Pittsburgh, 1973), ch.1.

19 David Hume, 'Of national characters', in *Essays Moral, Political, and Literary*, ed. Eugene F. Miller, rev. edn. (Indianapolis, 1987), pp. 198, 204; Robert Wokler, 'Apes and races in the Scottish Enlightenment: Monboddo and Kames on the nature of man', in Peter Jones (ed.), *Philosophy and Science in the Scottish Enlightenment* (Edinburgh, 1988), pp. 145–68.

20 Adam Smith, 'Letter to the *Edinburgh Review*', in W. P. D. Wightman et al. (eds.), *Essays on Philosophical Subjects* (Oxford, 1980), p. 248.

21 P. B. Wood, 'Buffon's reception in Scotland: the Aberdeen connection', *Annals of Science*, 44 (1987), pp. 169–90.

22 Smith, *Essays*, pp. 161–6.

23 Adam Ferguson, *Institutes of Moral Philosophy. For the use of Students in the College of Edinburgh*, 2nd edn. (Edinburgh, 1773), p. 11.

24 Adam Ferguson, *Principles of Moral and Political Science; being Chiefly a Retrospect of Lectures delivered in the College of Edinburgh*, 2 vols. (Edinburgh, 1792), vol. I, pp. 1–2, 5.

25 John Millar, *The Origin of the Distinction of Ranks*, 4th edn. (Edinburgh and London, 1806), pp. 11–12. Millar also devoted considerable attention to the status of women.

26 Kames, *Sketches*, 4 vols., 2nd edn. (Edinburgh, 1778), vol. I, pp. vii, 1.

27 Scottish responses to Rousseau are surveyed in R. A. Leigh, 'Rousseau and the Scottish Enlightenment', *Contributions to Political Economy*, 5 (1986), pp. 1–21; see also Peter France, 'Primitivism and Enlightenment: Rousseau and the Scots', *The Yearbook of English Studies*, 15 (1985), pp. 64–79.

28 [James Burnett, Lord Monboddo], *Of the Progress and Origins of Language*, 6 vols., 2nd edn. (Edinburgh, 1774–92), vol. I, p. 444.

29 Gregory, *Comparative View*, pp. 69–70.

30 J. C. Stewart-Robertson, 'Reid's anatomy of culture: a Scottish response to the eloquent Jean-Jacques', *Studies on Voltaire and the Eighteenth Century*, 205 (1982), pp. 141–63.

13   The natural history of the earth

1 J. C. Schaeffers, *Vorschlaege zu einer gemeinnuetzigen Ausbesserung und Befoerderung der Naturwissenschaft* (Ulm, *c.* 1760), p. 3.

2 R. Laudan, *From Mineralogy to Geology: The Foundations of a Science, 1650–1830* (Chicago, 1987), pp. 70–86.

3 Further details are available in T. M. Porter, 'The promotion of mining and the advancement of science: the chemical revolution in mineralogy', *Annals of Science*, 38 (1981), pp. 543–70, and in R. Laudan, *From Mineralogy to Geology*.

4 There is an extensive literature, especially in German, devoted to Werner and the Freiberg school. For an introduction see the article by A. Ospovat in C. C. Gillispie (ed.), *Dictionary of Scientific Biography*, 14 (1976), pp. 256–64.

5 For Woodward, see J. M. Levine, *Dr Woodward's Shield: History, Science, and Satire in Augustan England* (Berkeley and Los Angeles, 1977).

6 Translated by J. M. Winter as *The Prodromus of Nicholas Steno's Dissertation Concerning a Solid Body Enclosed by Process of Nature within a Solid* (New York, 1916).

7 For an edition in English see *The Lying Stones of Dr Beringer*, trans. M. E. Jahn and D. J. Woolf (Berkeley and Los Angeles, 1963).

8 J. Scheuchzerus, *Piscium querelae et vindiciae* (Tiguri, 1708).

9 For this and other debates about fossils see M. J. S. Rudwick, *The Meaning of Fossils: Episodes in the History of Palaeontology* (New York, 1976).

10 Illustrations from this work are reproduced in M. J. S. Rudwick, *Scenes from Deep Time: Early Pictorial Representations of the Prehistoric World* (Chicago, 1992), pp. 4–17.

11 J. Roger produced an excellent critical edition of Buffon's *Les Epoques de la nature* (Paris, 1962).

12 J. G. Lehmann, *Versuch einer Geschichte von Flötz-Gebürgen* (Berlin, 1756).

13 A. G. Werner manuscripts, vol. II, p. 206, Bibliothek der Bergakademie, Freiberg.

14 D. Dean, *James Hutton and the History of Geology* (Ithaca, 1992) surveys these publications and Hutton's later reputation.

15 J. E. Fichtel, *Mineralogen gegen das Ende des achtzehnten Jahrhunderts* (1792), in *Freiberger Forschungshefte* (Leipzig and Stuttgart, 1993), D199, p. 5.

14   *Naturphilosophie* and the kingdoms of nature

I am indebted for much help to Emma Spary and other members of the Cabinet of Natural History; also to Bill Clark, Marina Frasca-Spada, Sue Morgan, Betty Jackson, and Elinor Shaffer for guidance.

1 *Rapport historique sur les progrès des sciences naturelles* (Paris, 1810), pp. 235–6.

2 *De l'Allemagne* (orig. 1810), ed. J. de Pange and S. Balayé (Paris, 1959), vol. IV, p. 242.

3 *Diary, Reminiscences and Correspondence of Henry Crabb Robinson*, ed. T. Sadler (London, 1869), vol. I, p. 134.

4 Annotations in B.L.C. 43.b.12 (pp. 1–2).

5 For evocations of the moods and values of Romanticism see M. H. Abrams, *The Mirror and the Lamp* (Oxford, 1953), and J. L. Koerner, *Caspar David Friedrich and the Subject of Landscape* (London, 1990).

6 See R. Marks, *Konzeption einer dynamischen Naturphilosophie bei Schelling und Eschenmayer* (Munich, 1985).

7 On Romantic historiography see D. von Engelhardt, *Historisches Bewußtsein in der Naturwissenschaft* (Munich, 1979), pp. 103–58.

8 'Grundzüge allgemeiner Naturbetrachtung', *Zur Morphologie*, ed. Goethe, vol. II (1823), pp. 84–95, p. 85.

9 *Sämmtliche Werke*, ed. K. F. A. von Schelling (Stuttgart, 1856–61), vol. III, pp. 6–8.

10 *Alt und Neu* (Breslau, 1821), vol. II, p. 102.

11 See W. Riese, 'The impact of Romanticism on the experimental method', *Studies in Romanticism*, 2 (1962), pp. 12–22; and B. Lohff, *Die Suche nach Wissenschaftlichkeit der Physiologie in der Zeit der Romantik* (Stuttgart, 1990); and M. J. Trumpler, 'Questioning nature: experimental investigation of animal electricity in Germany, 1791–1810', unpubl. Ph.D. thesis (Yale, 1992).

12 On Goethean morphology see, for example H. Bräuning-Oktavio, 'Vom Zwischenkieferknochen zur Idee des Typus: Goethe als Naturforscher in den Jahren 1780–1786', *Nova Acta Leopoldina*, NS 18 (1956); T. Lenoir, 'The eternal laws of form: morphotypes and the conditions of existence in Goethe's biological thought', in F. Amrine et al., *Goethe and the Sciences: A Reappraisal* (Dordrecht, 1987), pp. 17–28.

13 On Schelling's philosophy see K. Fischer's classic *Schellings Leben, Werke und Lehre*, 2nd edn. (Heidelberg, 1899); A. Bowie, *Schelling and Modern European Philosophy* (London, 1993). On Schelling and natural history see D. von Engelhardt, 'Die organische Natur und die Lebenswissenschaften in Schellings Naturphilosophie', in R. von Heckmann et al. (eds.), *Natur und Subjektivität* (Stuttgart, 1985), pp. 39–57.

14 On Kant's genealogical natural history see works cited in n. 31 below.

15 *Sämmtliche Werke*, vol. III, p. 63. The reference is to Kant's *Von den verschiedenen Racen der Menschen*, in *Vorkritische Schriften*, ed. E. Buchenau (Berlin, 1912), vol. II, p. 451.

16 On Steffens's life and works see O. Liebemann, 'Steffens', *Allgemeine deutsche Biographie*, vol. XXXV (Leipzig, 1893), pp. 555–8; F. Paul, *Henrich Steffens. Naturphilosophie und Universalromantik* (Munich, 1973).

17 Cited in A. M. Ospovat's notes to his translation of Werner's *Kurze Klassifikation und Beschreibung der verschiedenen Gebürgsarten* (orig. 1787; New York, 1971), p. 102. On Werner and Steffens see A. M. Ospovat, 'Romanticism and German geology: five students of Abraham Gottlob Werner', *Eighteenth Century Life*, 7 (1981–2), pp. 105–17.

18 On Kielmeyer see D. Kuhn, 'Uhrwerk oder Organismus: Karl Friedrich Kielmeyers System der organischen Kräfte', *Nova Acta Leopoldina*, NS 36 (1970), pp. 156–67; K. T. Kanz, 'Einführung' to Kielmeyer's *Ueber die Verhältnisse der organischen Kräfte*, fac. (Marburg, 1993).

19 On Esenbeck see J. Proskauer, *D.S.B.*, vol. X, pp. 11–14.

20 Trans. B. Mueller, *Goethe's Botanical Writings* (Honolulu, 1952).

21 On Oken's life see A. Ecker, *Lorenz Oken* (Stuttgart, 1880), trans. A. Tulk (London, 1883). The only detailed account of Oken's *Naturphilosophie* known to me is G. Busse's excellent *Philosophische und geistesgeschichtliche Grundzüge der Lehre Lorenz Okens*, unpubl. D.Phil. thesis (Freiburg, 1950).

22 *Gesammelte Schriften*, ed. J. Schuster (Berlin, 1939), pp. 258–9.

23 Ecker, *Lorenz Oken*, p. xx.

24 Based on *Lehrbuch der Naturphilosophie*, 2nd edn., p. 40.

25 A. Tulk's trans. of *Principles of Physiophilosophy*, 3rd edn. (London, 1849), p. 45.

Similar parallelisms are invoked by many other German anatomists of the period: see O. Temkin, 'German concepts of ontogeny and history around 1800', *Bulletin of the History of Medicine* 24 (1950), pp. 227–46.

26 See Oken's account of his *Systemkunde* in *Lehrbuch der Naturgeschichte*, vol. I, pp. 13–18. On Oken's systematics see J. B. Stallo, *General Principles of the Philosophy of Nature* (Boston, MA, 1848), pp. 230–330; H. Querner, 'Ordnungsprinzipien und Ordnungsmethoden in der Naturgeschichte der Romantik', in R. Brinkmann (ed.), *Romantik in Deutschland* (Stuttgart, 1978), pp. 214–25.

27 See H. Bräuning-Oktavio, *Oken und Goethe im Lichte neuer Quellen* (Weimar, 1959).

28 See E. S. Russell's classic *Form and Function: A Contribution to the History of Animal Morphology* (London, 1916, repr. Chicago, 1982); P. F. Rehbock, *The Philosophical Naturalists* (Madison, WI, 1983).

29 M. Foucault, *Les Mots et les choses* (Paris, 1966), trans. as *The Order of Things* (London, 1970), chs. 7–8. A special issue of *Revue d'histoire des sciences et leurs applications*, 23 (1970), is devoted to Foucault's claims about the transformation of natural history; see also W. R. Albury and D. R. Oldroyd, 'From Renaissance mineral studies to historical geology, in the light of Michael Foucault's *The Order of Things*', *British Journal for the History of Science*, 10 (1977), pp. 187–215.

30 T. Levere, *Poetry Realized in Nature: Samuel Taylor Coleridge and Early Nineteenth-Century Science* (Cambridge, 1981), considers Coleridge's use of Schelling and Steffens; and E. Richards, 'A question of property rights: Richard Owen's evolutionism reassessed', *British Journal for the History of Science* 20 (1987), 129–71, discusses the reactions of Owen and others to Oken's works.

31 See J. L. Larson, 'Vital forces: regulative principles or constitutive agents? A strategy in German physiology, 1786–1802', *Isis* 70 (1979), pp. 235–49; T. Lenoir, *The Strategy of Life: Teleology and Mechanics in Nineteenth-Century German Biology* (Dordrecht, 1982); N. Jardine, *The Scenes of Inquiry: On the Reality of Questions in the Sciences* (Oxford, 1991), ch. 2.

32 C. E. McClelland, *State, Society and University in Germany, 1700–1940* (Cambridge, 1980), pts. 1–2.

33 Oken, *Lehrbuch der Naturgeschichte*, vol. I, n. 19; Steffens, *Beyträge*, pp. 96–7; Carus, *Versuch einer Darstellung des Nervensystems* (Leipzig, 1814), Vorrede.

34 I owe this observation to William Clark; a fine model for such investigation is R. L. Kremer, 'Building institutes for physiology in Prussia, 1836–1846', in A. Cunningham and P. Williams (eds.), *The Laboratory Revolution in Medicine* (Cambridge, 1992), pp. 72–109.

35 For example, C. E. McClelland, *State, Society and University*, ch. 5; W. Clark, 'On the ironic specimen of the doctor of philosophy', *Science in Context*, 1 (1992), pp. 97–137.

36 C. A. Culotta, 'German biophysics, objective knowledge and Romanticism', *Historical Studies in the Physical Sciences* 4 (1975–6), pp. 3–38; A. R. Cunningham and N. Jardine, 'Introduction: the age of reflexion', in *Romanticism and the Sciences* (Cambridge, 1990), pp. 1–9.

37 See N. Jardine, 'The laboratory revolution in medicine as rhetorical and aesthetic accomplishment', in Cunningham and Williams (eds.), *The Laboratory Revolution*, n.34, pp. 304–23; and B. Lohff, *Die Suche nach Wissenschaftlichkeit*.

38 For the impact of *Naturphilosophie* on the teaching of zoology at Jena see G. Uschmann, *Geschichte der Zoologie und der zoologischen Anstalten in Jena, 1779–1919* (Jena, 1959). G. McOuat's 'Cataloguing powers', forthcoming, looks at the impact of systems of classification derived from *Naturphilosophie* on the cataloguing, arrangement, and display of specimens at the British Museum in the 1820s–1860s.

15 New spaces in natural history

1 Library of the Institut de France, Paris, Fonds Cuvier 3159; quoted in Dorinda Outram, *Georges Cuvier: Vocation, Science and Authority in Post-Revolutionary France* (Manchester, 1984), pp. 62–3.

2 Sophie Forgan, 'Context, image and function: a preliminary enquiry into the architecture of scientific societies', *British Journal for the History of Science*, 19 (1986), pp. 89–113; 'The architecture of science and the idea of a university', *Studies in History and Philosophy of Science*, 20 (1989), pp. 405–34; Thomas A. Markus, *Buildings and Power: Freedom and Control in the Origins of Modern Building Types* (London and New York, 1993), pt. 3, 'Buildings and knowledge'. For some exploration of domestic space in science see P. Abir-Am and D. Outram, *Uneasy Careers and Intimate Lives* (New Brunswick, 1987).

3 M. Milner, *Eternity's Sunrise* (London, 1987); see also the psychoanalytical theory discussed in Klaus Theleweit, *Male Fantasies*, 2 vols. (Oxford, 1987); Henri Lefebvre, *The Production of Space* (Oxford, 1991).

4 K. Zacher, *Curiosity and Pilgrimage* (Baltimore, 1976).

5 'These gardens are an Elysium, which the friend of nature cannot approach without awe', quoted in D. Outram, *Georges Cuvier*, p. 161.

6 See Outram, *Georges Cuvier*, pp. 161ff., on the way the arrangement of the galleries represented conflicts over classification systems, and themselves represented an effort to *display* spatially the structure of nature which rendered them very different from early modern 'cabinets'.

7 Outram, *Georges Cuvier*, pp. 176, 251; Georges Cuvier, *Recueil des Éloges Historiques*, 3 vols. (Paris and Strasbourg, 1819–27), vol. III, p. 436.

8 Walter Benjamin, 'Paris, capital of the nineteenth century', in C. Hoare (ed.), *Charles Baudelaire: A Lyric Poet in the Era of High Capitalism* (London, 1973), pp. 155–76, p. 159.

9 Quoted in Outram, *Georges Cuvier*, pp. 62–3.

10 For a forceful description of field natural history writing, see Larzar Ziff, *Writing in the New Nation: Prose, Print and Politics in the Early United States* (New Haven, 1991), pp. 34–54, which examines field natural history writing as a way of controlling and defining terrain. The connection between manliness and field science is stressed in J. Secord, 'The Geological Survey of Great Britain as a research school, 1839–1855', *History of Science*, 24 (1988), pp. 223–75; Michael Shortland, 'Darkness visible: underground culture in the Golden Age of geology', *History of Science*, 32 (1994), pp. 1–61.

11 J. L. Jarret, 'On psychical distance', *Person*, 42 (1971), pp. 61–9; M. H. Nicolson, *Mountain Gloom and Mountain Glory: The Development of the Aesthetics of the Infinite* (New York, 1963).

12 L. Marin, *Portrait of the King* (London, 1988); Jacques Revel, 'Knowledge of the territory', *Science in Context*, 4 (1991), pp. 133–61. The literature on objectivity is vast; see Zeno G. Swijtink, 'The objectification of observation', in L. Kruger, L. Daston, and M. Heidelberger (eds.), *The Probabilistic Revolution* (Cambridge, MA, 1987), vol. I, pp. 261–86.

13 D. Outram, *The Body and the French Revolution* (New Haven and London, 1989); Norbert Elias, *The Civilising Process*, 2 vols. (New York, 1978); Jean de la Salle, *Traité de la civilité* (Paris, 1774), p. 23: 'Children like to touch clothes and other things that please them with their hands. This urge must be corrected, and they must be taught to touch all that they see only with their eyes.'

16 Minerals, strata and fossils

I am grateful to Jane Camerini, Rhoda Rappaport and Kenneth Taylor for helpful criticism of a draft of this essay. I have also gratefully borrowed the notion of 'proxies' from Mark Hineline, 'The visual culture of the earth sciences, 1863–1970' (Ph.D. dissertation, University of California, San Diego, 1993). The research for this essay was supported by the National Science Foundation (grant no. DIR–9021695).

1 Many such volumes were costly, and could only be owned by the rich or by institutions; hence it was important to most naturalists either to belong to a scientific institution with a good library, or to have access to the library of an affluent patron with scientific interests.

2 There were some speculations about the origins of animal, plant and mineral species, but rarely as more than marginal discussions of possible common origins of similar varieties, or – at most – of related species within a genus; it would be anachronistic to regard such theories as truly evolutionary.

3 Thin-sectioning techniques and polarized light microscopy were not developed until the mid-nineteenth century; before that time it was difficult to distinguish specimens of basalt from those of greywacke, for example, since little could be discerned with a hand-lens, and the results of gross chemical analysis were enigmatic.

4 Desmarest's study of the extinct volcanoes of Auvergne in central France (first published in 1774) proved literally decisive: a steady stream of other naturalists, going there in the subsequent decades to see for themselves, came away convinced of the volcanic origin of at least *these* basalts.

5 On Werner and the Neptunist–Vulcanist controversy, see Martin Guntau's chapter in this volume.

6 Saussure's four-volume *Voyages dans les Alpes* (1779–96), was a particularly influential model in this respect. By the early 1800s, demonstrable fieldwork experience had become an essential qualification for any author wishing to be taken seriously in this kind of science.

7 This kind of science therefore flourished first in countries, notably France, where accurate maps had already been made for political or economic reasons of state. See, for example, Josef Konvitz, *Cartography in France 1660–1848: Science, Engineering and Statecraft* (Chicago, 1987).

8 The distinction corresponds more closely to that between 'hard-rock' and 'soft-rock' terrains, in the colloquial usage of modern geologists, than to any modern stratigraphical or age-related terms. 'Primary' rocks included, in modern terms, igneous and metamorphic rocks of almost any age; 'Secondary' rocks included many sediments now regarded as Tertiary or Cenozoic in age, but also some that would be assigned to the Palaeozoic or even the Precambrian.

9 Werner's little book entitled *Kurze Klassifikation* ('Short Classification', Dresden, 1787), and the students from many countries who attended his lectures at Freiberg, spread the idea widely: see Guntau's chapter in this volume. Werner later interpolated a category of 'Transition' rocks between Primary and Secondary.

10 For animal and plant species, diagnostic features described from specimens were sometimes supplemented by those of habitat and geography; but there was no separate classification based *primarily* on fieldwork, until the development of biogeographical categories in the nineteenth century.

11 The geological map of England and Wales published in 1820 by the Society's first president, George Greenough, had become in effect the cartographical model; it had largely superseded the similar map by William Smith (1815).

12 Such 'theories' thus corresponded closely to the modern concept of scientific models. The genre was founded, in effect, by Thomas Burnet's *Sacred Theory of the Earth* (London, 1681–9), which ingeniously combined Cartesian with biblical motifs.

13 The nineteenth-century meaning of 'geology' was even wider than the modern, because it usually embraced the older 'mineralogy', and also such specialties as palaeontology and geophysics (not named as such until later): in effect, it was close to the modern phrase 'earth sciences'.

14 Desmarest's study of the extinct volcanoes of Auvergne (1779) was an influential example of this kind of geohistorical analysis. Specifically, he argued that basalts now capping some of the plateaux in Auvergne represented lavas that had flowed down valleys at a remote epoch in the past; but that subsequent erosion had left them high and dry above the present valleys, some of which were occupied by lavas of a far more recent epoch.

15 The introduction to Cuvier's great four-volume *Recherches sur les ossemens fossiles* (Paris, 1812) was later reprinted as a short popular book, which made his argument known throughout the Western world. Cuvier regarded his 'catastrophist' view as the minority position at the time, and the actualist interpretation (as it has since been termed) as the more usual opinion. An influential example of the latter, far

more widely read than Hutton's book, was the *Illustrations of the Huttonian Theory* (Edinburgh, 1802) by the Scottish mathematician John Playfair.

16 Agassiz's theory of an Ice Age was so extreme in form (he argued that most of the earth's surface had been covered in ice, and all forms of life wiped out) that it was met with great scepticism. On the other hand, geologists soon accepted a more moderate version, with extensive glaciers and drift ice, conceding that the climate in the geologically recent past had been much colder than at present, contrary to what had been assumed from the model of a steadily cooling earth.

17 An excellent account of the British side of that later development is in Bowdoin Van Riper, *Men among the Mammoths: Victorian Science and the Discovery of Human Prehistory* (Chicago and London, 1993).

17 Humboldtian science

I am grateful for permission to quote manuscript material from the following: Staatsbibliothek Preußischer Kulturbesitz, Berlin (SBPK) (Humboldt Nachlaß); Wellcome Institute for the History of Medicine, London; Conservatoire Botanique de Genève (Fonds Decandolle).

1 Susan Faye Cannon, *Science in Culture: The Early Victorian Period* (New York, 1978), ch. 4: 'Humboldtian science'.

2 See also Jack Morrell and Arnold Thackray, *Gentlemen of Science* (Oxford, 1981), pp. 512–31, for extensive use of 'Humboldtian science' to describe the activities of the British Association for the Advancement of Science.

3 'Beobachtungen über das Gesetz der Wärmeabnahme in den höhern Regionen der Atmosphäre, und über die untern Gränzen des ewigen Schnees', *Annalen der Physik*, 24 (1806), pp. 1–2.

4 Humboldt to David Friedländer (11 April 1799). *Die Jugendbriefe Alexander von Humboldts*, ed. Ilse Jahn and Fritz G. Lange (Berlin, 1973), p. 657.

5 Alexander von Humboldt, *Essai sur la géographie des plantes* (Paris, 1807), p. 1.

6 Humboldt, *Essai*, pp. 41–2.

7 M. Norton Wise (with the collaboration of Crosbie Smith), 'Work and waste: political economy and natural philosophy in nineteenth-century Britain', *History of Science*, 27 (1989), pp. 263–301, 391–449, discuss the types of equilibrium employed by natural philosophers around 1800 and their relation to secular ideals of reason and progress, with specific application to the calorimetric work of Laplace and Lavoisier and the dynamics of Lagrange. Bernadette Bensaude-Vincent, 'The Balance: between chemistry and politics', *The Eighteenth Century*, 33 (1992), pp. 217–37, extends Wise's analysis of Lavoisier's equilibrium ideal to his political economic views.

8 Humboldt to Karsten, Freiberg (26 November 1791). *Jugendbriefe*, pp. 161–2.

9 Humboldt to Lichtenberg (21 April 1792). *Jugendbriefe*, pp. 183–5. For German debates over anti-phlogistic chemistry, see Karl Hufbauer, *The Formation of the German Chemical Community* (Berkeley, 1982).

10 Alexander von Humboldt, 'Versuch über einige physikalische und chemische Grundsätze der Salzwerkskunde', *Bergmännisches Journal*, 5 (1792), pp. 1–45, 97–141; pp. 100–2.

11 Alexander von Humboldt, *Bergbau in den preußischen Fürstentümern Ansbach–Bayreuth* (Freiberg, 1959), pp. 214–16. Humboldt refers to experiments of the French chemist Guyton de Morveau, one of Lavoisier's chief colleagues and promoters, which showed that steel was a combination of 'the basis of fixed air' (carbon) and iron. Amalgamation was the recently imported process by which silver was dissolved out of silver ore by mercury. Its original promoter, Ignaz von Born, interpreted the process as the phlogistication of silver by mercury; Lavoisians saw it as the reduction of silver oxides by mercury metal.

12 'Lettre de M. Von Humboldt à M. Van-Mons, sur le procédé chimique de la vitalité', *Annales de chimie*, 22 (1797), p. 70.

13 *Magasin encyclopédique*, vol. I (1796), pp. 462–72.

14 'Die Entbindung des Wärmestoffs als geognostisches Phänomen betrachtet', in *Versuche über die chemische Zerlegung des Luftkreises* (Braunschweig, 1799).

15 Alexander von Humboldt, 'Von den isothermen Linien und der Verteilung der Wärme auf dem Erdkörper', *Schriften zur physikalischen Geographie*, ed. Hanno Beck (Darmstadt, 1989), pp. 18–96; pp. 22–3. This was Humboldt's own translation of the original memoir on isotherms, published in the *Mémoires de physique et de chimie de la Société d'Arcueil*, 3 (1817); pp. 462–602.

16 MS note, undated, in folder marked 'Lignes isothermes: généralités', Humboldt Nachlaß, großer Kasten 1, Mappe 1, Nr. 4. SBPK, Berlin.

17 Alexander von Humboldt, 'Versuch, die mittlere Höhe der Continente zu bestimmen', *Annalen der Physik und Chemie*, 57 (1842), pp. 407–19; pp. 409–410.

18 Alexander von Humboldt, 'Sur le nivellement barométrique de l'Espagne et le tracé des sections verticales qui représentent des grandes étendues de pays', Humboldt Nachlaß, großer Kasten 5, Nr. 15, p. 9, SBPK, Berlin.

19 Humboldt, 'Sur les lois que l'on observe dans la distribution des formes végétales', *Dictionnaire des sciences naturelles*, ed. F. Cuvier, vol. XVIII (Strasbourg, 1820), p. 431.

20 Alexander von Humboldt, *De distributione geographica plantarum secundum coeli temperiem et altitudinem montium*, (Paris, 1817).

21 Alexander von Humboldt, 'Nouvelles recherches sur la distribution des formes végétales', *Annales de chimie et de physique*, 16 (1821); pp. 267, 279–280.

22 Humboldt, *Kosmos*, vol. I, p. 82.

23 Humboldt to F.-C. Levrault (18 July 1820), Fonds Decandolle, Conservatoire Botanique de Genève. Levrault evidently did not take up Humboldt's offer; Humboldt supplied the article on 'Indépendence des formations' and an appendix on the numerical laws of plant geography to Decandolle's article on 'Géographie botanique', but no general article on 'Lignes'.

24 Humboldt, 'Über Steppen und Wüsten', 'Ideen zu einer Physiognomik der Gewächse', *Ansichten der Natur* (Tübingen, 1808); *Vues des Cordillères, et monumens des peuples indigènes de l'Amérique* (Paris, 1814), p. i.

25 Humboldt, 'Skizze einer geologischen Schilderung des südlichen Amerikas', *Allgemeine geographische Ephemeriden*, 9 (1802), pp. 310–29, 389–420, and *Essai sur le gisement des rochers dans les deux hémisphères* (Paris, 1823), especially appendix on 'algebraic' methods in geognosy; 'Sur les lois', *Dictionnaire des sciences naturelles*; 'Von den isothermen Linien', p. 54.

26 Humboldt mentions Darwin's offering in a letter to ?, MS 241, Wellcome Institute for the History of Medicine, London.

27 Beaufort to Humboldt (London, 29 November 1826). Großer Kasten 5, Nr. 40. Humboldt Nachlaß, SBPK, Berlin.

28 Helen Maria Williams, 'Translator's preface', *Personal Narrative of Travels to the Equatorial Regions of the New Continent* (London, 1814), vol. I, pp. vii–viii.

29 Humboldt's Orinoco diary, quoted in Paul Kanut Schafer (ed.), *Die Wiederentdeckung der Neuen Welt* (Berlin, 1989), p. 86. For an introduction to the colonial independence struggles, see Leslie Bethell (ed.), *The Independence of Latin America* (Cambridge, 1987), especially the essay by John Lynch.

30 Alexander von Humboldt, *Lateinamerika am Vorabend der Unabhängigkeitskrieg*, ed. Margot Faak (Berlin, 1982), pp. 286–8; correspondence of Humboldt with Thomas Jefferson, in *Briefe aus Amerika 1799–1804*, ed. Ulrike Moheit (Berlin, 1993), pp. 296, 307–8; 'Tablas geograficas del Reino de la Nueva Espana (1803)', *Boletìn de geogràfia y estadìstica*, 1 (1869), pp. 635–7. Humboldt supplied the government of Columbia with six Fortin barometers in 1823, and the 'politic administration' he foresaw involved interior canals connecting Buenos Aires to Angostura, steamships moving up the Orinoco, and a trans-isthmus canal. *Relation historique du voyage* (Paris, 1825), vol. III, p. 127.

31 *Selections from the Works of Baron de Humboldt, Relating to the Climate, Inhabitants, Productions and Mines of Mexico*, ed. John Taylor (London, 1824). The original title of Humboldt's essay was *Essai politique sur le royaume de la Nouvelle Espagne*,

2 vols. (Paris, 1811). For Humboldt's plans to travel India and the Himalaya on the back of British East India Company interests see *Briefe Alexander von Humboldts an seinen Bruder Wilhelm* (Berlin, 1923), p. 126; for the planned institute in Mexico sponsored by a French mining consortium see Alexander to Wilhelm von Humboldt (Verona, 17 October 1822); Humboldt to J.-B. Boussingault (Paris, 5 August 1822). *Lettres américaines d'Alexandre de Humboldt 1798–1807*, ed. E.-T. Hamy (Paris, 1905), pp. 294, 286.

32 Mary Louise Pratt, *Imperial Eyes: Travel Writing and Transculturation* (London, 1992), ch. 6.

33 Humboldt, *Ansichten der Natur*, p. 4; Pratt, *Imperial Eyes*, p. 124.

34 Humboldt, *Personal Narrative of Travels*, trans. Williams, vol. III, pp. 35–6.

18   Biogeography and empire

This paper is reprinted in a shortened form with permission from *Revue de l'histoire des sciences*.

1 Valuable work in this area includes Margaret Deacon, *Scientists and the Sea, 1650–1900* (London, 1971); Roy MacLeod, 'On visiting the moving metropolis: reflections on the architecture of imperial science', *Historical Records of Australian Science*, 5 (1982), pp. 1–16; John M. MacKenzie, *The Empire of Nature* (Manchester, 1988); and P. Petitjean, C. Jami, and A. M. Moulin (eds.), *Science and Empires: Historical Studies about Scientific Development and European Expansion* (Boston, 1992). Listings of literature on imperialism are in C. A. Bayly, *Imperial Meridian: The British Empire and the World, 1730–1830* (London, 1989) and Colin C. Eldridge (ed.), *British Imperialism in the Nineteenth Century* (Basingstoke, 1984).

2 The term 'biogeography' is a twentieth-century term. I use it here loosely, as a shorthand expression to include both animal and plant distribution studies.

3 Janet Browne, *The Secular Ark: Studies in the History of Biogeography* (New Haven, 1983), pp. 1–57. See also Frans A. Stafleu, *Linnaeus and the Linnaeans* (Utrecht, 1971); William T. Stearn, 'Botanical exploration to the time of Linnaeus', *Proceedings of the Linnean Society of London*, 169 (1958), pp. 173–96; Malcolm Nicolson, 'Alexander von Humboldt, Humboldtian science and the origins of the study of vegetation', *History of Science*, 25 (1987), pp. 167–94; and John Prest, *The Garden of Eden: The Botanic Garden and the re-creation of Paradise* (New Haven, 1981).

4 D. E. Allen, *The Naturalist in Britain: A Social History* (Harmondsworth, 1978).

5 Laurence Jameson, 'Biographical memoir of the late Professor Jameson', *Edinburgh New Philosophical Journal*, 57 (1854), pp. 1–49.

6 E. H. Ackerknecht, *Medicine at the Paris Hospital, 1794–1848* (Baltimore, 1967); Toby A. Appel, *The Cuvier–Geoffroy Debate: French Biology in the Decades before Darwin* (Oxford, 1987); and J. J. Keevil, C. C. Lloyd, and J. L. S. Coulter, *Medicine and the Navy, 1200–1900*, 4 vols. (Edinburgh, 1957–63).

7 Norman G. Brett-James, *The Life of Peter Collison* (London, 1925).

8 Two exemplary sources are Ray Desmond, *Dictionary of British and Irish Botanists and Horticulturists, Including Plant Collectors and Botanical Artists* (London, 1977), and Ron Cleeveley, *World Palaeontological Collections* (London, 1983). See also D. E. Allen, 'The early professionals in British natural history', in A. C. Wheeler and J. H. Price (eds.), *From Linnaeus to Darwin: Commentaries on the History of Biology and Geology* (London, Society for the History of Natural History, Special Publications no. 3, 1985), pp. 1–12.

9 Ray Desmond, *The India Museum, 1801–1879* (London, 1982); William T. Stearn, *The Natural History Museum at South Kensington* (London, 1981); and Albert E. Gunther, *A Century of Zoology at the British Museum Through the Lives of Two Keepers, 1815–1914* (Folkestone, 1975).

10 Richard Startin Owen (ed.), *The Life of Richard Owen* (London, 1894), and Adrian Desmond, *Archetypes and Ancestors: Palaeontology in Victorian London, 1850–1875* (London, 1982). For Gould, see Gordon Sauer, *John Gould, the Bird Man: A Chronology and Bibliography* (London, 1982), and Adrian Desmond, 'The making

of institutional zoology in London', *History of Science*, 23 (1985), pp. 153–85, 223–50.

11 F. H. Burkhardt and Sydney Smith (eds.), *The Correspondence of Charles Darwin*, 9 vols. (Cambridge, 1983–94), vol. II, p. 18.

12 Michael E. Hoare (ed.), *The Resolution Journal of Johann Reinhold Forster, 1772–1775*, 4 vols. (London, 1982), Introduction, vol. I, pp. 1–122.

13 See Richard Altick, *The English Common Reader: A Social History of the Mass Reading Public, 1800–1900* (Chicago, 1957). A close study of one such journal is given by Susan Sheets-Pyenson, 'From the North to Red Lion Court: the creation of and early years of the *Annals of Natural History*', *Archives of Natural History*, 10 (1981), pp. 221–49.

14 L. H. Brockway, *Science and Colonial Expansion* (New York, 1979), pp. 103–67, and Henry Hobhouse, *Seeds of Change: Five Plants that Transformed Mankind* (London, 1985).

15 J. C. Beaglehole (ed.), *The Journals of Captain James Cook*, 5 vols. (Cambridge, 1955–74) and H. B. Carter, *Sir Joseph Banks* (London, 1988).

16 Robert Hughes, *The Fatal Shore: A History of the Transportation of Convicts to Australia, 1787–1868* (London, 1987); Bernard Smith, *European Vision and the South Pacific*, 2nd edn. (London, 1985); and David Mackay, *In the Wake of Cook: Exploration, Science and Empire, 1780–1801* (London, 1985).

17 R. W. Russell (ed.), *Matthew Flinders: The Ifs of History* (Adelaide, 1979) and David Mabberley, *Jupiter Botanicus: Robert Brown of the British Museum* (Braunschweig/London, 1985).

18 Janet Browne and Michael Neve (eds.), *Voyage of the* Beagle *by Charles Darwin* (orig. London, 1839), edited with an introduction (Harmondsworth, 1989).

19 Joseph Dalton Hooker, Introductory essay to *Flora Novae-Zelandiae* (London, 1853–5), and Introductory essay to Joseph Dalton Hooker and Thomas Thomson, *Flora Indica* (London, 1855).

20 Charles Lyell, *Principles of Geology* (orig. 1830–3), reprinted with an introduction by M. J. S. Rudwick, 3 vols. (Chicago, 1991), vol. II, p. 66.

21 Burkhardt and Smith (eds.), *Correspondence of Charles Darwin*, vol. II, p. 311.

22 Browne, *Secular Ark*, pp. 58–85.

23 M. Vicziany, 'Imperialism, botany and statistics in early nineteenth century India: the surveys of Francis Buchanan (1762–1829)', *Modern Asian Studies*, 20 (1986), pp. 625–60.

24 Browne, *Secular Ark*, pp. 65–8. See also D. E. Allen, *The Botanists: A History of the Botanical Society of the British Isles Though 150 Years* (Winchester, 1986), and F. N. Egerton, 'Hewett C. Watson, Great Britain's first phytogeographer', *Huntia*, 3 (1979), pp. 87–102.

25 Arthur H. Robinson, *Early Thematic Mapping in the History of Cartography* (Chicago, 1982), and Nicholas Alfrey and Stephen Daniels (eds.), *Mapping the Landscape: Essays on Art and Cartography* (Nottingham, 1990).

26 Jane Camerini, 'Evolution, biogeography, and maps: an early history of Wallace's line', *Isis*, 84 (1993), pp. 700–27; and Martin Fichman, 'Wallace: Zoogeography and the Problem of Land Bridges', *Journal of the History of Biology*, 10 (1977), pp. 45–63.

27 A listing is given in F. N. Egerton, 'The history of ecology: achievements and opportunities', *Journal of the History of Biology*, 16 (1983), pp. 259–310. See also Joel B. Hagen, *An Entangled Bank: The Origins of Ecosystem Ecology* (New Brunswick, 1992), and Peter Bowler, *The History of the Environmental Sciences* (London, 1992).

28 Charles Darwin, *On the Origin of Species by Means of Natural Selection* (orig. London, 1859), edited with an introduction by J. W. Burrow (Harmondsworth, 1968), pp. 352–9.

29 Harriet Ritvo, *The Animal Estate: The English and Other Creatures in the Victorian Age* (Cambridge, MA, 1987).

30 Lyell, *Principles of Geology*, vol. II, p. 156.

19   Travelling the other way

1 The first form in which the book we now know as *The Voyage of the Beagle* appeared was as vol. III of *Narrative of the Surveying Voyages of his Majesty's Ships* Adventure *and* Beagle, *between the years 1826 and 1836, describing their Examination of the Southern Shores of South America, and the Beagle's Circumnavigation of the Globe* (London, 1839). Vol. I was by Captain P. Parker King and covers the earlier expeditions; vol. II is by Captain Robert Fitzroy, captain of the *Beagle*, and his account overlaps temporally with that of vol. III by Charles Darwin. The publisher Henry Colburn produced Darwin's volume as a separate book in the following year as *Journal of Researches into the Natural History and Geology of the Countries visited during the voyage of HMS.* Beagle *Round the World*, and in 1845 a second revised and expanded edition appeared.

20   Ethnological encounters

1 For example, J. C. Prichard, *The Natural History of Man* (London, 1842), p. 132.
2 For a discussion of the relationship between civil and natural history in the eighteenth century in Scotland, see Paul Wood's chapter in this volume.
3 E. Richards, 'The "moral anatomy" of Robert Knox: the interplay between biological and social thought in Victorian scientific naturalism', *Journal of the History of Biology*, 22 (1989), pp. 373–436.
4 [J. Barrow], 'Review of *Further Papers relating to the Slave Trade, Nos. III, and IV, 1821 and 1822, and Sixteenth Report of the Directors of the African Institution . . . May 1822*', *Quarterly Review*, 28 (1823), pp. 161–79. Authorship is attributed by Murray's 'Register of authors and articles' according to H. & H. C. Shine, *The Quarterly Review under Gifford. Identification of Contributors, 1809–1824* (Chapel Hill, NC, 1949).
5 J.-M. Degérando, *Considerations on the Diverse Methods to Follow in the Observation of Savage Peoples* (first published in French in 1800), trans. F. C. T. Moore (London, 1969).
6 J. Herschel (ed.), *A Manual of Scientific Enquiry*, 2nd edn. (London, 1851). For J. C. Prichard's instructions to naval officers and travellers, see 'Ethnology' (section 13).
7 H. Aarsleff, *The Study of Language in England, 1780–1860* (Princeton, 1967).
8 W. E. Parry, *Journal of a Second Voyage . . . North-West Passage . . . 1821–1823* (London, 1824); J. Ross and J. C. Ross, *Narrative of a Second Voyage . . . North-West Passage . . . 1829–1833* (London, 1835).
9 This example is taken from the standard account of the African Association. R. Hallett, *The Penetration of Africa: European Enterprise and Exploration . . . Vol. 1 to 1815* (London, 1965), pp. 204–9.
10 J. Rennell, 'Construction of the map of Africa', *Proceedings of the African Association* (London, 1790), p. 213.
11 Parry, *Journal* (1824), pp. 199–203; M. Vicziany, 'Imperialism, botany and statistics in early nineteenth-century India: the surveys of Francis Buchanan (1762–1829)', *Modern Asian Studies*, 20 (1986), pp. 625–60.
12 The canonical text for Humboldtian science is W. F. Cannon, 'Humboldtian science', in *Science in Culture: The Early Victorian Period*, (New York, 1978), pp. 73–110). For further discussion of the metaphorical mapping into natural history, see M. Dettelbach in this volume. For the emergence of a 'science of vegetation', see M. Nicolson, 'Alexander von Humboldt, Humboldtian science, and the origins of the study of vegetation', *History of Science*, 25 (1987), pp. 167–94. The most thorough study of phrenology is R. Cooter, *The Cultural Meaning of Phrenology and the Organisation of Consent in Nineteenth Century Britain* (Cambridge, 1984). A very suggestive study on Francis Galton is R. Cowan, 'Francis Galton's statistical ideas', *Isis*, 63 (1972), pp. 509–28. For a full-length history of biogeography, see J. Browne, *The Secular Ark: Studies in the History of Biogeography* (New Haven, 1983).

13 This analysis of the contact between the British naval officers and the Aivilingmiut and Iglulingmiut is derived from my doctoral dissertation. See M. T. Bravo, 'Science and discovery in the search for a North-west Passage, 1815–1825', Ph.D. dissertation (Cambridge University 1992).

14 J. Belich, *The New Zealand Wars and the Victorian Interpretation of Racial Conflict* (Auckland, 1986).

15 *The Colonial Intelligencer: or Aborigines' Friend*, 1 (1847), p. 7.

16 *The Colonial Intelligencer*, p. 202.

17 *The Colonial Intelligencer*, p. 8.

18 J. Conolly, *Address to the Ethnological Society of London . . . 25th May 1855* and R. Cull, *A Sketch of the Recent Progress of Ethnology* (London, c.1855), p. 4.

21 Equipment for the field

1 David E. Allen, *The Naturalist in Britain: A Social History* (London, 1976), p. 145.

2 Ken Arnold, 'Cabinets for the curious: the practice of science in early modern museums', Ph.D. dissertation (Princeton University, 1991, order no. DA 9217381), chs. 1 and 7.

3 John C. Lettsom, *The Naturalist's and Traveller's Companion* (London, 1772; 3rd ed. 1799), p.xiii.

4 George Graves, *Naturalist's Pocket-Book or Tourist's Companion* (London, 1818), pp. 126–7.

5 Graves, *Naturalist's Pocket-Book*, p. 56.

6 Allen, *Naturalist*, p. 152.

7 J. E. Taylor, 'Geological specimens', in J. E. Taylor (ed.), *Collecting and Preserving Natural-History Objects* (London, 1876), pp. 6–7.

8 *Ibid.*, pp. 7–11.

9 Allen, *Naturalist*, p. 58.

10 Allen, 'Some further light on the history of the vasculum', *Proceedings of the Botanical Society of the British Isles*, 6 (1965), p. 107.

11 Allen, *Naturalist*, p. 157.

12 Worthington G. Smith, 'Fungi', in Taylor (ed.), *Notes on Collecting and Preserving Natural-History Objects*, p. 179.

13 William Swainson, *Naturalist's Guide* (London, 1822), p. 31.

14 Graves, *Naturalist's Pocket-Book*, p. 128.

15 Allen, *Naturalist*, pp. 197ff.

16 Allen, 'The origin of sugaring', *Entomological Record*, 77 (1965), p. 117.

17 Quoted in George Samouelle, *The Entomologist's Useful Compendium* (London, 1819), pp. 315–16.

22 Artisan botany

Shortened version of 'Science in the pub: artisan botanists in early nineteenth-century Lancashire', *History of Science*, 32 (1994), pp. 269–315. Permission to quote from manuscripts has been granted by the Royal Botanic Gardens, Kew, and Cheshire County Council.

1 See James Cash, *Where There's a Will, There's a Way! or, Science in the Cottage: An Account of the Labours of Naturalists in Humble Life* (London, 1873).

2 Morag Shiach, *Discourse on Popular Culture: Class, Gender and History in Cultural Analysis, 1730 to the Present* (Cambridge, 1989).

3 Brian Harrison, *Drink and the Victorians: The Temperance Question in England 1815–1872* (Pittsburgh, 1971), p. 46.

4 Eileen Yeo, 'Culture and constraint in working-class movements, 1830–1855', in Eileen Yeo and Stephen Yeo (eds.), *Popular Culture and Class Conflict 1590–1914: Explorations in the History of Labour and Leisure* (Brighton, 1981), pp. 155–86.

5 Richard Buxton, *A Botanical Guide to the Flowering Plants, Ferns, Mosses, and*

Algae, found Indigenous within Sixteen Miles of Manchester . . . Together with a Sketch of the Author's Life (London, 1849), pp. iii–v.

6 Unless otherwise noted, quotations are taken from articles in the *Manchester Guardian* of 24 April 1850, p. 6; 14 December 1850, Suppl., p. 5; 21 December 1850, p. 5.

7 John Holt, *General View of the Agriculture of the County of Lancaster* (London, 1795), p. 229; *Manchester Guardian*, 31 December 1851, p. 3; James Middleton, *The Old Road: A Book of Recollections* (Manchester, 1985), p. 33.

8 *Manchester Guardian*, 31 December 1851, p. 3; L. H. Grindon, *Manchester Walks and Wild Flowers* (London, [1859]), p. 126.

9 Buxton, *Botanical Guide*, p. vii.

10 Middleton, *The Old Road*, p. 23.

11 Jack Goody, *The Interface Between the Written and the Oral* (Cambridge, 1987), pp. 177–9, points out that learning by exact repetition in this way is rarely a feature of oral societies which have no texts to serve as a corrective. Rather, it is a procedure intrinsic to the literate tradition.

12 For a discussion of co-operation between groups placing different meanings on the scientific objects that pass between them, see Susan Leigh Star and James R. Griesemer, 'Institutional ecology, "translations" and boundary objects: amateurs and professionals in Berkeley's Museum of Vertebrate Zoology, 1907–39', *Social Studies of Science*, 19 (1989), pp. 387–420.

13 *Rules and Regulations to be Observed by a Society Established at the House of Mr. Henry Rayner, the Sign of the Woodman, George-Street, Hyde, on Monday, July 2nd, 1849, to be called Hyde Faithful Botanical Society* (Hyde, [1849]), pp. 10–11.

14 William Wilson, diary entry (12 January 1833), Warrington Library, *Wilsoniana*, MS 72. Olivia Smith, *The Politics of Language 1791–1819* (Oxford, 1984), p. 13, points out that Samuel Johnson used the word 'illiterate' to signify ignorance of Latin and Greek.

15 Letters from William Wilson to W. J. Hooker (23 June 1831 and 19 July 1831), Royal Botanic Gardens Kew, Directors' Correspondence, vol. VI, letters 344 and 346.

16 I. J. Prothero, *Artisans and Politics in Early Nineteenth-Century London* (Folkestone, 1979), p. 336. For the decline in the status of handloom weavers, see E. P. Thompson, *The Making of the English Working Class* (London, 1980), ch. 9.

17 Letter from John Martin to William Wilson (19 June 1831), Warrington Library, William Wilson Correspondence, MS 53.

18 W. J. Hooker, *The English Flora of Sir James Edward Smith* (London, 1833), vol. V, pt. 1, p. 12.

19 John Rule, 'The property of skill in the period of manufacture', in Patrick Joyce (ed.), *The Historical Meanings of Work* (Cambridge, 1987), pp. 99–118.

20 David Vincent, *Literacy and Popular Culture, England 1750–1914* (Cambridge, 1989), p. 108.

21 Rule, 'The property of skill', p. 108.

22 Letter from William Bentley to W. J. Hooker (20 February 1843), Royal Botanic Gardens Kew, Directors' Correspondence, vol. XIX, letter 86.

23 Rule, 'The property of skill', p. 108.

24 John Horsefield, 'Notice of the Prestwich Botanical Society, and the Bury Botanical and Entomological Society, preceded by some critical remarks on a passage in the account of the conductor's tour in France', *The Gardener's Magazine*, 6 (1830), pp. 392–5, p. 393.

25 John Belchem, *Industrialization and the Working Class: The English Experience, 1750–1900* (Aldershot, 1990), p. 57.

26 Thomas Laqueur, 'The cultural origins of popular literacy in England 1500–1850', *Oxford Review of Education*, 2 (1976), pp. 255–75, p. 270.

27 Iain McCalman, *Radical Underworld: Prophets, Revolutionaries and Pornographers in London, 1795–1840* (Cambridge, 1988), pp. 234–5.

28 Shiach, *Discourse on Popular Culture*, p. 33.

29 Edward Forbes, *An Inaugural Lecture on Botany, Considered as a Science, and as a Branch of Medical Education* (London, 1843), pp. 18–19.

30 Letter from William Bentley to W. J. Hooker (21 January 1846), Royal Botanic Gardens Kew, Directors' Correspondence, vol. XXIV, letter 62; W. J. Hooker to W. Wilson (26 November 1846), Royal Botanic Gardens Kew, 'Letters from W. J. Hooker', fo. 101.

31 Steven Shapin and Simon Schaffer, *Leviathan and the Air-Pump: Hobbes, Boyle, and the Experimental Life* (Princeton, 1985), pp. 76–9.

32 For arguments that science should be regarded as sets of practices rather than a single conceptual network, see Andrew Pickering, 'From science as knowledge to science as practice', in Andrew Pickering (ed.), *Science as Practice and Culture* (Chicago, 1992), pp. 1–26.

33 J. Ginswick (ed.), *Labour and the Poor in England and Wales 1849–1851*, 3 vols. (London, 1983), vol. I, p. 40.

34 Shiach, *Discourse on Popular Culture*, p. 33.

23 Tastes and crazes

1 For a review of this and related concepts see my essay, 'Fashion as a social process', *Textile History*, 22 (1991), pp. 347–58.

2 J. Evans, *Pattern: A Study of Ornament in Western Europe from 1180 to 1900* (Oxford, 1931), vol. II, p. 92.

3 Letter from the Revd William Borlase (15 February 1752), cited in S. Savage, *Catalogue of the Manuscripts in the Library of the Linnean Society of London. Part IV. Calendar of the Ellis Manuscripts* (London, 1948), p. 8.

4 J. Ellis, *An Essay Towards a Natural History of the Corallines* (London, 1755), pp. v–vii; J. Groner, 'Some aspects of the life and work of John Ellis', Ph.D. dissertation (Loyola University of Chicago, 1987), p. 143.

5 J. B. Nichols (ed.), *Illustrations of the Literary History of the Eighteenth Century* (London, 1817–58), vol. IV, p. 753.

6 Now in the Department of Entomology Manuscripts Library, Natural History Museum, London.

7 J. Hardy (ed.), *Selections from the Correspondence of Dr George Johnston* (Edinburgh, 1892), p. 474.

8 Mrs A. Gatty, *British Sea-Weeds* (London, 1872), vol. I, p. viii.

9 For a detailed account, with references, see D. Allen, *The Victorian Fern Craze* (London, 1969).

10 Allen, 'Fashion', p. 354.

11 S. H. Ward, 'On the growth of plants in closely-glazed cases', *Proceedings of the Royal Institution*, 1 (1854), pp. 407–12; S. Hibberd, *Rustic Adornments for Homes of Taste* (London, 1856), p. 7.

12 For a detailed account, with references, see C. Hamlin, 'Robert Warington and the moral economy of the aquarium', *Journal of the History of Biology*, 19 (1986), pp. 131–54.

13 He had already done this a few months earlier in a more specialized place: 'On keeping marine animals and plants alive in unchanged sea-water', *Annals and Magazine of Natural History*, ser. 2, 10 (1852), pp. 263–8.

24 Nature for the people

We are grateful to Michael Clark for his revision of our English.

1 C. C. Gillispie, 'The *Encyclopédie* and the Jacobin philosophy of science: a study in ideas and consequences', in Marshall Clagett (ed.), *Critical Problems in the History of Science* (Madison, 1959), pp. 255–89.

2 J.-M. Drouin, 'L'Histoire naturelle à travers un périodique', in A. Corvol (ed.), *La Nature en révolution* (Paris, 1993), pp. 175–81.

3 See G. White, *The Natural History of Selborne*, ed. Richard Mabey (Harmondsworth, 1977).

4 For Henri Lecoq and his work as a popularizer and hybridizer, see Jean-Marc Drouin and Robert Fox, 'Corolles et crinolines, le mélange des genres dans l'oeuvre d'Henri Lecoq', paper presented to the Symposium *Pratiques et théorie de l'hybridation*, Clermont-Ferrand (October 1992), forthcoming.

5 Harriet Ritvo, 'Learning from animals: natural history for children in the eighteenth and nineteenth centuries', *Children's Literature*, vol. XIII (New Haven and London, 1985), pp. 72–93; Daniel Raichvarg and Denis Legros, 'Le chêne, l'os et la goutte d'eau: aventures et mésaventures du récit scientifique', *Romantisme. Revue du XIXe siècle*, 65 (1989), pp. 81–92.

6 See, for instance, Louis Figuier, *La Terre avant le déluge*, vol. 1 of *Tableaux de la nature* (Paris, 1863), pp. i–iv; J. H. Fabre, *La Science de l'oncle Paul* (Paris, 1926), pp. 7–8.

7 The specificity of the French case is related to the existence of a number of state and provincial institutions, like the Muséum d'Histoire Naturelle, the Institut de France (later Académie des Sciences), the secondary schools, and the provincial botanical gardens which provided naturalists with research, teaching or curator positions. Thus there was a number of careers for naturalists earlier than in other European countries: see, for instance, Dorinda Outram, *Georges Cuvier: Vocation, Science and Authority in Post-Revolutionary France* (Manchester, 1984).

8 In eighteenth-century academic life, descriptive naturalistic practices were closely associated with experimental practices, as can be seen from the monumental 'History of plants' project launched by the Paris Académie des Sciences which included the collection and description of plants, examination of their medicinal properties, and their chemical analysis: see F. L. Holmes, 'Argument and narrative in scientific writing', in Peter Dear (ed.), *The Literary Structure of Scientific Argument: Historical Studies* (Philadelphia, 1991), pp. 164–81, p. 167.

9 Georges Cuvier, 'Institut national', *La Décade philosophique* 23(6) (1799), pp. 327–8.

10 F. E. Guérin (dir.), *Dictionnaire pittoresque d'histoire naturelle et des phénomènes de la nature* (Paris, 1834–40), vol. 1, Introduction, pp. v–viii.

11 See L. Figuier, *La Terre*, p. viii.

12 [R. Chambers], *Vestiges of the Natural History of Creation* (London, 1844). Reprint edited by James A. Secord (Chicago, 1994).

13 G. Pancaldi, *Darwinism in Italy* (Bloomington, IN, 1991).

14 Armand de Quatrefages, *Souvenirs d'un naturaliste* (Paris, 1854), vol. 1, p. xiii.

15 *Taraxacum officinale* is named 'dandelion' in English, 'pissenlit' in French, 'diente de leon' in Spanish, 'dente di leone' in Italian, 'Kuhblume' or 'Löwenzahn' in German. *Grus grus* (or *Grus cinerea*) is 'crane' in English, 'grue cendrée' in French, 'grulla' in Spanish, 'Kranich' in German, 'gru' in Italian.

16 See for instance M. L. [Narcisse] Mauroy, *Dictionnaire d'agriculture*, Paris, *Encyclo-pédie théologique de Migne* (1852) and Emmanuel Le Maout, *Leçons élémentaires de botanique fondées sur l'analyse de cinquante plantes vulgaires* (Paris, Langlois, 1844). Le Maout also wrote, with Joseph Decaisne, a *Flore élémentaire* (Paris, 1855) with both scientific and common names.

17 Gaston Bonnier and Georges de Layens, *Petite Flore, contenant les plantes les plus communes, ainsi que les plantes utiles et nuisibles* (Paris, Dupont, 1889). The other floras by the same writers give Latin and French names. De Layens, a cousin of Bonnier, was a beekeeper.

18 Alphonse de Candolle, *Géographie botanique raisonnée* (Paris, Masson), vol. 1, p. xviii.

19 Alphonse de Candolle, *Géographie*, p. xix.

20 Alphonse de Candolle, *Géographie*, p. xix.

21 Edith Holden, *The Country Diary of an Edwardian Lady* (Exeter, 1977).

22 See J.-M. Drouin and R. Fox, 'Corolles et crinolines'.

23 Ann B. Shteir, 'Linnaeus's daughters: women in British botany', in Barbara J. Harris and JoAnn K. McNamara (eds.), *Women and the Structure of Society* (Durham, NC, 1984), pp. 67–73.

24 For instance, E. M. C., *Popular Geography of Plants* (London, 1855), was written by Maria E. Catlow.

25 Ann B. Shteir, 'Botany in the breakfast room: women and early nineteenth-century British plant study', in Pnina G. Abir-Am and Dorinda Outram (eds.), *Uneasy Careers and Intimate Lives: Women in Science 1789–1979* (London, 1987), pp. 31–44.

26 Jean-Jacques Rousseau, *Lettres sur la botanique* (1771–3), *Oeuvres complètes*, vol. IV (Paris, 1989), pp. 1149–95.

27 A.-P. de Candolle, 'Phytologie' in Bory de Saint-Vincent (ed.), *Dictionnaire classique d'histoire naturelle* (Paris, 1828), vol. XIII, p. 486.

28 Cf. J.-M. Drouin, 'Une société locale de sciences naturelles: la Société Linnéenne de Provence', *Revue d'histoire des sciences* 44(2) (1991), pp. 219–34.

29 Pascal Duris, *Linné et la France* (Paris, 1993).

30 A.-P. de Candolle, *Mémoires et souvenirs* (Geneva, 1862), pp. 138, 161.

31 A.-P. de Candolle, 'Jardin de botanique', in F. Cuvier (ed.), *Dictionnaire des sciences naturelles*, table 24 (Paris, 1822), pp. 359–422. Frédéric Cuvier (1773–1838) was Georges Cuvier's brother. See J.-M. Drouin, 'Une espèce de livre vivant: le rôle des jardins botaniques d'après Augustin-Pyramus de Candolle', *Saussurea* (1993), pp. 37–46.

32 Hugh Edwin Strickland, 'On geology, in relation to the studies of the University of Oxford', in William Jardine (ed.), *Memoirs of Hugh Edwin Strickland* (London, 1858), pp. 217–19.

33 Alexander von Humboldt, *Tableaux de la nature* (traduction française) (Paris, 1808), pp. ix–x.

34 Alexander von Humboldt, Letter to Arago (20 August 1827), in E. T. Hamy, *Correspondance d'Alexandre de Humboldt avec François Arago* (Paris, 1908), p. 31.

35 A. de Quatrefages, *Souvenirs d'un naturaliste* (Paris, 1854), table 1, p. xii.

36 Louis Figuier, *La Terre avant le déluge*, p. xv. For reproductions from this work see M. J. S. Rudwick, *Scenes from Deep Time: Early Pictorial Representations of the Deep Past* (Chicago, 1992).

37 Jéhan de Saint Clavien, entry 'Noisetier', *Dictionnaire de botanique*, in Migne (ed.), *Encyclopédie théologique* (Paris, Mame, 1851). See J.-M. Drouin, 'La botanique et ses usages', in C. Langlois and F. Laplanche (eds.), *La Science catholique* (Paris, 1992), pp. 143–57.

38 Camille Flammarion, *Dieu dans la nature*, 3rd edn. (Paris, 1867).

39 George E. Hutchinson, 'Homage to Santa Rosalia, or why are there so many kinds of animals?', *American Naturalist*, 93 (1959), pp. 145–59, esp. p. 146, n. 1. The number of known coleoptera exceeds 300,000 species. Hutchinson gives the anecdote as 'possibly apocryphal'.

40 'On the educational value of the natural history sciences', in T. H. Huxley, *Lay Sermons, Addresses and Reviews* (orig. 1854; London, 1887), pp. 78–9.

41 On French publishers, see Bruno Beguet (ed.), *La Science pour tous, 1850–1914* (Paris, 1990), and Elisabeth Parinet, *La Librairie Flammarion 1875–1914* (Paris, 1992).

42 Charles Elton, *Animal Ecology* (London, 1927), p. 1.

43 Aldo Leopold, in *A Sand County Almanac* (New York, 1970), p. 207.

25 Natural history and the 'new' biology

1 My discussion focuses exclusively on studies of animals. The story of late nineteenth-century German botany holds important parallels; see Eugene Cittadino, *Nature as the Laboratory: Darwinian Plant Ecology in the German Empire, 1880–1990* (Cambridge, 1990).

2 'Prospectus', *Archiv für Naturgeschichte*, 1 (1835), n. p.

3 Quoted in Ernst Ehlers, 'Carl Theodor Ernst von Siebold. Eine biographische Skizze', *Zeitschrift für wissenschaftliche Zoologie*, 42 (1885), pp. i–xxiii, p. xiii.

4 On the shifting relationships between scientific zoology, systematics and morphology

discussed below, see Lynn K. Nyhart, *Biology Takes Form: Animal Morphology and the German Universities, 1800–1910* (Chicago, 1995).

5 Carl Theodor von Siebold to Ernst Ehlers (18 February 1879), Universitätsbibliothek Göttingen, Cod. MS. E. Ehlers 1815.

6 Wilhelm Roux, 'Die Entwicklungsmechanik der Organismen, eine anatomische Wissenschaft der Zukunft' (1889), reprinted in Roux, *Gesammelte Abhandlungen über Entwickelungsmechanik der Organismen*, 2 vols. (Leipzig, 1895), vol. II, pp. 24–54. The American Thomas Hunt Morgan made similar claims in his book *Experimental Zoology* (New York, 1907), p. 5.

7 T. H. Huxley, 'On the study of biology' (1876), in *Huxley, American Addresses, with a Lecture on the Study of Biology* (New York, 1877), p. 135.

8 Joseph A. Caron, '"Biology" in the life sciences: a historiographical contribution', *History of Science*, 26 (1988), pp. 223–68.

9 Charles Darwin, *On the Origin of Species* (orig. London, 1859; fac. repr. Cambridge, MA, 1964), pp. 358–9, 385.

10 See, for example, Rudolf Leuckart, 'Helminthologische experimentaluntersuchungen', *Nachrichten der Georg-August Universität und der Königl. Gesellschaft der Wissenschaften zu Göttingen* (1862), pp. 13–21; (1865), pp. 219–32; Richard Greeff, 'Untersuchung über Bau und Entwicklungsgeschichte von Echinorhynchus miliarius Zenk.', *Archiv für Naturgeschichte*, 30 (1864), part I, pp. 98–140.

11 Carl Theodor Ernst von Siebold, 'Ueber die Parthenogenesis der Polistes gallicam', *Zeitschrift für wissenschaftliche Zoologie*, 20 (1870), pp. 236–42. August Weismann, 'Ueber den Saison-Dimorphismus der Schmetterlinge', *Annali del Museo civico di Storia Naturale di Genova*, 6 (1874), pp. 209–307: reprinted as the first part of his *Studien zur Descendenz-Theorie* (Leipzig, 1875), English translation: *Studies in the Theory of Descent* (London, 1882). Karl Möbius, 'Wo kommt die Nahrung für die Tiefseethiere her?' *Zeitschrift für wissenschaftliche Zoologie*, 21 (1871), pp. 294–304.

12 This relationship has been best studied in the American context, although until recently the discussion has centred more on the relationship between morphological and experimental studies than on natural history and experiment. See Garland Allen, *Life Science in the Twentieth Century* (New York, 1978); and the articles by Keith Benson, Jane Maienschein, Ronald Rainger, Garland Allen and Frederick Churchill in the 'Special section on American morphology at the turn of the century', *Journal of the History of Biology*, 14 (1981), pp. 83–191. More recently, historians have begun exploring the role of experiment in evolutionary and natural historical studies in early twentieth-century America. See, for example, Sharon Kingsland, 'The battling botanist: Daniel Trembly MacDougal, mutation theory, and the rise of experimental evolutionary botany in America, 1900–1912', *Isis*, 82 (1991), pp. 479–509, and David Magnus, 'Down the primrose path: competing epistemologies in early twentieth century biology', *Isis* (forthcoming).

13 Keith R. Benson, 'From museum research to laboratory research: the transformation of natural history into academic biology', in Ronald Rainger, Keith R. Benson and Jane Maienschein (eds.), *The American Development of Biology* (New Brunswick, NJ and London, 1991), pp. 49–83.

14 On the MCZ, see Mary P. Winsor, *Reading the Shape of Nature: Comparative Zoology at the Agassiz Museum* (Chicago, 1991). On the persistence of museums at American universities more generally, see Benson, 'Museum research', esp. p. 77; and Sally G. Kohlstedt, 'Museums on campus: a tradition of inquiry and teaching', in Rainger et al. (eds.), *American Development*, pp. 15–48.

15 Wilhelm Arndt, 'Statistisches über die Verteilung der Reichsdeutschen Museen', *Museumskunde*, NF 2 (1930), pp. 149–65, cited in Ilse Jahn, Rolf Löther and Konrad Senglaub (eds.), *Geschichte der Biologie* (Jena, 1982), p. 456.

16 Sir William Henry Flower, 'Modern museums' (orig. 1893), in Flower, *Essays on Museums and Other Subjects Connected with Natural History* (London, 1898), pp. 30–53.

17 Herwig, 'Die Thätigkeit der Sektion für Küsten- und Hochseefischerei seit ihrer

Gründung (1885) bis zum Sommer 1890', *Mittheilungen der Section für Küsten- und Hochseefischerei*, Sonderbeilage (1891), pp. 23–4.

18 Charles A. Kofoid, *The Biological Stations of Europe*, US Bureau of Education Bulletin, 1910, no. 4, whole number 440 (Washington, 1910), pp. 221–3. Inland stations are discussed on pp. 233–46. A similar coalition sponsored marine research in Britain: see David E. Allen, *The Naturalist in Britain* (London, 1976), pp. 211–13.

19 See, for example, Robert A. Lovely, 'Mastering nature's harmony: Stephen Forbes and the roots of American ecology', Ph.D. dissertation, University of Wisconsin, forthcoming 1995. See also the essays collected in Frank N. Egerton (ed.), *History of American Ecology* (New York, 1977), especially the essays by Egerton and J. L. McHugh.

20 Edmund B. Wilson, 'Aims and methods of study in natural history', *Science*, NS 13 (1901), pp. 14–23, on p. 19.

26 The crisis of nature

I am grateful to Patricia Fara, Nick Hopwood, Nick Jardine, Bernard Lightman, Martin Rudwick, Simon Schaffer, Anne Secord and Emma Spary for reading drafts of this essay, and to colleagues and friends in the systematics community for conversation and helpful suggestions.

1 M. P. Winsor, *Reading the Shape of Nature: Comparative Zoology at the Agassiz Museum* (Chicago, 1991), p. 176; N. A. Rupke, *Richard Owen: Victorian Naturalist* (New Haven, 1994), pp. 97–101. More generally, see T. Richards, *The Imperial Archive: Knowledge and the Fantasy of Empire* (London, 1993).

2 E. Cittadino, *Nature as the Laboratory: Darwinian Plant Ecology in the German Empire, 1880–1900* (Cambridge, 1991); R. M. Wood, *The Dark Side of the Earth* (London, 1985); E. L. Mills, *Biological Oceanography: An Early History, 1760–1960* (Ithaca, New York, 1991).

3 *Nature*, 346 (19 July 1990), p. 213.

4 D. Outram, *Georges Cuvier: Vocation, Science and Authority in Post-Revolutionary France* (Manchester, 1984).

5 J. Cullen, 'Libraries and herbaria', in V. H. Heywood and D. M. Moore (eds.), *Current Concepts in Plant Taxonomy* (London, 1984), pp. 25–38, p. 31.

6 I. Kant, *Critique of Judgement* (Oxford, 1952), sect. 75; for heroes and disciplines, see S. Schaffer, 'Scientific discoveries and the end of natural philosophy', *Social Studies of Science*, 16 (1986), pp. 387–420.

7 E. O. Wilson, 'In the queendom of the ants: a brief autobiography', in D. A. Dewsbury (ed.), *Studying Animal Behavior: Autobiographies of the Founders* (Chicago, 1989), pp. 465–84, pp. 474–5.

8 D. L. Hull, *Science as a Process* (Chicago, 1988); J. Dean, 'Controversy over classification: a case study from the history of botany', in B. Barnes and S. Shapin (eds.), *Natural Order: Historical Studies of Scientific Culture* (Beverly Hills, 1979), pp. 211–30; P. F. Stevens, 'Evolutionary classification in botany, 1960–1985', *Journal of the Arnold Arboreteum*, 67 (1988), pp. 313–39.

9 S. J. Gould, *Wonderful Life* (New York, 1989) provides an accessible (if somewhat tendentious) account of this kind of research.

10 G. R. de Beer, 'Preface', *Lectures on the Development of Taxonomy* (London, 1950), p. iii.

11 C. Holden, 'Entomologists wane as insects wax', *Science*, 246 (10 November 1989), pp. 754–6, p. 755.

12 F. Graham, *Audubon Ark: A History of the National Audubon Society* (Austin, TX, 1992); D. E. Allen, *The Botanists* (London, 1986).

13 H. T. Clifford, R. W. Rogers and M. E. Dettmann, 'Where now for taxonomy?', *Nature*, 346 (16 August 1990), p. 602; see also the correspondence in *Nature*, 347

(20 September 1990), pp. 222–4; 347 (25 October 1990), p. 704; and S. M. Walters, 'Herbaria in the 21st century: why should they survive?', *Webbia*, 48 (1993), pp. 673–82.

14 B. Mullan and G. Marvin, *Zoo Culture: A Book about Watching Man Watching Animals* (London, 1987); C. Tudge, *Last Animals at the Zoo* (London, 1993).

15 *Natural History Museum. Corporate Plan 1990–95* (London, 1990).

16 P. D. Gingerich, *Nature*, 345 (14 June 1990), p. 568.

17 P. Alberch, 'Museums, collections and biodiversity inventories', *Trends in Ecology and Evolution*, 8 (October 1993), pp. 372–5. The case for a global survey is made most prominently in E. O. Wilson, *The Diversity of Life* (Cambridge, MA, 1992).

18 B. Latour, 'Give me a laboratory and I will raise the world', in K. Knorr-Cetina and M. Mulkay (eds.), *Science Observed* (London, 1983), pp. 141–70.

19 M. L. Smith, 'Selling the moon: the US manned space program and the triumph of commodity scientism', in R. W. Fox and T. J. J. Lears (eds.), *The Culture of Consumption* (New York, 1983), pp. 175–209, 233–6; and S. Dunwoody, 'The science writing inner club: a communication link between science and the lay public', in *Scientists and Journalists: Reporting Science as News* (New York, 1986), pp. 155–69. More generally, see B. V. Lewenstein, 'The meaning of "public understanding of science" in the United States after World War II', *Public Understanding of Science*, 1 (1992), pp. 45–68.

20 Three collections of essays provide contrasting perspectives on these issues: J. Durant (ed.), *Museums and the Public Understanding of Science* (London, 1992); P. Boylan (ed.), *Museums 2000: Politics, People, Professionals and Profit* (London, 1992); P. Vergo, *The New Museology* (London, 1989).

21 W. Bodmer, 'The museum that has to change', *Nature*, 345 (14 June 1990), pp. 569–70, p. 569. The best book on Disney remains R. Schickel, *The Disney Version* (1968; new edn. 1985).

22 M. Lynch, 'Discipline and the material form of images: an analysis of scientific visibility', *Social Studies of Science*, 15 (1985), pp. 37–66. The origins of the cult of the real are analysed in M. Orvell's stimulating *The Real Thing: Imitation and Authenticity in American Culture, 1880–1940* (Chapel Hill, 1989), and the classic essay by W. Benjamin, 'The work of art in the age of mechanical reproduction', in *Illuminations* (New York, 1968), pp. 217–51.

23 For a stimulating analysis, see S. J. Gould, 'Dinomania', *New York Review of Books*, 12 August 1993, pp. 51–6.

24 H. Collins and S. Shapin, 'Experiment, science teaching, and the new history and sociology of science', in M. Shortland and A. Warwick (eds.), *Teaching the History of Science* (Oxford, 1989), pp. 67–79.

25 For a balanced discussion of these issues, see R. Silverstone, *Framing Science: The Making of a BBC Documentary* (London, 1985), esp. pp. 4–5 and 170–1.

26 A helpful discussion of these films is in A. Wilson, *The Culture of Nature* (1992), pp. 117–55.

27 A. Anderson, 'Source–media relations: the production of the environmental agenda', in A. Hansen, (ed.), *The Mass Media and Environmental Issues* (Leicester, 1993), pp. 51–68.

28 The prospects for environmental history, a large and fast-developing field, are best surveyed in 'A round table: environmental history', *Journal of American History*, 76 (March 1990), pp. 1087–1147; for current examples in practice, see W. Cronon, *Nature's Metropolis: Chicago and the Great West* (New York, 1991); A. Crosby, *Ecological Imperialism: The Biological Expansion of Europe, 900–1900* (1986); D. Worster, *The Wealth of Nature* (1993).

Index